BIOTECHNOLOGY :

PERSPECTIVES AND PROSPECTS

BIOTECHNOLOGY :
PERSPECTIVES AND PROSPECTS

Edited By

C P MALIK
CHITRA WADHWANI
BHAVNEET KAUR

School of Life Sciences
Jaipur University, Jaipur

MD Publications Pvt Ltd
New Delhi

Published by
MD Publications Pvt Ltd
"MD House" 11, Darya Ganj
New Delhi-110002
Tel. : +91-11-41563325, 41562846
E-mail : contact@mdppl.com
Website : www.mdppl.com

ISBN : 978-81-7533-116-7

Price : Rs. 1995/-

Published and Printed by **MD Publications Pvt Ltd** at Times Press, 910, Jatwara Street, Darya Ganj, New Delhi.

CONTENTS

PREFACE

This volume entitled "Biotechnology: Perspectives and Prospects" has taken months of hard work, diligence and an untiring effort to organize the contents so as to present the subject matter in the present form.

"Education", we would describe as the purest form of religion and is practised by maximum number of people globally. It is an invaluable possession borne by an individual that contributes in shaping his/her personality.

Biotechnology was born in the year 1976 with the decision of Robert Swanson, a 27-year old venture capitalist and Herb Boyer, a professor at University of California, to begin the first Biotech Company in the name of Genentech (Genetic Engineering Technology). It has been more than thirty years since then and tons of experimental data and knowledge has been created by the latest group of knowledgeable people styled to be biotechnologists or molecular biologists. There is infinite growth in this subject and attempts to measure the potential of this technology in units would surely dwarf its essence.

We have tried to compile the latest offered by this technology. Our efforts began with our previous book "Current Topics in Biotechnology" and will continue to provide study material for graduate and post graduate students. The present volume is a compilation of the most recent trends offered by this awesome technology and each chapter is resonant with the major research issues and sophisticated techniques that enrich the repertoire of this potential science.

A number of topics that in the past had taken a backseat and seemed to be of peripheral interest have been reignited, and fallen in place and taken on a new relevance.

We would like to thank the contributing authors without whom; the present volume would never have been a reality.

The present volume incorporates ten chapters, written by experts in their respective disciplines and provides a detailed account of the technical aspects involved as well as a review of the latest developments. The selected topics are most sought after by both researchers and students at the graduate as well as post-graduate level. Each chapter provides succinct review on different topics.

Chapter 1 by Jeff Schahczenski and Katherine Adam (USA) is a laudable description of transgenic crops in a new perspective. A substantial part of the chapter is devoted to statistical figures dealing with every aspect of transgenic crops. The write-up also highlights the importance of transgenic and gives an account of the risks and fears associated with their cultivation. The chapter concludes with a discussion on the regulatory aspects involved in commercializing their cultivation.

Chapter 2 by Lining Tian (Canada) is an extension of the previous chapter. It provides an overview of "Selectable Gene Excision from Transgenic Plants."

Chapter 3 elucidates a significant breakthrough by Indian-American Scientist, Keerti Rathore at the TAMU (USA). This wonderful invention has opened up the door for saving millions of lives by providing plentiful nutrition in the form of high protein in cotton seed. Rathore and co-workers have achieved the silencing of "gossypol" the toxic substance in cotton by employing RNA interference (RNAi) and a seed specific promoter. Resultantly, this toxic substance (essential for driving away pests and predators) is expressed in the entire plant save the seed. The cotton seed with high protein content and devoid of gossypol is now fit for human consumption and may become an engineered saviour of lives of millions of children who suffer from malnutrition.

Chapter 4 on "Gene Therapy" by Chitra Wadhwani and CP Malik conveys basic information about the types and vectors used along with the advancement achieved in this technique. The applications of this technology in correcting dreadful disease are indeed unbelievable. The chapter is student-savvy and clearly explains both basic and advanced concepts by diagrams suitably placed in the body of the text.

"RNA interference: Small and silent biotechnological tool" is the title of chapter 5, authored by Chitra Wadhwani, CP Malik and Bhavneet Kaur. RNAi has been aptly nicknamed and this chapter gives an extensive treatment to its varied aspects and applications.

The authors, Martías Zurbriggen, Vanesa B. Tognetti, Estela M. Valle and Néstor Carrillo (Argentina) have presented their protocol by employing Cyanobacterial flavodoxin in engineering Multiple Stress Tolerance in Chapter 6.

Chapter 7 by Bhavneet Kaur, CP Malik and Chitra Wadhwani is a concise expression of attainment of varied breakthroughs in different areas of biotechnology. The selected findings of past one and a half year have been listed along with the details. The objective of writing this chapter was to acquaint and enthuse young readers who will be in future specialists of this subject as well as laymen, with the power of this amazing technology.

Chapter 8 by William Laing and Sean Bulley (New Zealand) is aimed primarily at providing a comprehensible outline to University students of specialised processes such as "Control of Vitamin C levels in Plants."

Chapter 9 on "Catharanthus roseus: Molecular Biology and Biotechnology" by Chitra Wadhwani is a detailed description of every possible molecular aspect of this plant. It is a beautiful connotation of the author's own research work as well as findings of contemporary workers on different realms pertaining to molecular biology studies in this plant..

Chapter 10 by Chandan Agarwal summarizes pertinent information on "Mycobacterium", the causal agent of Tuberculosis. This chapter critically analyzes techniques available for sampling and screening of these clinical samples as well as the importance and implication of planning and surveillance to manage various samples.

We are thankful to Dr Ruth Irwin, Editor ISB News Report for his generosity and full cooperation whenever requested for. We extend our thanks to all the learned contributors for their cooperation in compiling very useful information on various facets of Biotechnology. Heartfelt thanks go to various authors, each of who have endeavoured to discuss up-to-date information of their area (s) and expertise and have kindly spared valuable time and knowledge. We are sure the information compiled in this volume will prove helpful to those interested in biotechnology.

We apologize unreservedly for any mistakes, omissions or failure to acknowledge fully. Finally, we would like to thank our families (CPM his spouse, Surakhsha; BK and CW their parents) for their continued support and encouragement throughout the completion of this book.

We wish to record our deep sense of gratitude and respect to our worthy Chancellor, Mr Sandeep Bakshi who always inspired and encouraged us. To Profs K L Sharma, Vice-Chancellor and H N Verma, ProVice - Chancellor, Jaipur National University for helpful suggestions and support. This reflects their incredible zest and zeal to promote academic growth in this portal of learning.

Thanks are due to Mr Pranav Gupta and Ms Pragati of MD Publishers, New Delhi, who supported us fully for completing this project on time, with patience, care and support.

CP Malik
Chitra Wadhwani
Bhavneet Kaur

1

TRANSGENIC CROPS

JEFF SCHAHCZENSKI AND KATHERINE ADAM

Introduction

The transference of genetic material between two dissimilar species for agricultural purposes and crop production is the intent of this discussion. Around 1983 the science and methods took birth to produce transgenic crops. There then technological movement began to modify organisms for economic, medical, military, and other general human needs. The modification of organisms or raising transgenic evoked ethical issues. The capacity to produce transgenic crops has resulted in great consternation among government agencies, business consortia, researchers, and certain nonprofit organizations. The genetically modified foods where parameters were quantifiable seem less in dispute. Understandably, much remains to be done in these areas. In 2001, the Experiment Station Committee on Organization and Policy (ESCOP) and the Extension Committee on a report on critical issues in agricultural biotechnology and recommended responses. While calling for education of the public in regard to transgenic technologies, the report also called for land-grant research on transgenic crops to address four substantive concerns raised by the environmental community. (Appendix 2) To date, little of this type of research has been conducted, since it has never been adequately funded.

As of June 30, 2005, the U.S. Department of Agriculture and the National Agriculture Statistics Service (USDA/NASS) reported that transgenic varieties comprised 87 per cent of all soybean acreage planted in the United States (up from 60 per cent in 2001, and 85 per cent in 2004). As of the same date, transgenic corn acreage planted was 52 per cent (up from 47 per cent in 2004). Transgenic upland cotton was 79 per cent (up from 76 per cent in 2004). No acreage was reported for transgenic varieties of other U.S. crops (Table 1).

Table 1. Acreage planted to transgenic varieties, as percentage of total corn, soybeans, cotton acreage by state. USDA/NASS (June 30th 2005)

	Corn	*Soybeans*	*Cotton*
Arkansas		92	96
California			53
Georgia			95
Illinois	36	81	
Indiana	26	89	
Iowa	60	91	
Kansas	63	90	
Louisiana			95
Michigan	40	76	
Minnesota	66	83	
Mississippi		96	96
Missouri	55	89	
Nebraska	60	91	
North Carolina			95
Ohio	18	77	
South Dakota	83	95	
Texas			63
Wisconsin	46	84	
Other States	52	84	91
US	52	87	79

Three types of transgenic produce have been commercialized—sweet corn, winter squash, and papaya. As of January 6, 2006, the following fruit and vegetable crops have been granted deregulated status by the USDA, Animal and Plant Health Inspection Service (APHIS): papaya (two varieties), potato, squash, sugar beet, sweet corn, and tomato. Except for papaya and a small amount of sweet corn, transgenic fresh produce is currently unavailable to American consumers. Fruits and vegetables for processing may be available very soon—perhaps by fall 2006—as seed has been released to contract growers. (Hagen, 2006). The European Union is debating the question of permitting transgenic crop production alongside its well-established organic production in order to avoid World Trade Organization (WTO) sanctions against trade barriers. India is conducting field trials of transgenic maize (corn), mustard (oilseed crop), sugarcane (ethanol production), sorghum (ethanol and animal feed), pigeonpea, chickpea, rice (staple food grain), tomato, brinjal (eggplant or aubergine), banana, papaya, soybean, and medicinal plants. China anticipates commercializing transgenic rice varieties by 2008 (Dansby, 2006).

The Benbrook Report: Genetically engineered crops and pesticide use in the United States: 1996–2004

The major genetically engineered (GE) crop varieties commercialized since 1996 in the United States have been designed to help control a damaging class of insects and simplify herbicide-based weed management systems. Over the first nine years of commercial use, 670 million acres of crops expressing GE traits have been planted, or about 23 per cent of the total 2,970 million acres of crops harvested across the country during this period.

Crops engineered to tolerate applications of herbicides, or so-called "herbicide-tolerant" crops (HT), account for the largest share of GE acres. About 487 million acres have been planted since 1996, or 73 per cent of total GE crop acres. Herbicide-tolerant soybeans are the most widely planted GE crop technology and account for more than half the total acres planted to GE varieties since 1996. The vast majority of HT crops are engineered to tolerate glyphosate (trade name "Roundup," or referred to as "Roundup Ready"), the herbicide introduced to the market in 1972, by Monsanto.

Corn and cotton have been genetically engineered to express the bacterial toxin *Bacillus thuringiensis*, or Bt. This transgenic trait allows plants to manufacture within their cells a crystalline protein that is toxic to most Lepidopteran insects (moths and butterflies). Some 183 million acres of Bt transgenic corn and cotton have been planted since 1996, representing 27 per cent of total GE crop acreage (Benbrook, 2004).

The top five countries growing transgenic crops in 2005, according to The International Service for the Acquisition of Agro-Biotech Applications (ISAAA) were the United States, Argentina, Brazil, Canada, and China (Table 2). Fourteen countries were ranked in the first tier as major adopters of the technology (ISAAA, 2006).

Table 2. Global Status of Commercialized Transgenic Crops (2005)

Rank	*Area*	*Area (mil. Ha/A.)*	*Crop*
1	USA	49.8/723.0	soybean, maize (corn), cotton,canola, squash, papaya
2	Argentina	17.1/42.2	soybean, maize, cotton
3	Brazil	9.4/23.2	soybean

4	Canada	5.8/14.3	canola, maize, soybean
5	China	3.3/8.2	cotton
6	Paraguay	1.8/4.4	soybean
7	India	1.3/3.2	cotton
8	So. Africa	0.5/1.2	maize, soybean, cotton
9	Uruguay	0.3/.7	soybean, maize, cotton
10	Australia	0.3/.7	cotton
11	Mexico	0.1/.2	cotton, soybean
12	Romania	0.1/.2	soybean
13	Philippines	0.1/.2	maize
14	Spain	0.1/.2	maize
15	Colombia	<0.1/.2	cotton
16	Iran	<0.1/.2	rice
17	Honduras	<0.1/.2	maize
18	Portugal	<0.1/.2	maize

1 ha = 2.47 a. (results rounded to .0)

WHAT ARE TRANSGENIC CROPS?

No uniformly accepted definition of biotechnology exists, according to the National Center for Agricultural Law Research and Information (NCALRI). The center provides several definitions and commentary. Under the broadest definition, the use of biological sciences to develop products—conventional plant and animal breeding techniques, conducted since the dawn of civilization—fall under biotechnology. In the popular press, biotechnology generally refers to newly developed scientific methods used to create products by altering the genetic makeup of organisms and producing unique individuals or traits that are not easily obtained through conventional breeding techniques. These products are often referred to as transgenic, bioengineered, or genetically modified because they contain foreign genetic material. Agriculture is one of the first industries radically affected by this new technology on both a fundamental production level and a legal level (NCALRI, 2000).

The focus of this publication is on crop varieties created through transgenic modification or genetic modification (GM). The products of transgenetic engineering are often called genetically modified organisms, or GMOs. All these terms refer to methods by which biologists splice genes from one or more species into the DNA of crop plants in an attempt to transfer chosen genetic traits. The method is known as recombinant DNA technology. Genes are segments of DNA that contain information that in part determines the end function of a living organism. Genetic engineers manipulate DNA, typically by taking genes from one species—an animal, plant, bacterium, or virus—and inserting them into another species, such as an agricultural crop plant. An intermediate organism or virus can be used to "infect" the host DNA with the desired genetic material. Microparticle bombardment technology is also widely used to deliver exogenous nucleic acids (DNA from another species) into plant cells. The desired genetic material is precipitated onto micron-sized metal particles and placed within one of a variety of devices designed to accelerate these "microcarriers" to velocities required to penetrate the plant cell wall. In this manner, transgenes can be delivered into the cell's genome. New DNA can also be inserted into a host cell using electroporation, in which a jolt of electricity is applied to cells to create openings in the plasma membrane that surrounds a cell. A (typically antibiotic-resistant) marker gene is included in the package to verify degree of effectiveness in introducing the foreign DNA. Gene stacking is becoming more common, adding a whole array of traits at once into the host organism (Stierle, 2006). The whole process can be illustrated by its application in the engineering of transgenic rice, using electroporation.

Steps in electroporation and other methods of gene transfer

1) The DNA sequence for the gene that will be altered is identified and obtained from a donor organism (bacterium). This can be done by referring to known information pertaining to the sequence of the gene which is to be selected, followed by the removal of the gene from the donor organism.

2) The desired gene is removed from the donor organism through the use of site-specific enzymes known as restriction enzymes.

3) The desired gene is then subject to polymerase chain reaction (PCR), a method to amplify DNA and produce a workable amount of the gene.

4) Once acquired, there are several ways to transfer the donor gene into the cells of the target organism. In rice, a somewhat advanced process is utilized. This process is electroporation, wherein special wall-denaturing enzymes remove the plant cell wall. The cells become protoplasts, which are plant cells stripped of the cell wall but still encapsulated in the cellular membrane. In the next step of electroporation, a very high voltage electric charge is sent through the protoplast-containing solution. This charge causes the membrane to temporarily deteriorate, forming small pores.

   Through these temporary pores, the donor gene's DNA is injected. The DNA is injected in the form of transfer plasmids that migrate to the chromosome and become incorporated in the plant's DNA. Shortly after the charge and injection, the cell membrane reforms. The cell wall also reforms in a reverse process.

5) The newly altered cells are then placed in a culture to reproduce the unique cell types that compose the organism.

6) The resulting cells are then transferred to a regular growth environment where the newly incorporated gene will be expressed (Bromley, no date).

With the advent of genetic engineering of plants around 1983, it appeared that transgenic manipulation might benefit and even revolutionize agriculture. The transfer of desirable genetic traits across species barriers offered potential promises to solve problems in the management of agricultural crops, provide new possibilities to improve human and animal health, and provide a new revenue stream for farmers through contract production of pharmaceutical and industrial crops (ESCOP/ECOP, 2000). Potential environmental benefits included reduced toxic pesticide use, improved weed control resulting in less tillage and soil erosion, and water conservation. Furthermore, the new technology promised increased yields.

Transgenic crops were also patentable. Technology agreements or engineering would insure that seed could not be saved over for planting the next year. The developer's intellectual property rights were thereby protected, which offered the potential to increase profits and theoretically garner a monopoly over the transgenic seed supply.

UNINTENDED EFFECTS

Current methods of gene transfer are not precise. While scientists can control with relative exactness the "trait gene" or its synthesized analog to be inserted into a host plant genome, they cannot entirely control its location, or the number of copies that get inserted. Location of genetic material is important because it controls the expression of biological traits, just as genes themselves do. Also, inserted DNA frequently contains multiple stacked genes for different traits (eight in the New Leaf potato), increasing chances of undesirable interactions.

A common and unpredictable occurrence is "silencing" of either the inserted genetic material or adjacent native genes. Present scientific knowledge is still a long way from being able to precisely control the traits the host plant will express and to guarantee genetic stability in subsequent generations (Ryan and Ho, 2001). This potential for instability can lead to unpredictable and undesirable effects, examples of which include plant infertility, production of toxins and allergens, and reductions in yield and plant fitness. The transgenic seed industry consistently counters that since genes from no known allergens are incorporated, adequate care has been taken to guard against this contingency (USDA/OIG, 2005).

Transgenetic engineers who rely on the simple model of gene expression—the position that one gene equals one effect—harbor an outdated interpretation of genetic theory, and one that could have serious implications. Pleiotrophy is the understanding that one gene may control multiple traits in an organism. Pleiotrophy multiplies the uncertainty surrounding transgenic crops. A gene identified as controlling a desirable trait may in fact control multiple traits in a variety of ways. Pleitrophy is common, and the interactions of genes with each other and with the environment add complexity. To accurately predict the effects of new genetic combinations is nearly impossible. The introduction of a novel life form into ecology can trigger effects perhaps too great to be understood during our time. While it is true most mutations don't survive, those that do can profoundly affect human and other life forms.

For instance, transgenic soy strains appear to exhibit unintended effects. Field observations reported to the University of Georgia (New Scientist, 1999) and University of Missouri (UM press release, 2001) noted physiological problems affecting yields. Research published by University of Arkansas scientists in 2000 noted that glysophate disrupts the nitrogen fixation process in Roundup Ready soy (King *et al.*, 2001).

The current marker and promoter genes of choice also may create new hazards. The antibiotic-resistant marker genes carry the potential to increase the variety of bacteria resistant to antibiotics (Sheldon, 1993). The viral promoter genes could combine with other infecting viruses, or be scrambled by the plant, to create new viral proteins. The cauliflower mosaic virus (CaMV) is a very powerful promoter and is commonly used. The CaMV can potentially cause the inserted DNA package to be expressed out of proportion with the rest of the genetic code. When inserted with a particle gun, the CaMV promoter can jump out of the DNA package and land somewhere else in the host genome, causing disruption. The bacterial and viral vector genes could recombine to form active pathogens—either new ones, or old ones with renewed virulence, or with broader host specificity (Stierle, 2006).

The ESCOP and ECOP report mentioned in the Introduction, while advocating and offering specific advice for an extensive education campaign in support of biotechnology, at the same time called for research studies to be carried out on several key safety issues raised by the public. An 18-member Biotechnology Implementation Task Force convened and issued an update in July, 2001. However, nothing more has been heard from this committee (ESCOP/ ECOP, 2000).

A main regulating agency for transgenic technology in the U.S. is the Animal and Plant Health Inspection Service (APHIS), a division within the Department of Agriculture. Rather than conduct safety studies, APHIS appears to accept risk-management tools such as "performance-based regulatory standards" and "science-based risk assessment policies and procedures." This approach allows for an acceptable level of possible collateral damage, as long as it is far enough down the road (USDA/OIG, 2005).

Each piece of the inserted gene package described above carries with it the potential to disrupt non-target portions of the host plant's DNA, to create instability in the new genetic construct, or to result in unpredictable combinations that can create new substances, viruses, or bacteria. What this adds up to is the possibility, again, of unintended effects—particularly in subsequent generations of the engineered plant. To date, no known replicated studies have been conducted that confirm or disprove potential long-term effects on human health. No known mechanism was proposed or included to identify undesirable side effects of the engineering process.

A December, 2005 USDA assessment of APHIS protocols for monitoring GE trial crops criticized oversight lapses. APHIS countered that it was relying on an accepted risk/benefit assessment process, while USDA took the position that oversight should be strengthened, on the assumption there is significant risk—until the new technology has been proven safe beyond doubt.

COMMERCIAL TRANSGENIC CROPS AND THEIR TRAITS

While increased yields and improved nutritional value are among the promised benefits of transgenic crops, most now planted worldwide are designed either 1) to survive exposure to certain herbicides (called herbicide-tolerant, or HT), or 2) to kill certain insect pests (called pesticidal or insecticidal). The transgenic tomato was designed for long shelf life. It is unclear whether the increased beta-carotene in transgenic "Golden Rice" (derived from the daffodil) is in a usable form for human nutrition, especially in the absence of dietary fats and proteins (Grains of Delusion, 2001).

Transgenic herbicide-tolerant crops have been altered to withstand being sprayed with broad-spectrum herbicides, with the idea that one application will take care of most types of weeds without killing the crop. Insecticidal crops contain genes of the soil bacterium *Bacillus thuringiensis* (Bt). These Bt genes cause the plants to produce a chemical toxic to the European corn borer, the cotton bollworm, and other caterpillars. (Caterpillars are the larvae of insects in the Lepidoptera order, which includes moths and butterflies.)

As of 2005, about 87 per cent of world transgenic acreage was in the U.S. (Table 2). Herbicide-tolerant crops accounted for about three quarters of the acreage planted, worldwide, to genetically engineered crops in 2005. Pesticidal crops, or a combination of pesticidal and herbicide-tolerant crops, accounted for most of the remaining acreage. Acreage devoted to crops with stacked genes intended to express a variety of traits is increasing (USDA/NASS, 2005).

With an overwhelming amount of U.S. commodity program crop acreage devoted to transgenic versions, seed for conventional varieties is becoming scarce for those who choose not to plant transgenic crops. Traditional seed scarcity can affect farmers who wish to return to non-transgenic corn, soy, or cotton (Holden, 2002). Cotton seed is controlled by two large suppliers working with a large public research institution. Development of the non-transgenic organic/specialty cotton sector, which accounts for the 37 per cent non-transgenic cotton acreage in Texas (Table 1), has been hampered by concerns about cross-pollination and boll-weevil control. Soybeans and corn (often planted in rotation in the Upper Midwest) cover the most transgenic acres. There may be some new evidence that field workers working with Bt cotton are developing allergic reactions (Bernstein *et al.*, 1999).

One other large-acreage North American transgenic crop is canola (a low erucic acid form of European rapeseed). Canola is a major oilseed crop in Canada, but only a minor crop in the U.S. However, until recently, it was thought that acreage of both canola and rapeseed would increase in the near future in the Pacific Northwest. On May 9, 2006, a proposed large production facility at Gray's Harbor, Washington, announced that it would produce biodiesel from Asian palm oil, thus bypassing the "seed crushing hassles" of canola/rapeseed (Montana Department of Environmental Quality, 2006).

Proposals to plant substantial acreages of canola and rapeseed (*Brassica napus, B. rapa*)— much of it transgenic varieties—in Oregon's Willamette Valley to produce raw material for biodiesel production caused considerable concern among small-acreage vegetable seed producers. A preliminary 2006 Oregon State University Extension study predicted a high potential for gene flow between *B. napus* canola and other *B. napus* crops (rutabaga and Siberian kale). Likewise, *B. rapa* rapeseed holds the potential for gene flow with its closely related vegetable crops (Chinese cabbage, pai-tsai, mizuna, Chinese mustard, broccoli raab, and turnip). Potential for crossbreeding between the two oilseed crop types was rated high, as well. Potential of crossbreeding with wild (*Raphanus raphanistrum*) and cultivated (*R. sativum*) forms of radish was considered low. More study was called for regarding outcrossing of canola with *B. oleracea* vegetables (cabbage, cauliflower, Brussels sprouts, kohlrabi, collards, and kale).

Table 3. Percentages of U.S. 2005 crop acreage planted to insecticidal, herbicidal and stacked-gene varieties

	Insect resistant	*Herbicide resistant*	*Stacked-gene*
Soy	0	87	0
Corn	26	17	9
Cotton	18	27	34

Oregon Extension concluded that "genetically modified canola [and rapeseed] present the greatest risk to vegetable crucifer seed crops. The presence of the gene would make the seed crop unsuitable for markets that have strict tolerance on GMO contamination"—i.e., organic, identity preserved (IP), and European exports. Furthermore, "transgenes are relatively easy to detect at very low levels, so it is likely that their presence could be detected even if only a few interspecific hybrids were found in a vegetable seed lot" (Myers, 2006).

While acknowledging the risks to the producers of the nation's garden seed crops located in the Willamette Valley, researchers suggested that the vegetable seed producers could pack up and move (Myers, 2006).

Most transgenic cotton is herbicide-tolerant, though some varieties have the Bt trait; transgenic canola is herbicide-tolerant. The first transgenic wheat, initially planned for commercial introduction in 2003, is Roundup-tolerant. On May 10, 2004, Monsanto announced that it was discontinuing all research and field trial activities on Roundup-Ready wheat. After seven years of development, the release said, efforts to win over farmers and the international wheat market had failed.

A 2005 study published by the Western Resource Council showed that introduction of genetically modified wheat would lower income for wheat growers and the wheat industry. The report projects costs per bushel and per acre for farmers adopting Roundup-Ready wheat and for nonadopters under best-case and worst-case scenarios. Either way, farmers were projected to lose money from introduction and use of the Roundup-Ready wheat (Benbrook, 2005). Other traits engineered into commercial transgenic varieties include disease resistance, high pH tolerance, and several nutritional, taste, texture and shelf-life characteristics (BIO, 2000) — primarily through gene stacking.

In the absence of transgenic labelling, the average U.S. consumer may not realize that ingredients derived from transgenic corn, soya, and oilseed are in 70 per cent of the foods found in U.S. retail food outlets. Most prevalent is high-fructose corn syrup, which is replacing other sweeteners in a wide variety of mass-produced food products. The Biotech Industry Organization agrees that transgenic oils and ingredients derived from corn and soya are pervasive in conventional processed foods. Now that transgenic horticultural crops are in the market place, no one will know for sure—in the absence of labelling— whether fresh produce or processed shelf products contain engineered crops.

Five years ago introduction of transgenic fresh produce appeared imminent. Winter squash and a limited amount of sweet corn are now being retailed. However, after the Flavr-Savr® tomato was withdrawn and Starlink® feed corn caused a recall of taco shells, the subsequent paths of crops such as tomatoes, potatoes, sunflowers, peanuts, and sweet peppers diverged. Field trials were conducted from 1993–2001 on transgenic peanuts, all in the U.S. Field trials were conducted from 1993–2002 on sunflowers—in Australia, three European countries, and the U.S. Sweet bell peppers have been joined by rice, alfalfa, cabbage, carrots, cauliflower, sweet corn, cucumber, lettuce, mustard—and most recently, eggplant—on the list under development for commercial release. Transgenic fruits for which field trials are currently underway (some in the U.S.) are apples, cherries, cranberries, grapefruit, kiwi fruit, pears, persimmons, pineapple, plum, and strawberries. Transgenic papaya, raised in Hawaii, has been commercialized for several years, and plum has recently been deregulated by APHIS (Plum is, of course, the source of prunes).

Transgenic Potato in the U.S.

More than 700 field trials of transgenic Bt potatoes were conducted in the U.S. from 1989–2002 by a single company. In 1996, Bt potatoes were made available to commercial growers, but after 2000 the Bt potato program was abandoned due to lack of consumer acceptance.

Large fast food chains, snack food manufacturers, and potato processing conglomerates eliminated transgenic potatoes from their products. There are no other types of transgenic potatoes currently approved for sale in the U.S.

(www.truefoodnow.org/crop/pipeline_rdfruit.html).

One variety of transgenic flax was approved in the U.S. in 1999, but transgenic flax is reportedly not being grown because of consumer resistance and market rejection. Flax seed oil and flax seed are popular nutraceutical products. Transgenic rice trials in Missouri were halted by public protests. So far Iran is the only known country producing transgenic rice for human consumption (Table 2).

Despite indications in 2002 that lack of public acceptance of transgenic food would cause transgenic firms to change course, it has turned out that transnational corporations have changed tactics—conducting trials overseas, keeping U.S. trials strictly secret (perhaps even from regulatory oversight by APHIS). The companies also lobby industry groups, such as the wheat boards, and seek to develop indirect markets such as processing aids and minor ingredients. Transgenic processing aids—enzymes and ingredients used to improve the color, flavor, texture, and aroma of manufactured foods—and preservatives, stabilizers, vitamin additives, and a vast number of minor ingredients are currently being derived from transgenic corn or soy (Non-GMO Source, 2002).

The industry currently takes the position that the public has been consuming highly processed, transgenic foods for several years and that this large-scale experiment with the American food supply has been a success. Corn, oilseeds, cotton, and wheat are the North American crops with the most acreage and profit potential. For a more complete list of current and future commercial transgenic crops and their traits, see the APHIS list (Appendix 1).

Many disturbing unanswered questions remain about transgenic crops and their potential benefits, costs, and risks. In fact, according to an independent survey of research data on transgenic crops, conducted by the Winrock Foundation's Henry A. Wallace Center for Agricultural and Environmental Policy, "The varieties and uses of genetically altered crops have grown much more rapidly than our ability to understand them." This study reveals that only four per cent of total federal agricultural biotech funding is dedicated to environmental assessment (Wallace Center Report, 2001).

It should also be noted that there is even less research dedicated to human and animal health impacts of the technology.

Unresolved Issues of Concern

Unresolved issues in transgenic agriculture:

- Food safety
- Farm management
- Crop yield, costs, and profitability
- Marketing and trade
- Organic industry impacts
- Influence on public research
- Industry concentration and farmers' right to save seed
- Regulation of transgenic crops and apportionment of liability

ISSUES FACING FARMERS AND RANCHERS

Since 2001 ecological risks of transgenic crops have become evident.

Flow to Neighboring Crops and to Related Wild Species

Gene flow from transgenic fields into conventional crops and related wild plants has occurred. This issue is of special concern to farmers because of the potential to cause herbicide resistance. For example, in western Canada, three different herbicide-resistant canola varieties have cross-pollinated to create canola plants that are resistant to all three types of herbicide. This new triple resistance has turned volunteer canola into a significant weed problem (Ellstrand, 2001).

Gene flow from transgenic crops to wild relatives causes wild plants to acquire traits that improve their fitness, turning them into "super weeds." For example, jointed goatgrass—a weedy relative of wheat—can acquire the herbicide-tolerant trait of Roundup Ready wheat, and can therefore thrive in crop fields unless applications of other herbicides are made. Frank Young and his colleagues at Washington State University found that imidazolone-resistant wheat (not a transgenic variety) outcrossed to goatgrass in one season (Stierle, 2006). Other traits that wild plants could acquire from transgenic plants that will increase their weediness are insect and virus resistance (Ervin *et al.*, 2001). Alfalfa, a popular hay crop, can easily cross with black medic, an invasive species prevalent in the western U.S. The Federal Register of June 27, 2005, announced that genetically modified alfalfa was unrestricted and that seed has been released for sale to farmers (Moore, 2005; Non-GMO Source, 2005). The Biotechnology Industry counters that resistant weeds can be controlled by "other herbicides." Research done at Iowa State University's Leopold Center found that the increased cost negates any advantage to the farmer of using transgenic seed (Benbrook, 2001).

Because of potential effects on pest management, crop marketability, and liability, more research needs to be done to determine the conditions under which gene flow from transgenic plants is likely to be significant.

Pharmacrops

After the year 2000, reorganizations in the agri-chemical/pharmaceutical industry led to a new emphasis on development of bioengineered products for enhancement of human and animal health. Between 1999 and 2002, 315 trials of pharmaceutical crops were conducted in the U.S., and such trials are ongoing. Corn is by far the most popular pharmacrop, accounting for more than two-thirds of the biopharm plantings. Other crops engineered for biopharmaceutical production include soybeans, rice, barley, wheat, canola, and tobacco.

Kentucky farmers report that transgenic tobacco has become the long-sought replacement crop after the tobacco buyout. Same was being trialed as a source of an AIDS medication. As of 2001, biopharm field trials had been conducted on at least 900 acres, probably closer to 1,600. The exact figure is not known because the USDA classifies these field trials as "confidential business information." The December, 2005 Office of Inspector General (OIG) report criticized the regulatory agency charged with monitoring company field trials for lax reporting and inadequate monitoring, especially of "high-risk pharmaceutical and industrial crops" and called for "science-based risk assessment" (USDA/OIG, 2005).

Pesticide Resistance in Insect Pests

Bacillus thuringiensis, or Bt, has been widely used as a microbial spray because it is toxic only to caterpillars. In fact, it is a pest management tool that organic farmers partially depend on—one of the few insecticides acceptable under organic rules. Unlike the commercial insecticide spray, the Bt engineered into transgenic crop plants is reproduced in all, or nearly all, the cells of every plant, not just applied on the plant surface for a temporary toxic effect. As a result, the possibility that transgenic Bt crops will accelerate development of pest resistance to Bt is of serious concern. Such resistance would remove this valuable and environmentally benign tool from the pest control toolbox of farmers and forest managers.

Antibiotic Resistance

As described in the earlier section on how gene transfer is accomplished, the use of antibiotic-resistant marker genes for the delivery of a gene package into a recipient plant carries the danger of spreading antibiotic-resistant bacteria. The implications for creation of antibiotic-resistant diseases are disturbing. Research is needed on antibiotic resistance management in transgenic crops (ESCOP/ECOP, 2000). The European Commission's new rules governing transgenic crops stipulated phasing out antibiotic-resistant marker genes by the end of 2004. Because of potential effects on pest management, crop marketability, and liability, more research needs to be done to determine the conditions under which gene flow from transgenic plants is likely to be significant. By the end of 2005, no such research was underway and implementation of the EU rule has been complicated by imminent publication of a WTO ruling against EU trade restrictions on transgenic crops. The ruling is certain to be appealed (Kiplinger, 2006).

Effects on Beneficial Organisms

Evidence continues to increase that transgenic crops—either directly or through practices linked to production—are detrimental to beneficial organisms. New studies show that Bt crops exude Bt in concentrations high enough to be toxic to some beneficial soil organisms. University of Arkansas agronomists found impaired "root development, nodulation, and nitrogen fixation" in Roundup-Ready soy (King *et al.*, 2001). Disruption of beneficial soil organisms can interfere with plant uptake of phosphorus, an essential plant nutrient (Massey, 2000). Beneficial insects that prey on insect pests can be affected by insecticidal crops in two ways. First, the Bt in transgenic insecticidal crops has been shown in some laboratory studies to be toxic to ladybird beetles, lacewings, and monarch butterflies (Ervin *et al.*, 2001). The extent to which these beneficials are affected in the field is a matter of further study. Second, because the insecticidal properties of Bt crops function even in the absence of an economic threshold of pests, Bt crops potentially can reduce pest populations to the point that predator species are negatively affected.

Reduced Crop Genetic Diversity

As fewer and larger firms dominate the rapidly merging seed and biotechnology market, transgenic crops may continue the trend toward simplification of cropping systems by reducing the number and type of crops planted. In addition, seed-saving, which promotes genetic diversity, is discouraged. In Europe, seed-saving traditionally practiced by a majority of farmers has been heavily restricted through registration requirements and subsidy payments. To be certified, seed must exhibit "distinctiveness, uniformity, and stability," called "DUS registration" (Toledo, 2002). A traditional landrace can be held uncertifiable (and effectively outlawed by billings for royalties and denial of subsidy payments) by being declared insufficiently distinct from a variety described in the EU Catalogue of Common Varieties. In an interview, Nancy Arrowsmith, founder of Arche Noah, (Arrowsmith, 1987) noted that traditional European landraces and seed-saving practices are being squeezed out in Common Market countries.

Seed legislation is quite restrictive. In order to be distributed, seeds have to be registered. There has to be extensive testing— up to seven years—and the registration fee is quite high. [Germany, Switzerland,] and all of the countries that belong to the Common Market have adopted what they call the Common Catalogue. Only the vegetable varieties listed in this Catalogue can be sold.

In Austria [Arrowsmith's home] many varieties are protected. ... In the catalogue, it will say that these cannot be reproduced in any way.

Outlawing landraces by legislative fiat (most recently in Iraq) was thwarted in the U.S. by organizations like the Seed Savers Exchange, which mobilized support for strong protection of the rights of seed savers in Plant Variety Patent legislation passed in the late 1980s. Traditional open-pollinated varieties are still vulnerable to genetic contamination by cross-pollination.

Following is a brief discussion of some of the remaining risk issues :

Food Safety

Food safety issues, except as they impact domestic marketing and exports, are beyond the scope of this publication. Five years ago, the major publicized concerns were environmental. Since then, the environmental community has stalled some transgenic crops. Food safety concerns include:

- Possibility of toxins in food
- Possibility of new pathogens
- Reduced nutritional value
- Introduction of human allergens
- Transfer of antibiotic resistance to humans
- Unexpected immune-system and genetic effects from the introduction of novel compounds

It is in part because of these concerns that domestic consumer demand for organically grown crops continues to increase. There are other marketing problems that reflect religious dietary and general religious (sometimes dismissed as "cultural") sensibilities, as well as ethical/philosophical concerns.

Farm Management Issues

The most widely planted transgenic crops on the market today can simplify short-term pest management for farmers and ranchers. In the case of herbicide-tolerant crops, initially farmers hoped to use a single broad-spectrum herbicide for all their crop weeds. It has turned out that they need more than one application in most seasons. By planting insecticidal crops, farmers can eliminate the need to apply pesticides for caterpillar pests like the European corn borer or the cotton bollworm, though they still have to contend with other crop pests.

While these crops offer simplified pest control features, they may complicate other areas of farm management. Farmers who grow both transgenic and conventional varieties of the same crop will need to segregate the two during all production, harvesting, storage, and transportation phases if they sell into differentiated markets or plan to save their own seed from the conventional crops.

To minimize the risk of gene flow from transgenic to adjacent conventional crop fields, federal regulations require buffer strips of conventional varieties around transgenic fields. Different transgenic crops require different buffer widths. Because the buffer strips must be managed conventionally, producers have to be willing to maintain two different farming systems on their transgenic fields. Crops harvested from the buffer strips must be handled and marketed as though they are transgenic.

Plant refuges—where pest species can live outside the fields of insecticidal and herbicide-tolerant transgenic crops—are also required to slow the development of weed and insect pest resistance to Bt and broad-spectrum herbicides. These refuges allow some individuals in the pest population to survive and carry on the traits of pesticide susceptibility. Requirements governing the size of refuges differ according to the type of transgenic crop grown, but a 2006 report in AgBioForum, based on a survey of Indiana farmers, states the requirements are misunderstood by farmers and routinely ignored. For some crops they are unworkable (Alexander and Van Melior, 2005).

Farmers growing herbicide-tolerant crops need to be aware that volunteer crop plants in the following year will be herbicide-resistant. Such resistance makes no-till or direct-seed systems difficult because volunteers can't be controlled with the same herbicide used on the rest of the crop. In a no-till system that relies on the same broad-spectrum herbicide that the volunteer plants are resistant to, these plants will contaminate the harvest of a following conventional variety of the same crop—a situation farmers tend to avoid for two reasons. First, the contamination means a following conventional crop will have to be sold on the transgenic market. This leads to the second reason. If farmers grow and market a transgenic crop for which they do not have a technology agreement and did not pay royalty fees, they may face aggressive collection by the company that owns the transgenic variety. Hundreds of U.S. farmers have already been charged with "theft" of a company's patented seed as a result of contamination in the field (Altieri, 2000).

Farmers growing insecticidal crops need to recognize that insect pressure is difficult to predict and may not warrant the planting of an insecticidal variety every year. In a year when pest pressure is low, the transgenic seed becomes expensive insurance against the threat of insect damage (Hillyer, 1999). Farmers growing transgenic crops need to communicate with their neighbors to avoid contaminating neighboring fields and to ensure that buffers are adequate. In Maine, farmers growing transgenic crops are now required by law to be listed with the state agriculture department, to help identify possible sources of cross-contamination when it occurs. The law also "requires manufacturers or seed dealers of genetically engineered plants, plant parts, or seeds to provide written instructions to all growers on how to plant, grow, and harvest the crops to minimize potential cross-contamination of non-genetically engineered crops or wild plant populations" (AgBioTech, 2001).

Farm management issues common to all transgenic crops include yield, cost, price, profitability, management flexibility, sustainability, market acceptance, and liability. Yield and profitability, as well as market acceptance, are discussed in separate sections below. Liability is discussed under regulation.

CROP YIELD, COSTS, AND PROFITABILITY

Some farmers will get higher yields with a particular transgenic crop variety than with their conventional varieties, and some will get lower yields. Yield variability is related to many factors, including choice of the conventional analog of the transgenic variety, making it very difficult to analyze how any one feature impacts yield. Costs of various inputs are also constantly changing; and the ability of farmers to adjust to changing costs, particularly rapid changes, is limited and affects profitability. However, some yield, cost, and profitability trends do appear to be emerging from the growing body of research data for transgenic crops. As noted in the Wallace Center report, Roundup Ready soybeans were designed simply to resist a particular chemical herbicide, not to increase yields. In contrast, Bt corn and cotton, by resisting insect pests, may result in higher yields from reduced pest pressure (Wallace Center, 2001).

Yield: Herbicide Tolerant Crops— Soybeans, Cotton, Canola

Herbicide-tolerant soybeans appear to suffer what's referred to as "yield drag." Again, in some areas and on some farms this tendency of Roundup Ready soybean varieties to yield less than their comparable, conventional counterparts varies, but overall, they appear to average yields that are five to ten per cent lower per acre. As described earlier, impaired root development, nodulation, and nitrogen fixation likely account for this yield drag. Drought conditions worsen the effects. The bacterium that facilitates nodulation and nitrogen fixation in the root zone is apparently sensitive to both Roundup and drought. University of Missouri scientists reported problems with germination of Roundup Ready soybeans in the 2001 crop year (UM press release, 2001). Yields of herbicide-tolerant cotton are reportedly not significantly different from those of conventional cotton (Benbrook, 2001; Wallace Center, 2001, summarizing research by Klotz-Ingram *et al.*, 1999).

Herbicide-resistant transgenic canola varieties yield less on average than conventional canola varieties. Transgenic canola costs less than conventional canola to produce, but because of its higher yields, conventional canola returns more profit per acre (Fulton and Keyowski, 1999).

Yield: Insecticidal Crops— Corn, Cotton

Insecticidal Bt corn and cotton generally yield higher "in most years for some regions" according to USDA Economic Research Service data from 1996 to 1998. Bt cotton, especially, outpaces yields of conventional cotton by as much as 9 to 26 per cent in some cases, though not at all in others. Yield increases for Bt corn have not been as dramatic (Fulton and Keyowski, 1999). Time will tell whether farmers can expect yield increases or decreases in the long run with these and other transgenic crop varieties.

Changes in Chemical Pesticide Use

One of the promises of transgenic technology is that it will reduce pesticide use and thereby provide environmental benefits while reducing farmers' costs. The herbicide-tolerant and insecticidal varieties are designed specifically to meet these goals. Studies estimate a two to three per cent decrease in U.S. pesticide use, but the effects vary widely by crop, region, and year. Increased future pesticide use resulting from the buildup of resistance to heavily used herbicides is a long-term concern (Ervin *et al.*, 2001) acknowledged by the transgenic crop industry. Pesticide use depends on the crop and its specific traits; weather, severity of pest infestations; farm management; geographic location of the farm; and other variables. As a result, conclusions drawn by various studies analyzing pesticide use on transgenic crops remain controversial. According to the Wallace Center report, in a review of the data available up through 2000, crops engineered to contain Bt appear to have decreased the overall use of insecticides slightly, while the use of herbicide-resistant crops has resulted in variable changes in overall herbicide use, with increases in use of some herbicides in some places and decreases in others (Wallace Center, 2001). The crop for which studies are showing the largest decrease in pesticide use is Bt cotton, with Bt corn resulting in only small changes. Herbicide-tolerant cotton has also resulted in little change in herbicide use (Ervin *et al.*, 2001).

The data for herbicide-tolerant soybeans seems harder to interpret. A recent study of herbicide use data on Roundup Ready soybeans by Charles Benbrook, Ph.D, former executive director of the National Academy of Sciences Committee on Agriculture and now with the Northwest Science and Environmental Policy Center, concludes that the use of herbicides has actually increased because the weeds have become resistant to Roundup (Benbrook, 2004). While another recent study by scientists in The Netherlands shows a decrease in herbicide use on transgenic soybeans, it is clear that weed resistance to Roundup may lead to increased herbicide use and to the need to shift to more toxic compounds in the future (Ervin *et al.*, 2001), and this is acknowledged by the industry. American Soybean Association president Tony Anderson agrees that the developing resistance of weeds to herbicides such as Roundup is a problem (Environmental News Service, 2001)..

The Wallace Center report emphasizes the importance of ongoing monitoring of pesticide use data. If farmers abandon integrated pest management, which utilizes a variety of pesticide and cultural control methods, in favor of the simplified control offered by herbicide-resistant and insecticidal transgenic crops, then early findings of reduced pesticide quantities and toxicity may not hold over the long run. Refer to chapter one of the Wallace Center report (Wallace Center, 2001) for USDA pesticide use data comparisons between transgenic and conventional crops, broken down by crop.

Profitability

Farmers need to consider all the factors that determine profitability. No single factor can tell the whole story. Transgenic crop seeds tend to be more costly, and farmers have the added expense of a substantial per-acre fee charged by the owners of transgenic varieties. These costs have to be considered along with input cost changes—whether herbicide or insecticide use and costs go down, go up, or stay the same. Market price is another factor. Prices for some transgenic crops in some markets are lower than prices for comparable conventional crops, though rarely they are higher. Farmers need to watch the markets. Some buyers will pay a premium for a non-transgenic product, though as transgenic seeds find their way into conventional transportation, storage, and processing steams, these premiums may disappear along with confidence that "GMO-free" products are in fact truly free of engineered genes. Future availability of conventional seed is another issue. Once farmers try transgenic crops, they have reported becoming locked into the technology, as alternate conventional seed supplies dry up. Also the potential liability of transgenic plants coming up in a conventional planting the next year is important to farmers. Transgenic seed suppliers aggressively pursue legal cases against any farmer using transgenic seed without having a signed technology agreement.

Marketing and Trade

Buyer acceptance is a significant marketing issue for farmers raising transgenic crops. Farmers need to know before they plant what their particular markets will or won't accept. Since most grain handlers cannot effectively segregate transgenic from non-transgenic crops in the same facility, many companies are channeling transgenic crops into particular warehouses. Farmers need to know which ones and how far away those are.

Many foreign markets have tended to be more leery of transgenic products than domestic markets, although this may change. World Trade Organization (WTO) directives can force dropping of trade barriers, but consumer acceptance cannot be forced (when choice is possible). Africa is a special case, as authority to accept or reject transgenic products was retained by governments, and several have banned GMOs in any form—even relief grain shipments. India has now developed its own transgenic industry and is producing transgenic cotton, while actively resisting attempts by others to patent its indigenous crop genetics. Brazil, Argentina, and China rank among the top five countries in acreage of transgenic soybeans, maize, and cotton. Even two European countries — Spain and Romania — are producing transgenic crops for animal feed (Table 2).

Eighty-six countries and the European Union have agreed on implementation steps for the UN's Cartagena Protocol on Biosafety, which came into force in September 2003. A rigorous system for handling, transporting, packing, and identifying transgenic crops was part of the agreement. All bulk shipments of genetically engineered crops intended for food, animal feed, or processing are to be labeled "May Contain LMOs," (Living Modified Organisms) according to the UNEP. Major producers of transgenic crops, including Canada, Argentina, and the U.S., did not sign the protocol (Agence France Presse, 2004).

Trade in transgenic livestock feed is more liberal than trade in transgenic human food. The rapid and widespread dissemination of the Cry9C Bt transgene (StarLink), which is not approved for human consumption but was detected in tacos, shows how easily transgenic material can spread from animal feed to human food products. The widespread publicity has resulted in even further resistance on the part of buyers to purchasing transgenic products for human food. According to a report in Britain's *The Guardian*, "No new transgenic crops have been approved by the European Union (EU) since April 1998, and a defacto moratorium on further approvals has been in place since June 1999" (Osborn, 2001). However, trials of food crops already approved continued, and the European Union officially lifted its moratorium on the introduction of new transgenic crops in 2004, although during the debate over labeling and traceability regulations the moratorium remained in effect (Evans, 2001).

If approved, the new regulations will complicate the export of U.S. farm products to the EU because the U.S. does not require traceability or labeling of transgenic crops.

Under the proposed new EU requirements, "all foods and animal feed derived from GMOs have to be labeled and, in the case of processed goods, records have to be kept throughout the production chain allowing the GMO to be traced back to the farm" (Evans, 2001). Spain and Romania rank in the top 14 countries growing biotech crops. Both grow transgenic animal feed crops. Portugal, Germany, France, and the Czech Republic grow small amounts of feed corn (maize)—less than 10,000 hectares (24,700 acres), probably much less.

While the U.S. does not require mandatory labeling of processed food containing transgenic ingredients, the EU, Russia, Japan, South Korea, Taiwan, Australia, New Zealand, and Ecuador do have such requirements, as of 2001 (Schrade and Raabe, 2001). The degree to which the Cartagena protocol (Agence France Presse, 2004) will be implemented by other signatories (see above) is unknown. Because many domestic merchandisers of agricultural commodities do not segregate transgenic from conventional crop varieties, it is impossible for them and the farmers that supply them to serve these food markets.

Twenty per cent of corn and 35 per cent of soybeans produced in the U.S. are exported (USDA/AMS, 2006), and more than 80 per cent of these crops are used in animal feed. Few, if any, animal feeding trials were carried out before transgenic crops were released. In 2005, grain exports were down 5 per cent overall from the previous year and 26 per cent at Gulf Coast ports (due to the hurricanes).

In contrast, in a dramatic increase from 2001, 45 per cent of U.S. wheat is exported (USDA/AMS, 2006) Exports to countries that are resistant to buying transgenic food—particularly Japan and European nations—are dropping, but being supplanted by increased demand from Nigeria and Iraq (USDA/AMS, 2006). Because wheat producers are so dependent on exports, they have vigorously resisted introduction of the first transgenic wheat, originally slated for 2003, now on hold. The Japanese milling industry has made it clear that it does not want transgenic products. As a result, Monsanto promised not to introduce Roundup Ready wheat until Japan gave its approval (Hord, 2001). North Dakota and Montana considered legislation that would place a state moratorium on the introduction of transgenic wheat. Recent federal regulation, under the Homeland Security Act, would nullify any such local or state food laws.

In addition to national and international policies on the use and importation of transgenic crops, processors and retailers in many countries have set their own corporate policies. Major retail chains in Europe and the U.S. have declared their commitment to avoiding the purchase of transgenic products, both feed and food. But, in the absence of labeling, most have been willing to accept a pervasive presence of transgenic corn, soy, and canola in processed products. Although European Union rules effectively barring U.S. corn imports have been recently relaxed, since 1997 the European Union ban has cost American farmers access to a $200 million annual market (Shadid, 2001) and the U.S. government billions in agricultural price supports.

ORGANIC INDUSTRY

Organic farmers face even bigger marketing and trade risks, since their buyers expect no transgenic contamination. Currently, organic production is process-oriented, not testing oriented—except for exports. The organic industry has a system for segregation, but recent tests for transgenic material in organic products demonstrate that it is not immune to contamination from conventional systems (Callahan, 2001). New technologies can reliably detect minute amounts of transgenic material. Published reports from Europe and the U.S. confirm a high degree of accuracy for detection methods (Non-GMO Source, 2004). European export markets organic farmers might have enjoyed, and those that producers of non-GE conventional crops could have built upon, have proven unstable in the presence of possible transgenic contamination. In 2005, U.S. exports of agricultural products fell below imports, for the first time in 20 years.

Seed testing for genetically modified traits

Selecting the appropriate test for seed will ultimately depend on the end use of the results. Are you looking for the absence or presence of a trait, or do you need quantitative data? The best approach to testing for genetically modified traits is to understand the ultimate use of the tests and then to talk with the laboratory or technologist that will be performing the test. The technologist will be able to describe the four types of tests commonly used by the seed industry to test for traits:

- Herbicide bioassay
- Immunoassay (ELISA, lateral flow strips)
- Electrophoresis (PAGE, IEF, starch-gel)
- Polymerase chain reaction (PCR)

The technologist can help you select the test that will best fulfill your needs. Five years ago, the Society of Commercial Seed Technologists (SCST) created an accreditation program for technologists in these four areas. The program ensures that the technologist is proficient in both the theory and practical application of the genetic purity tests currently utilized by the seed industry. Using a laboratory with a certified or registered genetic technologist ensures that GM tests are conducted by an experienced person. The SCST Genetic Technology Committee and working groups are extremely active in providing training and education to keep members up-to-date in this rapidly evolving area of seed testing.

Hall, Anita, executive director, Society of Commercial Seed Technologists, Inc.

INFLUENCE ON PUBLIC RESEARCH

While transgenic crop varieties are generally the property of private corporations, those corporations often contract with public-sector agricultural research institutions

To increase the genetic diversity of U.S. corn, the Germplasm Enhancement for Maize (GEM) project seeks to combine exotic germplasm, such as this unusually colored and shaped maize from Latin America, with domestic corn lines.

Photo by: Keith Weller, USDA ARS

A retooled gene in Endless Summer tomatoes controls ripening to give better flavor and shelf-life.

Photo by: Jack Dykinga, USDA ARS

for some of their development work. In fact, private investment in agricultural research, including germplasm development, has surpassed public investment in recent years (ESCOP/ECOP, 2000). With this shift in funding priorities, the following questions become important: Is the private sector unduly influencing the public research agenda? Are corporations directing public research in socially questionable directions while research on, for instance, sustainable agriculture wanes? Are the outcomes of corporate-funded transgenic research and development by our public institutions equitable across the food and agricultural sectors? Is equity even a consideration of our public institutions when they accept this work?

When intellectual property rights (patents) apply to living organisms, making them private property, the free flow of scientific information that has historically characterized public agricultural research is inhibited. What are the implications for the future of agriculture and society of the secrecy that now surrounds so much of what was formerly shared public knowledge?

These and other questions need to be addressed by citizens and their public institutions. These issues are of particular concern to farmers and consumers who would benefit from research into alternative technologies that are less costly (in every way), less risky, and more equitable. Equity requires that the economic benefits and risks of technology be fairly distributed among technology providers, farmers, merchants, and consumers.

For long-term sustainability, farmers need research that focuses on farms as systems, with internal elements whose relationships can be adjusted to achieve farm management goals (Union of Concerned Scientists, 2000). In contrast, transgenic crop research so far has focused on products that complement toxic chemical approaches to control of individual pest species. These are products that can be commercialized by large agribusiness or agri-chemical interests and that farmers must purchase every year. This research orientation only perpetuates the cost-price squeeze that continues to drive so many out of farming.

INDUSTRY CONCENTRATION AND FARMERS' RIGHT TO SAVE SEED

The broadening of intellectual property rights in 1980 to cover living organisms, including genes, has resulted in a flurry of mergers and acquisitions in the seed and biotech industries. According to the Wallace Center report, "Relatively few firms control the vast majority of commercial transgenic crop technologies". These firms have strategically developed linkages among the biotechnology, seed, and agri-chemical sectors to capture as much market value as possible. However, these tightly controlled linkages of product sectors raise serious issues of market access, product innovation, and the flow of public benefits from transgenic crops (Wallace Center, 2001).

Unlike Plant Variety Protection—which does not allow for the patenting of individual genes, but only of crop varieties—Intellectual Property Rights prohibit farmers from saving seed and undertaking their own breeding programs, and prohibit plant breeders from using the material to create new generations of varieties adapted to specific regions or growing conditions (Guebert, 2001). The Intellectual Property Rights have recently been upheld by the U.S. Supreme Court, which relied on a 1795 General Patents statute.

By 2000, agri-chemical giants DuPont and Monsanto together owned 73 per cent of the corn seed producers in the U.S. (Massey, 2000). Although some have recently been divested, this kind of corporate control and concentration raises the question of whether there remains enough competition in the seed industry for seed pricing to remain competitive. As additional concentration occurs, how affordable will seed—all seed, not just transgenic seed—be for farmers? This question takes on added gravity as an increasing number of seed varieties become proprietary and seed production, especially for gardeners, is pushed out of California and Oregon into Nevada and Idaho. Farmers can't save proprietary seed for planting and so must purchase new seed every year. In addition, farmers choosing transgenic varieties must sign a contract with the owner of the variety and pay a substantial per-acre technology fee, or royalty.

The anticipated commercial introduction of transgenic wheat represents a dramatic shift in an industry in which farmers still widely save their own seed. As non-transgenic varieties become contaminated with transgenic ones, even those farmers who choose to stick with conventional varieties will lose the right to replant their own seed. This loss has already occurred in Canada's canola industry, with Monsanto winning its court case against farmer Percy Schmeiser for replanting his own canola variety that had become contaminated with Monsanto's Roundup Ready canola.

The adoption of transgenic crop varieties has brought with it an increasing prevalence of contract production. While contract production can lead to increased value and reduced risk for growers, farmers are justified in their concern about their loss of control when they sign a contract with a private company. Issues associated with contract production of transgenic crops must be considered within the broader context of a sustainable agriculture to include ownership, control, and social equity.

REGULATION OF TRANSGENIC CROPS AND APPORTIONMENT OF LIABILITY

Much of the controversy over transgenic crops, both internationally and in the U.S., is in part a result of how the U.S. regulates transgenic crops. The federal government has determined that the commercial products of agricultural biotechnology are "substantially equivalent" to their conventional counterparts and that therefore no new regulatory process or structure is needed for their review and approval. Currently, three federal agencies regulate the release of transgenic food crops in the U.S.: the U.S. Department of Agriculture's Animal and Plant Health Inspection Service (USDA- APHIS), the U.S. Environmental Protection Agency (EPA), and the U.S. Food and Drug Administration (FDA).

USDA–APHIS: APHIS looks at how a transgenic plant behaves in comparison with its unmodified counterpart. Is it as safe to grow? The data it uses are supplied largely by the companies seeking a permit for release of the new crop. Under "fast-track" approval, a process in place since 1997, companies introducing a crop similar to a previously approved version need give only 30 days' advance notice p nor to releasing it on the market. According to the Wallace Center report, APHIS staff estimate that by 2000, 95 to 98 per cent of field tests were taking place under simple notification rules rather than through permitting (Wallace Center, 2001). The Office of Inspector General (OIG) report has called for much stricter tracking—in light of the industry shift to industrial and pharmacrops.

EPA: The EPA regulates the pesticides produced by transgenic crops, such as the Bt in Bt corn and cotton. It does not regulate the transgenic crops themselves. In contrast with its regulation of conventional pesticides, the EPA has set no tolerance limits for the amount of Bt that transgenic corn, cotton, and potatoes may contain (Wallace Center, 2001).

FDA: The FDA focuses on the human health risks of transgenic crops. However, its rules do not require mandatory pre-market safety testing or mandatory labeling of transgenic foods. Initially, the U.S. regulatory process for transgenic food crops required product-by-product reviews. Now, however, to simplify and speed up the process, new products can be approved based on the experience gained in reviewing earlier products. According to the Wallace Center report, the implication is that "some crops might be approved, or disapproved, without actual field testing" (Wallace Center, 2001). The regulatory process, in fact, may not answer most questions about the environmental and human health risks of commercial production of these crops (Advisory Committee on Biotechnology, 2003).

Central to the policy of substantial equivalence is the assumption that only the end product of transgenic technology is of concern—not the process of genetic modification. Canada has adopted a similar approach. Europe and other U.S. trading partners, however, have taken a more conservative approach. They focus on the process of genetic modification—the source of many of the environmental and human health risks of greatest concern.

How these different approaches play out in reality can be summed up simply. The U.S. and Canada assume a product is safe until it is proven to carry significant risk; the European Union, which follows the "precautionary principle," assumes the same product may carry significant risk until it can be proven safe. The science used by the two approaches is not fundamentally different. The difference is in the level of risk the different societies and political systems are willing to accept (ESCOP/ECOP, 2000).

Liability

Farmers who choose not to grow transgenic varieties risk finding transgenic plants in their fields anyway, as a result of cross-pollination via wind, insects, and birds bringing in pollen from transgenic crops planted miles away. Besides pollen, sources of contamination include contaminated seed and seed brought in by passing trucks or wildlife. Those farmers whose conventional or organic crops are contaminated, regardless of the route, risk lawsuits filed against them by the companies that own the proprietary rights to seed the farmer didn't buy. Likewise, farmers who grow transgenic crops risk being sued by neighbors and buyers whose non-transgenic crops become contaminated.

Because contamination by transgenic material has become so prevalent in such a short time, all farmers in areas of transgenic crop production are at risk. Insurance, the most common recourse for minimizing potential losses because of liability, is not available to the nation's farmers for this risk because insurance companies do not have enough experience for gauging potential losses.

Most of the farmers who have been accused by transgenic seed companies of illegally growing and harvesting their proprietary transgenic varieties have paid fines to the companies rather than go to court to defend themselves. One Canadian grower of non-transgenic canola, Percy Schmeiser, did go to court against Monsanto, and lost. The initial ruling required him to pay Monsanto the approximately U.S. $85,000 value of his crop and $13,000 in punitive damages. The amount was eventually reduced, upon appeal. This farmer's case has implications for other farmers, especially those who traditionally save their own seed for planting the next year's crop.

In another example, a Texas organic farm's corn, assumed to be GE-free, was purchased by a processor who made it into organic tortilla chips. Only after the product had been sold and shipped to European retailers was it discovered to be contaminated with transgenic corn. The processor had to recall its product at a cost of over $150,000. The processor chose not to sue the organic farmer, but could have (Shadid, 2001). The retailer, in turn, apparently did not sue the processor. Had it done so, the liability for the retailer's loss could have fallen on both the processor and the farmer.

Until laws or legal precedent clarify the extent of farmer liability, farmers would do well to avoid making assumptions or claims about the purity of their non-transgenic products. (Some export crops, as well as Identity Preserved crops, are now being tested before shipping.) Furthermore, producers of transgenic crops need to take all possible precautions against spreading pollen and seed to their own and others' nontransgenic fields and markets. A full risk assessment and legal clarification of the distribution of liability among farmers, seed companies, grain handlers, processors, and retailers is needed before farmers can rest assured that transgenic crops won't result in lawsuits against them. For farmers who have adopted transgenic technology—and for those who have not—in the words of Brian Leahy, executive director of California Certified Organic Farmers, "This technology does not respect property rights" (Shadid, 2001).

In essence, farmers who grow transgenic crops on some of their fields, and farmers who grow none, risk bearing tremendous liability. This situation won't change until farmers either gain legal protection or stop growing transgenic crops entirely.

Precautionary Principle

The principle of precaution was developed specifically for issues involving ethics and risk. It is described by Katherine Barrett, project director with the Science and Environmental Health Network, as: "... a process for decision-making under conditions of uncertainty. The principle states that when there is reason to believe that our actions will result in significant harm, we should take active measures to prevent such harm, even if cause-and-effect relationships have not been proven conclusively... It has been invoked in many international laws, treaties, and declarations on a range of environmental issues including climate change, marine dumping of pollutants, and general efforts towards sustainability— [including the 2000 international Cartagena Biosafety Protocol on the transfer of 'living modified organisms'] ... There is growing consensus that the precautionary principle has reached the status of international customary law" (Barrett, 2000).

The primary elements of the precautionary principle are identified by Barrett as:

1. Avoid harms to the environment that are "irreversible, persistent, bioaccumulative, or otherwise serious".
2. "Anticipate and prevent potential harms at the source," rather than relying on reactive measures of mitigation, clean-up, or compensation.
3. Recognize the limits of scientific knowledge and don't expect a full and conclusive understanding of potential consequences before taking precautionary action, especially when the potential consequences are long-term, unconfined, and broad-scale.
4. Shift the burden of proof to the developers of potentially hazardous technologies.
5. Finally, some versions of the precautionary principle include analysis of the costs and benefits of precautionary action, though cost-benefit analysis "is not a sufficiently robust decision-making framework to stand alone" (Barrett, 2000).

Implications for Sustainable Agriculture

In contrast to the ecological approach of sustainable agriculture, "the current generation of transgenic crops follows a pest management model like that employed for chemical pesticides— through interventions that are toxic to pests," the Wallace Center report points out. This "single-tool" approach is likely to fail in the long run because pests will successfully develop resistance that allows them to thrive (Wallace Center, 2001).

Standard plant-breeding methods can potentially solve many of the same problems in agriculture that genetic engineers are working on, though there are areas in which genetic engineering can enhance traditional plant breeding. Armed with the map of an organism's genetic code, scientists can test which genes are in a plant to select more easily which ones to cross-breed. "Before we knew where the genes were, we were still breeding in the dark," according to Steven Briggs, head of genomics for Syngenta, a Swiss biotechnology giant, as quoted in the New York Times (Pollack, 2001).

Gauging a Technology's Impact on Agricultural Sustainability

The sustainability of any agricultural technology can be gauged in part by answering a series of questions that emerge from the principles of sustainable agriculture. Farmers and ranchers can ask themselves these questions in the context of their own operations to help determine whether adoption of the technology will move them away from, or toward, increased sustainability.

1. Does the technology increase genetic diversity?
2. Does it maintain a positive balance of pests and predators?
3. Does it protect or enhance soil biota?
4. Does it decrease the quantity or concentration of toxins released into the environment?
5. Does it decrease soil erosion?
6. Does it protect non-target organisms?
7. Does it help protect natural habitats?
8. Does it reduce pest populations and viability?
9. Does it increase farmers' yields? Decrease farmer's costs?
10. Does it increase farmers' market control? Management flexibility? Time?
11. Does it provide benefits to consumers? Will consumers accept it?
12. Does it help citizens globally gain better access to food?
13. Does it protect the public's access to information and improve public trust in agriculture?

If the answer to any of the above questions is no, a cautious approach to the adoption of the technology in question would seem in the interest of agricultural sustainability.

Conclusion

Evelyn Fox Keller, author of *The Century of the Gene* (Fox Keller, 2000), describes the scientific understanding of genetics that originated with the discovery of DNA in 1953, and on which the current generation of transgenic crops is still based :

For almost fifty years, we lulled ourselves into believing that, in discovering the molecular basis of genetic information, we had found the "secret of life"; we were confident that if we could only decode the message in DNA's sequence of nucleotides, we would understand the "program" that makes an organism what it is.

Recent scientific discoveries no longer support this theory. Outdated as it has become, the view that each genetic message comes from a distinct gene continues to drive promises of feeding the world and curing incurable diseases—a view Keller calls a now "utterly fantastic premise." Keller, a professor of History and Philosophy of Science at Massachusetts Institute of Technology, insists that the most recent scientific understanding of genetics has more to tell us about biological organization than about how to modify individual traits. In fact, the new leading-edge biotechnology will strive to capture the benefits of genetic engineering without the costs, risks, and potential genetic instability.

Molecular biologists and breeders are beginning to utilize the emerging knowledge of gene location and function to guide them in the application of conventional and innovative breeding techniques that do not rely on a transgene with promoter and marker genes (AgBioTech, 2001). This is a good news for sustainable agriculture, which is based on understanding how natural systems work in order to fit human enterprises into them. According to Fred Kirschenmann, organic farmer and former director of Iowa State University's Leopold Center for Sustainable Agriculture, "The real benefit of genetics seems to be derived not from the manipulation of a few genes, but from our enhanced understanding of how nature works" (Kirschenmann, 2001).

Regardless of the future direction of transgenic technology, one thing remains certain: Many of the unresolved issues for farmers, ranchers, and the general public will not be settled through the use of biological or natural sciences alone.

REFERENCES

Advisory Committee on Biotechnology & 21st Century Agriculture (AC21) (2003) Summary: First Plenary Meeting of AC21 June 16–17 p 21

AgBioTech InfoNet (2001) Maine GE cross contamination bill signed into law June 4

Agence France Presse (2004) Victory over US claimed as rules agreed on GM exports February 27

Alexander CE Thuy Van M (2005) Determinants of corn rootworm resistant corn adoption in Indiana *AgBioForum* **8** (4) 9–10

Altieri MA (2000) The ecological impacts of transgenic crops on agroecosystem health Ecosystem Health **6** 13–23

Arrowsmith Nancy (1987) Nancy Arrowsmith—personal interview [by Kent Whealey] at Heritage Farm Decorah IA July 31 Harvest Edition Seed Savers Exchange p 79–92 quotation p 85

Barrett K (2000) Risk and precaution in agricultural biotechnology: A role for science and scientists Inquiry in Action, the newsletter of the Consortium for Sustainable Agriculture Research and Education No 4647 Spring–Summer p 16–19

Benbrook C (2001) When does it pay to plant Bt corn? Farm-level economic impacts of Bt corn, 1996–2001

Benbrook C (2004) Excerpt: Executive Summary: Genetically Engineered Crops and Pesticide Use in the United States: The First Nine Years October 25

Benbrook C (2005) Harvest at risk? Impacts of Roundup Ready Wheat in the Northern Great Plains. Western Organization of Resource Councils

Bernstein I *et al.*, (1999) Immune responses in farm workers after exposure to *Bacillus thuringiensis* pesticides Environmental Health Perspectives **107** 575–582

Biotechnology Industry Organization (BIO) (2000) Guide to Biotechnology Ag Produce on the Market, and On the Market in 6 Years

Bromley M No date Genetically Modified Rice

Callahan P (2001) Genetic draft affects more than biology: US farmers stand to lose millions Boston Globe April 4.

Dansby A (2006) China & India: The next seed super powers? Seed World February

Duffy M (2001) Who benefits from biotechnology? American Seed Trade Association Meeting December

Ellstrand NC (2001) When transgenes wander, should we worry? *Plant Physiology* **125** 1543–1545

Environment News Service (ENS) (2001) Herbicide resistant weeds spring up in bioengineered soy fields May 4

Ervin DE Sandra SB Chantal LC Rick W 2001 Public research for U.S. biosafety regulation of transgenic crops Paper prepared for Biotechnology, Science and Modern Agriculture: A New Industry at the Dawn of the Century. 5th International Conference organized by the International Consortium on Agricultural Biotechnology Research (ICABR), Ravello, Italy, June 15–18. p. 1

Evans D (2001) EU presents tough rules on gene labels, tracing. Reuters. July 24

ESCOP, ECOP (Experiment Station and Extension Committees on Organization and Policy) (2000) Agricultural Biotechnology: Critical Issues and Recommended Responses from the Land-Grant Universities p 20

Fox K Evelyn (2000) The Century of the Gene Harvard University Press p 192

Fulton M Keyowski L (1999) The producer benefits of herbicide-resistant canola. University of Saskatchewan *AgBioForum* **2** (2) 85–93

Grains of Delusion (2001) Jointly published by BIOTHAI (Thailand)f; CEDAC (Cambodia), DRCSC (India), GRAIN< MASIPAG (Philippines), PANIndonesia, and UBINIG (Bangladesh) February

Guebert A (2001) Supreme Court blesses plant patents: Bye-bye bin-run seed The Land (Minnesota) December 21 p 3

Hagen K (2006) Biotech fruits, vegetables and flowers. What's on the market. Seed World. February p 12

Hillyer G (1999) Biotechnology offers U.S. farmers promises and problems *AgBioForum* **2** (2) 99–102

Holden P (2002) Report: Seeds of Doubt: North American farmer's experiences of GM Crops Soil Association, Bristol, UK p 58

Hord B (2001) Wheat industry is cautious on biotech introduction World Herald May 29

International Service for the Acquisition on Agro-Biotech Applications (ISAAA) ISAAA AmericCenter. 417 Bradfield Hall, Cornell University, Ithaca, NY

King C Purcell L Vories E (2001) Plant growth and nitrogenase activity of glyphosate-tolerant soybeans in response to foliar application *Agronomy J* **93** 179–186

Kiplinger editors (2006) WTO ruling will help exports The Kiplinger Agriculture Letter February 17 p 2

Kirschenmann F (2001) A common ground to discuss genetics *Leopold Letter* **13** (2) 6 Massey R (2000) Sustainability and *ag biotech* Rachel's Environment and Health News No. 686, Feb. 10. www.rachel.org Environmental Research Foundation, Annapolis, MD

Moeller DE (FLAG) and Sligh M (RAFI-USA) Karen R Krub (FLAG) (ed.) (2004) Farmers' Guide to GMOs 5–6

Montana Department of Environmental Quality (2006) Biodiesel production in the Northwest Staff e-mail to NCAT program staff May 6

Moore, Tam (2005) Monsanto fee doubles cost of alfalfa seed. Capital Press [agriculture weekly]. Salem, Oregon. August 5. p. 5.

Myers, James R(2006). Outcrossing Potential for Brassica Species and Implications for Vegetable Crucifer Seed Crops of Growing Oilseed Brassicas in the Willamette Valley. Special Report 1064. (PDF / 286K) January. Oregon State University Extension, Corvallis, OR.

NCALRI (2000) Biotechnology: Overview. The National Center for Agricultural Law Research and Information. Fayetteville, Arkansas.

New Scientist. (1999) Splitting headache. November 20, No. 2213.

Non-GMO Source (2002) Manufacturers face GMO challenges with minor ingredients. September. (Ken Roseboro).

Non-GMO Source (2004) Genetic ID Augsburg receives perfect scores in ISTA proficiency test. August. p. 15.

Non-GMO Source (2005) GMO news: GM alfalfa seed now on sale in U.S. October. p. 12.

Osborn, A (2001) GM multinationals are thrown a lifeline with new regulations. The Guardian (Britain). Feb. 15.

Pest Management at the Crossroads

See for more information on Bt effects on beneficial insects and on pest resistance. This site is sponsored, designed, and managed by Benbrook Consulting Service (BCS) for biointensive Integrated Pest Management.

Pollack, A (2001) Mapping genes may help breed better food, bypassing genetic engineering.

New York Times March 7

Ryan, Angela, and Mae-Wan Ho (2001) Europe's New Rules Could Sink All GMOs. A report from the Institute of Science in Society. London. August 28.

Schrade, Ann, and Steve R (2001) States join global fight over biotech labeling. Denver Post. May 29.

Shadid, A (2001) Genetic drift affects more than biology; U.S. farmers stand to lose millions. Boston Globe. April 4.

Sheldon, A (FDA microbiologist) to James Maryanski, FDA Biotechnology Coordinator. 1993. Internal memo: Use of Kanamycin resistance marker gene in tomatoes. March 30.

Stierle, Andrea. Personal communication. May 9, 2006.

Toledo Alvaro (2002) Excerpt: Saving the seed: Europe's challenge.

South Bulletin. No. 35. May 15. p. 17–20; p. 18.

Union of Concerned Scientists (2000) Biotechnology and Sustainable Agriculture.

United States Department of Agriculture, Office of Inspector General (Southwest Region)(USDA/ OIG-A/506-8-Te). 2005. Audit Report: Animal and Plant Health Inspection Service [APHIS] Controls Over Issuance of Genetically Engineered Organism Release Permits. Audit 50601-B-Te. December. Washington, DC. 63 p. See p. 56.

USDA/AMS(2006) Grain Transportation Report. Feb. 2. p. 1, 13.

USDA/NASS (2005) Biotechnology varietiesAcreage, 06.30.05. USDA, p. 22–24 (of 46).

University of Missouri press release (2001) Bad news beans – A year of challenges confronts soybean growers. July 27.

Wallace Center for Agricultural and Environmental Policy. Winrock International (2001) Transgenic Crops: An Environmental Assessment. Policy Studies Report No. 15. p 52

APPENDICES

APPENDIX 1

Petitions for Nonregulated Status Granted

Applicant Documents						APHIS Documents		
Petition	**Extension of Petition No. *****	**Institution**	**Regulated Article**	**Transgenic Phenotype**	**Transformation Event or Line**	**Preliminary EA ******	**FR Notice**	**Final EA & Determination**
92-196-01p		Calgene	Tomato	Fruit ripening altered	FLAVR SAVR	92-196-01p_ea	92-196-01p_fr	92-196-01p_com
92-204-01p		Upjohn	Squash	WMV2 & ZYMV resistant	ZW-20	92-204-01p_ea	92-204-01p_fr	92-204-01p_com
93-196-01p		Calgene	Cotton	Bromoxynil tolerant	BXN	93-196-01p	93-196-01p_fr	93-196-01p_com
93-258-01p		Monsanto	Soybean	Glyphosate tolerant	40-3-2	93-258-01p	93-258-01p	93-258-01p_com
94-090-01p		Calgene	Rapeseed	Oil profile altered	pCGN3828-212/86- 18 & 23	94-090-01p	94-090-01p	94-090-01p_com
94-227-01p	92-196-01p	Calgene	Tomato	Fruit ripening altered	Line N73 1436-111	94-227-01p	94-227-01p_fr	94-227-01p_com
94-228-01p		DNA Plant Tech	Tomato	Fruit ripening altered	1345-4	94-228-01p_ea	94-228-01p_fr	94-228-01p_com
94-230-01p	92-196-01p	Calgene	Tomato	Fruit ripening altered	9 additional FLAVR SAVR lines	94-230-01p	94-230-01p	94-230-01p_com
94-257-01p		Monsanto	Potato	Coleopteran resistant	BT6, BT10, BT12, BT16, BT17, BT18, BT23	94-257-01p_ea	94-257-01p_fr	94-257-01p_com
94-290-01p		Zeneca & Petoseed	Tomato	Fruit polygalacturonase level decreased	B, Da, F	94-290-01p	94-290-01p	94-290-01p_com
94-308-01p		Monsanto	Cotton	Lepidopteran resistant	531, 757, 1076	94-308-01p	94-308-01p	94-308-01p_com
94-319-01p		Ciba Seeds	Corn	Lepidopteran resistant	Event 176	94-319-01p	94-319-01p	94-319-01p_com
94-357-01p		AgrEvo	Corn	Phosphinothricin tolerant	T14, T25	94-357-01p	94-357-01p	94-357-01p_com
95-030-01p		Calgene	Tomato	Fruit ripening altered	20 additional FLAVRSAVR lines	95-030-01p	95-030-01p	95-030-01p_com
95-045-01p		Monsanto	Cotton	Glyphosate tolerant	1445, 1698			95-045-01p_com
95-053-01p		Monsanto	Tomato	Fruit ripening altered	8338			95-053-01p_com
95-093-01p		Monsanto	Corn	Lepidopteran resistant	MON 80100			95-093-01p_com
95-145-01p		DeKalb	Corn	Phosphinothricin tolerant	B16			95-145-01p_com
95-179-01p	2-196-01p	Calgene	Tomato	Fruit ripening altered	2 additional FLAVRSAVR lines			95-179-01p_com
95-195-01p		Northrup King	Corn	European Corn Borer resistant	Bt11			95-195-01p_com
95-228-01p		Plant Genetic Systems	Corn	Male sterile	MS3			95-228-01p_com
95-256-01p		Du Pont	Cotton	Sulfonylurea tolerant	19-51a			95-256-01p_com
95-324-01p		Agritope	Tomato	Fruit ripening altered	35 1 N			95-324-01p_com
95-338-01p		Monsanto	Potato	CPB resistant	SBT02-5 & -7, ATBT04-6 & -27, -30, -31, -36			95-338-01p_com

Continued: Petitions for Nonregulated Status Granted

Applicant Documents						APHIS Documents		
Petition	**Extension of Petition No. *****	**Institution**	**Regulated Article**	**Transgenic Phenotype**	**Transformation Event or Line**	**Preliminary EA ******	**FR Notice**	**Final EA & Determination**
95-352-01p		Asgrow	Squash	CMV, ZYMV, WMV2 resistant	CZW-3			95-352-01p_com
96-017-01p	95-093-01p	Monsanto	Corn	European Corn Borer resistant	MON809 &MON810			96-017-01p_com
96-051-01p		Cornell U	Papaya	PRSV resistant	55-1, 63-1			96-051-01p_com
96-068-01p		AgrEvo	Soybean	Phosphinothricin tolerant	W62, W98, A2704- 12, A2704-21, A5547-35			96-068-01p_com
96-248-01p	92-196-01p	Calgene	Tomato	Fruit ripening altered	1 additional FLAVRSAVR line			96-248-01p_com
92-196-01p		Calgene	Tomato	Fruit ripening altered	FLAVR SAVR	92-196-01p-ea	92-196-01p-fr	92-196-01p_com
96-291-01p		DeKalb	Corn	European Corn Borer resistant	DBT418			96-291-01p_com
96-317-01p		Monsanto	Corn	Glyphosate tolerant & ECB resistant	MON802			96-317-01p_com
97-008-01p		Du Pont	Soybean	Oil profile altered	G94-1, G94-19, G-168			97-008-01p_com
97-013-01p		Calgene	Cotton	Bromoxynil tolerant & Lepidopteran resistant	Events 31807 & 31808			97-013-01p_com
97-099-01p		Monsanto	Corn	Glyphosate tolerant	GA21			97-099-01p_com
97-148-01p		Bejo	Cichorium intybus	Male sterile	RM3-3, RM3-4, RM3-6			97-148-01p_com
97-204-01p		Monsanto	Potato	CPB & PLRV resistant	RBMT21-129 &RBMT21-350			97-204-01p_com
97-205-01p		AgrEvo	Rapeseed	Phosphinothricin tolerant	T45			97-205-01p_com
97-265-01p		AgrEvo	Corn	Phosphinothricin tolerant & Lep. resistant	CBH-351			97-265-01p_com
97-287-01p		Monsanto	Tomato	Lepidopteran resistant	5345			97-287-01p_com
97-336-01p		AgrEvo	Beet	Phosphinothricin tolerant	T-120-7			97-336-01p_com
97-339-01p		Monsanto	Potato	CPB & PVY resistant	RBMT15-101,SEMT15-02,SEMT15-15			97-339-01p_com
97-342-01p		Pioneer	Corn	Male sterile & Phosphinothricin tolerant	676, 678, 680			97-342-01p_com
98-014-01p	96-068-01p	AgrEvo	Soybean	Phosphinothricin tolerant	A5547-127			98-014-01p
98-173-01p		Novartis Seeds &Monsanto	Beet	Glyphosate tolerant	GTSB77			98-173-01p_com
98-216-01p		Monsanto	Rapeseed	Glyphosate tolerant	RT73			98-216-01p_com
98-238-01p		AgrEvo	Soybean	Phosphinothricin tolerant	GU262			98-238-01p_com
98-278-01p		AgrEvo	Rapeseed	Phosphinothricin tolerant& Pollination control	MS8 & RF3			98-278-01p_com
98-329-01p		AgrEvo	Rice	Phosphinothricin tolerant	LLRICE06,LLRICE62			98-329-01p_com
98-335-01p		U. of Saskatchewan	Flax	Tolerant to soil residues of sulfonyl urea herbicide	CDC Triffid			98-335-01p_com
98-349-01p	95-228-01p	AgrEvo	Corn	Phosphinothricin tolerant and Male sterile	MS6			98-349-01p_com
99-173-01p	97-204-01p	Monsanto	Potato	PLRV & CPB resistant	RBMT22-82			99-173-01p_com
00-011-01p	97-099-01p	Monsanto	Corn	Glyphosate tolerant	NK603			00-011-01p_com
00-136-01p		Mycogen c/o Dow & Pioneer	Corn	Lepidopteran resistant phosphinothricin tolerant	Line 1507			00-136-01p_com
00-342-01p		Monsanto	Cotton	Lepidopteran resistant	Cotton Event15985			00-342-01p_com
01-121-01p		Vector	Tobacco	Reduced nicotine	Vector 21-41			01-121-01p_com
01-137-01p		Monsanto	Corn	Corn Rootworm Resistant	MON 863			01-137-01p_com

Petitions of Nonregulated Status Granted or Pending by APHIS as of 5 February 2007

Abbreviations:

CMV-cucumber mosaic virus

CPB-colorado potato beetle

PLRV- potato leafroll virus

PRSV-papaya ringspot virus

PVY-potato virus Y

WMV2- watermelon mosaic virus 2

ZYMV-zucchini yellow mosaic virus

Appendix 2

Excerpt: Agricultural Biotechnology: Critical Issues and Recommended Responses from the Land-Grant Universities—A report to the Experiment Station Committee on Organization and Policy (ESCOP) and the Extension Committee on Organization and Policy (ECOP). (PDF / 79K). January 21, 2000.

Environment (p. 8)

Public discussions have raised a number of concerns about potential environmental effects of the use of crops derived from agricultural biotechnology. Among the most prominent are concerns

- that the flow of genetic material from genetically engineered crops to weed species will improve weed fitness.
- about the validity of industry claims that genetically engineered varieties can help improve environmental quality by reducing the use of chemical pesticides.
- that the development of resistance to introduced genetic material will undermine the efficacy of widely used pest control products; and
- that the use of marker genes will accelerate the spread of antibiotic resistance.

The Land-Grant University (LGU) system can help inform discussions of these issues through research and education. In particular, research is sorely needed on all these topics because the current information base is inadequate.

Transgenic Crops

By Jeff Schahczenski and Katherine Adam

NCAT Program Specialists

Paul Driscoll, Editor

Sherry Vogel HTML Production

IP189

Slot 172

ATTRA - National Sustainable Agriculture Information Service is managed by the National Center for Appropriate Technology (NCAT) and is funded under a grant from the United States Department of Agriculture's Rural Business-Cooperative Service. Visit the NCAT Web site for more information on our sustainable agriculture projects.

Summary

Transgenic Crops describes the basics of genetic modification for agricultural purposes and a brief history of the technology and the governing policies surrounding it. This publication offers a brief overview of the main agricultural crops that have been genetically modified, the characteristics they express, and the market roles they play. Unintended consequences, economic considerations, and safety concerns surrounding the cultivation and dissemination of transgenic crops are also discussed. Biopharmaceutical aspects of transgenic crops are also briefly addressed. Economic, legal, and management concerns associated with these types of crops are addressed, as well as political and regulatory aspects. Implications of transgenic technologies for sustainable agriculture are briefly addressed along with concluding remarks. References and resources follow the narrative.

2

SELECTABLE MARKER GENE EXCISION FROM TRANSGENIC PLANTS

LINING TIAN

Selectable markers are commonly used in plant genetic transformation; however, marker genes usually are not needed once transgenic plants have been identified. The presence of marker genes, especially antibiotic markers and herbicide resistance markers in transgenic crop plants, may elicit environmental and consumer concerns. In addition, the existence of marker genes in transgenic crops could evoke additional, lengthy risk assessments for release of crops that contain useful novel traits. Markers that promote plant regeneration have been recently described. However, continuous expression of these markers may interfere with normal plant growth and development. Removal of this type of marker from plant tissues is necessary unless expression is under good control. Furthermore, current transformation technologies permit only the introduction of a very limited number of genes into plants. Retransformation of the same line is needed for multiple trait modifications. New selectable markers are thus needed with each transformation to pyramid the same crop variety with different desirable traits. The number of selectable marker genes that are suitable for each crop species is usually very limited. This is especially true for transformation of recalcitrant species. Marker excision can allow reuse of a marker after each transformation step. Marker elimination will not only appease some potential environmental and consumer concerns, it will also remove technical barriers for plant genetic transformation (Miki & McHugh, 2004; Tian, 2006).

Co-transformation and technology development

Transformation selection uses a marker gene to identify the transformants that contain the gene of interest or to eliminate non-transformed tissues. Conventionally, the gene of interest and the marker gene are tightly linked in the transformation vectors and are transferred and integrated into the plant genome together to ensure selection. Nevertheless, the marker gene can permit selection even if it is distant from the gene of interest. Co-transformation was developed for marker removal. The gene of interest and the marker gene are transferred separately via different approaches, and the two genes are inserted at loci that are not linked. The marker is segregated later in the next generation.

Co-transformation can be conducted by different approaches, such as the use of separate transformation vectors in the same *Agrobacterium*, or the use of different *Agrobacterium* strains. Co-transformation using mixed *Agrobacterium* strains or vectors is technically simple and is effective for transformation and marker removal. Nevertheless, this approach may not be suitable for species in which transformation efficiency is low. Various technologies have been developed for co-transformation improvement. For example, two copies of T-DNA can be placed on the same transformation vector. In this case, the transgene and the marker gene are located independently on each of the T-DNAs, which are separated by intervening DNA sequences. The two T-DNAs can also be located adjacent to each other using two extra T-DNA border regions, creating two almost contiguous T-DNAs. Instead of using multiple T-DNAs, a single T-DNA with a unique design can also be used for co-transformation. Lu and co-workers (Lu *et al.* 2001) developed a system called double right-border binary vector. The vector contained only one copy of T-DNA but carried two copies of the T-DNA right border sequences, which flanked a selectable marker gene. The gene of interest was inserted next, and then one copy of the left border sequence was placed at the end. The vector enabled two separate insertions: one developed from the first right border and contained the selectable gene along with the target gene, and the other came from the second right border carrying only the target gene.

Another system, described by Huang *et al* (Huang *et al.* 2004.) broke the traditional T-DNA based principle of gene transfer. Contrary to conventional vector design, they repositioned the selectable marker in the backbone of a regular binary vector, leaving only the gene of interest within the T-DNA region. They reasoned that the marker gene would be transferred into the plant cell either along with the T-DNA that was initiated at the first border but failed to terminate at the second border, or as part of the T-strand that was initiated at the second border. The attempt was based on previous observations that vector backbone sequences frequently transfer along with T-DNA into plant cells. They tested this design in maize and found that co-transformation efficiency was higher than when the vector was comprised of multiple T-DNAs containing the gene of interest and the marker separately. Marker-free transgenic plants were successfully recovered in subsequent generations.

Co-transformation is technically simple and is an effective approach for segregating the transgene and the marker gene, and for subsequent marker removal.

Nevertheless, co-transformation efficiency should be reasonably high and the gene locations should be distantly separated. Such demands may not always be possible.

A major limitation with co-transformation is that marker excision occurs only in the next generation, and thus the time course is relatively long. Also, the methodology is not suitable for woody tree plants and for vegetatively propagated plants.

Site specific excision and the system control

The Cre-*lox* site-specific recombination system of bacteriophage P1 can be used as a strategy for marker gene removal. The system consists of two components: the Cre recombinase and two 34 bp *lox*P recognition sequences. The expression of the Cre protein causes recombination between the two *lox*P sites and results in excision of the DNA sequence flanked by the sites (Dale & Ow, 1991).

Cre recombinase is introduced into the transgenic plants carrying the selectable marker gene either by a second round of transformation or by out-crossing. In addition to Cre/*lox*, several other site-specific DNA excision systems derived from other organisms may be adapted for marker removal in transgenic plants. These include the R/*RS* system from *Zygosaccharomyces rouxii*, Flp/ *frt* from *Saccharomyces cerevisiae*, and Gin/*gix* from bacteriophage Mu.

Site-specific excision is an effective approach for marker gene removal. However, the system has some limitations. For instance, re-transformation is laborious, expensive, and time consuming. Out-crossing is not suitable for vegetatively propagated plants and woody tree species. In addition, the continuous and high level of expression of the *cre* recombinase gene may result in phenotypic aberrations in some plant species.

Some new techniques that aim to control recombinant systems have been developed. The *cre* gene expression system can be controlled by specific gene promoters, such as inducible gene promoters, including chemical inducible promoters and heat shock protein systems. The marker gene is allowed to process for selection. Once selection is complete and upon induction of the promoter, the *cre* gene expresses, which then activates the *lox* sequences and the marker gene is excised. This new technology is effective for marker gene removal in different crop species and can also be applied to generate marker-free woody tree species via vegetative propagation.

Expression control of recombinases no doubt has significantly extended the site-specific excision system and made the system more useful and practical. In addition to using a specific promoter to control the *cre* gene, transient expression of *cre* recombinase in some approaches may be sufficient to activate the excision and lead to marker gene removal. However, the system still needs improvement for wider use.

Transposon elements and marker gene removal

Transposition systems such as *Ac/Ds* have been used as a strategy to eliminate marker genes. The technique is based on the fact that the DNA sequence located within *Ds* repeats can be mobilized to excise along with the *Ds* mobile element. The gene of interest or the marker sequence can be placed within the mobile sequence. Upon activation by *Ac* transposase, the transgene or marker gene residing inside of the transposable *Ac* element can be excised and subsequently reinserted elsewhere in the plant genome. Thus, the transgene and the marker gene may be separated by transposon elements. The use of transposition to relocate the transgene to a new chromosomal locus is a useful approach for generating marker-free plants. However, genetic crossing or segregation is required to separate the gene of interest and the marker gene, and thus it is time consuming and entails a long process.

In rare cases due to transposition errors, the excised DNA may fail to reinsert into the host genome and be lost. Based on this observation, Ebinuma *et al.* (Ebinuma *et al.* 2005) described a vector in which the marker gene was constructed within the *Ac/Ds* transposable system. Marker-free transgenic tobacco and woody aspen plants were obtained via the reinsertion failure. Although the system was effective and marker elimination occurred in the same generation, marker removal efficiency via this strategy is poor, due to the low incidence of occurrence.

Marker gene removal from chloroplasts

Chloroplast DNA, in general, is maternally inherited. However, there exists some concern that marker genes may transfer to relevant weeds in the environment. Marker genes may also move to microorganisms, and the marker gene may function in bacteria, as the expression system is directed by prokaryotic regulatory elements. Therefore removal of marker genes after transgenic plants are recovered is still highly desirable in transplastomic plants. The well characterized site-specific Cre/*lox* system has also been used for plastid marker gene removal. In a study by Corneille *et al.* in 2001(Corneille,. 2001) the negative selectable marker *cod*A gene, which was used to identify marker removal events, was flanked by two *lox* sites placed in plastids. Nuclear-encoded but plastid targeted Cre recombinase was introduced via either *Agrobacterium*-mediated retransformation or by pollination. Efficient elimination of the marker gene from plastids occurred via either approach. Plastid marker removal using the Cre/*lox* system was also demonstrated in other studies. It has been noticed that marker removal in plastids occurs rather rapidly, which is in contrast to the slow process of marker establishment in plastids during transformation. Another approach for marker removal from plastids is via homologous DNA recombination. Iamtham and Day (Iamtham and Day, 2000) described a system in which several marker genes, including *addA*, *bar*, and *gus*, were conventionally flanked by different recombination sites. These marker genes were sequentially eliminated later via loop-out recombination. This system, however, is difficult to regulate because selection of homoplastomic transformants appeared to be unpredictable. Klaus *et al.*9 described a system in which the marker gene is placed outside of the flank regions used for homologous recombination. Recombination via either left or right flanks resulted in co-integration of the vector, but a further round of recombination resulted in the loss of the marker gene. Thus, the marker gene was only transiently co integrated into the genome. The marker gene expression for transplastomic selection was still provided effectively via this manner (Klaus, 2004).

The advantage of this method is that marker-free plants can be generated directly in the first generation.

Summary

Marker removal is an important and integral aspect of plant transgenic research today, and many different approaches and technologies have been demonstrated. Routine marker gene removal from transgenic plants will eventually become more feasible in a wider array of plant species as research in the area continues to proceed rapidly.

REFERENCES

Corneille S (2001) Efficient elimination of selectable marker genes from the plastid genome by the CRE-lox site-specific recombination system *Plant Journal* **27**, 171-178

Dale EC & Ow DW (1991) Gene transfer with subsequent removal of the selection gene from the host genome *Proceedings of the National Academy of Sciences* **88**, 10558-10562

Ebinuma H *et al* (2005) Elimination of marker genes from transgenic plants using MAT vector systems *Methods in Molecular Biology* **286**, 237-253

Huang SS *et al* (2004) Generation of marker-free transgenic maize by regular two-border *Agrobacterium* transformation vectors *Transgenic Research* **13**, 451-461

Iamtham S Day A (2000) Removal of antibiotic resistance genes from transgenic tobacco plastids *Nature Biotechnology* **18**, 1172-1176

Klaus SMJ (2004) Generation of marker-free plastid transformants using a transiently cointegrated selection gene *Nature Biotechnology* **22**, 225-229

Lu HJ *et al* (2001) Generation of selectable marker-free transgenic rice using double right-border (DRB) binary vectors *Australian Journal of Plant Physiology* **28**, 241-248

Miki B & McHugh S (2004) Selectable marker genes in transgenic plants: applications, alternatives and biosafety *Journal of Biotechnology* **107**, 193-232

Tian L (2006) Marker gene removal from transgenic plants In: *Floriculture, Ornamental and Plant Biotechnology Advance and topical issues* (1st Edition), Jaime A Teixeira da Siva (Ed) Pp 26-29 (and the references therein)

3

REDUCING GOSSYPOL IN COTTONSEED MAY IMPROVE HUMAN NUTRITION

KEERTI S. RATHORE

As world population continues to rise, there is a growing requirement for food and fiber, especially in developing countries. In addition, agriculture is increasingly tapped to supplement world energy needs. Since the area of arable land is limited, meeting these needs must involve improving crop yields sustainably and utilizing global agricultural output more efficiently. The terpenoid class of secondary metabolites produced by many plant species has important ecological functions either as attractants (e.g., linalool) or as defense compounds (Aharoni *et al.*, 2005; Stipanovic *et al.*, 1999) (e.g., bitter triterpenoid cucurbitacins, pungent diterpenoid polygodial, gossypol, and related compounds in cotton). However, some terpenoids and other defense compounds are toxic to humans and animals. Over the course of human history, man has learned to either avoid consumption of toxic plants/parts, or to inactivate/neutralize toxic compounds before ingestion (e.g., cassava, kidney beans). In some cases, the inedible plant product is used as feed for those domestic, ruminant animals that suffer little or no ill effects from consuming these products. Cottonseed, produced in abundance as a byproduct of fiber production, represents such a case.

Cotton is a cash crop for more than 20 million farmers in developing countries in Asia and Africa. For every 1 kg of fiber, cotton plants produce approximately 1.65 kg of seeds: e.g., global cotton cultivation in 2005 yielded 23.5 million metric tons (MMT) of lint and over 40 MMT of seed. Cottonseed contains 21% oil and 23% protein of relatively high quality. The global annual production of cottonseed could potentially provide the total protein requirements of nearly 500 million people for a year (50 g/day rate) if the seed were safe for human consumption. However, the utility of cottonseed for food is hampered by the presence of gossypol, located in seed glands. Gossypol is a terpenoid that is cardio- and hepatotoxic to humans and monogastric animals; unfortunately, this toxicity limits cottonseed to use primarily as ruminant cattle feed, either as whole seeds, or as meal following oil extraction. Even though cottonseed does find some indirect use in human nutrition in the form of cattle feed, ruminant animals are highly inefficient in terms of feed conversion. Thus, the potential of cottonseed to help meet food requirements of the burgeoning world population provides a great impetus for eliminating gossypol from cotton seed.

Classical breeding and its limitations

Sub-epidermal, lysigenous glands containing one or more terpenoids are present in all parts of the cotton plant except stele tissue, young roots, xylem, and seed coat (Bell *et al.*, 1994). Gossypol is the predominant terpenoid in glands of seed-kernel and petal, while glands in other organs contain varying levels of gossypol and related terpenoids. Discovery of a glandless trait in a mutant cotton variety (Hopi Moencopi) provided plant breeders with a tool to eliminate glands and therefore gossypol from seed (Lusas & Jividen, 1987). Several national and international programs were launched to transfer this useful trait into commercial varieties to produce gossypol-free cottonseed. These programs yielded cottonseed that could be fed to more efficient feed-utilizing, monogastric animals and was even deemed safe for human consumption.

In nutrition studies, glandless cottonseed compares favorably as a source of protein with other traditional food sources. However, these glandless cotton varieties were a commercial failure. Under field conditions, glandless plants were more prone to insect damage, because they constitutively lacked gossypol and other protective terpenoids (Jenkins *et al.*, 1966) consequently, these cotton plants were not accepted by farmers. A few wild, diploid *Gossypium* species exhibit a glandless-seed and glanded-plant phenotype. Attempts to introgress this highly desirable trait into commercially popular tetraploid cotton have had limited success. Consequently, the potential of cottonseed to contribute to human nutrition has not yet been realized.

Biotechnology to the rescue

As mentioned, glands containing various terpenoids in aerial plant parts protect against herbivory by insects. Many terpenoids are also induced in response to fungal or bacterial challenge and are believed to function as phytoalexins (Stipanovic *et al.*, 1999). These compounds are derived from the primary sesquiterpene skeleton, (+)-δ-cadinene. (+)-δ-cadinene synthase (δ-CS), a cyclase enzyme, catalyzes conversion of farnesyl diphosphate to (+)-δ-cadinene, the first committed step in biosynthesis of gossypol and related terpenoids (Martin *et al.*, 2003). The ability to block this committed step solely in seeds could produce cottonseeds with reduced gossypol levels while maintaining wild-type levels of gossypol and other defensive terpenoids in the rest of the plant. Attempts to use the antisense gene suppression mechanism targeting δ-CS to eliminate gossypol from cottonseed were unsuccessful, or resulted in a small reduction in gossypol levels, or provided ambiguous results (Townsend *et al.*, 2005). It took a more powerful gene silencing mechanism and a strong, highly seed-specific promoter to achieve a meaningful reduction in seed-gossypol.

Sunilkumar *et al.* (2006) used RNA interference (RNAi)-mediated suppression of the δ-CS gene under control of a cotton-seed promoter to significantly reduce δ-CS activity in developing embryo, which resulted in a seed-specific reduction of toxic gossypol. The efficacy, strength, and tissue specificity of the α-globulin promoter had been demonstrated in an earlier study. This promoter becomes active at 15 days post-anthesis (dpa) in developing cotton embryos. At this stage, promoter activity is low and localized in the middle portion of the zygotic embryo. There is a slow increase in activity until 20 dpa; however, after this stage, it rises rapidly and spreads throughout the embryo. The promoter remains active until maturity. On the other hand, δ-CS transcripts are first detected at 23 dpa and reach steady-state level from 30 to 50 dpa. Thus, the α-globulin promoter is active for quite some time prior to appearance of δ-CS transcripts in developing embryo and continues to be active until embryo maturity. This promoter was therefore used to direct generation of hairpin transcripts targeting the δ-CS gene in cotton.

An efficient means to induce heritable, RNAi in plants is by stable integration and expression of a transgene construct encoding self-complementary transcripts that can fold back to generate double-stranded RNA molecules ('hairpin' RNA or hpRNA). The efficiency with which a target gene is silenced is significantly improved when the loop portion of these hpRNAs consists of a functional intron (ihpRNA). To generate ihpRNA transcripts, a 604 bp sequence from a δ-cadinene synthase cDNA clone obtained from *Gossypium hirsutum* was used as the trigger sequence to assemble the transformation construct. This sequence bore a high degree of homology to several previously published sequences of δ-CS genes from diploid and tetraploid cottons and was expected to silence most members of the δ-CS, multigene family.

Following *Agrobacterium*-mediated transformation, several lines were obtained that produced seeds containing significantly reduced levels of gossypol. This reduced seed gossypol phenotype was correlated with presence of transgene, as well as with substantial reductions in levels of target transcripts and enzyme activity in the developing transgenic embryo. Most relevantly, levels of gossypol and related terpenoids involved in defense against insects and diseases were not reduced in foliage, floral organs, and roots of transgenic plants, as compared to wild-type controls. Three different transgenic lines were monitored up to the T2 generation, and the reduced seed-gossypol trait was found to be stable and heritable. One of the lines produced seeds with gossypol levels as low as 0.2 μg mg^{-1}, showing approximately a 98% reduction in the level of the toxin (seeds from greenhouse-grown, wild-type, cv. Coker 312 cotton plant contain approx. 10 µg gossypol mg^{-1} of kernel). The United States Food and Drug Administration (FDA) and United Nations Food and Agriculture Organization and World Health Organization (FAO/WHO) permit up to 0.45 µg mg^{-1} (450 ppm) and 0.6 µg mg^{-1} (600 ppm), respectively, of free gossypol in edible cottonseed products. Thus, by using tools of modern molecular biology, gossypol was reduced in cottonseed to a level considered safe for human consumption.

An intriguing property of RNAi is that it can spread from cell-to-cell and long distances from its point of origin. This phenomenon, initially observed in *C: elegans*, is also believed to occur in plant systems. On the basis of these observations, doubts have been cast on the feasibility of achieving tissue- or cell-specific silencing of desired gene(s) in plants. If the RNAi-induced silencing signal were to spread out of developing embryos, or if the components of the silencing mechanism remaining in mature seed were to spread to the plant upon its germination, the seed (kernel)-specificity offered by the α-globulin promoter would be lost, and the plant would suffer from the same weakened resistance that was observed in glandless cotton.

However, results presented in the PNAS report indicate strict confinement of silencing (Sunilkumar *et al.* 2006). Single seed analysis showed that null-segregant seeds, which developed side-by-side with RNAi silenced embryos in the boll of a (Smith *et al.* 2000; Wang & Waterhouse 2001); transgenic line, had wild-type levels of gossypol. Thus, they were not affected by the silenced status of adjacent embryos. In addition, leaves, floral organs, and roots of transgenic plants, grown from silenced seeds, exhibited wild-type levels of terpenoids. This observation suggests that spreading of the RNAi-induced silencing signal does not occur in cotton, and consequently, gossypol reduction remains confined to cottonseed.

The study also provided additional evidence from unrelated research showing that RNAi mediated silencing targeting the expression of GFP in cotton does not spread and is strictly limited to tissues expressing hairpin transcripts. These results suggest that RNAi-mediated gossypol reduction will be restricted to the developing embryo and the kernel in mature seed. Thus, defensive capabilities offered to remaining plant organs by gossypol and related terpenoids should not be compromised in RNAi lines.

Conclusions

The availability of reduced-gossypol cottonseed offers the possibility that the cotton plant, in addition to providing fiber, can also be a source of nutrition, either directly as food in the form of seed/seed products or as feed for more efficient feed-utilizing, monogastric animals. Of course, RNAi lines will have to be field tested, and pass the regulatory approval process, before the technology becomes available and is accepted by cotton farmers. Plant breeding will continue to help meet requirements of humanity for food, fiber, and fuel. The results from the RNAi study suggest that tools of modern molecular biology offer an alternative means to address problems that are not always possible to solve using traditional breeding methods. This research also shows that biotechnology, in addition to improving productivity of agricultural crops, can help us better utilize agricultural products by either modifying their nutritional composition or eliminating harmful compounds from edible organs of the plant.

REFERENCES

Aharoni A *et al* (2005) Volatile science? Metabolic engineering of terpenoids in plants *Trends Plant Sci* **10** 594-602

Stipanovic RD *et al* (1999) Cotton pest resistance: the role of pigment gland constituents In: *Biologically Active Natural Products: Agrochemicals,* HG Cutler & SJ Cutler (eds), CRC Press, Boca Raton, pp 211-220

Bell AA *et al* (1994) Genetic manipulation of terpenoid phytoalexins in Gossypium: Effects on disease resistance In: *Genetic engineering of plant secondary metabolism* B E Ellis et al (eds), Plenum Press, New York pp 231-249

Lusas EW Jividen GM (1987) Glandless cottonseed: A review of the first 25 years of processing and utilization research *J Amer Oil Chem Soc* **64** 839-854

Jenkins JN *et al* (1966) The comparative preference of insects for glanded and glandless cottons *J Econ Entomol* **59** 352-356

Martin GS *et al* (2003) Reduced levels of cadinene sesquiterpenoids in cotton plants expressing antisense (+)-ä-cadinene synthase *Phytochem* **62** 31-38

Townsend BJ *et al* (2005) Antisense suppression of a (+)-d-cadinene synthase gene in cotton prevents the induction of this defense response gene during bacterial blight infection but not its constitutive expression *Plant Physiol* **138** 516-528

Sunilkumar G *et al* (2006) Engineering cottonseed for use in human nutrition by tissue-specific reduction of toxic gossypol *Proc National Acad Sci,* USA **103** 18054-18059

Smith NA *et al* (2000) Total silencing by intronspliced hairpin RNAs *Nature* **407** 319-320

Wang M – B Waterhouse PM (2001) Application of gene silencing in plants *Current Opinion Plant Biol* **5** 146–150

4

GENE THERAPY: A BLESSING OVERCOMES CONVENTIONAL THERAPEUTICS

CHITRA WADHWANI AND C.P. MALIK

Imagine that you happen to break one of your neighbor's window pane. What would you do? You could:

-Stay silent: no one will ever know that you are guilty, but the window pane doesn't get repaired.

-Try to repair the cracked window pane: not the best long-term solution.

-Put in a new window pane: you solve the problem, and do the honorable thing.

How is it related with gene therapy?

You can simulate a medical condition or illness to a "broken window pane." Many medical conditions result from mutations, in one or more of a person's genes. Mutations cause the protein encoded by that gene to malfunction. When a protein malfunctions, cells that rely on that protein's function don't behave normally, causing problems for whole tissues or organs. Medical conditions related to gene mutations are called genetic disorders.

So, if a mutated gene caused our "broken window pane," can you "fix" it? What are your options?

-Stay silent: ignore the genetic disorder and nothing gets fixed.

-Try to treat the disorder with drugs or other approaches: depending on the disorder, treatment may or may not be a good long-term solution.

-Put in a normal, functioning copy of the gene: if you can do this, it may solve the problem!

If it is successful, gene therapy provides a solution to a problem at its source. Adding a corrected copy of the gene may help the affected cells, tissues and organs function properly. Gene therapy differs from traditional drug-based approaches, which may treat the problem, but which do not repair the underlying genetic defect.

But gene therapy is not a simple solution-it's not a molecular bandage that will automatically fix a disorder. Though scientists and physicians have made progress in gene therapy research, much remains to be done before its full potential is harnessed. In this chapter, we'll explore several approaches to gene therapy and find out why creating successful gene-based therapies is so challenging.

BACKGROUND

In the 1980s, advances in molecular biology had enabled human genes to be sequenced and cloned. Scientists looking for methods of easily producing proteins—such as insulin, the protein deficient in diabetes mellitus type 1—investigated introducing human genes to bacterial DNA. The modified bacteria then produce the corresponding protein, which can be harvested and injected in people who cannot produce it naturally.

On September 14, 1990 at the U.S. National Institutes of Health W. French Anderson and his colleagues R. Michael Blaese, and Kenneth Culver, performed the first approved gene therapy procedure on a four-year old Ashanthi DeSilva. Born with a rare genetic disease called severe combined immunodeficiency (SCID), she lacked a healthy immune system, and was vulnerable to every passing germ. Children with this illness usually develop overwhelming infections and rarely survive to adulthood; a common childhood illness like chickenpox is life-threatening. Ashanthi led an isolated existence—avoiding contact with people outside her family, remaining in the sterile environment of her home, and battling frequent illnesses with massive amounts of antibiotics. In Ashanthi's gene therapy procedure, doctors removed white blood cells from the child's body, let the cells grow in the lab, inserted the missing gene into the cells, and then infused the genetically modified blood cells back into the patient's bloodstream. Laboratory tests have shown that the therapy strengthened Ashanthi's immune system; she no longer had recurrent colds, she has been allowed to attend school, and she was immunized against whooping cough. This procedure was not a cure; the white blood cells treated genetically only work for a few months, after which, the process must be repeated (VII, Thompson [First] 1993).

Although this simplified explanation of a gene therapy procedure sounds like a happy ending, it is little more than an optimistic first chapter in a long story; the road to the first approved gene therapy procedure was rocky and fraught with controversy. The biology of human gene therapy is very complex, and there are many techniques that still need to be developed and diseases that need to be understood more fully before gene therapy can be used appropriately. The public policy debate surrounding the possible use of genetically engineered material in human subjects has been equally complex. Major participants in the debate have come from the fields of biology, government, law, medicine, philosophy, politics, and religion, each bringing different views to the discussion. Scientists took the logical step of trying to introduce genes straight into human cells, focusing on diseases caused by single-gene defects, such as cystic fibrosis, hemophilia, muscular dystrophy and sickle cell anemia. However, this has been much harder than modifying simple bacteria, primarily because of the problems involved in carrying large sections of DNA and delivering it to the right site on the comparatively large human genome.

GENE THERAPY-AN OVERVIEW

In 1977 scientists began the task of mapping, isolating and decoding every gene in the human genetic make-up. This task was accelerated by the establishment of the Human Genome Project in 1995 and was declared complete in 2003. Work now turns towards discovering how each gene functions and the protein or proteins for which it codes. As a result of the Human Genome Project, it is expected that there will be a dramatic increase in both our knowledge of and ability to treat (and in some cases, cure) conditions with a genetic basis. One of the most exciting and highly publicized areas in biomedical research today is human gene therapy-the replacement of a person's faulty genetic material with normal genetic material to treat or cure a disease or abnormal medical condition. Over time and with proper oversight, human gene therapy might become an effective weapon in modern medicine's arsenal to help fight diseases such as cancer, diabetes, high blood pressure and heart disease. Also, for patients with presently incurable diseases, such as cystic fibrosis, hemophilia A and B, and other genetic disorders, human gene therapy offers hope of a cure, or at least a measure of relief.

Gene and its role in humans

You look a little like your mother and a little like your father because of the genes they gave to you. They only determine certain physical characteristics, such as eye color, and are passed on from parents to children. Genes which are carried on chromosomes are the basic physical and functional units of heredity that tells the body how to operate. Genes are those conceptual units composed of deoxyribonucleic acid—DNA with specific sequences of bases that encode instructions on how to make proteins. The body buries genes deep in the heart of every cell, the nucleus, and organizes them in the chromosomes that hold the DNA. Although genes get a lot of attention, it's the proteins, the building blocks of our bodies that perform most life functions and even make up the majority of cellular structures. When genes are altered so that the encoded proteins are unable to carry out their normal functions, genetic disorders can result. Genes play a role in some cancers, either by stimulating the growth of cancer cells or by failing to stop them before they grow out of control. For better understanding how human gene therapy works, it may be worthwhile to review some facts about human genes and gene delivery techniques:

- Body of every person is made up of trillions of cells, which are small and microscopic. Every cell has a nucleus, which is also visible under a microscope.
- The nucleus contains set of chromosomes, which can only be detected with special laboratory techniques.

- Every chromosome contains DNA, the genetic material unique for each living creature.
- DNA comprises genes, which are still smaller.

Genes can be detected by advanced analytical techniques such as gene sequencing methods that have been used to map the entire human genome. In future, gene doctors would use a variety of techniques, some similar to current vaccinations, to inject normal genes into a person's body. Other approaches might put genes into mechanical devices to deliver them into the body. Even nanotechnology-the use of very small, even microscopic particles could eventually be part of a gene therapy delivery system.

Altered Genes

Each one of us carries nearly half a dozen defective genes and remains blissfully unaware of this fact unless we are amongst the many millions who suffer from a genetic disease. About one in ten people has, or will develop at some later stage, an inherited genetic disorder, and approximately 2,800 specific conditions are known to be caused by defects (mutations) in just one of the patient's genes. Some single gene disorders are quite common-cystic fibrosis is encountered in one out of every 2,500 babies born in the Western World-and in total, diseases that can be traced to single gene defects account for about 5% of all admissions to children's hospitals. In the U.S. and Europe, the exciting new programs have helped to 'map' the entire human genome-all of our genes. This information has enabled scientists and doctors to understand the genes that control all diseases to which the human race is prone, and hopefully develop new therapies to treat and predict diseases.

Diseases of Genetic Origin

Most of us do not suffer any harmful effects from our defective genes since we carry two copies of nearly all genes, one derived from our mother and the other from our father. The only exceptions to this rule are the genes found on the male sex chromosomes. Males have one X and one Y chromosome, the former from the mother and the latter from the father, so each cell has only one copy of the genes on these chromosomes. In the majority of cases, one normal gene is sufficient to avoid all the symptoms of disease. If the potentially harmful gene is recessive, then its normal counterpart will carry out all the tasks assigned to both. Only if we inherit from our parents two copies of the same recessive gene will a disease develop.

Definition of gene therapy

Gene therapy is the insertion of genes into the cells and tissues of an individual to treat a disease, and hereditary diseases in which a defective mutant allele is replaced with a functional one. Though the technology is still in its infancy, it has been used with some success. Antisense therapy is not strictly a form of gene therapy, but is a genetically-mediated therapy and is often considered together with other methods.

Human Gene Therapy

Much attention has been focused on the so-called genetic metabolic diseases in which a defective gene causes an enzyme to be either absent or ineffective in catalyzing a particular metabolic reaction effectively. A potential approach to the treatment of genetic disorders in man is gene therapy. Gene therapy is correcting functional gene loss by delivering genes to human tissues or simply expression of a "therapeutic gene" in a target tissue to achieve benefit. Thus the absent or faulty gene is replaced by a functional gene, so that the body can make the correct enzyme or protein and consequently eliminate the root cause of the disease. Often DNA viruses engineered to be safe or nonviral DNA are used to help deliver a healthy gene to the tissue cells. In general, non-viral vectors are very inefficient at delivering genes to cardiovascular cells. Viruses are more efficient but also have a higher safety risk. Scientists are also able to adjoin three genes together to use the "triple whammy" method to cut off and kill cancerous cells in our body. When using this later method it becomes very effective. Naturally, our body has a response to seek out the cancerous cells but is not strong enough to kill the tumor. With this modification there is a first attack on the tumor then with the other two antibodies that strengthen the first, the first becomes strong enough to fight back and finally kill the tumor. So far this has shown promise on mice for breast cancer, renal cancer and sarcomas, and colon cancer.

In brief, Gene therapy is a technique for correcting defective genes responsible for disease development. Researchers may use one of several approaches for correcting faulty genes:

- A normal gene may be inserted into a nonspecific location within the genome to replace a nonfunctional gene. This approach is most common.
- An abnormal gene could be swapped for a normal gene through homologous recombination.
- The abnormal gene could be repaired through selective reverse mutation, which returns the gene to its normal function.
- The regulation (the degree to which a gene is turned on or off) of a particular gene could be altered.

How does gene therapy work?

In most gene therapy studies, a "normal" gene is inserted into the genome to replace an "abnormal," disease-causing gene. To reverse disease caused by genetic damage, researchers isolate normal DNA and package it into a vector, a molecular delivery truck usually made from a disabled virus which must be used to deliver the therapeutic gene to the patient's target cells. Doctors then infect a target cell—usually from a tissue affected by the illness, such as liver or lung cells—with the vector. The vector unloads its DNA cargo, which then begins producing the missing protein and restores the cell to normal (See color plate: Fig 1).

Currently, the most common vector is a virus that has been genetically altered to carry normal human DNA. Viruses have evolved a way of encapsulating and delivering their genes to human cells in a pathogenic manner. Scientists have tried to take advantage of this capability and manipulate the virus genome to remove disease-causing genes and insert therapeutic genes. Target cells such as the patient's liver or lung cells are infected with the viral vector. The virus then unloads its genetic material containing the therapeutic human gene into the target cell. The generation of a functional protein product from the therapeutic gene restores the target cell to a normal state (See color plate: Fig 2).

Recently, French researchers have reported dramatic results in treating a disease called severe combined immune deficiency (SCID), the disorder suffered by David (The Boy in the Bubble). A broken gene eliminates the production of an enzyme essential for the development of a normal immune system. Scientists isolated the normal copy of the gene and packaged it into a vector. In the laboratory, they then used the vector to transport the gene into the patient's own bone marrow cells. Bone marrow cells create the immune system. The treated bone marrow cells are then given back to the patient in a germ-free isolation room, where they reconstitute a normal, functioning immune system, freeing the patient from the need to remain in isolation.

A. Gene Therapy Inserts Healthy Genes

1. This includes procedures to provide patient healthy genes to make up for a faulty gene.
2. It also includes use of genes to treat genetic disorders and various human illnesses.
3. Currently about 1,000 patients are enrolled in 200 gene therapy trials in U.S.
4. There are *ex vivo* (outside body) and *in vivo* (inside body) methods of gene therapy.

B. Genetic Disorders

1. Children with severe combined immunodeficiency syndrome (SCID) underwent *ex vivo* gene therapy.
   a. Lacking an enzyme ADA involved in maturation of T and B cells, they faced life-threatening infections.
   b. Bone marrow stem cells are removed, infected with retrovirus that carries a normal gene and returned.
   c. Use of bone marrow stem cells allow them to divide and produce more cells with same genes; also can be tested *ex vivo* to ensure gene transfer has occurred.
2. Gene therapy trials include treatment of familial hypercholesterolemia where liver cells lack a receptor for removing cholesterol from blood.

a. High levels of blood cholesterol make patient subject to fatal heart attacks when young.

b. A small portion of liver is surgically removed and infected with retrovirus with normal gene for receptor.

c. Chemotherapy in cancer cells kills off healthy as well as cancer cells; genes given to cancer patients make healthy cells more tolerant or make tumors more vulnerable.

3. Cystic fibrosis patients lack a gene for trans-membrane chloride ion carriers; patients die from respiratory tract infections.

a. Liposomes, microscopic vesicles that form when lipoproteins are in solution, are coated with healthy cystic fibrosis genes and sprayed into a patient's nostrils.

b. Due to limited gene transfer, this method is not yet successful.

4. New research may allow direct correcting of the base sequence of patients with a genetic disorder.

C. Cancer

1. Chemotherapy for cancer patients often kills healthy cells as well as cancer cells.

a. In an *ex vivo* trial, bone marrow stem cells from women with ovarian cancer were infected with a virus carrying a gene to make them more tolerant to chemotherapy.

b. Another strategy could make cancer cells more vulnerable to chemotherapy.

c. Injecting a retrovirus containing a normal *p*53 gene into tumors of a lung cancer patient stopped growth or shrank the tumor; *p*53 helps regulate cell cycle and promotes apoptosis in damaged cells.

d. Adenoviruses are easier to use because they can be produced in large quantities and are active even when not dividing.

e. The virus will kill tumor cells that lack *p*53 but not healthy cells that have the *p*53 gene.

D. Other Illnesses

1. Coating a balloon catheter (used to open up coronary arteries) with VEGF may promote growth of extra blood vessels.

2. It may be possible to use *in vivo* therapy to cure hemophilia, diabetes, Parkinson disease, or AIDS.

a. Hemophilia patients could get regular doses of cells with normal clotting-factor genes.

b. Organoids (artificial organs) could be implanted in the abdominal cavity.

c. To cure Parkinson disease, dopamine-producing cells could be grafted directly into the brain.

Different types of viruses used as gene therapy vectors

- **Retroviruses-**A class of viruses that can create double-stranded DNA copies of their RNA genomes. These copies of its genome can be integrated into the chromosomes of host cells. Human immunodeficiency virus (HIV) is a retrovirus.
- **Adenoviruses-**A class of viruses with double-stranded DNA genomes that cause respiratory, intestinal, and eye infections in humans. The virus that causes the common cold is an adenovirus.
- **Adeno-associated viruses-**A class of small, single-stranded DNA viruses that can insert their genetic material at a specific site on chromosome 19 (See color plate: Fig 3).
- **Herpes simplex viruses-**A class of double-stranded DNA viruses that infect a particular cell type, neurons. Herpes simplex virus type 1 is a common human pathogen that causes cold sores.

Besides virus-mediated gene-delivery systems, there are several non-viral options for gene delivery. The simplest method is the direct introduction of therapeutic DNA into target cells. This approach is limited in its application because it can be used only with certain tissues and requires large amounts of DNA.

Another non-viral approach involves the creation of an artificial lipid sphere with an aqueous core. This liposome, which carries the therapeutic DNA, is capable of passing the DNA through the target cell's membrane. Therapeutic DNA also can get inside target cells by chemically linking the DNA to a molecule that will bind to special cell receptors. Once bound to these receptors, the therapeutic DNA constructs are engulfed by the cell membrane and passed into the interior of the target cell. This delivery system tends to be less effective than other options.

Researchers also are experimenting with introducing a 47th (artificial human) chromosome into target cells. This chromosome would exist autonomously alongside the standard 46—not affecting their workings or causing any mutations. It would be a large vector capable of carrying substantial amounts of genetic code, and scientists anticipate that, because of its construction and autonomy, the body's immune systems would not attack it. A problem with this potential method is the difficulty in delivering such a large molecule to the nucleus of a target cell.

The current status of gene therapy research

The Food and Drug Administration (FDA) has not yet approved any human gene therapy product for sale. Current gene therapy is experimental and has not proven very successful in clinical trials. Little progress has been made since the first gene therapy clinical trial began in 1990. In 1999, gene therapy suffered a major setback with the death of 18-year-old Jesse Gelsinger. Jesse was participating in a gene therapy trial for ornithine transcarboxylase deficiency (OTCD). He died from multiple organ failures 4 days after the treatment and his death is believed to have been triggered by a severe immune response to the adenovirus carrier.

Another major blow came in January 2003, when the FDA placed a temporary halt on all gene therapy trials using retroviral vectors in blood stem cells. FDA took this action after it learned that a second child treated in a French gene therapy trial had developed a leukemia-like condition. Both this child and another who had developed a similar condition in August 2002 had been successfully treated by gene therapy for X-linked severe combined immunodeficiency disease (X-SCID), also known as "bubble baby syndrome."

FDA's Biological Response Modifiers Advisory Committee (BRMAC) met at the end of February 2003 to discuss possible measures that could allow a number of retroviral gene therapy trials for treatment of life-threatening diseases to proceed with appropriate safeguards. FDA has yet to make a decision based on the discussions and advice of the BRMAC meeting.

Some recent developments in gene therapy research

- University of California, Los Angeles, research team inserts genes into the brain using liposomes coated in a polymer call polyethylene glycol (PEG). The transfer of genes into the brain is a significant achievement because viral vectors are too big to get across the "blood-brain barrier." This method has immense potential for treating Parkinson's disease.
- RNA interference (RNAi) or gene silencing may be a new way to treat Huntington's. Short pieces of double-stranded RNA (short, interfering RNAs or siRNAs) are used by cells to degrade RNA of a particular sequence. If a siRNA is designed to match the RNA copied from a faulty gene, then the abnormal protein product of that gene will not be produced.
- New gene therapy approach repairs errors in messenger RNA derived from defective genes. Technique has potential to treat the blood disorder thalassaemia, cystic fibrosis, and some cancers.
- Gene therapy for treating children with X-SCID (sever combined immunodeficiency) or the "bubble boy" disease is stopped in France when the treatment causes leukemia in one of the patients.

- Researchers at Case Western Reserve University and Copernicus Therapeutics are able to create tiny liposomes 25 nanometers across that can carry therapeutic DNA through pores in the nuclear membrane.
- Sickle cell is successfully treated in mice.

Some of the ethical considerations for using gene therapy

—Some Questions which need to be considered...

- What is normal and what is a disability or disorder, and who decides?
- Are disabilities diseases? Do they need to be cured or prevented?
- Does searching for a cure demean the lives of individuals presently affected by disabilities?
- Is somatic gene therapy (which is done in the adult cells of persons known to have the disease) more or less ethical than germline gene therapy (which is done in egg and sperm cells and prevents the trait from being passed on to further generations)? In cases of somatic gene therapy, the procedure may have to be repeated in future generations.
- Preliminary attempts at gene therapy are exorbitantly expensive. Who will have access to these therapies? Who will pay for their use?

THERAPEUTIC APPROACHES

There are several major therapeutic approaches that may ultimately be useful in the treatment of various diseases. The majority-enzyme replacement therapy, bone marrow transplantation; neural stem cell therapy, hormonal replacement therapy-are aimed at restoring enzyme activity. These are currently more theoretical approaches to therapy and much remains to be learnt.

Bone marrow transplantation/Stem cell transplantation

Bone marrow or stem cell transplantation makes use of the fact that certain brain cells, as well as blood cells, arise from bone marrow or stem cells isolated from umbilical cord blood. Researchers theorize that such transplantation from a healthy donor would introduce healthy stem cells into the brain and multiply. However, it is not known if the number of healthy cells would be large enough to provide enough of the missing enzyme to make a clinical difference. If not, healthy stem cells could be genetically manipulated prior to transplantation to increase the production of the missing enzyme. Other factors which could limit the potential of bone marrow transplantation for the treatment of the lysosomal storage and allied diseases are: (1) the need to find an immunologically matched healthy bone marrow donor and (2) the risks associated with the bone marrow transplantation procedure. However finding a match with cryopreserved fetal cord blood is much easier and the risks associated with the transplant procedure are somewhat less than with conventional bone marrow transplantation.

Stem cells are immature cells that have the capability to develop into all of the different types of cells. The two leading sources of cells include mesenchymal cells (from blood) and neural stem cells (from brain). In mice, stem cells have been shown to spread throughout the brain, a necessity when it comes to treating diseases that affect the entire brain. Stem cells can also be genetically modified to produce and secrete high levels of a missing enzyme, such as Hex-A. For example, when neural stem cells genetically engineered to produce Hex-A were inserted into the brains of mice with Tay-Sachs they traveled throughout the brain and produced therapeutic levels of the needed enzyme. The enzyme then broke down the GM2 ganglioside, preventing its accumulation.

Replacement therapy with immunoglobins (Also known as IVIG, SCIG, Antibody Replacement Therapy)

"Gammaglobulin" and *"Immunoglobulin"* are terms used to describe antibodies, which are vital proteins produced by the immune system to deal with infections. *"Plasma"* is the liquid part of blood that remains when all cells have been removed.

Significance of Immunoglobulin therapy

Many people with immunodeficiencies have insufficient antibodies to adequately fight infections. Immunoglobulin replacement therapy is one of the most important and successful therapies for these people, and can be life saving. Immunoglobulin replacement therapy can:

- Treat existing infections
- Prevent new infections from occurring and
- Prevent the long-term damage of chronic infections (for example bronchiectasis in the lung)

Immunoglobulin therapy

Immunoglobulin products are purified from pooled plasma of many healthy blood donors. When donors give blood samples, the red cells and plasma are separated. The plasma is pooled together and processed in highly specialized and regulated facilities to produce the product. Each preparation therefore contains a wide variety of antibodies.

Delivered methods

There are two common ways that immunoglobulin replacement therapy can be delivered; intravenous (into the vein) and subcutaneous (under the skin).

1. *Intravenous injections*–Infusions are delivered directly into the vein, usually in a hospital day-stay clinic or sometimes in the patient's home. The infusion takes approximately two to four hours. The dose and frequency vary, and depend on the person's weight and immunoglobulin levels. Most people receive doses once every month.

2. *Subcutaneous injections*–The last few years has seen a move toward subcutaneous immunoglobulin (SCIG) injections, especially throughout Europe. This involves slowly infusing the antibody preparation directly under the skin, which can be done at home using a special pump. This is usually done once or more each week as only 10-15mL can be infused into any one site. When beginning SCIG therapy, red lumps may form under the skin. These usually disappear quite quickly and after a few weeks of therapy usually stop appearing.

Side effects of Immunoglobulin Replacement Therapy

Immunoglobulin replacement therapies are normally very well tolerated and serious side effects are very rare. There are some side effects that a person needs to be aware of:

1. Risk of blood-borne infections

As immunoglobulin preparations are derived from blood donors, there will always be a theoretical risk of blood borne infections. There has never been a case of HIV transmitted by this means. However, in the past, there have been cases of transmission of Hepatitis C and other viral illnesses. Current preventative measures have been greatly enhanced so that the risk of infection from antibody therapy is now close to zero.

2. Other side effects

Some patients experience minor side effects such as low-grade fevers or headaches which can usually be reduced by a slower infusion rate or treated with paracetamol. Occasionally patients experience hives or wheezing or rarely severe headaches. Rare cases of severe allergic-type reactions and abnormal kidney function have been described.

3. IgA reactions

In rare cases, where the patient lacks the antibody called IgA, reactions similar to severe allergy may occur when receiving blood products containing IgA. However, the great majority of IgA deficient patients receive blood products without difficulty. These reactions are less likely with current IgA-depleted Immunoglobulin products.

4. Pathogen specificity

Most antibody therapies are pathogen-specific. This is a disadvantage in dealing with mixed infections. Mixed infection with multiple *S. pneumoniae* serotypes was recognized as a cause for the failure of serum therapy.

5. Development of resistance in microbes

Widespread use of antibody-based therapies could produce selective pressure on microbial populations for the emergence of antibody-resistant variants. Antibody-resistant mutants of *Borrelia burdorferi* have been produced in the laboratory, and may be selected in patients who undergo antibody-based therapies. Microorganisms may become resistant to antibodies by acquiring mutations that change the antigenic site recognized by the antibodies or by producing proteases that destroy immunoglobulins. The IgA protease genes of *N. gonorrhoeae* can be transferred between strains and the widespread use of antibody therapies may select for protease-producing strains.

6. Only preventive measure

Antibodies are more effective in preventing infection than in treating established infection. Antibody-based therapies have been most useful when administered early in the course of disease: serum therapy for pneumococcal pneumonia was most effective if serum was administered within 3 days of the onset of clinical symptoms. Because antibodies are proteins, therapy for invasive infections is likely to require systemic administration. This is a serious disadvantage in developing countries where access to medical care is limited. For enteric pathogens, oral antibody administration may be most effective. Efforts to develop MAbs for cancer therapy have run into unexpected pharmacologic problems.

7. Allergic responses

Antibodies can, in principle, elicit neutralizing antibody, allergic responses, or both. Administering rodent MAbs to patients elicits human antibodies to the rodent MAb. Antibody therapies against emergent pathogens, if attempted, are likely to use human, human-animal chimeric, or humanized antibodies. Mouse-human chimerics and humanized antibodies are less immunogenic than heterologous antibodies; therefore, the likelihood that the patient will mount a neutralizing antibody response to the therapeutic antibody may be reduced. Nevertheless, anti-idiotypic responses have been observed in patients receiving humanized antibody therapy. The clinical importance of such anti-idiotypic responses is uncertain. Many infections are single life-threatening episodes in the life of a person, and the occurrence of anti-idiotypic antibodies following therapy may require repeated or long-term antibody administration.

8. Blood brain barrier

Antibodies are largely excluded from the central nervous system by the blood-brain barrier. Nevertheless, antibody treatment of brain infections is feasible. In some brain infections, the blood-brain barrier is more permeable to serum components because of inflammation, and systemic antibody therapy was used successfully to meningococcal meningitis.

9. High cost factor

Antibody therapies are also costly to develop and hence expensive for the patient. For example, antibody prophylaxis for cytomegalovirus infections can cost several thousand dollars per patient.

Limitations of Immunoglobulin Replacement Therapy

Antibody preparations contain essentially only one of the important components of the immune system's response to infection. This therapy does not cure the antibody deficiency, and will seldom reverse long-term organ injury from chronic infections.

Growth hormone replacement therapy

Growth hormone (GH) or somatotropin (STH) is a protein hormone that is synthesized and secreted by cells called *somatotrophs* in the anterior pituitary and it stimulates growth and cell reproduction in humans and other animals. It is a 191-amino acid, single chain polypeptide hormone which is synthesised, stored, and secreted by the somatotroph cells within the lateral wings of the anterior pituitary gland. Growth hormone is also of considerable interest as a drug used in both humans and animals.

Physiologic Effects of Growth Hormone

A critical concept in understanding growth hormone activity is that it has two distinct types of effects:

- Direct effects are the result of growth hormone binding its receptor on target cells. Fat cells (adipocytes), for example, have growth hormone receptors, and growth hormone stimulates them to break down triglyceride and supresses their ability to take up and accumulate circulating lipids.
- Indirect effects are mediated primarily by an insulin-like growth factor-I (IGF-I), a hormone that is secreted from the liver and other tissues in response to growth hormone. A majority of the growth promoting effects of growth hormone is actually due to IGF-I acting on its target cells. (See color plate: Fig 4).

Keeping this distinction in mind, we can discuss two major roles of growth hormone and its minion IGF-I in physiology.

Effects on Growth

Growth is a very complex process, and requires the coordinated action of several hormones. The major role of growth hormone in stimulating body growth is to stimulate the liver and other tissues to secrete IGF-I. IGF-I stimulates proliferation of chondrocytes (cartilage cells), resulting in bone growth. Growth hormone does seem to have a direct effect on bone growth in stimulating differentiation of chondrocytes. IGF-I also appears to be the key player in muscle growth. It stimulates both the differentiation and proliferation of myoblasts. It also stimulates amino acid uptake and protein synthesis in muscle and other tissues.

Metabolic Effects

Growth hormone has important effects on protein, lipid and carbohydrate metabolism. In some cases, a direct effect of growth hormone has been clearly demonstrated, in others, IGF-I is thought to be the critical mediator, and some cases it appears that both direct and indirect effects are at play.

- Protein metabolism: In general, growth hormone stimulates protein anabolism in many tissues. This effect reflects increased amino acid uptake, increased protein synthesis and decreased oxidation of proteins.

- Fat metabolism: Growth hormone enhances the utilization of fat by stimulating triglyceride breakdown and oxidation in adipocytes.
- Carbohydrate metabolism: Growth hormone is one of a battery of hormones that serves to maintain blood glucose within a normal range. Growth hormone is often said to have anti-insulin activity, because it supresses the abilities of insulin to stimulate uptake of glucose in peripheral tissues and enhance glucose synthesis in the liver. Somewhat paradoxically, administration of growth hormone stimulates insulin secretion, leading to hyperinsulinemia.

Clinical problems: excess and deficiency

Growth hormone excess: (acromegaly and pituitary gigantism)

The most common disease of GH excess is a pituitary tumor consisting of somatotroph cells of the anterior pituitary. These somatotroph adenomas are benign and grow slowly, gradually producing more and more GH. For years, the principal clinical problems are those of GH excess. Eventually the adenoma may become large enough to cause headaches, impair vision by pressure on the optic nerves, or cause deficiency of other pituitary hormones by displacement. Prolonged GH excess thickens the bones of the jaw, fingers and toes. Resulting heaviness of the jaw and increased thickness of digits is referred to as acromegaly. Accompanying problems can include pressure on nerves (e.g., carpal tunnel syndrome), muscle weakness, insulin resistance or even a rare form of type 2 diabetes, and reduced sexual function. GH-secreting tumors are typically recognized in the 5th decade of life. It is extremely rare for such a tumor to occur in childhood, but when it does the excessive GH can cause excessive growth, traditionally referred to as pituitary gigantism. Surgical removal is the usual treatment for GH-producing tumors. In some circumstances focused radiation or a GH antagonist such as bromocriptine or octreotide may be employed to shrink the tumor or block function.

Growth hormone deficiency (GHD)

Deficiency of GH produces significantly different problems at various ages. In children, growth failure and short stature are the major manifestations of GH deficiency. In adults the effects of deficiency are more subtle, and may include deficiencies of strength, energy, and bone mass, as well as increased cardiovascular risk. There are many causes of GH deficiency, including mutations of specific genes, congenital malformations involving the hypothalamus and/or pituitary gland, and damage to the pituitary from injury, surgery or disease. Diagnosis of GH deficiency involves a multiple step diagnostic process, usually culminating in GH stimulation test(s) to see if the patient's pituitary gland will release a pulse of GH when provoked by various stimuli. GH deficiency is treated by replacing GH. All GH in current use is a biosynthetic version of human GH, manufactured by recombinant DNA technology. As GH is a large protein molecule, it must be injected into subcutaneous tissue (or muscle) to get it into the blood. When the patient has had a long-standing deficiency of GH, benefits of treatment are often dramatic and gratifying and side effects of treatment are rare. Increased growth in childhood can result in dramatically improved adult height. GH is used as replacement therapy in adults with GH deficiency of either childhood-onset (after completing growth phase) or adult-onset (usually as a result of an acquired pituitary tumor). In these patients, benefits have variably included reduced fat mass, increased lean mass, increased bone density, improved lipid profile, reduced cardiovascular risk factors, and improved psychosocial well-being.

Other GH uses and treatment indications

Many other conditions besides GH deficiency cause poor growth, but growth benefits (height gains) are often poorer than when GH deficiency is treated. Examples of other causes of shortness often treated with growth hormone are Turner syndrome, chronic renal failure, Prader-Willi syndrome, intrauterine growth retardation, and severe idiopathic short stature. Higher ("pharmacologic") doses are required to produce significant acceleration of growth in these conditions, producing blood levels well above physiologic. Sometimes GH is used for benefits other than increasing height. GH treatment improves muscle strength and slightly reduces body fat in Prader-Willi syndrome, benefits more important to these children than increased height. It has also been shown to help maintain muscle mass in AIDS wasting. GH can also be used in patients with short bowel syndrome to lessen the requirement for intravenous parenteral nutrition.

Risks and side effects of GH treatment

Side effects in adults may include fluid retention, joint pain, and nerve compression symptoms. There is theoretical concern and weak evidence that GH treatment may increase the risks of diabetes or cancer, especially in those with other predispositions treated with higher doses. One survey of adults, who had been treated with replacement cadaver GH, during childhood showed a mildly increased incidence of colon cancer, but linkage with the GH treatment was not established. Regular application of GH did show several negative side effects such as joint swelling, joint pain, carpal tunnel syndrome, and an increased risk of diabetes. However, proponents of GH use as an anti-ageing substance claim that these side effects will not happen if injections are done in physiological amounts.

Hormone Replacement Therapy (Estrogen replacement therapy, HRT, Menopausal hormone therapy)

During her reproductive years, a woman's ovaries produce the hormones estrogen and progesterone in the ovaries. Estrogen works to regulate a woman's monthly menstrual cycle and secondary sexual characteristics, and also rises at different times in the menstrual cycle to prepare the body for fertilization and reproduction. Progesterone also rises in a cyclical fashion to prepare the uterus for possible pregnancy and to prepare the breasts for lactation. As a woman reaches menopause, typically around 50 years old, her body produces less and less estrogen and progesterone. Menopause is the time in a woman's life when she stops menstruating, signaling the end of her reproductive life. It is a normal part of aging. Natural menopause typically occurs in women when they are in their late forties or early fifties. Before menstruation stops at menopause, a woman will go through premenopause and perimenopause. During premenopause, menstrual periods begin to become irregular. In the years before and during menopause, the levels of female hormones can go up and down. During perimenopause, a woman's menstrual periods will become highly irregular and she may experience symptoms of menopause, such as hot flashes, vaginal dryness, mood swings, sleep disturbances and decreased sexual desire. The dryness of tissue surrounding the vagina and urethra may cause pain during sexual intercourse or urination. Perimenopausal occurs a few years before a woman reaches menopause. Once a woman reaches menopause, menstruation will stop forever unless she takes HRT.

Some women take hormone replacement therapy (HRT) to relieve these symptoms. Hormone replacement therapy (HRT) is medication containing one or more female hormones, commonly estrogen plus progestin (synthetic progesterone). Some women receive estrogen-only therapy (usually women who have had their uterus removed) HRT may also protect against osteoporosis. This medication may be taken in the form of a pill, a patch, or vaginal cream.

- However, HRT also has risks. It can increase your risk of breast cancer, heart disease and stroke. Certain types of HRT have a higher risk, and each woman's own risks can vary depending upon her health history and lifestyle. You and your health care provider need to discuss the risks and benefits for you. Should you decide to take HRT, it should be the lowest dose that helps and for the shortest time needed.

Limitations

Based on early studies, many physicians used to believe that HRT might be beneficial for reducing the risk of heart disease and bone fractures caused by osteoporosis (thinning of the bones) in addition to treating menopausal symptoms. The results of the Womens' Health Initiative (WHI) has led physicians to revise their recommendations regarding HRT. The WHI, started in 1993, has enrolled 161,809 women, ages 50-79, in 40 different medical centers. Part of the study was intended to examine the health benefits and the risks of hormone replacement therapy, including the risk of developing breast cancer, heart attacks, strokes, and blood clots.

In 2002, one component of the WHI, which studied the use of estrogen and progestin in women who had a uterus, was stopped early because the health risks exceeded the health benefits. The main reason for stopping the estrogen-progestin study was because of a 26% increase in breast cancer. In 2004, a second component of the WHI, which studied estrogen-only therapy in women who no longer had a uterus (see hysterectomy) was stopped early. This was primarily because of an increase in the risk for strokes.

Side effects of HRT

As with all medicines, there are side effects associated with HRT. Some women taking HRT may experience water retention, bloating, nausea, breast soreness, mood swings, and headaches. Changing the dosage or the form of HRT may help to alleviate these side effects. Some women have irregular bleeding when they start taking HRT, but changing the dosage often eliminates this side effect (Table 1).

Table 1: The risks associated with various diseases.

Adverse event	*Relative risk (95% CI)*	*Change in number of events per 10,000 women in a year*
Breast cancer	1.26 (1.00-1.59)	8 more
Heart disease	1.29 (1.02-1.63)	7 more
Stroke	1.41 (1.07-1.85)	8 more
Pulmonary embolism	2.13 (1.39-3.25)	8 more
Colorectal cancer	0.63 (0.43-0.92)	6 fewer
Hip fracture	0.66 (0.45-0.98)	5 fewer

Summary of risks/benefits

The WHI study shows relatively small *absolute* increases in the risk of heart disease, breast cancer, blood clots, and stroke to an individual woman. However, when the entire population of postmenopausal women and the number of years a woman may be on HRT are considered—the number of strokes, heart attacks, breast cancer cases, and blood clots appears to outweigh the protective effect of HRT on bones. Some women may still wish to consider HRT for short-term treatment of menopausal symptoms. The key is to weigh the risks associated with taking HRT against a particular woman's risk of heart disease or osteoporosis without taking HRT. Every woman is different. Therefore, entire medical history should be evaluated individually when considering HRT.

Insulinotherapy

A general description of intensive or flexible insulin therapy

A working pancreas continously secretes small amounts of insulin into the blood to prevent the body from shifting into "starvation metabolism." This insulin is referred to as *basal insulin secretion*. Most insulin used each day is produced during the digestion of meals. Insulin levels rise immediately as we begin to eat, remaining higher than the basal rate for 1 to 4 hours. This meal-associated (*prandial*) insulin production is roughly proportional to the amount of carbohydrate in the meal.

Intensive or flexible therapy involves supplying a continual supply of insulin to serve as the *basal insulin*, supplying meal insulin in doses proportional to nutritional load of the meals, and supplying extra insulin when needed to correct high glucose levels.

These three components of the insulin regimen are commonly referred to as basal insulin, meal insulin, and high correction.

Methods

One method of intensive insulinotherapy is based on multiple daily injections (sometimes referred to in medical literature as *MDI*). Meal insulin is supplied by injection of rapid-acting insulin before each meal in an amount proportional to the meal. Basal insulin is provided as a once or twice daily injection of dose of long-acting insulin. In an MDI regimen, long-acting insulins are preferred for basal use. Older insulin used for this purpose is ultralente. *Levemir*, made by Novo Nordisk, is long-acting insulin in trials. Many people on MDI regimens carry insulin pens to inject their rapid-acting insulins instead of traditional syringes.

The other method of intensive/flexible insulin therapy is an insulin pump. It is a small mechanical device about the size of a deck of cards. It contains a syringe-like reservoir with about three days' insulin supply. This is connected by thin, disposable, plastic tubing to a needle-like cannula inserted into the patient's skin and held in place by an adhesive patch. The infusion tubing and cannula must be removed and replaced every few days. An insulin pump can be programmed to infuse a steady amount of rapid-acting insulin under the skin. This steady infusion is termed the basal rate and is designed to supply the background insulin needs. Each time the patient eats, he or she must press a button on the pump to deliver a specified dose of insulin to cover that meal. Extra insulin is also given the same way to correct a high glucose reading. Current pumps do not include a glucose sensor and cannot automatically respond to meals or to rising or falling glucose levels.

Both MDI and pumping can achieve similarly excellent glycemic control. Some people prefer injections because they are less expensive than pumps and do not require the wearing of a continually attached device. A primary advantage of pumps is the freedom from syringes and injections. Intensive/flexible insulin therapy requires frequent blood glucose checking. To achieve the best balance of blood sugar with either intensive/flexible method, the patient must check his or her glucose level with a meter monitoring of blood glucose several times a day. This allows optimization of the basal insulin and meal coverage as well as correction of high glucose episodes.

Advantages and disadvantages

For most people with diabetes, the two primary advantages of intensive/flexible therapy are greater flexibility of timing and amounts of meals and better glycemic control. Major disadvantages of intensive/flexible therapy are that it requires greater amounts of education and effort to achieve the goals, and it substantially increases the daily cost of diabetes care. The frequency of hypoglycemia increases with increasing effort to achieve normal blood glucoses with any insulin regimen. When traditional regimens are used aggressively enough to achieve near-normal hemoglobin A1c levels, hypoglycemia is as frequent as with flexible regimens. When used correctly, flexible regimens offer greater ability to achieve good glycemic control with easier accommodation to variations of eating and physical activity.

Enzyme replacement therapy

Genes code for proteins and proteins are essentially enzymes. So, when a gene is mutated and does not produce the gene product, an enzyme is missing or defective. One way to treat a genetic condition is to replace the enzyme that the gene is not producing. The treatment of enzyme deficiency states represents an obvious use of enzymes. Enzyme replacement therapy (ERT) is a therapeutic approach in which the specific enzyme that is inactive or absent in affected individuals is replaced with functional enzyme molecule isolated or produced in the lab. For example, oral pancreatic extracts have commonly been used in the treatment of cystic fibrosis and pancreatic insufficiency. The most commonly available preparations are of three types: (1) pancreatin an alcoholic extract of pig pancreas containing amylase and trypsin activity, (2) pancrealipase, a lipase-enriched extract, and (3) Viokase also enriched in lipase.

More intriguing is the treatment of the inborn errors of metabolism in which deficiency of a single enzyme leads to accumulation of abnormal amounts of substrate. With the recognition that many of these errors are owing to inadequacies of lysosomal enzymatic catabolism, it was reasoned that exogenously administered enzyme might react with and dispose of such accumulations.

The infusion of crude glucosidase from *Aspergillus niger* into patients with type 11 glycogenolysis, a condition attributed to a deficiency of this enzyme, was reported in the mid 1960s. Infusion of normal plasma, which contains galactosidase, the enzyme missing in Fabry's disease, reduced the amount of abnormal substrate in the plasma and tissues of two patients heterozygous for this disorder.

Gaucher disease is a condition for which enzyme replacement therapy has been developed. Patients with Gaucher disease have abnormal accumulations of a compound called glucocerebroside due to the deficiency of the enzyme glucocerebrosidase. This accumulation causes cells to form what is known as "Gaucher cells" which displace healthy cells in bone marrow, cause an enlargement of the liver and spleen, organ dysfunction, and deterioration of the skeleton. There are three different types of Gaucher disease which range in terms of severity and age of onset. An enzyme assay can confirm the diagnosis.

Enzyme replacement therapy (ERT) has been used in over 1000 patients for the treatment of gaucher disease—to reverse the symptoms and improve the quality of life. It must be taken throughout the life of the individual, otherwise the health problems from the gene defect return. Patients must be under the care of a physician to monitor the therapy for any adverse reactions and adjust dosages. ERT has been successful for the treatment of Gaucher Type 1, Fabry, MPS I and, most recently, has received approval for Pompe disease. ERT is effective in the non-neurological symptoms of Mucopolysaccharidosis Types I, II IV and VI, Pompe and Niemann-Pick B, but has not yet proven to be beneficial in storage diseases that primarily affect the central nervous system since the replacement enzymes do not efficiently cross the blood-brain barrier.

SIGNIFICANCE OF GENE THERAPY

As a result of recent advances in medical science, researchers believe that a gene can be transplanted into human who suffer from severe diseases. Such gene transplants may alleviate and / or even cure diseases for which no adequate treatment now exists. The treatment is called human gene therapy and is one of the emerging genetic techniques, commonly called genetic engineering, based on new knowledge about how genes work. It is expected that researchers will soon request approval for a human gene therapy experiment. Many benefits are foreseen. However, because of the novelty of the field, concerns about the discriminatory and eugenic misuse of the techniques, and the possible effects on future generations from some types of human gene therapy, important ethical questions will also be raised.

Diseases and their treatment

The human body is made up of about fifty trillion cells. Inside each cell is information that tells the cell what to do and how to work. This information is contained in genes, which are made up of a chemical called DNA. Through small differences in the structure of the DNA, the information is coded and stored, just as different letters combine to form words which are then stored in books. All the many activities that a cell does start with reading part of the information that is stored in the DNA of the genes. There are approximately one hundred thousand genes in each cell of the human body. Although the genes are the same in every cell, each type of cell reads only certain genes. In this way a muscle cell, for instance, looks and works differently from a skin cell or a liver cell.

There are two major types of cells in the human body, *somatic* (non-reproductive) cells and *germ line* (reproductive) cells. Most cells in the body are somatic. Somatic cells provide all the body structures and perform all the functions except for passing genetic information on to the next generation. Germ cells include eggs in women and sperm in men. The genes in sperm and egg cells store information that will go to the next generation, to one's children. The genetic information in somatic cells is not passed on to the next generation.

If the DNA information of a particular gene contains mistakes, the gene may not function properly. Sometimes the malfunction will not be serious, but other times it will cause a severe genetic disease. Examples of some genetic diseases are cystic fibrosis, sickle cell anemia, and hemophilia. Hemophilia, for instance, is caused by the malfunctioning of the gene that makes the factor that causes blood to clot. As more is learned about human genetics, it is becoming clear that diseases such as diabetes, cancer, heart disease, and some manic depressive illnesses also result in part from faulty DNA information.

For some genetic diseases, there are satisfactory therapies that do exist. Drugs, blood transfusions, changes in diet, introducing enzymes and hormones or transplantation of body organs can often help to compensate for the incorrect information from the malfunctioning gene. For example, clotting factor can be administered to patients with hemophilia.

Human gene therapy is a possible alternative approach to the treatment of some genetic diseases. The basic idea behind gene therapy is to insert normal genes with correct information into the DNA of the cells that contain malfunctioning genes. Adding genes in this way is called "gene insertion." The added genetic information would allow these cells to function properly and might reduce or eliminate the signs or symptoms of the disease. For example, instead of repeatedly treating a hemophiliac with clotting factor, one could insert the correct genetic information into his cells to allow those cells to make their own clotting factor.

There may be other uses of gene therapy, for example, to repair a wound, or to grow new blood vessels, during which the effect of the gene would only need to be active for a short time. Advances in human gene therapy may allow doctors to treat a disease or abnormal medical condition by turning off a faulty gene and stopping the growth of a cancerous tumor, for example. Or they may allow the body to begin producing a necessary protein or other substance, such as an enzyme, that the faulty gene cannot order the body to produce.

For example, a person with the disease cystic fibrosis (CF) has a faulty gene for handling lung development, which results in excess mucous in the lungs, leading to chronic coughing, choking and serious respiratory infections. Cystic fibrosis patients often die young from pneumonia or other life-threatening respiratory diseases. The current median life span for CF patients is now about 30 years. For these persons, a doctor might inject a normal CF gene into a patient's bloodstream. The hope is that the CF gene would find its way into the patient's lungs, replace the faulty gene and help the patient's lungs to function properly.

It seems likely that human gene therapy will also be used to combat certain diseases that may not be genetic. For example, malignancies are usually treated with surgery, radiation and/or chemotherapy. For cancer patients who are not helped by these therapies, researchers are now planning to treat the patients' disease with genetically-altered white blood cells. Scientists have developed methods for inserting genes into human somatic cells. The techniques for isolating human genes and making multiple copies of them in the laboratory are well established. Now scientists are studying how to insert those genes into cells and how to make those genes work properly once inside the cells. One method for inserting genes into cells is to link the genes with a virus that has been crippled and rendered harmless. As part of the modification, such a virus, sometimes called a vector or vehicle, has been deliberately altered so that it can carry genes into cells but cannot then escape to infect other cells. After the cells to be treated have been temporarily removed from a patient's body, the virus or vector is used to carry the desired gene into them. The final step will be to return the treated cells, which now contain the correct genetic information, to the patient's body. For example, bone marrow, liver cells, or white blood cells could be removed from the body of a patient, treated in the laboratory, and returned to the patient.

The most likely candidates for future gene therapy trials will be rare diseases such as Lesch-Nyhan syndrome, a distressing disease in which the patients are unable to manufacture a particular enzyme. This leads to a bizarre impulse for self-mutilation, including very severe biting of the lips and fingers. The normal version of the defective gene in this disease has now been cloned. If gene therapy does become practicable, the biggest impact would be on the treatment of diseases where the normal gene needs to be introduced into only one organ. One such disease is phenylketonuria (PKU). PKU affects about one in 12,000 white children, and if not treated early can result in severe mental retardation. The disease is caused by a defect in a gene producing a liver enzyme. If detected early enough, the child can be placed on a special diet for their first few years, but this is very unpleasant and can lead to many problems within the family.

The types of gene therapy described thus far all have one factor in common: that is, that the tissues being treated are somatic (somatic cells include all the cells of the body, excluding sperm cells and egg cells). In contrast to this is the replacement of defective genes in the germline cells (which contribute to the genetic heritage of the offspring). Gene therapy in germline cells has the potential to affect not only the individual being treated, but also his or her children as well. Germline therapy would change the genetic pool of the entire human species, and future generations would have to live with that change. In addition to these ethical problems, a number of technical difficulties would make it unlikely that germline therapy would be tried on humans in the near future.

Before treatment for a genetic disease can begin, an accurate diagnosis of the genetic defect needs to be made. It is here that biotechnology is also likely to have a great impact in the near future. Genetic engineering research has produced a powerful tool for pinpointing specific diseases rapidly and accurately. Short pieces of DNA called DNA probes can be designed to stick very specifically to certain other pieces of DNA. The technique relies upon the fact that complementary pieces of DNA stick together. DNA probes are more specific and have the potential to be more sensitive than conventional diagnostic methods, and it should be possible in the near future to distinguish between defective genes and their normal counterparts, an important development.

The Human Genome Program in the U.S. has been providing about $200 million each year to scientists in multidisciplinary research centers who are attempting to determine the makeup of all human genes. Together with similar programs in Europe, it is hoped that in 15 years time we shall be able to identify and treat all diseases to which humans are susceptible. This will revolutionize modern medicine, and hopefully improve the quality of life of all men, women, and children. Already, the genes for Duchenne muscular dystrophy, cystic fibrosis, and retinoblastoma have been identified, and more such information is emerging all the time.

The best outcome of human gene therapy would be a single treatment that would correct enough cells to provide a permanent cure for the patient's disease. This kind of complete success is unlikely in the beginning stages of human gene therapy but will remain the long-term goal of research scientists working in this field.

APPROACHES

The transfer of DNA is usually achieved either by mixing the DNA with a substance that enhances its uptake or by packaging it inside a disabled virus (a virus that can carry DNA into a cell but cannot cause a disease). Gene therapy may be carried out by first removing cells from the patient and then re-introducing those that have been appropriately modified. Alternatively, the direct introduction of DNA into the patient's body may be the only effective strategy. There are several approaches to gene therapy:

Gene augmentation

This is appropriate for the treatment of inherited disorders caused by the loss of a functional gene product. The aim is to add a functional copy of the lost gene back into the genome and express it at sufficient levels to replace the missing protein. It is only suitable if the pathogenic effects of the disease are reversible.

Gene inhibition

This is suitable for the treatment of infectious diseases, cancer and inherited disorders caused by inappropriate gene activity. The aim is to introduce a gene whose product inhibits the expression of the pathogenic gene or interferes with the activity of its product.

Killing of specific cells

This is suitable for diseases such as cancer that can be cured by eliminating certain populations of cells. The aim is to express within such cells a suicide gene, whose product is toxic. One approach is the expression of an enzyme that converts a harmless prodrug into a highly toxic molecule. Another is the expression of a protein that makes the cells vulnerable to attack by the immune system. It is very important to ensure that suicide genes are appropriately targeted otherwise the therapy would result in widespread cell death.

TYPES OF GENE THERAPY

In theory it is possible to transform either somatic cells (most cells of the body) or cells of the germline (such as sperm cells, ova, and their stem cell precursors). All gene therapy to date on humans has been directed at somatic cells, whereas germline engineering in humans remains controversial. For the introduced gene to be transmitted normally to offspring, it needs not only to be inserted into the cell, but also to be incorporated into the chromosomes by genetic recombination.

Originally known as 'genetic replacement therapy' during the early 1980s, 'gene therapy' has now outgrown its original definition and is applied to all manner of protocols that involve an element of gene transfer, either *in vivo* or *ex vivo*, and not necessarily a gene that is known to cause a disease. Gene therapy can be subdivided into somatic cell gene transfer (that is transfer to normal diploid cells), which is the focus of this review, and germline gene transfer (transfer to haploid sperm or egg cells of the reproductive system).

Somatic gene therapy can be broadly split in to two categories: *ex vivo*, which means exterior (where cells are modified outside the body and then transplanted back in again) and *in vivo*, which means interior (where genes are changed in cells still in the body). Recombination-based approaches *in vivo* are especially uncommon, because for most DNA constructs recombination has a very low probability. *In vivo* gene transfer is the introduction of genes to cells at the site they are found in the body, for example to skin cells on an arm, or to lung epithelial cells following inhalation of the gene transfer vector. *Ex vivo* gene transfer is the transfer of genes into viable cells that have been temporarily removed from the patient and are then returned following treatment (e.g. bone marrow cells).

The ethical issues associated with germline gene therapy are far more complex than those surrounding somatic cell gene transfer, because the genes are transferred not only to treated individuals but also to their progeny. Germline gene therapy is being widely used for the production of transgenic animals for research, and increasingly for agriculture and biotechnology, but the long-term effects of each transferred gene in animals will need to be carefully monitored and analysed, as well as the significance of any residual vector DNA if applicable. The benefits that the use of germline gene therapy in humans could bring are significant. The development of serious and distressing inherited genetic diseases could be prevented before birth and eliminated in subsequent generations. However, because of the potential for abuse and eugenics, gene therapy in humans needs to be widely discussed and the associated safety issues evaluated before this approach can be used for the treatment of diseases.

Somatic Cell Gene Therapy

Human somatic cell gene therapy is a form of medical treatment that is being developed as a genetic approach to disease management. Two basic gene therapy strategies have been investigated;

IN VIVO APPROACH

- *In vivo* gene therapy requires that the gene transfer vector be delivered in a cell-type selective manner, either through direct tissue injection, or perhaps someday, by receptor-mediated processes (See color plate: Fig 5).

EX VIVO APPROACH

- *Ex-vivo* gene therapy involves removing tissue from the patient, transfecting (or virally-infecting) the cells in culture, and then reimplanting the genetically altered cells into the patient (See color plate: Fig 6).

In vivo gene therapy is done by targeting the gene delivery system to the desired cell type in the patient using either physical means such as tissue injection (brain tumor) or biolistics (dermal DNA vaccination), or potentially using systemic infusion of cell-specific receptor-mediated DNA carriers (reconstructed liposomes or viruses). Importantly neither of these gene therapy strategies involve reproductive germline cells and therefore the genetic alteration will not be transmitted to the next generation. In many countries, human germline gene therapy is considered unethical or even illegal.

Ex-vivo gene therapy is performed by transfecting or infecting patient-derived cells in culture with vector DNA and then reimplanting the transfected cells into the patient. Two types of *ex-vivo* gene therapies under development are those directed at fibroblasts and hematopoietic stem cells.

Monogenic loss-of-function diseases have been used as paradigms to develop gene therapy strategies using animal models. In this context, gene therapy refers to genotypic alterations (chromosomal or episomal) of somatic cells, resulting in sustained gene expression. This is to be distinguished from transient DNA transfection approaches, such as repeated inhalation or infusion of functional nucleic acids, for example, drug therapies utilizing ribozymes or antisense oligonucleotides.

Table 2: An account of genes to be transfected for gene therapy of various diseases

Disease	*Target cells*	*Transfected gene(s)*
Hemophilia A Hemophilia B	liver, muscle, bone marrow cells, fibroblasts	Factor VIII Factor IX
Familial hypercholesterolaemia	liver	Low-density lipoprotein receptor
Severe combined immunodeficiency	bone marrow cells, T cells	Adenosine deaminase (ADA)
Haemoglobinopathies	red blood precursor cells	a-globin, b-globin
Cystic fibrosis	lung airway cells	Cystic fibrosis gene (CFTR)
Gaucher's	bone marrow cells, macrophages	glucocerebrosidase
Cancer	tumor cells	p53, Rb, interleukins, growth-inhibitory genes, apoptosis genes

In addition to being safe and cost-effective, the most important properties of an efficacious gene transfer system will be;

- Target cell selective
- Transcriptionally competent for the desired length of time

- Available in a highly concentrated active form
- Immunologically neutral

While it is still too early to know how useful somatic cell gene therapy will be as a routine treatment for human genetic diseases, most biomedical researchers agree that it will represent just one of the many new molecular genetic tools that 21st century physicians will have at their disposal to better prevent, diagnose and treat their patients.

Human Germline Gene Therapy

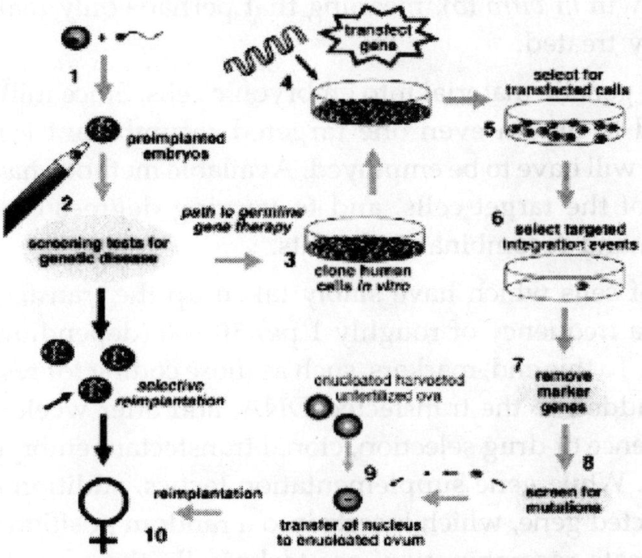

Fig A: Steps Involved in a Human Germline Gene Therapy Protocol

1. Isolation of totipotent embryonic cells at an undifferentiated stage S (a preimplanted embryo within four days of conception). Step 1 is achieved either by laparoscopic flushing of oviducts to recover a naturally-fertilized ovum before implantation, or *in vitro* fertilization. Both these strategies are currently possible in humans, but are expensive and often require months or years of repeated attempts (1). While it may be morally preferable to genetically manipulate gametes instead of embryos, this is not technically feasible, and this strategy does not seem to be an option (2).
2. Determination of the genetic state of the embryo. "Preimplantation diagnosis" is already available on an experimental basis (3). Note that if normal embryos can be differentiated from abnormal embryos, the simplest option by far is selective reimplantation, obviating the need for genetic manipulations (Steps 3-9). On the other hand, step 2 can be skipped in rare cases (both parents

homozygous for a recessive disorder, or one parent homozygous for a dominant disorder) where all embryos will be afflicted, and would not be a necessary step in "enhancement" procedures. (See Fig. A).

3. Expansion of embryonic stem cells in culture. While already achieved in mice and, more recently, in monkeys (4), this step would raise several issues if applied to humans. There is a poor success rate for establishing such cultures, and maintenance is expensive and labour-intensive. A normal karyotype has a limited lifespan in culture; in particular, second X chromosomes are frequently lost during growth *in vitro* (5), meaning that perhaps only male embryos could be successfully treated.

4. Transfer of genetic material into embryonic cells. Since millions of cells must be transfected to obtain even one targeted recombinant (6), mass transfection techniques will have to be employed. Available methods have limited efficiency, kill some of the target cells, and to varying degrees cause undesired DNA integration and recombination events.

5. Selection of cells which have stably taken up the transfected gene. These are present at a frequency of roughly 1 per 10 000 (depending on the transfection technique). To this end, markers, such as those coding for resistance to a cytotoxic drug, are added to the transfected DNA, and after weeks of careful culturing in the presence of drug selection, clonal transfectant embryonic cell lines can be grown out. While gene supplementation tactics, addition of a normal copy of the transfected gene, which inserts into a random position in the cell's genome by illegitimate recombination, are technically the easiest way to replace the function of a defective gene, they are inherently unpredictable in their downstream effects and can only be used for somatic gene therapy. Unfortunately, current technologies only permit the addition of a supplementary gene copy (7) causing human germline gene therapy to adopt the more difficult targeted gene replacement (8).

6. Targeted gene replacement. This step represents a true genetic cure, where the deleterious DNA base pair sequence in its native chromosomal position is changed to a normal sequence that permits proper function of the gene product. However, mammalian cells favour illegitimate recombination events by a factor of at least 1000 over the homologous recombination events required for a targeted gene replacement (6). Differentiating targeted gene replacement events from illegitimate insertions is especially tedious and difficult, as in both cases cells have taken up selectable marker tags attached to the transfected DNA. Molecular biologists have devised transfection vectors that increase the proportion of targeted recombinants (7, 9), but only by selecting against marker-

expressing illegitimate integration events; there is no way currently available to induce significantly higher absolute levels of homologous recombinations, which remain at the abysmally low rate of around 1 per 10 million cells transfected. While gene targeting techniques are probably obligatory for human germline manipulation, biological limitations make Step 6 very inefficient and difficult.

7. Marker removal. Leftover marker genes in the vicinity of the modified locus may influence neighbouring genes in undesired ways (10), and should not become part of every cell in a person, or passed on to future generations. Fortunately, molecular biologists have developed strategies to permit the removal of markers, leaving the genome completely unaltered except for the targeted sequence change (9); this requires reexpansion of the culture, and another round of drug selection.
8. Confirming genomic integrity. An embryonic cell line with the exact, desired genetic change has finally been established. However, genetic changes can occur during long term culture, and there is evidence to suggest that subtle mutations are frequently induced in regions treated by targeted homologous recombination (6). Thus, it may be necessary to recheck the genomic integrity of the cell line before attempting to reimplant an embryo. How this step could be accomplished is uncertain; it would require tedious, extensive, expensive, yet necessarily incomplete investigation. Months or years of growth in culture would be required to reach this point, but the cells should still be viable (11) for nuclear transfer.
9. Nuclear transfer. The well-publicized work of Wilmut *et al.*, who cloned sheep from cultured cells arrested in the G0 phase of the cell cycle (11), makes step 9 theoretically possible in humans, although it has never been tried. Unfertilized ova would need to be recovered from a superovulated woman, either the mother, or a donor (who would then donate the progeny's mitochondrial DNA). With the nucleus removed, and replaced with a nucleus from Step 8, an embryo could grow, carrying the modified human genome; again, a difficult, expensive and tedious step (12).
10. Reimplantation into the mother. Embryo loss is roughly 98% (13), and remains the limiting factor in human *in vitro* fertilizations. Subsequent spontaneous abortion is frequent; only about 15% of couples attempting *in vitro* fertilization ever have a successful pregnancy (3). To increase the chance of success, multiple embryos are reimplanted simultaneously, and if multiple embryos from Step 9 "take," identical twins or triplets will result.

There are some arguments in favour of germ line gene therapy (for example, it would allow the correction of disease-causing mutations that are certain to be passed on) but many more arguments against.

The principle objections are:

- The technology is imperfect. The effects of gene transfer are unpredictable and, even if the target disease was cured, further defects could be introduced into the embryo.
- Denial of human rights. Individuals resulting from germ line gene therapy would have no say in whether their genetic material should have been modified.

Potential abuse

Germ line gene therapy could by used not only to eliminate disease, but also to enhance favourable characteristics and suppress unfavourable ones. On a small scale, this would result in a generation of 'designer children', with traits chosen by their parents. On a large scale, gene therapy could result in eugenics-manipulation of the genetic properties of a population.

METHODS/TECHNIQUES FOR TRANSPORT OF RECOMBINANT GENES TO TARGET CELLS

In recent years many diseases have been shown, at least in part, to have a genetic component to their aetiology. In some cases, the causal component of a particular disease state results from a mutation in the DNA encoding a single gene. For example, a mutation in the cystic fibrosis transmembrane conductance regulator (CFTR) gene pre-disposes an individual to cystic fibrosis (Riordan *et al.*, 1989). However, the majority of genetic diseases arise from mutations in more than one gene (Brookes and Vile, 1998). Thus, an important first step in establishing a therapy for a particular genetic disease that involves manipulation of the disrupted or affected gene(s) is the identification of the genetic component of that disease. Fortunately, this task has been made easier for the gene therapist with the completion of the sequencing of the human genome. Equally, there are other important considerations that must be addressed to establish the therapeutic potential of gene therapy in the treatment of any disease. Thus, the therapeutic DNA might be required to restore the activity of a gene that has been abolished or altered, to control the expression of a gene that has become deregulated and/or to block the function of an affected gene.

All of these scenarios require the introduction of genetic material into affected cells to restore the original function of the mutated gene. The introduction of the therapeutic genetic material only into affected cells also might be necessary (e.g. in the treatment of cancers). Where available, the use of tissue or cell specific promoters to control the expression of the therapeutic DNA would ensure the expression of the introduced DNA only in the target cellos and not in normal, neighbouring cells. Considerations such as specific characteristics of the targeted cells, the required duration of the gene therapy and/or the preferred route of administration of the therapeutic gene might have an important influence on the therapeutic approach employed. A number of human inherited disorders have been corrected in cultured cells and several diseases (e.g. malignant melanoma and severe combined immunodeficiency disease, SCID) are currently being treated by gene therapy techniques indicating that gene therapy is likely to be a powerful therapeutic technique against a host of diseases in coming years.

GENE TRANSFER VECTORS

The most common technique utilized in gene therapy studies is the introduction of the corrected gene into bone marrow cells, skin fibroblasts or hepatocytes. All viruses attack their hosts and introduce their genetic material into the host cell as part of their replication cycle. This genetic material contains basic 'instructions' of how to produce more copies of these viruses, hijacking the body's normal production machinery to serve the needs of the virus. The host cell will carry out these instructions and produce additional copies of the virus, leading to more and more cells becoming infected. Some types of viruses actually physically insert their genes into the host's genome (a defining feature of retroviruses, the family of viruses that includes HIV, is that where the virus will introduce the enzyme reverse transcriptase into the host and thus use its RNA as the "instructions"). This incorporates the genes of that virus among the genes of the host cell for the life span of that cell.

Vectors are the vehicles that are used in gene therapy to transfer the gene(s) of interest [transgene(s)] to the target cells, which will then go on to express the therapeutic protein encoded by the transgene(s). The vectors most commonly utilized are derived from retroviruses and utilize only the transcriptional promoter regions of these viruses (the LTRs) to drive expression of the gene of interest. The advantage of retroviral-based vector systems is that expression occurs in most cell types. The first successful examples of this type of retro-viral gene therapy were reported in August.

Doctors and molecular biologists realized that viruses like this could be used as vehicles to carry 'good' genes into a human cell. Before viruses can be used to deliver therapeutic genetic material, they must be modified to prevent the uncontrolled replication of the engineered viral vector and the potential for reversion of the therapeutic virus into an actively pathogenic form. So first, a scientist would remove the genes in the virus that cause disease. Then they would replace those genes with genes encoding the desired effect (for instance, insulin production in the case of diabetics). This procedure must be done in such a way that the genes which allow the virus to insert its genome into its host's genome are left intact. This can be confusing, and requires significant research and understanding of the virus' genes in order to know the function of each. An example:

A virus is found which replicates by inserting its genes into the host cell's genome. This virus has two genes-A and B. Gene A encodes a protein which allows this virus to insert itself into the host's genome. Gene B causes the disease this virus is associated with. Gene C is the "normal" or "desirable" gene we want in the place of gene B. Thus, by re-engineering the virus so that gene B is replaced by gene C, while allowing gene A to properly function, this virus could introduce your 'good gene'-gene C into the host cell's genome without causing any disease.

The most important factor in any gene transfer protocol, apart from the gene of interest, is the choice of vector, which can result in either success or failure. Unfortunately, there is no such thing as a 'good universal vector'; all of the vectors that are currently available have both advantages and disadvantages. All this is clearly an oversimplification, and numerous problems exist that prevent gene therapy using viral vectors such as: trouble preventing undesired effects, ensuring the virus will infect the correct target cell in the body, and ensuring that the inserted gene doesn't disrupt any vital genes already in the genome. For example, one vector might be able to enter target cells very efficiently but once there invokes a strong immune response, resulting in that cell being killed by the immune system. Many factors must be taken into consideration when choosing a vector. The most import ones are: (1) the length of time that the transgene needs to be expressed, (2) the dividing state of the target cells, (3) the type of target cell, (4) the size of the transgene, (5) the potential for an immune response against the vector to be induced, and whether this is deleterious, (6) the ability to administer the vector more than once, (7) the ease of production of the vector, (8) the facilities available, (9) safety issues and (10) regulatory issues. However, this basic mode of gene introduction currently shows much promise and doctors and scientists are working hard to fix any potential problems that could exist.

Viral Gene Therapy Vectors

For millions of years, viruses have been transferring genes into all types of cells, including plant, animal and human cells. The experimental technique of viral gene delivery was developed from this natural ability, which offers many intrinsic advantages to scientists and clinicians: (1) specific cell-binding and entry properties, (2) efficient targeting of the transgene to the nucleus of the cell and (3) the ability to avoid intracellular degradation. The general principle involved in the development of most viral vector systems is that an intact wild-type virus is modified for safe use and effective gene transfer; for example, the specific genes that are involved in viral replication can be modified or deleted, thus rendering the new recombinant virus 'replication defective' and safer for use in gene therapy protocols. Usually, the transgene that is to be delivered by the virus must be inserted into the viral genome, using molecular biological techniques; transgenes are often inserted into the space created by the removal of viral replication genes. In general, the more severely attenuated the viral vector is from its wild-type state (i.e. the greater the number of virulence-associated genes that have been removed), the safer the virus is for use in gene therapy protocols. The size of the transgene has to be matched to the potential space in the viral genome; if the new viral genome is too large, it cannot be packaged into an infectious particle. Because many of the viruses that are used as vectors lack replication genes and therefore cannot replicate in normal cells, the recombinant virus with its transgene must be grown up to higher titres in a packaging cell line. This is a cell line that contains all of the complementary genes that the virus requires to replicate (i.e. those that were previously removed). The recombinant viral particles can then be purified as live infectious virus from the packaging cell line and used to infect (transduce) cells or tissues *in vivo* or *ex vivo*.

Retroviruses

Retroviruses have long been recognized as a powerful means of delivering foreign genes to target cells since they are relatively simple to construct as a vector-delivery system (See color plate: Fig 7). The Retroviridae is a large family of RNA viruses including spumavirus (foamy viruses), Moloney-murine-lentivirus-related viruses [e.g. Moloney murine leukaemia virus (MMLV) and the C family of human endogenous retroviruses (HERV-C)] and lentivirus [e.g. human immunodeficiency virus type 1 (HIV-1) and human immunodeficiency virus type 2 (HIV-2)]. The diameter of retroviral virions ranges from 80 nm to 130 nm, and their genomes consist of two identical positive-sense single-stranded RNA molecules, in the size range of 3.5–10 kb. The genomes are encased in a capsid, along with the integrase and reverse transcriptase enzymes. Retroviral vectors are the most widely used viral vectors in clinical trials at present.

The genetic material in retroviruses is in the form of RNA molecules, while the genetic material of their hosts is in the form of DNA. When a retrovirus infects a host cell, it will introduce its RNA together with some enzymes into the cell. This RNA molecule from the retrovirus must produce a DNA copy from its RNA molecule before it can be considered for part of the genetic material of the host cell. The process of producing a DNA copy from an RNA molecule is termed reverse transcription. It is carried out by one of the enzymes carried in the virus, called reverse transcriptase. After this DNA copy is produced and is free in the nucleus of the host cell, it must be incorporated into the genome of the host cell. That is, it must be inserted into the large DNA molecules in the cell (the chromosomes). This process is done by another enzyme carried in the virus called integrase.

Now that the genetic material of the virus is incorporated and has become part of the genetic material of the host cell, we can say that the host cell is now modified to contain a new gene. If this host cell divides later, its descendants will all contain the new genes. Sometimes the genes of the retrovirus do not express their information immediately.

Retroviruses will only transduce cells that are actively undergoing mitosis, and are therefore well suited for gene transfer protocols to pluripotent (i.e. able to differentiate into several different final differentiated cell types) HSCs. Retroviral vectors give good gene expression over the long term and are technically easy to produce. However, low viral titres are yielded (generally up to 1 x 10^7 colony-forming units per ml) and, although very rare, contamination with helper virus is a possibility, which needs to be monitored.

As with all DNA-delivery systems, the use of retroviral vectors has both advantages and disadvantages. Advantages of the system include the ability of retroviral vectors to integrate the therapeutic gene into the genetic material of targeted cells with high efficiency. Also, the introduced gene remains functional for the duration of the life of the cell and is inherited by any daughter cells. Additionally, retyroviral vectors have been widely used and studied, therefore, as a delivery system they are well characterised and relatively safe. To ensure safety, the viral genes involved in the replication and pathogenesis of the retrovirus are removed in the viral vector and this reduces the risk of reactivation of the virulent form of the virus when expressed in target cells.

One of the disadvantages, however is that the retroviral-encoded therapeutic DNA can only enter the nucleus and be integrated into the genetic material of the host cell when the nuclear membrane of the infected cell is disrupted in the process of cell division (Roe *et al.*, 1993; Lewis and Emerman, 1994). Another disadvantage of gene therapy using retroviruses is that the integrase enzyme can insert the genetic material of the virus in any arbitrary position in the genome of the host. If genetic material happens to be inserted in the middle of one of the original genes of the host cell, this gene will be disrupted (insertional mutagenesis). If the gene happens to be one regulating cell division, uncontrolled cell division (i.e. cancer) can occur. Also, loss of function of another essential gene may occur. This problem has recently begun to be addressed by utilizing zinc finger nucleases or by including certain sequences such as the beta-globin locus control region to direct the site of integration to specific chromosomal sites. Additionally, the size of the therapeutic DNA encoded by retroviral vectors is restricted to approximately 8 kb in size (Shin *et al.*, 2000) and this limits the use of such vectors for large genes, such as dystrophin (14 kb cDNA) (Koenig *et al.*, 1987), that cannot be delivered via this route.

Gene therapy trials to treat severe combined immunodeficiency (SCID) were halted or restricted in the USA when leukemia was reported in three of eleven patients treated in the French Therapy X-linked SCID (XSCID) gene therapy trial. Ten XSCID patients treated in England have not presented leukemia to date and have had similar success in immune reconstitution. Gene therapy trials to treat SCID due to deficiency of the Adenosine Deaminase (ADA) enzyme continue with relative success in the USA, Italy and Japan.

a) MMLV

Most of the retroviral vectors that are being used for gene therapy applications are based on the oncoretroviruses, such as the *Moloney murine leukaemia virus* (MMLV), such as the LNSX series of vectors from Miller's laboratory. Replication is prevented by removing the *gag, pol* and *env* gene regions. The *gag* region encodes the capsid proteins, the *pol* region encodes the reverse transcriptase and the integrase, and the *env* region encodes the proteins that are required for receptor recognition and envelope anchoring. The genome also includes long terminal repeats (LTRs) at either end, which play vital roles in initiating DNA synthesis and regulating transcription of the viral genes. For example, in the LNSX vector, the LTR drives the transcription of a neomycin-resistance marker gene (which is used to select transduced cells), and an internal Simian virus 40 (SV40) promoter drives the transcription of the transgene. The *gag, pol* and *env* gene products have to be supplied by a complementary packaging cell line, into which these genes have been transferred and are stably expressed. When a retroviral vector plasmid is introduced into a packaging cell line (such as pA317), viral RNA is produced, packaged into virions and secreted into the medium. Viral titres of up to 1 x 10^7 colony-forming units per ml can be obtained in this way. Because the resultant viral particles lack the *gag, pol* and *env* genes, each particle is only able to integrate itself into the genome of the host cell and is unable to produce more viral particles. The transduced DNA sequences are integrated stably into the chromosomal DNA of the target cells and thus transferred to the progeny of transduced cells.

The first generation retroviral vectors were largely derived from MMLV and were unable to transfer genes into non-dividing cells (Roe *et al.*, 1993; Lewis and Emerman, 1994). This limited the potential for their application as a delivery system in gene therapy since certain cell types that are targeted for such therapeutic approaches are non-dividing cells (e.g. cardiomyocytes and neurons). The utilisation of the lentivirus family of retroviruses, however, has overcome this shortcoming (Lewis and Emerman, 1994; Naldini *et al.*, 1996). Obviously, safety concerns arise when HIV-based retroviral vectors are considered for gene therapy approaches since although the viral genes that encode the viral pathogenesis and replicative capacity of the HIV-based viral vector are inactivated or deleted, the risk of re-activation of the pathogenic virus must be considered carefully.

b) Lentiviruses

The most recently discovered members of the retrovirus family are the human immunodeficiency viruses (HIVs), which belong to a subclass of retroviruses known as lentiviruses. Gene therapy vectors that have been derived from HIVs have several advantages over MMLV retrovirus vectors. Lentivirus vectors are able to transduce non-dividing cells, as well as those that are actively dividing, thereby considerably broadening their usefulness as gene transfer vehicles. Because they integrate their genetic material into the genome of the host cell, lentivirus vectors have the potential to result in the long-term, stable gene expression of transgenes. The prospects for using lentivirus vectors as gene therapy vectors for immunological purposes are very exciting because of their inherent tropism (affinity) for $CD4^+$ T cells, macrophages and HSCs; this makes them useful vehicles for gene therapeutic approaches to prevent or treat HIV infection and acquired immunodeficiency syndrome (AIDS). Genetic modifications, such as the introduction of vesicular stomatitis virus G protein into the lentiviral envelope, have widened the tropism of this vector. These vectors can now be used to target airway epithelial cells for the gene therapy of cystic fibrosis, or potentially to target fully differentiated neurones for the gene therapy of Parkinson's disease. Although this development is exciting, in practice, the use of an HIV-based vector for therapies targeting diseases other than HIV might be extremely difficult to introduce clinically, until it has been established that such a vector is safe to use.

Retroviral vectors used in gene therapy are replication deficient, such that they are unable to replicate in the host cell and can infect only one cell (Mann *et al.*, 1983: Cone and Mulligan, 1984). This characteristic, although essential for the safety of viral vectors in gene therapy, imposes restrictions on the amounts of virus that can safely be administered (Kim *et al.*, 1998: Sheridan *et al.*, 2000). Generation of the retroviral vector first involves insertion of the therapeutic gene into the replication deficient retroviral genome. Since retroviral vectors are unable to replicate, the retroviral vector genome encoding the gene of interest must then be packaged into a viral envelope. Propagation and packaging of the viral vector requires the transfection of the viral vector into a packaging cell line, which provides the viral packaging machinery required for the production of infectious replication deficient viral particles for *in vivo* or *ex vivo* administration (Mann *et al.*, 1983; Cone and Mulligan, 1984; Kim *et al.*, 1998; Sheridan *et al.*, 2000). Retroviral mediated delivery of therapeutic DNA has been widely used in clinical gene therapy protocols, including the treatment of cancers, such as melanoma (Fujii *et al.*, 2000) and ovarian cancer (Tait *et al.*, 1999), adenosine deaminase deficiency-severe combined immune deficiency (ADA-SCID) (Bordignon *et al.*, 1995; Onodera *et al.*, 1998) and Gaucher's disease (Dunbar *et al.*, 1998).

Adenoviruses

Adenoviruses are non-enveloped, icosahedral, double-stranded DNA viruses (See color plate: Fig B with a capsid diameter of 70–100 nm, and comprise 252 capsomeres (240 hexons and 12 pentons). They cause respiratory (especially the common cold), intestinal, and eye infections in humans. When these viruses infect a host cell, they introduce their DNA molecule into the host. They are not incorporated into the genome of the target cell (non-integrating) but remain as an extrachromosomal entity in the nucleus of the host cell.

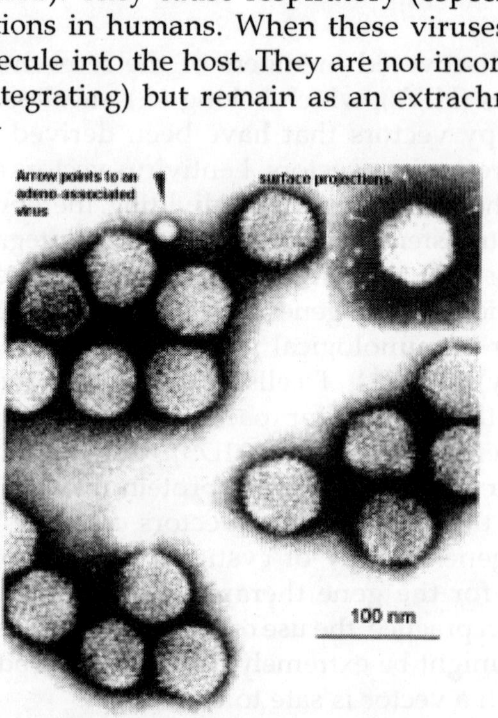

Fig B: Adenovirus micrograph

The instructions in this extra DNA molecule are transcribed just like any other gene. The only difference is that these extra genes are not replicated when the cell is about to undergo cell division so the descendants of that cell will not have the extra gene. As a result, treatment with the adenovirus will require readministration in a growing cell population although the absence of integration into the host cell's genome should prevent the type of cancer seen in the SCID trials. This vector system has shown real promise in treating cancer and indeed the first gene therapy product to be licensed is an adenovirus to treat cancer.

One of the major advantages of using adenoviral vectors for delivery of therapeutic DNA is that both dividing and non dividing cells can be affected readily and most cells are susceptible to infection. Construction of these vectors is slightly easier than for retroviruses and various kits are now available from a range of manufacturers that simplify the procedure still further. During adenovirus uinfection, adenovirus particles enter the host cells by receptor-mediated endocytosis and once inside the cytoplasm of the cell, the adenoviral DNA enters the host nucleus where viral reproduction occurs.

Replication-defective recombinant adenoviruses (See color plate: Fig 8) are the second most commonly used viral vectors in clinical trials today. Adenoviruses commonly infect humans, and were first isolated in 1953 by Rowe and colleagues from US army recruits who had acute respiratory symptoms. Primary (untransformed) cell cultures derived from the adenoids of these recruits were established, and the cells were seen to degenerate spontaneously in culture owing to the presence of the virus. To date, 47 serotypes of adenovirus have been characterised, and have been associated with a variety of symptoms ranging from a mild cold to acute febrile pharyngitis. Ad2 and Ad5 have been subjected to the most studies and are the main serotypes in use for gene therapy applications; they are not associated with severe disease, causing mainly only mild cold symptoms. The structure and replication cycle of adenoviruses has been recently reviewed by Lee and Shenk.

The 36-kb genome of adenoviruses can be divided into two main regions, early (E) and late (L), according to the time at which their genes are expressed during the replication cycle of the virus. There are four regions of early genes, which are termed E1, E2, E3 and E4, and one region of late genes, which comprise the five coding units termed L1, L2, L3, L4 and L5.

The E1 region of adenoviruses has been subdivided into E1A and E1B. The E1A gene product is a viral transcription unit, which activates the expression of other adenoviral transcription units by binding to viral promoters. The E1B region codes for a 55-kD protein that interacts with the cellular p53 tumour-suppressor protein. p53 regulates the progression of the host cells' cycle from the G_1 phase to the S phase; this phase is optimal for viral replication. E1B also binds to viral E4 proteins as well as p53, which together act to shut down host protein synthesis. The E2 region codes for viral DNA polymerase and the adenovirus single-stranded DNA-binding protein. The E3 region is not required by the adenovirus for in vitro replication but does, however, offer the virus some protection against host defence mechanisms by inhibiting infected cells being killed by CTLs or TNF-a. The E4 region codes for proteins that are known to be involved in (1) the regulation of viral and cellular protein expression, (2) viral DNA replication and (3) the switching off of the synthesis of host proteins. The late genes (L1–L5) are expressed at the onset of viral DNA replication, and code for structural polypeptides that are required for virion assembly. Disruption of the cellular cytoskeleton and membrane, owing to the accumulation of newly synthesized viral particles, results in the collapse of the cell and release of the virus.

The E1 region is essential for viral replication; thus, those adenoviruses that artificially lack the E1 region are considered replication defective. In a replication-defective adenovirus, the E1 region can be replaced with the transgene that is to be expressed. Further removal of genetic material from the vector, such as the deletion of the E3 region and even the E4 region, has been carried out to allow larger genes to be inserted in their place and also to reduce the immunogenicity of the virus; such recombinant viruses are usually referred to as 'gutless'.

For gene therapy in almost all cell types both *in vivo* and *ex vivo*, the transduction efficiency of adenoviruses is high compared with that of other viral vectors. For transplantation, adenovirus has the distinct advantage that it can bind to the surface of its target cells at low temperatures (e.g. 4°C). This allows pre-transplant gene transfer to be carried out during the process of cold preservation. Because of the structural stability of the capsid polypeptides of adenovirus, viral particles can be purified and concentrated to a very high titre of 1×10^{13} plaque-forming units (pfu) per ml, although a titre of 1×10^{10} pfu per ml is more usual. Retroviral titres are much lower ($\sim 1 \times 10^7$ pfu per ml) because their capsid is structurally more unstable and cannot be purified and concentrated on a caesium chloride gradient. Another advantage of adenoviruses for use in gene therapy is that the adenovirus genome does not integrate into the human genome, but remains in the nucleus of target cells as a non-replicating extrachromosomal entity; however, this means that there is a very low chance of activating human oncogenes or interrupting a human tumor-suppressor gene.

Additionally, the temporary expression of a therapeutic gene might be desirable in the treatment of acute disease states. For example, adenoviruses have been used to deliver therapeutic DNA to patients suffering from metastatic breast, ovarian and melanoma cancers (Stewart *et al.*, 1999; Alvarez *et al.*, 2000 a, b). Treatment of chronic disorders such as cystic fibrosis, however require repeated administration of adenoviral therapy that could cause complications due to the innate immune response that these vectors induce (Byrnes *et al.*, 1995; Yang *et al.*, 1996a, b; van Ginkel *et al.*, 1997). Indeed, the severe immune response of the host contributes to the limited survival of the adenoviral DNA in targeted cells and results in a transient expression of the therapeutic gene since the adenoviral DNA is lost over time (Byrnes *et al.*, 1995; Yang *et al.*, 1996 a, b; van Ginkel *et al.*, 1997; Michou *et al.*, 1997). However, it might be possible to extend the retention time of viral DNA in targeted cells by parallel immunosuppressant treatment and/or further manipulation of the viral genes involved in stabilizing the viral DNA in the host cell.

A further advantage of using adenoviral vectors for gene therapy is that they are able to accommodate large foreign genetic sequences in their viral genome. First generation adenoviral vectors were able to accommodate the introduction of therapeutic genes over

7kb long (but rarely larger) into targeted cells (Bett *et al.*, 1993). However the generation of gutless adenoviral vectors, which lack all viral genes, has facilitated adenoviral delivery of upto 30kb of a therapeutic DNA sequence (Clemens *et al.*, 1996; Kochanek *et al.*, 1996; Parks and Graham, 1997; Morsy *et al.*, 1998) with decreased toxicity (Schiedner *et al.*, 1998). Classically the therapeutic gene of interest could not be inserted directly into the adenoviral genome.

The construction of an adenovirus expressing a therapeutic gene for use in gene therapy

First, the therapeutic gene must be inserted into a shuttle vector, which contains adenoviral sequences on either side of the inserted gene. This shuttle vector and the replication-deficient adenoviral genome DNA then are introduced or co-transfected into a packaging cell line [e.g. 293 cells (Graham *et al.*, 1977)], which express genes that allow the production of infective viral particles. Once inside the cell, the viral sequences on either side if the therapeutic gene in the shuttle vector allow the transfer of the therapeutic gene into the adenoviral genome. This event is called homologous recombination. The adenoviral genome expressing the therapeutic DNA then is packaged into virus particles or virions. These particles then are amplified to generate large numbers of viral particles for *in vivo or ex vivo* administration. Because the generation of replication-competent recombinant adenoviruses is a concern when using traditional methods for generating therapeutic adenoviruses, methods have been developed that reduce the chance of this occurring (He *et al.*, 1998; Richards *et al.*, 2000).

Non replicating retroviral vectors contain all of the *cis* elements required for transcription of mRNA molecules encoding a gene of interest and packaging of these transcripts into infectious virus particles (See color plate: Fig 9). The vectors typically comprise an *E. coli* plasmid backbone containing a pair of 600-bp viral long terminal repeats (LTRs) between which the gene of interest is inserted. The LTR is divided into three regions: The U3 region contains the retroviral promoter/enhancer, and is flanked in the 3′ direction by the R region, which contains the viral polyadenylation signal (pA). The R region is followed by the U5 region, which, along with R, contains sequences that are critical for reverse transcription. Expression of the viral RNA is initiated within the U3 region of the 5′ LTR and is terminated in the R region of the 3′ LTR. The viral packaging signal (Ψ), which is required in *cis* for the viral RNA to be packaged into virion particles, resides between the 5′ LTR and the coding sequence for the gene of interest.

To generate infectious virus particles that carry the gene of interest, specialized packaging cell lines have been constructed that contain chromosomally integrated expression cassettes for viral gag, pol, and env proteins, all of which are required in *trans* to make virus. The *gag* gene encodes internal structural proteins; the *pol* gene encodes reverse transcriptase (RT) and integrase; and the *env* gene encodes the viral envelope protein, which resides on the viral surface and facilitates infection of the target cell by direct interaction with cell type-specific receptors. Thus, the host range of the virus is dictated not by the DNA vector but by the choice of the *env* gene used to construct the packaging cell.

The packaging cell line is then transfected with the vector DNA and, at this point, either stable viral producer cell lines may be selected (providing the vector has an appropriate selectable marker), or virus is produced transiently as mRNAs that are transiently transcribed from the vector are encapsidated, and bud off into the cell supernatant. These supernatants are collected and used to infect target cells. Upon infection of the target cell, the viral RNA molecule is reverse transcribed by RT (present in the virion particle), and the cDNA of the gene of interest, flanked by the LTRs, is integrated into the host DNA. Because the vector itself does not express viral proteins, once a target cell is infected, the LTR expression cassette is incapable of proceeding through another round of virus production. Thus, a new gene is inserted into an adenovirus vector, which is used to introduce the modified DNA into a human cell. If the treatment is successful, the new gene will make a functional protein (See color plate: Fig 10).

A major disadvantage of using adenovirus as a vector *in vivo* is the potential for patients to develop CTL response that is induced against capsid-derived peptides; this response can cause the destruction of vector-transduced cells and also leads to local tissue damage and inflammation (Byrnes *et al.*, 1995; Yang *et al.*, 1996a, b; van Ginkel *et al.*, 1997). Host cells presenting peptides of adenovirus-encoded foreign transgene products have also been shown to target the host cell for CTL-mediated destruction. A pre existing immune response to adenoviral infection could block adenoviral-driven gene therapy treatment completely or might result in an adverse reaction to the treatment, thereby threatening the life of the patient. Strategies for the development of adenoviral vectors that are less likely to stimulate an immune response are underway (Schiedner *et al.*, 1998; Morsy *et al.*, 1998). These second generation vectors either have more viral genes deleted or inactivated or are devoid of viral genes altogether (the so called 'gutless' adenoviral vectors) (Clemens *et al.*, 1996; Kochanek *et al.*,1996; Parks and Graham, 1997; Schiedner *et al.*, 1998; Morsy *et al.*, 1998)

Because adenovirus is a commonly occurring virus, most humans are pre-primed to at least one specific serotype. Using the same serotype in a gene therapy context will almost certainly result in a rapid and vigorous memory immune response, such that high levels of anti-adenovirus antibody become measurable in the sera of the recipients within days of administration of the adenovirus vector. It might help to screen recipients of such vectors to detect which serotypes they have previously encountered, and then use a different serotype as the vector. However, this approach would require the availability of a very large panel of recombinant vectors with different serotypes. Another potential problem is the strong secondary immune response that would be induced by the re-administration of a vector of the same serotype.

The period of expression of an adenovirus-encoded transgene is also relatively short. Expression is reported to last at a 'reasonable' level for ~14 days *in vivo*; however, manipulation of the immune response has resulted in expression for longer periods. This short expression time is mainly due to the expression of viral polypeptides inducing a CTL response, and to some extent the transgene itself especially if it is not normally expressed in that individual (i.e. it is foreign). Because the adenoviral genome does not integrate into the genome of the target cell, only one of the daughter cells (if the target cells are dividing) will contain the transgene, therefore halving the total number of cells that contains the transgene. Adenoviral gene delivery is ideally suited to those situations that require only a one-off delivery of a transgene, for example in growth factor therapy, in which transient, as opposed to long-term, expression of growth factor is required. In protocols that are aimed at inducing transplantation tolerance, the delivery of an adenoviral vector to a recipient before transplantation should be sufficient to induce a regulatory population of T lymphocytes that will impart long-term immunological tolerance to that recipient.

Advantages:

- Unlike retroviral vectors, they do not integrate into the host genome and thus should not be able to disrupt host genes (thus avoiding the danger that resulted in the three little boys developing leukemia).
- They can infect nondividing cells with high efficiency.

Disadvantage: They elicit a powerful immune response, both by T cells and by B cells (antibodies) so repeated doses soon lose their effectiveness.

Adeno-associated viruses

Adeno-associated viruses, from the parvovirus family, are small viruses with a genome of single stranded DNA. These are not associated with any disease in humans (Rose *et al.*, 1969). The adeno-associated virus (AAV) vectors offer many of the same advantages as adenovirus vectors, including a wide host-cell range and relatively high transduction efficiency, in some situations. In addition, unlike adenovirus, which can cause a high degree of cell death (cytopathogenicity), AAV causes little damage in target cells. These viruses can insert genetic material at a specific site on chromosome 19 (Kotin *et al.*, 1990; Samulski *et al.*, 1991) which has the beneficial effect of a longer-lasting transgene expression. It is certain to go here. Therefore it provides an alternative to adenoviral vectors for gene therapy (See color plate: Fig 11) and a means for long term gene expression with a reduced risk of adverse reactions upon administration of the vector (Fisher *et al.*, 1997; Jooss *et al.*, 1998).

However, there is evidence to suggest that AAVs are significantly less efficient than retroviral vectors at transducing primary-cell cultures. In primary-cell transductions, most of the AAV vector DNA does not integrate into the host genome but remains extra chromosomal, and this inefficiency might limit its use for *in vivo* application. There are a few other disadvantages to using AAV, including the small amount of DNA it can carry (low capacity) and the difficulty in producing it. This type of virus is being used, however, because it is non-pathogenic (most people carry this harmless virus).

In contrast to adenoviruses, most people treated with AAV will not build an immune response to remove the virus and the cells that have been successfully treated with it. Several trials with AAV are on-going or in preparation, mainly trying to treat muscle and eye diseases; the two tissues where the virus seems particularly useful. However, clinical trials have also been initiated where AAV vectors are used to deliver genes to the brain. This is possible because AAV viruses can infect non-dividing (quiescent) cells, such as neurons in which their genomes are expressed for a long time. In recent human trials, CD8+ immune cells have recognized the AAV infected cells as compromised and killed these cells accordingly. This action appears to be triggered by part of the capsid or outer coat of the type 2 virus.

The AAV life cycle involves two distinct stages: initially, the AAV enters the cell by receptor-mediated endocytosis and is transported to the nucleus. In the nucleus of the host cell the single stranded viral DNA is synthesized to form a double stranded viral DNA fragment, which then is integrated into the genome of the host cell. The integrated virus then remains latent in the infected host cell until subsequent infection of the cell with another virus, such as an adenovirus, that provides the sequences required for the viral replication and packaging.

In the engineering of AAV vectors, most of the AAV genome can be replaced with the therapeutic gene (Samulski *et al.*, 1989), which significantly reduces potential adverse responses of the host to viral infection. However, the size of the therapeutic gene is limited to approximately 5 kb (Dong *et al.*, 1996; Hermonat *et al.*, 1997), potentially restricting the use of AAV vectors in gene therapy compared to adenoviral vectors. To propagate the recombinant AAV vector, a cell line must be infected with two DNA plasmids encoding AAV viral sequences. The first contains the therapeutic gene of interest and viral sequences necessary for DNA integration. The second plasmid encodes all other AAV sequences except those necessary for integration. With the AAV DNA integrated into the genetic material of the cell line, infection with a helper virus, such as an adenovirus, provides the necessary replicative and/or packaging capacity for AAV virus particles to be generated (Samulski *et al.*, 1989).

The need to utilize a helper virus in the production of AAV vectors increases the safety concerns regarding the use of AAVs. Contamination of the AAV vector with replication competent adenovirus or herpes simplex virus is difficult to eliminate with current methods available for AAV propagation. Although the AAV genome is relatively small and relatively simple, much less is known about the specific AAV sequences that are involved in infection and replication. Current research into the replication of AAVs might provide a means for generating AAV particles on a large scale without the use of helper viruses, thereby improving the safety of AAV vector use in human gene therapy.

Herpes simplex virus

Retroviral and adenoviral based vectors provide well characterized and highly efficient methods for DNA delivery into cells for gene therapy purposes. Both methods of gene delivery have advantages and disadvantages that affect their applications in gene therapy approaches. In view of their potential limitations, research into alternative viral vectors continues. For example, Herpes simplex virus (HSV) vectors are being developed for several applications, including for use in gene transfer protocols that target neuronal tissue, such as the treatment of Parkinson's disease, malignant gliomas (a type of brain tumour) and cerebral ischaemia (starving of brain tissue from essential nutrients). HSV is a large, relatively complex, enveloped, double stranded DNA virus that has the capacity to encode large therapeutic genes and like AAV can remain latent in infected cells providing the potential for long term expression of the therapeutic gene (Carpenter and Stevens, 1996). HSV is maintained as an extrachromosomal DNA element in the nucleus of host cells, and has an excellent ability to establish long-lived asymptomatic infections in the sensory neurones of the peripheral nervous system and in some central nervous tissue. This provides the opportunity for long-term gene expression in neuronal target tissue. HSV vectors also have a wide host range, can accept large gene inserts, and have had multiple deletions of immediate-early (IE) genes that are essential for replication to make the vectors less cytotoxic to target cells and reduce safety concerns. The main problem associated with the use of HSV as a gene therapy vector at present is concern about its safety for clinical use, owing to reports of the wild-type virus replicating lytically in the human brain and resulting in potentially serious encephalitis. Therefore although able to infect many cell types, HSV vectors currently are limited in their use by vector toxicity (Lowenstein *et al.*, 1994).

An alternative method of DNA delivery involves the use of ultraviolet light (UV)-inactivated Hemagglutinating virus of Japan (HVJ), which has been complexed with liposomes (a non viral method for DNA delivery discussed below) and the therapeutic DNA. The inactivated HVJ virus promotes fusion of the liposome-DNA complex with the cellular membrane of the target cells and delivers the liposome-DNA complex directly into the cell (Tomita *et al.*, 1994). This method has been used successfully to deliver therapeutic DNA *in vitro, ex vivo* and *in vivo* (Tomita *et al.*, 1994; Yoshida *et al.*, 1999; Kawauchi *et al.*, 2000; Morishita *et al.*, 2000).

The development of hybrid viral delivery system to be utilized whilst overcoming the problems associated with the negative features of individual delivery systems. Examples of chimaeric delivery systems that have been investigated in this way include recombinant adenoviruses expressing retroviral sequences that permit the stable integration of the therapeutic DNA without the disadvantages associated with retroviral mediated delivery (Feng *et al.*, 1997; Caplen *et al.*, 1999) and adenovirus/AAV hybrid vectors that are simple to propagate and integrate stably into the genome of infected host cells (Ueno *et al.*, 2000).

Vaccinia virus

Although vaccinia virus vectors are not currently in use for transplantation studies, they are in development for cancer gene therapy. Vaccinia virus has been used worldwide to eradicate smallpox and has been shown to be a safe live vaccine. Vaccinia virus vectors do not integrate into the genome of the host cell; they can, however, accommodate large transgenes and are extremely immunogenic. Vaccinia virus can be used to immunise patients against tumour antigens by cloning (inserting) into its large genome tumour antigen genes or genes encoding proteins that enhance the immune response (e.g. cytokine genes). Most transgenes are expressed at a high level *in vivo*, eliciting a specific immune response against the tumour antigen, which otherwise would not have occurred at levels that are sufficient to kill cancer cells. If required, more than one gene can be cloned into the vector, owing to its large capacity.

ENVELOPE PROTEIN PSEUDOTYPING OF VIRAL VECTORS

The viral vectors described above have natural host cell populations that they infect most efficiently. Retroviruses have limited natural host cell ranges, and although adenovirus and adeno-associated virus are able to infect a relatively broader range of cells efficiently, some cell types are refractory to infection by these viruses as well. Attachment to and entry into a susceptible cell is mediated by the protein envelope on the surface of a virus. Retroviruses and adeno-associated viruses have a single protein coating their membrane, while adenoviruses are coated with both an envelope protein and fibers that extend away from the surface of the virus. The envelope proteins on each of these viruses bind to cell-surface molecules such as heparin sulfate, which localizes them upon the surface of the potential host, as well as with the specific protein receptor that either induces entry-promoting structural changes in the viral protein, or localizes the virus in endosomes wherein acidification of the lumen induces this refolding of the viral coat. In either case, entry into potential host cells requires a favorable interaction between a protein on the surface of the virus and a protein on the surface of the cell.

For the purposes of gene therapy, one might either want to limit or expand the range of cells susceptible to transduction by a gene therapy vector. To this end, many vectors have been developed in which the endogenous viral envelope proteins have been replaced by either envelope proteins from other viruses, or by chimeric proteins. Such chimera would consist of those parts of the viral protein necessary for incorporation into the virion as well as sequences meant to interact with specific host cell proteins. Viruses in which the envelope proteins have been replaced as described are referred to as pseudotyped viruses. For example, the most popular retroviral vector for use in gene therapy trials has been the lentivirus Simian Immunodeficiency virus coated with the envelope proteins, G-protein, from Vesicular Stomatitus virus. This vector is referred to as VSV G-pseudotyped lentivirus, and infects an almost universal set of cells. This tropism is characteristic of the VSV G-protein with which this vector is coated. Many attempts have been made to limit the tropism of viral vectors to one or a few host cell populations. This advance would allow for the systemic administration of a relatively small amount of vector. The potential for off-target cell modification would be limited, as well as many concerns from the medical community. Most attempts to limit tropism have used chimeric envelope proteins bearing antibody fragments. These vectors show great promise for the development of "magic bullet" gene therapies.

PROMOTER ATTENUATION

It is essential for the success of gene therapy vectors that an appropriate promoter is linked to the gene of interest. A promoter is a regulatory sequence of DNA that is located upstream of a gene and to which proteins (transcription factors and RNA polymerase) bind, to initiate the synthesis of mRNA and, subsequently, protein. Most experimental expression vectors and gene therapy vectors use promoter elements that have been derived from pathogenic viruses, because of the high level of constitutive (constant) gene expression that they will induce. Various gene transfer studies have used promoter and enhancer elements from cytomegalovirus (CMV), Rous sarcoma virus (RSV) and SV40, and have reported encouraging success; however, the levels of expression do depend on many factors including the vector used, its route of administration and the type of cell transduced.

One of the common problems encountered by investigators is that transgenes are expressed only at a low level and only transiently. Although the molecular mechanisms that are responsible for this poor expression are poorly defined, it could be mainly due to attenuation of the promoter.

Considering the importance of promoter attenuation to the field of gene therapy, remarkably few studies have been carried out to look at this problem directly. It has been shown in experimental systems that the application of adenoviral vectors *in vivo* induces cytokine production through specific or non-specific immune responses. These cytokines can then act on the adenovirus-infected cell that is carrying the transgene, and initiate cytokine-mediated cellular signals, which will modulate transgene expression. Qin and colleagues showed that transgene expression, controlled by many viral promoters, was inhibited by IFN-γ and TNF-α, and that both of these cytokines had synergistic effects. Promoters derived from CMV and RSV were found to be the most sensitive to this cytokine treatment. In a different mouse model that also used recombinant adenovirus, Harms and Splitter demonstrated that the administration of a neutralising anti-IFN-γ monoclonal antibody *in vivo* resulted in enhanced expression of the transgene.

At the molecular level, promoters derived from SV40, CMV and RSV all have an interferon response sequence. Nuclear factors, produced as a result of IFN-γ interacting at the cell surface, bind to elements in these viral promoters, which inhibits transcription of the transgene. Strong constitutive viral promoters have been used (and work successfully) *in vitro* in mammalian expression vectors where there are no inflammatory cytokines present. The use of these strong viral promoters was naturally assumed to be ideal for the development of clinical gene therapy protocols. However, low levels of transgene expression are usually seen, and this is thought to result from the design of the vector as a whole, as opposed to a particular component of the vector. Common to most gene therapy expression systems are the viral promoter and enhancer elements. Viruses and isolated viral promoters that are used in both in vitro expression vectors and *in vivo* gene therapy vectors can be adversely affected by cytokines produced by infected cells.

It therefore makes sense to choose promoters for gene transfer that can be upregulated by factors in the environment that will be present where the vector is delivered and when and where transgene expression is desired. For example, the MHC class I promoter might be a better choice as a promoter for immunoregulatory gene therapy applications because inflammatory cytokines, such as IFN-γ, actually act on this promoter to enhance transcription.

Identifying Where Genes Can Successfully Integrate

A small number of the treated individuals developed leukemia-like diseases because the retrovirus integrated into the genome at a site(s) that caused the cells to divide uncontrollably.

Analysis of the sites of retroviral integration in individuals treated with gene therapy is therefore of utmost importance to the further development of such gene therapy approaches and three studies have appeared in the Journal of Clinical Investigation (August 2007 issue). The three studies—reported by researchers from the National Center for Tumor Diseases, Heidelberg, Germany; University College London, London, United Kingdom; and the San Raffaele Telethon Institute for Gene Therapy, Milan, Italy—indicate that the retroviruses used to deliver the genes of interest in these studies, gammaretroviruses, integrated near the 5" ends of genes that are active in the treated cell population.

Although the diversity of the detected retroviral integration sites decreased following transplantation into patients because the treated cells differentiated, the sites detected remained at the 5" end of genes active in the cell populations analyzed. Importantly, the retrovirus integration sites observed in cells from successfully treated patients were not significantly different from those observed in cells from patients that went on to develop leukemia-like disease. These studies are critical if researchers are to determine whether the dangers of retroviral gene therapy can be predicted and/or eliminated.

Non-viral Gene Therapy Methods

Scientists have used viruses as a gene delivery mechanism for more than 20 years because of their adeptness at getting inside cells and inserting themselves in DNA. Non-essential genes can be removed from viral vectors to make more room for transgenes, to reduce inflammatory responses or to increase their safety; this involves the virus being simplified, sometimes to an extreme. What remains can be an artificial 'vector shell', which has been designed to allow the gene of interest to be expressed at high levels, in a highly regulated specific manner and for a controlled period of time (either short term or long term). But efficiency comes at a price.

Gene therapy trials have been halted because of major complications, including deaths. As examples, one patient died because of his immune response to an adenovirus and three children in another study developed leukemia because the virus inserted itself upstream of a cancer-causing gene. "With viruses, you don't have control," says Dr.

Kaminski. "People have tried to modify viruses for site-specific integration and have not been very successful. Once they get into the cell, they can insert wherever they want."

Another approach to achieve the same result is to produce a system that can simply introduce genetic material to the nucleus of cells. This has been the focus of intensive research over the past few years, and has resulted in the development of several non-viral vector systems. A jumping gene first identified in a cabbage-eating moth may one day provide a safer, target-specific alternative to viruses for gene therapy, researchers say. They compared the ability of the four best-characterized jumping genes, or transposons, to insert themselves into a cell's DNA and produce a desired change, such as making the cell resistant to damage from radiation therapy. They found the piggy Bac transposon was five to 10 times better than the other circular pieces of DNA at making a home and a difference in several mammalian cell lines, including three human ones.

"If we want to add a therapeutic gene, we can put it within the transposon and use it to deliver the gene into the cell," says Dr. Joseph M. Kaminski, radiation oncologist at the Medical College of Georgia Cancer Center and a corresponding author on research published the week of Sept. 25 in the online Proceedings of the National Academy of Sciences Early Edition. "You can use these wherever retroviruses have been used."

Non-viral methods of DNA delivery into cells have been used for the delivery of DNA *in vitro* in the laboratory for many decades. However, many of these approaches, such as the chemical methods of DNA delivery [e.g. calcium phosphate (Graham and Eb, 1973) or 2-(diethylamino) ether (DEAE)-dextran-mediated DNA delivery (Vaheri and Pagano, 1965)], cannot readily be employed for *in vivo* use mainly because of problems associated with low delivery efficiency and high cytotoxicity. Despite these limitations, non-viral delivery systems do offer many advantages over viral based technologies, and, therefore, the improvement and development of non-viral DNA-delivery technology has been the focus of intense research. Non viral methods of DNA delivery are simple to use and the synthetic components of these systems, such as cationic lipids or polymers, can be produced in large quantities with relative ease. Perhaps the most desired feature of non-viral vectors is their safety for use *in vivo* as a specific host immune response to the vector is not encountered. Additionally, non-viral mechanisms of DNA delivery do not disrupt the genetic material of the targeted host cell and provide short-term expression of the therapeutic gene, a feature suitable for the treatment of acute conditions, such as the inhibition of vascular smooth muscle cell proliferation following balloon angioplasty. Improvements in the efficiency and longevity of non-viral DNA delivery will make these vectors a competitive choice over viral-based systems in the future.

Non-viral methods also present certain other advantages over viral methods; simple large scale production and low host immunogenicity being just two. Previously, low levels of transfection and expression of the gene held non-viral methods at a disadvantage, however recent advances in vector technology has yielded molecules and techniques with transfection efficiencies similar to that of viruses.

Naked DNA

One of the simplest ideas for non-viral gene delivery techniques is the use of purified DNA in the form of plasmids. This is the direct transfer of naked DNA (i.e. a therapeutic gene sequence incorporated into a plasmid vector) that encodes a potentially therapeutic gene. This approach is being used for DNA vaccines, among other protocols, and has been tried in many situations for gene therapy. The injection of plasmid DNA encoding an MHC class I antigen derived from a donor rat strain into the thymus of a recipient rat strain, in conjunction with a dose of anti-lymphocyte serum, induced donor-specific tolerance to subsequently transplanted liver grafts. Donor DNA was detected in the thymus up to 4 days after injection and in the spleen 7 days after injection.

Naked DNA can be administered via two possible routes, either by *ex vivo* delivery or by *in vivo* delivery. The *ex vivo* method of naked DNA delivery provides a precise, but time-consuming, method for transfer of therapeutic DNA to target cells. In this procedure, target cells are removed from the patient and are grown in cell culture under sterile laboratory conditions. The therapeutic naked DNA then is introduced into these cultured cells, using a method such as microinjection, and cells are re-implanted into the patient. Although this methodology has been used successfully for the introduction of DNA into endothelial and smooth muscle cells (Nakamura *et al.*, 1998; Mann *et al.*, 1999a), its reliance on the culture of harvested cells renders it unsuitable for many cell types.

In vivo delivery of naked DNA is by far the simplest method for the administration of therapeutic genetic material to a patient. This method of gene delivery was first described in 1990 when it was demonstrated that direct injection of DNA into skeletal muscle was sufficient fro the successful introduction and expression of exogenous DNA encoding reporter genes such as chloramphenicol acetyltransferase, luciferase and β-galactosidase into myoblasts (Wolff *et al.*, 1990). Efficiency of the delivery of naked DNA can be improved when administered in a pressure-mediated fashion, for example, by controlled non-distending pressure in cardiovascular tissue (Mann *et al.*, 1999a) or by hydrodynamic force in intravenous injection (Liu *et al.*, 1999). Compared with the *ex vivo* delivery of DNA, the *in vivo* method of gene transfer results in less precise delivery of DNA to the target cells and exogenous DNA might be subject to degradation by serum nucleases. Several modifications of this method have been described that increase the efficiency of uptake of naked DNA into target cells, including the particle bombardment or "gene gun" approach (Yang *et al.*, 1990) and electroporation techniques (Rols *et al.*, 1998).

Clinical trials carried out of intramuscular injection of a naked DNA plasmid have occurred with some success, however despite the simplicity of this approach, studies have revealed that the transfection expression has been very low in comparison to other methods of transfection and will therefore limit its application. In addition to trials with plasmids, there have been trials with naked PCR product, which have had similar or greater success. This success, however, does not compare to that of the other methods, leading to research into more efficient methods for delivery of the naked DNA such as electroporation and the use of a "gene gun", which shoots DNA coated gold particles into the cell using high pressure gas.

Ballistic gene delivery/Particle bombardment

This physical method involves the use of microcarriers (usually gold particles or another inert substance like tungsten), which are coated with therapeutic DNA and 'fired' at high velocity using an explosive or helium gas-powered ballistic device called a gene gun (See color plate: Fig 12). Once the particles are inside the target cell, the DNA is slowly released from the microcarriers, and can result in useful levels of gene transcription and translation. This technology enables the localized delivery of DNA readily into skin or muscle (Fynan *et al.*, 1993) and, therefore, has applications as a method for delivery of DNA vaccines. This technique has been widely used experimentally, but its clinical use is restricted to exposable surfaces because the fired particles do not penetrate deeply into tissues. Possible clinical applications are gene transfer to the bladder urothelium, the cornea and skin epithelial cells.

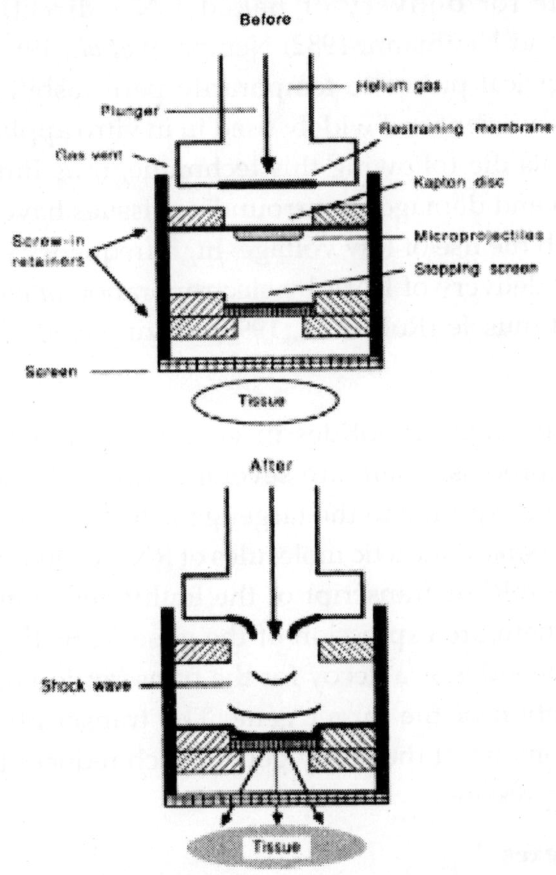

Fig C: Sketch of a hand-held helium powered gene gun works by pushing DNA coated particles into tissue. (Images from Williams, 1991)

There are several variables in these experiments that must be controlled in order to attain maximal transformation efficiency. Optimal responses have been shown to be dependent on the delivery of a sufficient number of DNA-coated particles, as well as how much DNA coats the metal particles. Temperature, number of cells, and their ability to regenerate also has an effect on the overall efficiency, as well as the type of gun used: helium powered vs. gun-powder (See Fig. C), hand-held vs. stand-alone, etc. It is also important to adjust the length of the flight path of the particles: fragile tissues cannot be bombarded at the same high speed as those which have more resistance to foreign particles entering. How to adjust these variables depends mainly on which metal particles are being used to transfer the genetic material, and what type of cells are being attempted to transfect.

Electroporation

Another technique for delivery of naked DNA directly into target cells is electroporation (Wong and Neumann, 1982; Neumann *et al.*, 1982; Rols *et al.*, 1998). The use of high voltage electrical pulses to temporarily permeasbilise cell membranes and deliver DNA into cells is efficient and widely used in in vitro applications. Unfortunately, a large percentage of cells die following this technique, thus limiting its application *in vivo*, although cell death and damage to surrounding tissues have been minimized for *in vivo* applications through the use of low voltage, high frequency electrical pulses (Rols *et al.*, 1998). The successful delivery of DNA by electroporation *in vivo* has been reported in tissues such as skin and muscle (Rols *et al.*, 1998; Rizzuto *et al.*, 1999).

Oligonucleotides

The use of synthetic oligonucleotides in gene therapy is to inactivate the genes involved in the disease process. There are several methods by which this is achieved. One strategy uses antisense specific to the target gene to disrupt the transcription of the faulty gene. Another uses small catalytic molecules of RNA called siRNA to cleave specific unique sequences in the mRNA transcript of the faulty gene, disrupting translation of the faulty mRNA, and therefore expression of the gene. A further strategy uses double stranded oligodeoxynucleotides as a decoy for the transcription factors that are required to activate the transcription of the target gene. The transcription factors bind to the decoys instead of the promoter of the faulty gene, which reduces the transcription of the target gene, lowering expression.

Lipoplexes and polyplexes

To improve the delivery of the new DNA into the cell, the DNA must be protected from damage and its entry into the cell must be facilitated. New molecules, lipoplexes and polyplexes, have been created that have the ability to protect the DNA from undesirable degradation during the transfection process (See color plate: Fig 13). The most common use of lipoplexes has been in gene transfer into cancer cells, where the supplied genes have activated tumor suppressor control genes in the cell and decrease the activity of oncogenes. Recent studies have shown lipoplexes to be useful in transfecting respiratory epithelial cells, so they may be used for treatment of genetic respiratory diseases such as cystic fibrosis. The use of cationic lipids to transfer DNA into cells was first described as an *in vitro* method of DNA delivery (Felgner *et al.*, 1987) and subsequently this technology has been widely adopted and shown to be suitable for the transfection of many different cell types.

In their most basic form, liposomes consist of two lipid species, a cationic amphiphile and a neutral phospholipid (typically, dioleoyl-phosphatidylethanolamine) both of which are commercially available. Liposomes bind to and condense DNA spontaneously to form complexes that have a high affinity for the plasma membranes of cells (See color plate: Fig 14); this results in the uptake of liposomes to the cytoplasm by the process of endocytosis. Many adaptations of this basic protocol have been tested and have resulted in varying levels of gene expression.

Plasmid DNA can be covered with lipids in an organized structure like a micelle or liposome. When the organized structure is complexed with DNA it is called a lipoplex. There are three types of lipids, anionic (negatively charged), neutral, or cationic (positively charged). Initially, anionic and neutral lipids were used for the construction of lipoplexes for synthetic vectors. However, in spite of the facts that there is little toxicity associated with them, that they are compatible with body fluids and that there was a possibility of adapting them to be tissue specific; they are complicated and time consuming to produce so attention was turned to the cationic versions. (See Fig. D & E).

Cationic lipids due to their positive charge, naturally complex with the negatively charged DNA. Also as a result of their charge they interact with the cell membrane, endocytosis of the lipoplex occurs and the DNA is released into the cytoplasm. The cationic lipids also protect against degradation of the DNA by the cell. This process is of non specific endocytosis. The therapeutic DNA must be then transported into the nucleus of the cell where the therapeutic gene can be expressed. Cationic liposomes have been used in clinical trials to deliver therapeutic DNA; for example, the liposomal-mediated delivery of the CFTR gene to the nasal epithelium in patients with cystic fibrosis (Caplen *et al.*, 1995; Hyde *et al.*, 2000; Noone *et al.*, 2000) and for the treatment of patients with melanoma (Nabel *et al.*, 1993, 1996).

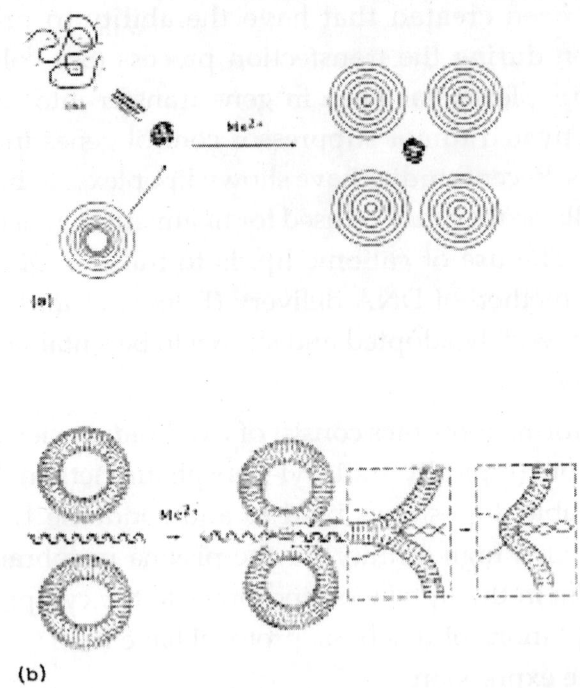

Fig D: Divalent metal cation bridged zwitterionic multilamellar (a) and unilamellar (b) vesicle-nucleic acid aggregation and fusion. Me^{2+} refers to inorganic divalent metal cations.

There are major barriers that must be overcome to improve the efficiency of cationic lipids as DNA-delivery vehicles. First, formation of the DNA/lipid complex can be a variable process and variation in such complexes might influence the effiiency of DNA delivery. Secondly, the efficiency by which the lipid/DNA complex is taken up by targeted cells must be improved.

Complexes of polymers with DNA are called polyplexes. Most polyplexes consist of cationic polymers and their production is regulated by ionic interactions. Synthetic polymers, such as protective interactive non-condensing polymers (PINC), poly (L-lysine), cationic polymers or dendrimers, offer an alternative to cationic lipids as a vehicle for DNA delivery into target cells (Boussif *et al.*, 1995; Wadhwa *et al.*, 1995; Kukowska-Latallo *et al.*, 1996; Tang and Szoka, 1997; Mumper *et al.*, 1998). One large difference between the methods of action of polyplexes and lipoplexes is that polyplexes cannot release their DNA load into the cytoplasm, so to this end, co-transfection with endosome-lytic agents (to lyse the endosome that is made during endocytosis, the process by which the polyplex enters the cell) such as inactivated adenovirus must occur as described by process stated below. However this isn't always the case, polymers such as polyethylenimine have their own method of endosome disruption as does chitosan and trimethylchitosan. Such polymers can be used to deliver the complex efficiently into a variety of cell types.

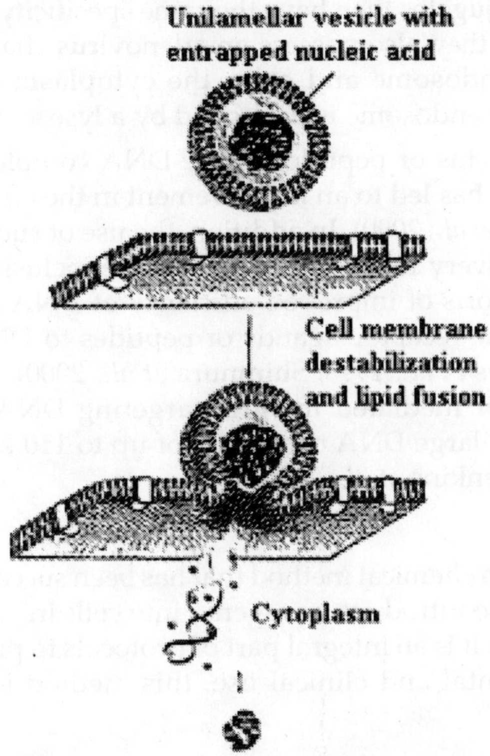

Fig E: Proposed mechanism for cellular internalization of divalent metal cation mediated DNA-induced zwitterionic lipoplex formulation.

Controlled chemical synthesis of the cationic polymers, however, ensures that the size and shape of the dendrimers is consistent and defined, thereby improving the reproducibility of DNA delivery. The use of cationic polymers also might permit the controlled expression of potentially therapeutic genes in gene therapy approaches. Thus, encapsulation of a DNA molecule or even a therapeutic viral vector within a biodegradable polymer has been demonstrated to permit the controlled release of the DNA in a targeted cell over a period of weeks or months (Naughton *et al.*, 1992; Singh *et al.*, 2000). Using such a system to deliver naked DNA has the advantage of providing a means of controlling the long term expression of an introduced therapeutic gene whilst employing a non viral DNA-delivery system.

DNA-Ligand conjugates

DNA–ligand conjugates have two main components: a DNA-binding domain and a ligand for cell-surface receptors. The transgene can therefore be guided specifically to the target cell, where it is internalised via receptor-mediated endocytosis. Once the DNA–ligand complex is in the endocytic pathway, the conjugate is likely to be destroyed when the endosome fuses with a lysosome. Curiel and colleagues have developed a method to avoid this, which incorporates an adenovirus-derived domain into the cell-surface receptor part of the ligand. The conjugates then have the same specificity as adenoviruses, binding to a wide host-cell range; they also possess an adenovirus characteristic that allows the conjugate to leave the endosome and enter the cytoplasm (by a process known as endosomolysis) before the endosome is destroyed by a lysosome.

The inclusion of proteins or peptides in the DNA complex that are recognised by receptors on targeted cells has led to an improvement in the efficiency of DNA uptake in several instances (Jenkins *et al.*, 2000). In addition, the use of such proteins could improve the targeting of DNA delivery to specific cells and the inclusion of other neighbouring cell. Indeed, there are reports of improved efficiency of DNA delivery by cationic lipid via the coupling of specific receptor ligands or peptides to DNA/Liposome complexes (Ellison *et al.*, 1996; Simoes *et al.*, 1998; Shinmura *et al.*, 2000). One such example is the recently reported receptor mediated integrin-targeting DNA-delivery system that is capable of delivering very large DNA constructs of up to 110 kb both in vitro (Compton *et al.*, 2000) and *in vivo* (Jenkins *et al.*, 2000).

$CaPO_4$ transfection

$CaPO_4$ transfection is a chemical method that has been successfully used by molecular biologists for many years to introduce transgenes into cells in vitro with a relatively good efficiency (10%). Although it is an integral part of protocols to produce many viral vectors for subsequent experimental and clinical use, this method is not suitable for *in vivo* application.

Other non-viral delivery methods

Another method of gene delivery suitable for gene therapy applications arose from the discovery of the family of transcription factors called the homeodomain proteins, which are able to translocate across the plasma membranes of living cells. This unusual property has formed the basis of the development of the penetratin family of polypeptides. These synthetic peptides retain the ability to translocate efficiently across the plasma membrane of the living cells and are capable of carrying therapeutic oligonucleotides or peptides into the cytoplasm and nucleus of the target cells (Derossi *et al.*, 1994). The mechanism of cellular internalization by penetratin is not fully understood. It is not cell type-specific, receptor-mediated nor concentration-dependent but is thought to result from interaction of the peptide with membrane lipids (Derossi *et al.*, 1996; Thoren *et al.*, 2000). The efficient and rapid internalization of penetratin and its cargo by all cell types at low concentrations, and with low cellular toxicity, are properties that make this non-viral DNA-delivery approach very attractive. Despite these desired characteristics, the penetratin system does have disadvantages, including the fact that when penetratin binds double-stranded DNA its translocation properties are lost (Derossi *et al.*, 1998). Thus, therapeutic DNA delivered via this system must be in the form of a single-stranded oligonucleotide. The size of the therapeutic DNA that can be delivered by penetratin is also limited to approximately 55 nucleotides (Derossi *et al.*, 1998). The route of administration of penetratin and its cargo also must be carefully considered since internalization of penetratin is highly efficient and the penetratin-DNA complex will translocate rapidly into cells at the point of administration only.

Penetratin is not the only polypeptide hat is internalized by living cells. Other proteins that have demonstrated similar internalization properties include the TAT transcription factor of HIV (Vives *et al.*, 1997) and VP22 of the Herpes simplex virus (Elliott and O'Hare *et al.*, 1997). High transfer efficiency and the low toxicity levels associated with gene-delivery systems that utilize translocating peptides make them attractive candidates for use in gene therapy.

Hybrid Viral/Synthetic Methods

Due to every method of gene transfer having shortcomings, there have been some hybrid methods developed that combine two or more techniques. Virosomes are one example; they combine liposomes with an inactivated HIV or influenza virus. This has been shown to have more efficient gene transfer in respiratory epithelial cells than either viral or liposomal methods alone. Other methods involve mixing other viral vectors with cationic lipids or hybridising viruses.

Fusigenic virosomes

More recently, some of the advantages of viral delivery vectors have been combined with the safety and 'simplicity' of the liposome to produce fusigenic virosomes. Virosomes have been engineered by complexing the membrane fusion proteins of haemagglutinating virus of Japan (HVJ, which is also known as Sendai virus) with either liposomes that already encapsulate plasmid DNA or oligodeoxynucleotides for antisense applications. The inherent ability of the viral proteins in virosomes to cause fusion with cell membranes means that these hybrid vectors can be very efficient at introducing their nucleic acid to the target cell, resulting in good gene expression. Each viral vector has a limit on the size of transgene that can be incorporated into its genome; no such limit exists for virosome or liposome technology. Genes of up to 100 kilobase pairs have been delivered by fusigenic virosomes to cells both *ex vivo* and *in vivo*.

Hybrid nanovectors

University of Illinois is developing a new class of gene delivery vectors by combining viral and synthetic components to produce hybrid, active nanostructures. Recombinant viruses by themselves are highly efficient vectors, but suffer from serious safety concerns, difficulty with redirecting cell specificity, and expensive production and purification. Polymeric vectors are potentially safer, cheaper and more versatile, but in their current form lack the efficiency needed for clinical application. Hybrid viral/synthetic nanovectors may be designed to exhibit several of the advantages (and possibly avoid disadvantages) of both types of vectors.

Targeting the delivery of therapeutic DNA

The ability to deliver therapeutic DNA specifically to target cells and not to normal, surrounding cells is a desirable attribute in any DNA-delivery system and might be an important consideration in the design of gene therapy protocols. A number of approaches can be taken to target the delivery of the therapeutic DNA and these vary depending upon the delivery system employed. For instance, cell type-specific viral-mediated DNA delivery can be achieved via the manipulation of the natural tropism of viruses (Douglas *et al.*, 1996). The viral vectors employed for DNA delivery, like the viruses upon which they are based, enter host cells using a receptor-mediated mechanism. Altering the viral proteins involved in the recognition of these receptors on the host cell and engineering them to recognize alternative receptors found only on the targeted cell type would enable therapeutic viral based DNA-delivery systems to target specific cells. Another method that can be used to ensure that a therapeutic gene only is expressed in a specific population of cells exploits the transcriptional control mechanisms present in all cells (Shering *et al.*, 1997). Transcriptional control of therapeutic gene expression can be achieved in gene therapy protocols that use either viral or non-viral-based delivery systems. Placing expression of the therapeutic gene under the control of a cell type-specific promoter will ensure that, although the therapeutic DNA may be delivered to many different cell types, the gene can be expressed only in cells possessing the specific transcriptional machinery that allows expression from that promoter. One example of such a promoter is the myosin light chain-2v promoter that is present only in cardiac myocytes (Chen *et al.*, 1998). Thus, therapeutic genes placed under the transcriptional control of this promoter will be expressed only in cardiac myocytes, thereby enabling genes to be specifically targeted to heart muscle cells.

Table 3: Each of these gene delivery strategies has their own advantages and disadvantages.

Gene delivery system	*Advantages*	*Disadvantages*
Retrovirus	integrates into host cell genome providing stable gene expression	random integration may cause insertional mutations
Adenovirus	contains >30 kb of non-viral DNA, infects non-dividing and dividing cells	does not provide long term gene expression; no integration
Adeno-associated virus	contains no viral genes, non-pathogenic, no immunity problems	small capacity for gene sequences, difficult to obtain large viral stocks
Liposomes	non-pathogenic, no immunity problems, no limit to size of gene	low transfection efficiency, low rate of stable integration
Biolistics	same as liposome-mediated transfer, promising as a vaccination method	limited to dermal tissue, low rate of stable integration, difficult to QC

Table 4: Comparison of viral and non-viral-based gene therapy delivery systems

Property	*Viral delivery*	*Non-viral delivery*
Efficiency	Highly efficient gene delivery. Potential for long-term gene expression (retroviral delivery)	Low efficiency of gene transfer
Size of therapeutic DNA	Limited by size that can be accomodated by viral genome and packaging of viral particles	Largely unlimited
Safety issues	Some viral vectors stimulate severe immune response in the patient. Concern regarding use of viral vectors based on HIV.	Low toxicity to host
Longevity of expression	Transient (adenoviral-based systems) or long-term (retroviral- or AAV-based systems) expression	Transient expression

*See color plate: Fig 16

Routes of administration and cell targets

Intrathymic administration

The application of the process of intrathymic T-cell development to transplantation and tolerance induction was first described by Posselt and colleagues. Self tolerance (the failure to respond to antigen borne on self tissue) develops as T-lymphocyte precursor cells that are $CD4^-$and $CD8^-$('double negative') pass through the thymus. Because the T cells are exposed to antigen on thymic epithelial cells, any T cells that have a high-affinity interaction with antigen in the thymus, and are therefore potentially autoreactive cells, are 'negatively selected' by the process of clonal deletion. Cells that have TCRs that have no (or an extremely low) affinity for intrathymic antigen, yet still have a high-affinity interaction with self MHC, are 'positively selected'; they can thus mature and go on to populate and expand into larger clonal populations in the periphery.

Knechtle and colleagues showed it was possible to induce tolerance using a gene therapy strategy in a rat model. First, they took syngeneic recipient muscle cells and then transfected them *in vitro* with an MHC class I gene derived from the donor. These cells were then injected into the thymus of the recipient. Next, the peripheral immune system of the recipient was depleted of potentially alloreactive T cells using anti-lymphocyte serum. This was then followed by a liver transplant from the donor, to which the recipient was found to be unresponsive. In a subsequent study, the MHC class I complementary DNA (cDNA) from the donor strain rat was introduced directly into the thymus to transfect recipient thymic cells *in situ*; analysis using the polymerase chain reaction (PCR) detected the transient expression of donor DNA in the thymus (and later in the spleen, which was probably due to the export of transfected thymocytes out of the thymus).

The above approaches have used either live cells transfected with DNA or naked DNA itself to deliver donor MHC genes. Adenovirus could be used to improve the efficiency of gene therapy that has been achieved using DNA transfection (See color plate: Fig 17). Adenovirus vectors are ideal for use in intrathymic applications because they can be generated at high titres and can transduce a range of cell types. Not only can genes be transferred to the antigen-presenting thymic epithelial cells and possibly to the developing thymocytes, but central tolerance (tolerance that is established in lymphocytes developing in central lymphoid organs such as the thymus, spleen and bone marrow) to the immunogenic adenoviral antigens can be induced, as shown by Ilan and colleagues. Their work demonstrated that intrathymic inoculation of the recombinant adenovirus inhibited the appearance of neutralising antibodies and CTLs against the recombinant adenovirus.

Liver

The liver has many interesting properties with respect to gene transfer and transplantation. Recipients of liver grafts have been shown in some situations to accept MHC-mismatched grafts spontaneously, without the requirement for post-transplant systemic immunosuppression. It has been hypothesised that this observation was due to the post-transplant solubilisation and release into the circulation of donor MHC molecules, which subsequently downregulated the alloreactive CTL response.

The portal vein, hepatic artery or both offer the best routes of access to the liver *in vivo* for delivering either a perfusate of a viral or non-viral gene therapy vector. In one study, Chia and colleagues demonstrated that efficient gene transfer could be achieved after perfusion with an adenovirus encoding a reporter gene via both the portal vein and the hepatic artery into clamped, cold-preserved liver grafts. This efficiency was said to have been due, in part, to the improved access to the hepatic microcirculation and thus virus–cell contact.

Retroviral vectors have also been used in mouse models for hepatic gene transfer; however, because these vectors are efficient at transducing only actively dividing cells, a partial hepatectomy must be carried out in most cases before retroviral transduction, to encourage the hepatocytes to divide.

Bone marrow cells

The importance of bone marrow cells, particularly the sub-population of haematopoietic stem cells (HSCs), has never been underestimated in the context of gene therapy. Their potential for self renewal and differentiation into all haematopoietic lineages makes them very attractive as targets for gene transfer, especially in those situations (such as genetic disorders) where long-term transgene expression is required. Unfortunately, because HSCs occur at extremely low frequencies in the bone marrow and peripheral blood, it is difficult to obtain sufficient numbers of cells to transduce *ex vivo* for a subsequent *in vivo* biological effect. The purification of HSCs for gene therapy relies mainly on mobilising the stem cells from the bone marrow into the peripheral circulation, using an agent such as granulocyte macrophage colony-stimulating factor; cells are then selected for, using methods such as fluorescent-activated cell sorting or antibody-coated magnetic beads. This type of positive-enrichment method requires the existence of cell-surface markers that are potentially specific for stem cells, such as c-kit (the receptor for stem-cell factor in the mouse) and CD38 in humans (although this is more controversial). Negative depletion (which excludes those cells that are definitely not stem cells) is another method, and is often used in combination with positive selection. The search for new stem-cell-specific markers is an extremely active area of research at present.

With regards to clinical transplantation, bone marrow itself is a frequently transplanted tissue, for example in patients who have undergone radical cytotoxic therapy for leukaemias and other haematological malignancies. The infusion of donor bone marrow has often been used to expose recipients before transplantation to the alloantigens of a mismatched organ. This approach is thought to be worthwhile despite the possibility that GvHD could occur. The exposure of recipients to MHC transgenes that have been derived from donors is a more specific and safer method; furthermore, it does not require the introduction of live donor lymphocytes and therefore poses no risk of the transfer of cells that induce GvHD. The transfer of MHC genes to syngeneic bone marrow *ex vivo* or *in vivo* could be carried out as a means of exposing recipients to alloantigen genes. Gene transfer to bone marrow could also be used to introduce genes that encode immunoregulatory molecules (such as cytokines) that would modulate the immune function in haematopoietic cells.

Sykes and colleagues demonstrated in irradiated mice, using a retroviral gene therapy vector, that the pre-transplantation transfer of a donor MHC class I gene to recipient strain bone marrow cells *ex vivo* prolonged the survival of mismatched skin grafts that were mismatched by a single alloantigen; however, fully allogeneic skin grafts with multiple mismatches were rejected.

Wong and colleagues studied a similar system that also used a retroviral vector encoding a donor MHC class I molecule. This time, fully allogeneic CBA mice with the MHC haplotype $H2^k$ were used as transplant recipients. First, they were pre-treated 28 days before transplant with two doses of an anti-CD4 monoclonal antibody and 5×10^6 recipient bone marrow cells. These cells had been transduced *ex vivo* with a retroviral vector carrying the donor-specific MHC class I gene K^b. As a result of this tolerisation regime, the mice were able to accept indefinitely a subsequent donor-specific [C57BL/10 ($H2^b$)] cardiac allograft. This study has important clinical implications, because it demonstrated that recipients do not need to be exposed to the full range of donor MHC molecules present on a graft for that graft to be accepted long term. This tolerogenic (or unresponsive) state does not reduce the potency of the immune system; it still reacts with full vigour against any other foreign or third-party antigen.

GENE THERAPY IN DISEASES

Gene therapy is "the use of genes as medicine". It involves the transfer of a therapeutic or correct gene into specific cells of an individual in order to repair a faulty gene. Thus it may be used to replace a faulty gene, or to introduce a new gene whose function is to cure or to favorably modify the clinical course of a disease.

The scope of this new approach to the treatment of disease is broad, with potential in the treatment of many genetic conditions, some forms of cancer and certain viral infections such as Acquired Immune Deficiency Syndrome (AIDS). At the present time however, gene therapy remains an experimental discipline and much research remains to be done before this approach to the treatment of disease will realise its full potential.

There are over 600 clinical gene therapy trials initiated or approved worldwide. Of these, the majority are being conducted in the United States and Europe, with only a modest number initiated in other countries, including Australia. Perhaps surprisingly, the majority of trials focus on treating acquired diseases such as cancer and AIDS, although an increasing number of inherited conditions are being targeted.

A form of immune deficiency called adenosine deaminase (ADA) deficiency was the first disease to be treated with a gene therapy approach in humans in the early 1990's, and was also the first condition for which therapeutic gene transfer into stem cells has been attempted in the clinical arena (Candotti F, 2001).

Another from of immune deficiency is due to a mutation in a gene located on the X chromosome and is called Severe Combined Immune Deficiency (SCID). The use of gene therapy in the treatment of this condition was done by a French research group (Cavazzana-Calvo M *et al.*, 2000). Gene therapy of human severe combined immunodeficiency (SCID-X1) disease was hailed as the first example of a genetic condition being successfully treated by gene therapy and is a milestone in medical history. Seven out of 10 infants treated to date have restored immune function but 2 of the children treated initially have developed a form of leukemia. The leukemia in these two patients was caused when the virus used to deliver the therapeutic gene activated a cancer-causing gene. After the first boy developed leukaemia in October 2002, and the second in January 2003, the clinical trials of the gene therapy being conducted in a number of countries were halted. These have now been resumed but only for patients with no other treatment options. Work is continuing to make the therapy as safe as possible.

Many human diseases are caused by defective genes

A gene is a linear sequence of DNA that codes for a particular protein. On rare occasions, usually during the division of the cell, the nucleotide sequence (the order of the DNA base pairs) of a gene can get jumbled up and mutated, so that the resultant protein is faulty. Such a mutation event is the root cause of genetic diseases such as cystic fibrosis, adenosine deaminase (ADA) deficiency and sickle-cell anaemia. For example, people who suffer from cystic fibrosis produce a faulty cellular transport protein called cystic fibrosis transmembrane conductance regulator, which results in the build-up of mucous in their lungs.

The earliest applications of gene therapy were based on the principle that a disease is caused by a faulty gene (or combination of genes), and if such genes can be replaced with 'correct' versions, the disease might be controlled, prevented or cured. Gene therapy is being applied to many different genetic diseases, both congenital (since birth) and acquired. However, most diseases involve multiple genetic factors (they are polygenic). Until the precise involvement of different genes (their regulation and expression) in the disease process and the proteins they encode is established, gene therapy is most likely to be clinically effective as a preventative or curative treatment for single-gene defects such as ADA deficiency, familial hypercholesterolaemia and cystic fibrosis. Several clinical trials employing gene therapy protocols have already been completed, with some success in patients who have cystic fibrosis and ADA deficiency, although the effectiveness of the protocols was not as dramatic as first envisaged, mainly owing to the inefficiency of the gene transfer vectors that were used.

Table 5: A few common examples of diseases based on genetic defects

Disease	*Genetic defect*
hemophilia A	absence of clotting factor VIII
cystic fibrosis	defective chloride channel protein
muscular dystrophy	defective muscle protein (dystrophin)
sickle-cell disease	defective beta globin
hemophilia B	absence of clotting factor IX
severe combined immunodeficiency (SCID)	any one of several genes fail to make a protein essential for T and B cell function

All of these diseases are caused by a defect at a single gene locus. (The inheritance is recessive so both the maternal and paternal copies of the gene must be defective.) Is there any hope of introducing functioning genes into these patients to correct their disorder? Probably yes. Other diseases also have a genetic basis, but it appears that several genes must act in concert to produce the disease phenotype. The prospects of gene therapy in these cases seem to be far more remote.

Heredity and cystic fibrosis

Cystic fibrosis (CF) is the most common, fatal genetic disease in the United States. About 30,000 people in the United States have the disease. Cystic fibrosis (CF) is a common genetic disorder characterized by defective epithelial chloride transport and progressive lung disease. The largest of the clinical trials of gene therapy have involved transfer of the cystic fibrosis transmembrane conductance regulator (CFTR) gene in cystic fibrosis, and over 100 patients have received such therapy to date (reviewed in crystal, 1995: Zeitlin, 2000).

Cystic Fibrosis: A Single Gene Disease

Cystic Fibrosis (CF) is a genetic disease caused by mutations in the cystic fibrosis transmembrane conductance regulator (CFTR) gene, located on chromosome 7 in humans. The gene was discovered in 1989. Since then, more than 900 mutations of this single gene have been identified. Being an autosomal recessive genetic disease this means that for a person to develop the disease they must inherit a mutation from both parents. A person who has a mutation in only one chromosome is considered a carrier and will remain healthy. Approximately 1 in 25 people are carriers of the disease. Around 7000 people are affected by CF in the UK.

CF causes the body to produce thick, sticky mucus that clogs the lungs, leads to infection, and blocks the pancreas, which stops digestive enzymes from reaching the intestine where they are required in order to digest food. Although great strides have been made in the treatment of CF, it remains lethal, often by early adulthood. The CFTR gene encodes an epithelial chloride ion channel which forms an ion channel that facilitates the movement of salt (electrically charged ions) and water across the cell membrane, from the cell cytoplasm to the extracellular space (ductal space) (See color plate: Fig 18). When this gene becomes defective, the cells do not release the chloride. As a result there is an improper salt balance in the cells which affects the secretion of mucus in the lungs and digestive tract. In CF patients, CFTR is unable to properly carry out these functions leading to dehydration of affected organs such as the sweat glands, lungs, pancreas and digestive system (See color plate: Fig 19). As you might imagine, this phenomenon creates an environment that favours bacterial growth and consequently people who suffer from CF are susceptible to repeated respiratory tract infections which lead to inflammation and progressive lung damage. Lung disease is the primary cause of morbidity and mortality in CF.

The normal CFTR protein is expected to exert a dominant function over the mutant CFTR, which makes it a good candidate disorder for treatment by gene therapy. Since the target epithelial cells are terminally differentiated and relatively short-lived, there is no necessity for integration of the transgene, which means that gene delivery can be achieved using non-integrating vectors. Preclinical and clinical research has lead to the rapid development of a variety of vectors designed to correct the basic defect in CF, including adenovirus, adeno-associated virus, and liposomes. Although both adenoviral vectors and liposomes have been shown to work effectively in *in vitro* gene therapy studies for cystic fibrosis, clinical studies delivering DNA to the nasal, and more recently lung, epithelia have shown only weakened short-lived correction (Alton *et al., 1999; Harvey et al.*, 1999). There is an added complication in that repeated administration of adenoviral vectors have reduced effectiveness and can cause adverse reactions in some patients (Harvey *et al., 1999).*

Researchers are focusing on ways to cure CF by correcting the defective gene, or correcting the defective protein. CF is one of the most extensively researched genetic diseases as a target for gene therapy development. It may also serve as an important model for gene therapy of other diseases. Clinical studies have identified the host immune response and low vector efficiency as key impediments to effective CF gene therapy. Further research promises to refine vector technology and bring CF gene therapy to the bedside.

Gene Therapy Research Offers Promise of a Cure for Cystic Fibrosis

Gene therapy offers great promise for life-saving treatment for CF patients since it targets the cause of CF rather than just treating symptoms. Gene therapy for CF had its start in 1990, when scientists successfully corrected faulty CFTR genes by adding normal copies of the gene to laboratory cell cultures. In 1993, the first experimental gene therapy treatment was given to a patient with CF. Researchers modified a common cold virus to act as a delivery vehicle-or "vector"-carrying the normal genes to the CFTR cells in the airways of the lung.

Subsequent studies have tested other methods of gene delivery, such as fat capsules, synthetic vectors, nose drops or drizzling cells down a flexible tube to CFTR cells lining the airways of lungs. Researchers are now testing aerosol delivery using nebulizers.

But finding the best delivery system for transporting normal CFTR genes is only one problem that scientists must solve to develop an effective treatment for CF. Scientists must also determine the life span of affected lung cells, identify the "parent cells" that produce CFTR cells, find out how long treatment should last and how often it needs to be repeated.

The first cystic fibrosis gene therapy experiments have involved lung cells because these cells are readily accessible and because lung damage is the most common, life-threatening problem in CF patients. But scientists hope that the technologies being developed for lung cells will be adapted to treat other organs affected by CF.

CF gene therapy-progress to date

Only a year after the gene for CF was identified scientists showed in cells in the laboratory that it was possible to correct the CF defect by adding a healthy copy of the gene. These findings have been consistently repeated by other groups using a variety of models and gene transfer agents. Studies in humans followed quickly. Gene therapy for CF has been tested in humans using both viruses and liposomes. Five of the liposome trials were undertaken by members of the UK CF Gene Therapy Consortium. These early studies were concerned mainly with safety issues. The amount of gene transfer achieved is similar for both systems and is probably still too small to have any real therapeutic benefit-although it is important to note that none of these trials actually measured therapeutic benefit. Also at present the effect seen (gene expression) lasts for only a few days.

Use of gene therapy to treat CF versus conventional treatment methods

Conventional treatments for CF include the use of physiotherapy, antibiotics and pancreatic supplements. Many patients require treatment four times daily, including a considerable time spent on physiotherapy. This combination of treatment has helped increase life expectancy considerably to the current median of approximately 30 years of age. Despite small further improvements, more recently it has become apparent that there is a need for a more effective and convenient therapy. The identification of the gene responsible for CF (the cystic fibrosis transmembrane conductance regulator or CFTR protein) in 1989, was responsible for the advent of potential new treatments for CF. Gene therapy, the transfer of a normal copy of the CFTR gene into the lungs of CF patients, was proposed as an attractive new option.

The future of CF gene therapy

The scientific principle of gene therapy for CF is a sound one and the technology already exists to transfer the CF gene into the airway cells in man but there are still challenges ahead of the field. The two main challenges are getting the gene into the cells more efficiently and to make gene expression last longer. However the field is moving forward rapidly and with the development of better new viruses and liposomes these problems are surmountable. The members of the UK Cystic Fibrosis Gene Therapy Consortium have played a significant role in bringing the field of gene therapy for CF to where it is today. After initial encouraging clinical trials most research for CF gene therapy has returned to the laboratory to solve the problems mentioned above. In the UK there are no trials ongoing at present. It is the goal of the Consortium to work together to meet the challenges and enhance progress to a Phase 3 (large-scale) study by 2009.

Gene Database Speeds Research

As the amount of information about CF grows, scientists have recognized the need to share their research findings. To facilitate this sharing of information, the Cystic Fibrosis Foundation funds Cystic Fibrosis Foundation Therapeutics, Inc.-National BioInformatics Center (CFFTI-NBC) located at The University of North Carolina Chapel Hill. The center is becoming a repository for data derived from gene expression studies. By pooling information, researchers hope to accelerate the process of finding a cure for CF.

SEVERE COMBINED IMMUNODEFICIENCIES (SCIDS)

Without doubt the greatest clinical experience with gene therapy is in use of murine retroviral vectors for the treatment of SCID, particularly for patients with adenosine deaminase deficiency (ADA) (Blaese *et al.*, 1995; Bordignon *et al.*, 1995; Kohn *et al.*, 1995; Hoogerbrugge *et al.*, 1996), though more recently success has been reported for patients with X-linked SCID (Cavazzana-Calvo *et al.*, 2000).

SCIDs are a group of inherited disorders characterized by a profound reduction or absence of T-lymphocyte function (reviewed in Fischer, 2000). The resulting deficits in both cell mediated and humoral immune responses invariably lead to premature mortality in the absence of hematopoetic stem cell transplantation. Since the first successful HLA-matched bone marrow transplants (BMT) were performed in two primary immunodeficiency disorders over 30 years ago, transplantation technology has advanced considerably. The cure rate for SCID using matched sibling donors is now over 95% and one recent study reported 100% cure (Buckley *et al.*, 1999). However, for only 30% of patients does such a donor exist, and for T cell depleted haplo-identical parental grafts, the success rate falls to 60%. Complications can relate to toxicity arising from any conditioning regimen, the presence of mature T cells in the marrow graft, pre-existing infection and delayed reconstitution of immune function post transplant (Fischer *et al.*, 1990; Buckley *et al.*, 1999; Fischer, 1999). *In utero* transplantation of haplo-identical cells has been attempted to pre-empt the development of clinical disease, and to utilize the proliferative and possibly tolerogenic fetal environment to facilitate engraftment (Flake *et al.*, 1996; Wengler *et al.*, 1996). The efficacy and safety of this approach awaits comparison with conventional transplantation, and remains limited to those families with previously affected children.

Gene Transfer Therapy for Severe Combined Immunodeficieny Disease (SCID) Due to Adenosine Deaminase (ADA) Deficiency and purine nucleoside phosphorylase deficiency

SCID can arise as a result of deficiencies in certain enzymes involved in purine metabolism.

SCID is a disease in which the patient has

- No cell-mediated immune responses
- Not able to make antibodies.

It is a disease of young children because, until recently, the absence of an immune system left them prey to infections that ultimately killed them.

About 25% of the cases of SCID are the result of the child being homozygous for defective genes encoding the enzyme adenosine deaminase (ADA), while a few cases (about 4%) are caused by defects in purine nucleoside phosphorylase (PNP). The normal catabolism of purines is deficient, and this is particularly toxic for T cells and B cells.

ADA deficiency in human beings results primarily in immunological dysfunction with T and B lymphopenia (reviewed in Hershfield and Mitchell, 1995). The highest enzymatic activity of ADA is found in immature thymocytes, and deficiency results in impaired intrathymic differentiation. Accumulation of deoxyadenosine in T cells is accompanied by preferential conversion to the toxic compound deoxyadenosine triphosphate (dATP), and the build up of other toxic metabolites. Despite ubiquitous constitutive expression of ADA, lymphocytes are particularly susceptible to toxicity from accumulating metabolites, possibly because of predominant inhibition of T cell specific molecules during intrathymic differentiation (Benveniste *et al.*, 1995).

Other autosomal recessive SCID

There are now many forms of SCIDs described which are individually quite rare, usually only a few families or individuals are affected. These are caused by the inheritance of defective autosomal genes. Many of these genes are involved in T-cell development, signaling or proliferation (reviewed in Fischer, 2000).

Jack-3 defects

The jack-3 kinase is known to associate with γc chain-containing cytokine receptors and jack-3 mutations have been identified in a small number of SCID patients with immunophenotypes equivalent to x-SCID, including block NK-cell differentiation.

Recombination activating genes (RAG1 and RAG2) defects

As many as half of T-B-SCID patients may be affected by mutations in the RAG1 and/or RAG2 genes. These proteins are involved in the process of VDJ rearrangements, which is essential for the production of functional immunoglobulin and T-cell receptor (TCR) proteins.

ZAP 70 defects

Mutations in ZAP 70, a TCR-associated protein tyrosine kinase that is associated with the ζ chains of the TCR-CD 3 complex produce a distinctive form of SCID characterized by absence of CD8+ and abundance of CD4+ peripheral T-cells (CD4+CD8-B+ SCID).

TCR-CD3 abnormalities

Rarely, the defective expression and function of the TCR-cd3 complex of T-cells can occur as a result of mutations in the γ or ε genes of the CD3 complex, resulting in a variable phenotype disease.

Treatment Options for ADA-SCID:

- The child is raised in a strictly germfree environment: all food, water, and air to be sterilized. David, the "bubble boy" from Houston, survived this way until he was 12 years old.
- The child is given a transplant of bone marrow from a normal, histocompatible, donor. Ideally, this would give the child a continuous source of ADA^+ T and B cells. However,
- Even though the child cannot reject the transplant (the child has no immune system), T cells in the transplant (unless the donor was an identical twin) can attack the cells of the child producing graft-versus-host disease.
- The donor cells may be infected with a virus which could overwhelm the recipient before his or her immune system was restored. (David received a

bone marrow transplant from his sister, but she, like many people, had been infected earlier with the Epstein-Barr virus (the cause of "mono"). The virus was still present in the cells she donated, and killed her brother.

- Injections of ADA are given (the enzyme is currently extracted from cows). When conjugated with polyethylene glycol (PEG) to delay its breakdown in the blood, ADA-PEG injections have kept SCID patients reasonably healthy. But just like the insulin injections of a diabetic, they must be repeated at frequent intervals. So,
- What about giving the patient functioning ADA genes; that is, gene therapy?

A new method for delivering gene transfer therapy to patients with severe combined immunodeficiency disease (SCID) due to a defective adenosine deaminase (ADA) gene has been evaluated. This gene codes for the adenosine deaminase enzyme, which is essential for the proper growth and function of infection-fighting white blood cells called T and B lymphocytes. Patients who lack this enzyme are vulnerable to frequent and severe infections.

Somatic gene therapy

Though rare, primary immunodeficiencies are in many ways ideally suited to the application of this technology, and have become model conditions on which to design and test gene therapy protocols. Rapid progress in understanding the molecular basis for some of these disorders, including ADA-SCID, X-SCID, Jak-3 kinase deficiency and ZAP 70 deficiency, has facilitated the development of therapeutic strategies based on genetic manipulation of autologus cells (Osborne *et al.*, 1990; Ferrari *et al.*, 1991; Candotti *et al.*, 1996; Hacein-Bey *et al.*, 1996; Taylor *et al.*, 1996; Bunting *et al.*, 1998; Cavazzana-Calvo *et al.*, 2000; Soudais *et al.*, 2000).

Some patients with this disease receive enzyme replacement therapy with weekly injections of the drug PEG-ADA (ADAGEN). This drug may increase the number of immune cells and reduce infections, but it is not a cure. Gene transfer therapy, in which a normal ADA gene is inserted into the patient's cells, attempts to correct the underlying cause of disease. This therapy has been tried in a small number of patients with varying degrees of success. In this, the gene is inserted into the patient's stem cells (cells produced by the bone marrow that mature into the different blood components-white cells, red cells and platelets).

Enzyme replacement therapy with bovine ADA conjugated to polyethylene glycol (PEG-ADA), has resulted in significant improvement in lymphocyte cell numbers and function in many ADA-SCID patients, but the response is variable, and may be transient (Hershefield, 1995). This probably reflects the main site of PEG-ADA activity which is extracellular, and which only partially complements the cellular deficiency. Complete absence of ADA results in lethality before two years of age. However, retention of only 1-5% of activity have been found to be immunologically normal, while levels of ADA 50 times above normal have also been associated with normal immunity. These considerations are important implications for the design of gene therapy vectors, but suggest that relatively simple expression systems may be efficacious for this condition. Furthermore, spontaneous reversion to normal of inherited mutations *in vivo* (resulting in somatic mosaicism) has recently been demonstrated in 2 patients with attenuated disease (one with ADA-SCID and another with X-SCID) suggesting that corrected cells have a distinct growth and differentiation advantage over mutants (Hirschhorn *et al.*, 1996; Stephan *et al.*, 1996).

Gene Therapy: requirements

- The gene must be identified and cloned. This has been done for the ADA gene.
- It must be inserted in cells that can take up long-term residence in the patient. So far, this means removing the patient's own cells, treating them in tissue culture, and then returning them to the patient.
- It must be inserted in the DNA so that it will be expressed adequately; that is, transcribed and translated with sufficient efficiency that worthwhile amounts of the enzyme are produced.

All these requirements seem to have been met for SCID therapy using a retrovirus as the gene vector. Retroviruses have several advantages for introducing genes into human cells.

- Their envelope protein enables the virus to infect human cells.
- RNA copies of the human ADA gene can be incorporated into the retroviral genome using a packaging cell.

Packaging cells are treated so they express:

- an RNA copy of the human ADA gene along with
  a) A packaging signal (P) needed for the assembly of fresh virus particles
  b) Inverted repeats ("R") at each end; to aid insertion of the DNA copies into the DNA of the target cell.

*An RNA copy of the retroviral gag, pol, and env genes but with no packaging signal (so these genes cannot be incorporated in fresh viral particles).

*See color plate: Fig 20

Target cells for SCID gene therapy

T cells

The first attempts at gene therapy for SCID children (in 1990), used their own T cells (produced following ADA-PEG therapy) as the target cells.

The T cells were:

- Placed in tissue culture

- Stimulated to proliferate (by treating them with the lymphokine, Interleukin 2 (IL-2)
- Infected with the retroviral vector
- Returned, in a series of treatments, to the child

The children developed improved immune function but:

- The injections had to be repeated because T cells live for only 6–12 months in the blood (But someday there might be a way to make genetically-engineered T cells live indefinitely by getting them to express telomerase).
- The children also continued to receive ADA-PEG so the actual benefit of the gene therapy was unclear

Stem cells

In 1993, ADA gene therapy was attempted with blood ("hematopoietic") stem cells drained from the umbilical cords of three newborn babies who were known to be homozygous for ADA deficiency.

Blood stem cells:

- Produce (by mitosis) all the types of blood cells, including T and B lymphocytes
- Produce (by mitosis) more stem cells, thus ensuring an inexhaustible supply

At last report, some of the circulating T cells in these children carried the ADA gene. But because the children continue to receive ADA-PEG, it is unclear how much gene therapy has contributed to their good health.

However, in June of 2002, a team of Italian and Israeli doctors reported on two young SCID patients that were treated with their own blood stem cells that had been transformed in vitro with the ADA gene. After a year, both children had fully-functioning immune systems (T, B, and NK cells) and were able to live normal lives without any need for treatment with ADA-PEG or immune globulin (IG). The doctors attribute their success to first destroying some of the bone marrow cells of their patients to "make room" for the transformed cells.

Retroviral vectors for gene therapy of ADA-SCID

A variety of techniques have been developed for transferring genetic material into human cells, each with its associated advantages and disadvantages. The most commonly used vectors tend to be virus based, because of the relatively high efficiencies of transduction that can be obtained. Where stable integration of transferred sequences is required, then retroviruses have most often been the vectors of choice. The murine oncoretrovirus *Moloney Murine leukaemia virus* (MoMLV) has formed the backbone of thye retroviral vectors that have so far been used in clinical trials for SCID. The single stranded RNA genome is reverse transcribed to double stranded DNA which integrates into the chromosome of the target cell during cell division, where it becomes a stable genetic element passed onto all subsequent progeny cells. Thus, although stability can be achieved it is dependent on the process of cell division furthermore integration is at random sites in the target cell genome which could potentially lead to unpredictable gene expression or insertional mutagenesis. Another potential problem is the generation of replication-competent recombinant viruses, though the risks are minimized. Gene transfer to human cells has been optimized using the murine amphotropic retrovirus envelope protein, though the receptor expression on human cells may be limited (Orlic *et al.*, 1996). Future clinical trials may benefit from more effective gene transfer mediated by alternative envelope proteins including the gibbon ape leukaemia virus (GALV) protein (Demaison *et al.*, 2000) and the feline endogenous viral envelope (RD114) protein (Porter *et al.*, 1996).

T lymphocyte gene therapy for ADA-SCID

The ability of HLA-matched genotypically identical stem cell transplants for severe ADA deficiency to selectively reconstitute the lymphoid component in the absence of cytoablative conditioning suggested that biochemical correction of autologous T-cells by somatic gene transfer could provide a similar therapeutic outcome. This was supported by preclinical studies which demonstrated a survival advantage of transduced ADA-SCID T-cells in immunodeficient mice (Ferrari *et al.*, 1991).

The first ever approved clinical trial of gene therapy began in 1990 at the National Institutes of Health (NIH), and involved repeated *ex vivo* transduction of autologous peripheral blood T cells with a MoMLV-based retroviral vector expressing human ADA, followed by expansion and reinfusion into two patients (Blaise *et al.*, 1995). Although both patients had attenuated disease, at the time of enrolment both were lymphopenic, anergic and had depressed immune reactivity to specific antigens despite receiving optimal PEG-ADA enzyme replacement therapy. Within five to six months of gene therapy, the lymphocyte count in one patient had risen into the normal range, and remained stable for at least 2 years after the last infusion of transduced cells. Levels of ADA in blood lymphocytes stabilized at roughly half that found in heterozygote carriers, and was associated with return of delayed type hypersensitivity (DTH) skin test reactivity, cytolytic responses and antibody production. The second patient also responded with an initial increase in peripheral T lymphocyte numbers to levels in the high normal range, although these fell over time to levels just above those at enrolment. In contrast to the first patient, ADA levels remained very low, correlating with low gene transfer efficiency (<10% as efficient as the other patient). Some return of immune reactivity was demonstrated, but could have arisen as a result of infusion of an expanded T-cell population, and was variable anyway prior to enrolment.

This trial demonstrates the potential efficacy of gene transfer to autologus T cells for therapy of ADA-SCID, and at the present time is suggestive of clinical benefit. The outcome is confused by the concurrent administration of exogenous PEG-ADA, which to a variable extent complements the cellular immunodeficiency. Discontinuation of this treatment would seem to be obvious test for efficacy of gene transfer, but normalization of T lymphocyte ADA levels alone may also be insufficient to correct the immunodeficiency fully, and may necessitate continued therapy. One concern with T lymphocyte gene therapy is the potential for insertional mutagenesis arising from multiple integration events during the transduction procedure. At present there is no evidence to suggest that this has occurred. Interestingly, attempts to utilize cytokine-mobilised peripheral blood progenitor cells from ADA-SCID patients for gene transfer have been limited by poor yield (Sekhsaria *et al.*, 1996).

Stem cell gene therapy for ADA-SCID

The optimal target cell for curative gene therapy of many haematopoeitic disorders is the pluripotent haematopoetic stem cell (PHSC). Much interest has therefore been directed towards attempts to achieve this goal. The first attempt in a clinical setting to transducer bone marrow cells from a patient with ADA deficiency was conducted by Bordignon and colleagues in Milan, and was combined with re-infusion of genetically modified peripheral blood lymphocytes (Bordignon *et al.*, 1995). In this study, two retroviral vectors distinguishable by molecular techniques, were used to transfer an ADA minigene into either bone marrow or lymphocytes from two patients with attenuated ADA-SCID who had become immunologically refractory to exogenous enzyme replacement. Administration of genetically modified cells resulted in normalization of lymphocyte counts, antigen-specific immune responses, and mitogen and antigen-specific proliferation. The T-cell repertoire (V_β) was also shown to normalize progressively. Significantly, one year after discontinuation of gene therapy, peripheral T cells derived from the transduced lymphocyte population were gradually replaced by T cells derived from bone marrow, suggesting a proliferative advantage. Both children enrolled in this trial continued to receive enzyme replacement therapy, although relative dosages were decreasing. Withdrawal of PEG-ADA should reveal the true efficacy of gene transfer, and may enhance the proliferative advantage of corrected cells.

In a separate study, CD34+ cells obtained from the bone marrow of three children with ADA deficiency by a retroviral vector encoding the human ADA cDNA (Hoogerbrugge *et al.*, 1996). Mononuclear cells and granulocytes retaining the vector genome were detectable by PCR in the peripheral circulation for up to three months after gene transfer, and in one patient, in the bone marrow at six months. Detection of the vector genome after this period of time was not possible. One patient did not receive enzyme replacement until three months after infusion in an attempt to maximize proliferative advantage, although it is likely that because gene transfer efficiency was very low, this was not a sufficient time period for genetically corrected clones to reveal themselves.

Limiting factors for stem cell gene therapy

The inefficiency of retroviral transduction of the human PHSC population has been highlighted. The requirement for the current generation of retroviral vectors to transduce actively dividing cells may be a limiting factor because PHSCs are relatively quiescent. It was thought that cells harvested from umbilical cord blood might be better targets for retrovirus-mediated gene transfer, and they were therefore used in a clinical trial for ADA-SCID. In this study, three prenatally diagnosed infants received infusions of transduced cord blood CD34+ cells shortly after birth and were commenced on enzyme replacement therapy. Eighteen months after transplantation the proportion of lymphocytes retaining the transgene and levels of ADA activity in unselected cells was extremely low (Kohn *et al.*, 1995). It was expected that gradual withdrawal of PEG-ADA might facilitate the proliferation of gene corrected cells, however this does not appear to be the case, cessation of PEG-ADA treatment in one patient four years after gene therapy treatment lead to a decline in immune function. This suggests that although there is evidence for long term engraftment of transduced stem cells and accumulation of peripheral T-cells containing the transgene, there is still a need for improved efficiency of gene transfer and expression before clinical benefit can be achieved for this disease (Kohn *et al*,. 1998).

Clinical gene therapy for X-SCID

The most exciting results in this field have recently been reported from a clinical trial of gene therapy in two infants with X-SCID (Cavazzana-Calvo *et al*,. 2000). Following successful preclinical studies for X-SCID (Candotti *et al*,. 1996; Hacine-Bey *et al*,. 1996; Taylor *et al*,. 1996; Soudais *et al*,. 2000), CD34+ bone marrow cells were harvested from two X-SCID patients aged 11 months and 8 months and preactivated and infected daily for 3 days with retroviral γc vector-containing supernatant. Following this, the CD34+ cells were infused into the patients without prior chemioablation. Two weeks after infusion, peripheral blood cells were found to be positive for the γc transgene by PCR. T-cell counts increased between one and two months and continued to rise over the next eight months. The children were able to leave protective isolation three months after treatment. After a ten month period, γc transgene expression and function in the appropriate cell types, and antigen-specific responses in the normal range, could be observed. It can be concluded that gene therapy has provided full correction of the disease phenotype and that these children are 'cured' of their disease. Similar results have now been obtained for two additional children.

The improved success of this clinical trial over those previously conducted for ADA-SCID is attributable to several factors. First, X-SCID is not the same disease as ADA-SCID, and the children were not on enzyme replacement therapy prior to or during the gene transfer. This substantiates the idea that gene transfer confers a selective advantage for the transduced cells in this disease. Secondly, since the relatively poor results obtained in the clinical trials for ADA-SCID were reported there has been a substantial improvement in methodology, with use of optimized cytokine cocktails and a fibronectin fragment for gene transfer to CD34+ cells. The application of these improved protocols to the treatment of ADA-SCID patients may well prove to be efficacious in future clinical trials.

Stem Cell Gene Therapy to Treat X-Linked Severe Combined Immunodeficiency (XSCID)

A clinical trial of gene therapy was done for X-linked severe combined immunodeficiency (XSCID), a genetic disease caused by defects in a protein called the common gamma chain, which is normally on the surface of immune cells called lymphocytes. XSCID patients cannot make T lymphocytes, and their B lymphocytes fail to make essential antibodies for fighting infections. Without T and B lymphocytes patients develop fatal infections in infancy unless they are rescued by a bone marrow transplant from a healthy donor. However, even transplanted patients may achieve only partial immune recovery and still suffer from many infections, auto-immunity and/or and poor growth. A recent, successful trial in France used gene therapy instead of bone marrow transplantation for infants with XSCID.

Gene therapy has been succeeded for 15 French baby boys who suffered from this form of severe combined immunodeficiency, called X-linked SCID, because it is caused by a mutated X-linked gene encoding a subunit—called γc (gamma-c)—of the receptor for several interleukins, including interleukin-7 (IL-7). IL-7 is essential for converting blood stem cells into the progenitors of T cells. Boys with X-linked SCID can make normal B cells, but because B cells need T-helper cells to function, these boys could make neither cell-mediated nor antibody-mediated immune responses and had to live in a sterile bubble before their treatment.

The French doctors

- Isolated blood stem cells from the bone marrow of each boy;
- Treated the cells with a retroviral vector containing the normal gene for the ãc interleukin receptor subunit;
- Returned the treated cells to each donor.

The results: Now after as long as four years, 14 of these boys

- Are able to live normal lives at home instead of inside a sterile "bubble";
- Have normal (with some exceptions) numbers of T cells of both the CD4 and CD8 subsets;
- Have responded to several childhood immunizations, including diphtheria, tetanus and polio by producing both T cells and antibodies specific for these agents.
- Antibody production is sufficiently good that they have no need for periodic infusions of immune globulin (IG).

* Three of the little boys have developed cancer (one has died):

- In one case caused by a proliferating clone of T cells in which the vector has inserted itself in a gene (on chromosome 11) implicated in some cases of acute lymphoblastic leukemia (ALL).
- In a second case, the leukemia was of T cells.

Haemophilia

Queen Victoria was a carrier of the defective gene responsible for hemophilia, and through her it was transmitted to the royal families of Russia, Spain, and Prussia. Minor cuts and bruises, which would do little harm to most people, can prove fatal to hemophiliacs, who lack the proteins (Factors VIII and IX) involved in the clotting of blood, which are coded for by the defective genes. Sadly, before these proteins were made available through genetic engineering, hemophiliacs were treated with proteins isolated from human blood. Some of this blood was contaminated with the AIDS virus, and has resulted in tragic consequences for many hemophiliacs. Use of genetically engineered proteins in therapeutic applications, rather than blood products, will avoid these problems in the future. In the last few decades dramatic improvements in the management of haemophilia patients have occurred. Haemophilia has moved from a fatal or disabling disease to a hereditary disorder with available treatment and much better clinical outcomes. The safety of antihaemophilic factor concentrates has been dramatically improved and, in a multidisciplinary environment including haematologists, orthopaedic surgeons, paediatrics, infectiologists, specialised nurses and physiotherapists, complications related to haemophilia are now limited, markedly improving the quality of life of haemophiliacs. One can even think that the cure of haemophilia through gene therapy might occur in the next decades. Keeping this ultimate aim in mind, efforts at present are mainly focused on bioengineered Factor VIII/Factor IX concentrates with increased efficacy or longer half-life or decreased immunogenicity. In addition, several preclinical and clinical studies are being carried out for optimising and individually tailoring the therapeutic regimens of antihaemophilic therapies using global haemostasis tests in combination with the routine coagulation assays.

*See color plate: Fig 21

Gene therapy and hemophilia

An example of *ex vivo* gene therapies used by Transkaryotic Therapies, Inc., to treat hemophilia A (lack of Factor VIII) was recently reported.

Non viral transfer of the gene encoding coagulation factor VIII in patients with severe hemophilia A

This study was conducted on six individuals with severe hemophilia A all of which received the gene therapy treatment knowingly. The results showed that a few of the patients did indeed express higher levels of Factor VIII and did not need as much exogenous treatment. However, the effect lasted for less than a year and in several patients there was not much benefit (Roth *et al.*, 2001). This study was published in New England Journal of Medicine.

Let us examine the data:

Justification for using the *ex vivo* gene therapy approach for hemophilia A (Factor VIII):

- Factor VIII production is not regulated in response to bleeding
- The broad therapeutic index of factor VIII minimizes risk of overdose
- Delivery of factor VIII into the bloodstream does not require cell-specific expression
- Even low levels of the protein can be beneficial to the patient

The fibroblasts from the skin-biopsy specimen were transfected with a plasmid containing the gene encoding human factor VIII from which the B domain had been deleted (See color plate: Fig 22). Prior basic science studies had shown the the B domain is glycosylated and is involved in secretion. Recombinant factor VIII lacking the B domain still functions to stimulate coagulation and has a longer half-life because it is more resistant to protein degradation. The function of the B domain is not known.

Conclusion from this small pilot study is that it shows the benefit of a non-viral *ex vivo* approach. The cells come from the patient and the site of placement is relatively simple. The down side of course is that it didn't work that well and many more studies are needed.

TREATMENT OF AIDS

Safety and Efficacy of T cell Genetic Immunotherapy for HIV

The aim of gene therapy is to modify the genetic material of living cells for therapeutic purposes (Amado and Chen, 1999). Gene therapy involves the insertion of a functional gene or another molecule that contains and information sequence into a cell to achieve a therapeutic effect. Thus, the gene serves as a drug (Lasic, 1997). There are two types of gene therapy: somatic cell and germ line. Somatic cell gene therapy is the only technique now in use. The purpose of the procedure is to eliminate the clinical consequences of a disease and the inserted gene is not passed on to the patient's offspring. In germ line gene therapy a healthy gene is inserted into the fertilized egg of an animal that has a genetic effect. Every cell that develops from this egg, including the reproductive cells, will have the new gene. However, there are very serious social and ethical considerations with this type of gene therapy (Nichols, 1998).

Before 1996 scientists relied mainly on modified retroviruses such as Moloney murine leukemia virus when gene transfer into the chromosomes of target cells was needed, and adenovirus vectors when such integration was not needed. However, there has been little success in gene transfer with such virus vectors because even though the vectors can enter into their target cells, the cells need to be dividing, so that their nuclear membrane are broken down, for the gene to enter and integrate into the chromosome (Sikorski and Peters, 1998; CFAR at UC San Diego). However, scientists soon realized that members of the subfamily lentivirus, such as the retrovirus human immunodeficiency virus (HIV), would have the same ability to transfer genetic material into the genomes of cells, but could do this with non-dividing, dormant cells *in vivo* and growth-arrested cells *in vitro*

(Amado and Chen, 1999; CFAR at UC San Diego). Exploring this new method of gene therapy has been the work of many labs in the past few years.

Role of Lentiviral vectors in HIV treatment

Lentiviral vectors are a type of retrovirus that can infect both dividing and nondividing cells because their preintegration complex (virus "shell") can get through the intact membrane of the nucleus of the target cell. Lentiviruses can be used to provide highly effective gene therapy as lentiviruses can change the expression of their target cell's gene for up to six months. They can be used for nondividing or terminally differentiated cells such as neurons, macrophages, hematopoietic stem cells, retinal photoreceptors, and muscle and liver cells, cell types for which previous gene therapy methods could not be used. HIV is a very effective lentiviral vector because it has evolved to infect and express its genes in human helper T cells and other macrophages. The only cells lentiviruses cannot gain access to are quiescent cells (in the G0 state) because this blocks the reverse transcription step (Amado and Chen, 1999). To understand how HIV is a good vector for gene therapy, we must understand the structure of HIV and how it functions and infects its host.

HIV is a good vector for gene therapy

The preintegration complex of the human immunodeficient virus (HIV), which allows the vector assess inside human cells, dividing or non-diving, is composed of the enzyme integrase, the product of the *vpr* gene (an accessory gene), and a protein encoded by the *gag* gene (an essential structural gene) called matrix. This matrix protein contains a localization sequence which is recognized by the import machinery of the nucleus of a cell. The virus is surrounded by a lipid bilayer with protruding membrane proteins (See color plate: Fig 23). One of these proteins, gp120, is recognized by the host helper T cell CD4 receptor protein. Then HIV binds to a secondary receptor (CCR5 or CXCR4) and triggers a membrane fusion-mechanism with the gp41 transmembrane protein. This allows the virus asses to the cell interior and the virus content is released into the cytoplasm of the cell (Adler *et al.*; Schmidt, The HIV Page). Once inside of the cell in the cytoplasm, the matrix protein of the HIV contains a localization sequence that is recognized by the nuclear import machinery, which docks the complex at a nuclear membrane pore (See color plate: Fig 24). This enables the preintegration complex of the HIV lentiviral vector to pass into the nucleus (Amado and Chen, 1999).

Components of HIV Provirus

It is useful to understand the components of HIV and how it affects its host cell. The major protein components of the HIV virus can be seen in Table.

Table 6: The major protein components which are expressed by all retroviruses and are necessary for virus replication. They are encoded by three major transcripts: *gag, pol, env*. These proteins are synthesized as fusion proteins, which are post-translationally cleaved by the virus-encoded protease. HIV has some additional genes (From Schmidt, The HIV Page).

Name	*Protein*	*Function*
MA	Matrix	Matrix protein (gag gene); lines envelope
CA	Capsid	Capsid protein (gag gene); protects the core; most abundant protein in virus particle
NC	Nucleocapsid	Capsid protein (gag gene); protects the genome; forms the core
PR	Protease	Essential for gag protein cleavage during maturation
RT	Reverse transcriptase	Reverse transcribes the RNA genome; also RNAseH activity
IN	Integrase	Encoded by the pol gene; needed for integration of the provirus
SU	Surface glycoprotein	The outer envelope glycoprotein; major antigen
TM	Transmembrane protein	The inner component of the mature envelope glycoprotein

How HIV Infects Its Target Cell

Lentiviruses are the only type of virus that are diploid; they have two strands of RNA. Thus, HIV contains a diploid single stranded positive sense RNA-genome that is approximately 10 kb long. The ends are flanked with long terminal repeats (LTRs). A Psi-sequence is found near the 5′ end of the RNA-genome which is necessary for packaging viral RNA into virus capsids to continue the infection of HIV in its host (Schmidt, The HIV Page). However, the HIV's genetic information is integrated into the DNA of the host cell, so its RNA must be converted into DNA inside of the host for viral replication to be successful. This is done by reverse transcription of the RNA into DNA, and some of the proteins described in Table 8 are essential for this process. Reverse transcriptase synthesizes the first strand of DNA from the RNA template, and the host DNA polymerase synthesizes the second strand to produce dsDNA. Thus, quiescent cells do not have the ability to perform this second step in the reverse transcription process, so the RNA is not turned into DNA in cells in the G0 state. This is the reason for the limitation on gene therapy with HIV vectors. The DNA copy just made, which contains the genes *gag, env,* and *pol*, is inserted by integrase into the host genome (Adler *et al.*). LTRs are also necessary for integration of the dsDNA into the host chromosome. LTRs also serve as part of the promoter for transcription of the viral genes (Schmidt, The HIV Page). Thus, the virus is protected from attack by the immune system. It is this ability of the HIV to integrate its genetic material into a host cell which scientists would like to harness to put towards gene therapy. It has been shown that the HIV vector has an even higher rate of expression in its host cells than other retroviruses. HIV gene therapy vectors also do not trigger immune reactions, making them very attractive delivery systems (Adler *et al.*).

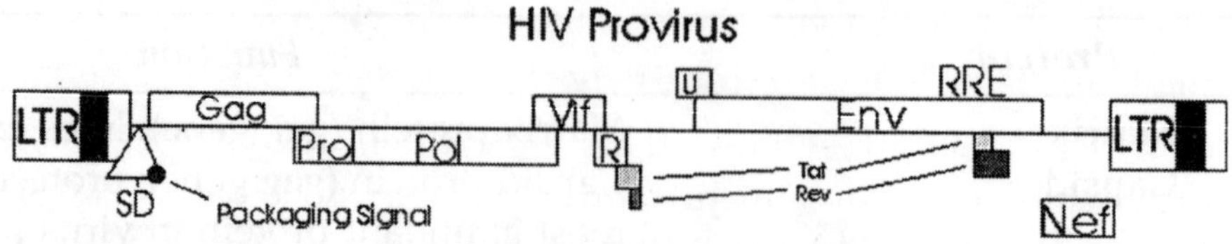

Fig F: HIV Provirus Used to Construct HIV Based Gene Therapy Vector

With the new genes from the HIV vector, DNA copy duplication, excision, and integration of the virus can take place. After infection and integration of the virus into its host regulatory proteins let the retroviral DNA exist in three stages—the latent period with inactivity, the stage where the virus gradually infects helper T cells, and then rapid production of infective viral particles that are released into the blood by the host cell lysis to infect other cells (Adler *et al.*). Researchers must curtail these second and third phases of HIV infection or HIV cannot be used as a gene therapy vector as patients would be infected with not only the therapeutic gene product but also the AIDS disease.

Production of HIV lentiviral vectors

To obtain a lentiviral gene therapy vector, a reporter gene or therapeutic gene is cloned into a vector sequence that is flanked by LTRs and the Psi-sequence of HIV. The LTRs are necessary to integrate the therapeutic gene into the genome of the target cell, just as the LTRs in HIV integrate the dsDNA copy of the virus into its host chromosome. The Psi-sequence acts as a signal sequence and is necessary for packaging RNA with the reporter or therapeutic gene in virions. Viral proteins which make virus shells are provided in the packaging cell line, but are not in context of the LTRs and Psi-sequences and so are not packaged into virions. Thus, virus particles are produced that are replication deficient, so are designed to be unable to continue to infect their host after they deliver their therapeutic content (Schmidt, HIV as a Vector for Gene Therapy).

*(See color plate: Fig 25)

HIV Vectors Have A Three Plasmid Expression System

Lentiviral vectors are usually created in a transient transfection system in which a cell line is transfected with three separate plasmid expression systems. These include the transfer vector plasmid (portions of the HIV provirus), the packaging plasmid or construct, and a plasmid with the heterologous envelop gene (ENV) of a different virus (Amado and Chen, 1999). The three plasmid components of the vector are put into a packaging cell which is then inserted into the HIV shell (Kalapana, 1999). The virus portions of the vector contain insert sequences so that the virus cannot replicate inside the cell system (Adler *et al.*).

Transfer Vector Plasmid

The transfer vector plasmid contains cis-acting genetic sequences necessary for the vector to infect the target cell and for transfer of the therapeutic (or reporter) gene and contains restriction sites for insertion of desired genes. The 3′ and 5′ LTRs, the original envelop proteins, and *gag* sequence promoter have been removed (Adler *et al.*; Naldini *et al.*, 1996).

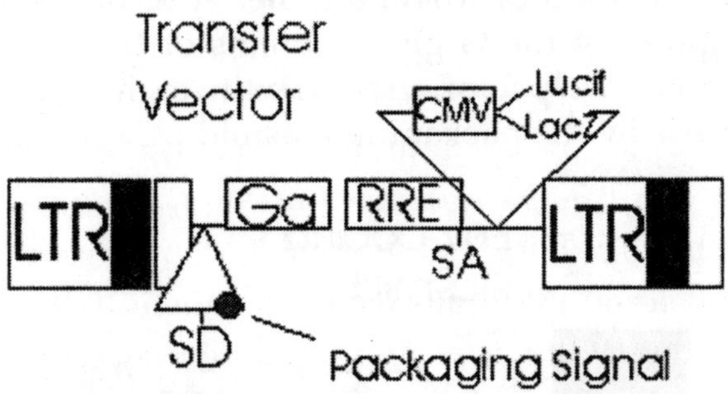

Fig G: Transfer Vector

Packaging Plasmid

The packaging plasmid is the backbone of the virus system and is also known as pCMVAR9. In this plasmid are found the elements required for vector packaging such as structural proteins, HIV genes (except the gene env which codes for infection of T cells, or the vector would only be able to infect these cells), and the enzymes that generate vector particles (Amado and Chen, 1999). Also contained is the human cytomegalovirus (hCMV) which is responsible for the expression of the virus proteins during translation. The packaging signals and their adjacent signals are removed so the parts responsible for packaging the viral DNA have been separated from the parts that activate them. Thus, the packaging sequences will not be incorporated into the viral genome and the virus will not reproduce after it has infected the host cell (Adler *et al.*; Naldini, 1996). Previous HIV vectors used two plasmids as the packaging plasmid contained the viral envelop gene. However, in the newer, better vectors the packaging plasmid lacks a viral envelop gene because this has been shown to be more desirable in terms of titer (minimum volume needed to cause a particular result in titration), stability, and broad range of target cells (CFAR at UC).

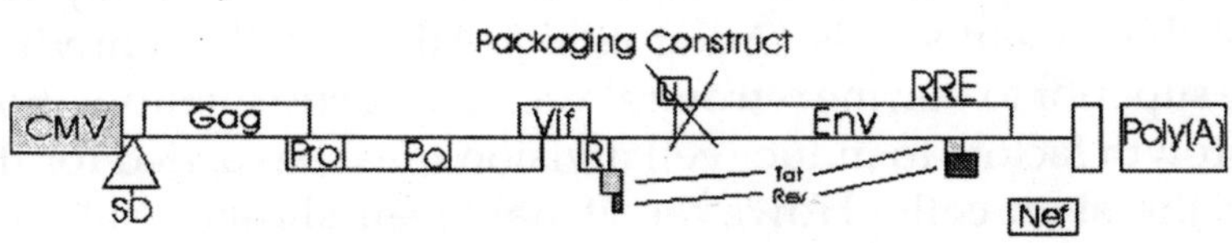

Fig H: Packaging Construct pCMVAR9

Envelop Gene of Third Plasmid

The third plasmid's envelope gene of a different virus specifies what type of cell to target and infect instead of the T cells (Amado and Chen, 1999). Normally HIV can infect only helper T-cells because they use their gp120 protein to bind to the CD4 receptor. However, it is possible to genetically exchange the CD4 receptor-binding protein for another protein that codes for the different cell type on which gene therapy will be performed (Schmidt, HIV as a Vector for Gene Therapy). This gives the HIV lentiviral vector a broad range of possible target cells. There are two types of heterologous envelope proteins. The amphoteric envelop of MLV, another type of vector, is transcribed first followed by the transcription of the G glycoproteins of the vesicular stomatitis virus, known as VSV-G. Both of these help to provide stability to the vector by bringing together the particles that were made by the packaging plasmid pCMVAR9 (Adler *et al.*; Naldini, 1996).

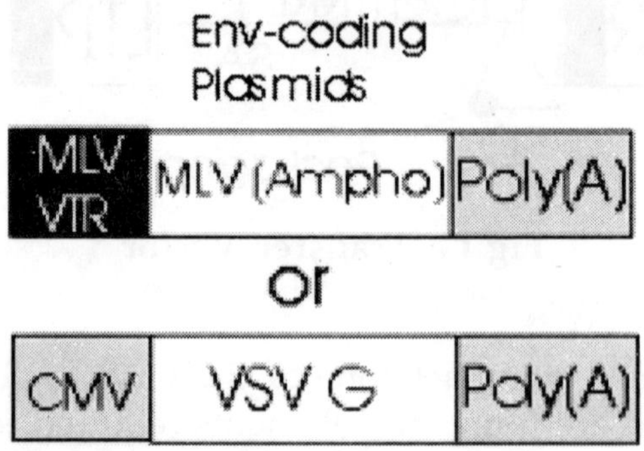

Fig I: Envelop Genes in the Third Plasmid

Scientists are challenged when making efficient packaging lines of HIV gene therapy vectors because expression of the VSV-G envelope and a number of HIV proteins is toxic to cells. They are dealing with this problem by designing vectors whose expression of the packaging genes and VSV-G can be turned on at will. Thus, the toxic genes can be turned off to produce more vectors without toxicity. This cell line can produce high titer vector without generating HIV vectors that can self-replicate and infect the patient with disease (Amado and Chen, 1999).

Use of HIV lentiviral vectors

Delivery into Patients' Target Cells

The HIV-based vector can be delivered directly into the body without *in vitro* manipulations of the patient's cells (Adler *et al.*). Additionally, lentiviral vectors have been shown to be superior to murine retroviral vectors. *Ex vivo* manipulations that activate stem cells with growth factors to induce cell division must be carried for the retrovirus to be able to enter the stem cells. However, it has been shown that *ex vivo* stem cell stimulation is not necessary with lentiviral vectors, so the vectors can be inserted directly into the patient and will find their way to the target cell (Amado and Chen, 1999).

Previous gene therapy using retroviral vectors required that cells be dividing, limiting therapy to proliferating cells *in vivo* or *ex vivo*. In the *ex vivo* method, the target cells are removed from the patient's body, treated to stimulate replication and then transduced with the vector before being returned to the patient. However, with lentiviral vectors there is no need for *ex vivo* treatment, and the target cells need not be dividing. The HIV-based vector is simply injected into a patient, upon which it seeks out its target cells based on cell membrane receptor proteins. Immune responses to the lentiviruses have not been found (Peel, 1998).

Uses for HIV Lentiviral Vectors

Scientists have recently been using the HIV lentiviral vector to repair neurons. HIV is also being developed as a delivery system to provide successful gene therapy in many diseases such as metabolic diseases, cancer, viral infection, cystic fibrosis, muscular dystrophy, hemophilia, retinitis pigmentosa, and maybe even Alzheimer's disease (Adler *et al.*; Naldini *et al.*; Amado and Chen, 1999; Planelles).

Concerns With Using HIV Lentiviral Vectors

There is still concern with using lentiviral vectors for safety reasons. One concern involves the possibility that the HIV could self-replicate and could be produced during manufacture of the vector in the packaging cell line or in the target cells by a process of recombination. Thus, the person undergoing gene therapy would also be infected with HIV in addition to the new therapeutic gene. A self-replicating infectious vector could cause cancer by inserting itself into the host genome and activate a neighboring proto-oncogene, thus causing mutagenesis (Amado and Chen, 1999).

Current research involving lentiviral vectors

Because scientists have shown that lentiviruses, such as HIV, are successful and efficient gene delivery vehicles, the field has now turned its attention to producing vectors with built-in safety features to prevent the development of replication competent lentiviruses (RCL). However, even the earliest studies with HIV lentiviral vectors did not generate RCL *in vitro* or *in vivo* (Amado and Chen, 1999), but precautions are still very important.

Safety Modifications

HIV lentiviral vectors are being produced whose packaging plasmid does not contain the necessary HIV genes. This does not interfere with efficient vector production and is a great increase in safety because potential RCL's cannot use the HIV genes necessary for replication of HIV in humans. The drawback to these vectors is that they cannot transduce macrophages because the accessory gene vpr is needed for HIV infection of this type of cells. Thus, scientists are showing that the type of lentiviral vector necessary is dependent on the type of cell chosen as target, so the HIV vectors will be made with different accessory genes (Amado and Chen, 1999).

Researchers at the Salk Institute are creating HIV lentiviral vectors that are self-inactivating. The scientists are working on packaging a defective HIV genome that contains only the necessary elements for gene transduction into a virion that has a broad host range. HIV normally targets human CD4 (helper T cells) through interactions with membrane-bound target proteins, but to broaden the host cell targets a surrogate targeting molecule (VSV-G) was inserted into the viral membrane. The HIV genome was modified to produce a minimal construct and the cytomegalovirus promoter and green fluorescence protein as a marker were added (Sikorski and Peters, 1998). A deletion in the LTR region at the end of the virus genome is also created. These are unique cis-acting sequences that are essential to the virus life cycle. The deletion inactivates the LTR promoter and eliminates the production of vector RNA. The gene to be transferred by the vector is expressed from an exogenous viral or cellular promoter that is inserted into the lentiviral vector. Inactivation of the promoter activity of the LTR reduces the possibility of insertional mutagenesis as the lentiviral products are integrated into the host genome. Also, as expression of the vector RNA is eliminated, the potential for RCL production in the target cell is further minimized (Amado and Chen, 1999).

Other safety methods include using specific internal promoters that regulate gene expression either temporally or with tissue or cell specificity so as to prohibit gene expression that would cause replication of HIV in the gene therapy target cell (Amado and Chen, 1999).

Use of Non-Human Lentiviral Vectors

By using non-human lentiviruses, scientists hope to bypass the issue of host infection by the gene therapy vector. Researchers are developing non-human lentiviruses such as the feline immunodeficiency virus (FIV) to be used in gene therapy (Amado and Chen, 1999). FIV infects two to twenty per cent of domestic cats worldwide and causes a disease similar to human AIDS. While humans have been exposed to this virus through cat bites, humans have never been shown to be infected by the virus. It has been shown that evolutionarily FIV diverged early on from HIV and other lentiviruses. Researchers at the University of San Diego, though, have found that while nonprimate lentiviruses may provide safer alternatives to HIV they have highly restricted host range of infection. However, promoter substitution of FIV enabled an env-deleted, three plasmid, human cell-FIV lentiviral vector system to express high levels of FIV proteins and FIV vectors in human cells. The researchers replaced the U3 element within the 5′ LTR of FIV with the human cytomegalovirus early gene promoter. Pseudotyped FIV vectors were shown to be able to efficiently transduced dividing, growth-arrested, and postmitotic human targets. The researchers also showed that human cells supported mechanisms of the FIV life cycle needed for efficient lentiviral vector transduction. It is the U3 element in FIV that is the only restriction to the productive phase of FIV replication in human cells. The researchers concluded that lentivirus-specific properties of FIV vectors are retained in human cells, and they speculate that eventually FIV vector will have advantages in human clinical use. Additionally, vectors derived from FIV may represent a safer alternative to HIV vectors, even those with deleted nonstructural proteins, because they cannot induce HIV-reactive antibodies in recipients. Overall, FIV has experimental advantages over HIV (Poeschla *et al.*, 1998).

Researchers at the University of North Carolina at Chapel Hill are working with equine infectious anemia virus (EIAV) to be used as a lentiviral vector in humans. EIAV is a lentivirus that normally infects horses, donkeys, and mules. It has been shown to be able to infect mature macrophages, and thus has the potential to infect quiescent cells, and has relatively simple genome organization. The researchers constructed separate plasmids encoding EIAV proteins, a viral envelop, and an EIAV vector. They attempted to broaden the host range of the vector to human cells by using non-EIAV enhancer/ promoter elements to drive expression and a non-EIAV envelope glycoprotein. They succeeded in transducing up to about 60 to 70 percent of human CFT1 cells which were placed in a culture dish. This is still quite a bit lower than the transduction level obtained using murine retroviruses, but more work with EIAV will hopefully increase the efficiency of this procedure. In addition, the fact that both EIAV-based and HIV vector can mediate gene transfer and expression to non-dividing human cells suggests that nuclear targeting mechanisms of equine and human lentiviruses are functionally conserved (Olsen, 1998).

Lentiviral Vectors for Hematopoietic Stem Cells

Many recent studies with lentiviral vectors have focused on modifying the hematopoietic stem cell which has the capacity to self-renew and to differentiate into all of the mature cells of the blood and immune systems. Thus, by introducing therapeutic genes into stem cells many diseases that affect these systems could be treated (Amado and Chen, 1999).

Liver-Directed Gene Therapy

Initial research aimed at delivering genes to the liver *in vivo* with HIV-based lentiviral vectors showed promising results, reported Ganjam Kalpana of Albert Einstein College of Medicine this year. This scientist developed a crippled version of HIV and used it as a vehicle for *in vivo* gene therapy on low-density lipoprotein receptor-deficient Watanabe heritable hyperlipidemic rabbits. A eukaryotic humanized gene fluorescent protein gene was cloned into the transfer vector to act as the reporter gene for successful cell transduction. The HIV vector was highly superior to previous methods of gene therapy using retroviral vectors which were highly invasive to the patient. There was also no host mediated cellular immune response to the lentiviral vector (Kalpana, 1999). This is another application to HIV-based gene therapy vectors that has been shown to be successful.

Therapy Against Retinitis Pigmentosa

Retinitis pigmentosa is an inherited genetic disease which causes the retina to degenerate leading to loss of visual field and night blindness. Genetic defects of photoreceptor cells of the visual system are the cause of this disease. A vector for gene therapy of retinitis pigmentosa should only target photoreceptor cells, which are located in the outer nuclear layer of the retina. Miyoshi, Takahashi, Gage, and Verma conducted an experiment using an HIV-based vector with a gfp-gene (green fluorescent protein) as a reporter. The vector was injected into rat retina. It was shown that the HIV-based vector did achieve long-term gene expression in the photoreceptor cells when a rhodopsin-promoter was used in the vector. This is only active in the photoreceptor cells, so the vector only targets these cells and not others in the retina. Thus, the researchers were successful in performing gene therapy on their rat patients (Schmidt, HIV as a Vector for Gene Therapy).

Gene therapy: potential applications in clinical transplantation

Gene therapy continues to offer much hope for the future treatment of a variety of clinical conditions. The development of tailored, novel gene transfer vectors will improve the efficiency and stability of therapeutic gene expression in the many settings of gene therapy. In the context of tissue and organ transplantation, gene therapy is being harnessed to prevent the acute and chronic rejection of transplanted tissues by introducing either new genes that are important in preventing rejection (e.g. co-stimulatory blocking molecules or immunosuppressive cytokines) or antisense nucleic acids to block the production of rejection-associated molecules such as adhesion molecules. The delivery of genes by gene therapy vectors that encode foreign donor antigens (alloantigens) might also be an effective means of inducing donor-specific unresponsiveness (immunological tolerance) in the recipient, perhaps eliminating the requirement for potentially harmful whole-body immunosuppression.

Study of diseases over several hundreds of years has resulted in the wide range of sophisticated techniques that is used today for diagnosis, treatment and research. During the 1960s, research to determine the underlying cause of diseases was limited to analysing the biochemistry of diseased cells and investigating various protein interactions. Although this research was valuable, scientists at the time lacked the specific technology and reagents to break down disease processes into their constituent parts in order to understand them more thoroughly. During the 1970s, restriction enzymes that cut DNA at specific sequences were first discovered and then used in molecular biology. Using these restriction enzymes to cut, remove and join genes together, researchers were able to start to understand the important roles that genetic factors play in disease. Currently, as the Human Genome Project nears completion, we can attempt to interpret the wealth of data available to us and thus make new links between particular genes and diseases. Once established, this information should accelerate the progress in the use of gene therapy as a therapeutic strategy.

Allograft rejection and immunological tolerance

For the foreseeable future, patients will continue to be transplanted with allogeneic organs which, by definition, are not identical at the major histocompatibility complex (MHC) locus. Without any other treatment, such as the administration of immunosuppressive drugs, the immune response generated by the recipient, which is mainly T-cell mediated, will reject such a graft. Self tolerance (the inability of self T cells to react against self tissue) is acquired as the immature T cells develop and pass through the thymus. This occurs because the majority of any potentially autoreactive T cells are 'negatively selected' by the process of clonal deletion, although clonal anergy (the existence of viable T cells that are unresponsive to antigen) and the generation of a population of regulatory T cells might have a role to play. The ultimate aim of transplant immunologists is to induce long-term transplantation tolerance to alloantigens in graft recipients. Such an immunological state would allow patients to continue to respond as normal to foreign antigens (e.g. bacteria, viruses and arising malignant cells), but tolerate, and not reject, a transplanted graft. In such an 'ideal' situation, there would be no need for systemic immunosuppressive drugs (with all the disadvantages that they bring), and transplanted patients would have a fully functioning, healthy immune system.

Gene therapy applied to transplantation

In 1958, the first recorded attempts to use DNA in transplantation research were carried out by Haskova and colleagues and Medawar, who were investigating whether the administration of DNA from a donor strain could result in immunisation to a subsequent transplant (that is its rapid rejection). In the experiments carried out by Medawar, DNA from the spleen of donor mouse strain A was purified and 5 mg injected into the peritoneal cavity of a previously unmanipulated recipient mouse (CBA strain). The recipient mice were transplanted 3–5 days later with skin from donor A strain mice and the grafts monitored over time. The grafts were rejected at the same rate as that in mice that had not received DNA, and no heightened response was measured. In another experiment by Medawar, following on from his success in injecting donor strain cells into neonates to induce transplantation tolerance, newborn mice were repeatedly injected with 'high doses' of donor strain DNA in an attempt to induce transplantation tolerance; however, this approach did not prolong the acceptance of skin grafts after subsequent transplantatation. The negative results of these early experiments were attributed by Medawar to impure DNA preparations and contamination with polysaccharides, although, with hindsight, quite different results could probably have been achieved if a different route of injection, such as intramuscular, had been employed, as was recently demonstrated by Geissler and colleagues.

Organ transplantation is now well established as a therapy for end-stage organ failure. Remarkable advances in the development of immunosuppressive drugs (e.g. cyclosporin, corticosteroids and rapamycin) have increased the survival rate of 1-year and 5-year renal grafts to 85% and 70%, respectively. Although this is an impressive achievement, healthy grafts can still be rejected, and the use of systemic immunosuppressive drugs is associated with significant risks, such as an increased risk of the development of cancers, infections and ischaemic heart disease, even in patients with long-term functioning grafts.

Gene therapy is a good strategy for approaching existing problems associated with transplantation, but will most likely be used only as a complementary approach. For example, grafts themselves could be targeted to reduce their immunogenicity by the introduction of genes to block T-cell activation, or donor-specific MHC antigens could be introduced into the recipient before transplantation to induce transplantation tolerance. Both strategies are potentially powerful and their prospects are discussed later.

Genes of interest in transplantation

MHC

The MHC is a highly conserved yet polymorphic gene locus. MHC molecules are surface proteins that present intracellularly processed peptides in a helical groove to their ligand, the T-cell receptor (TCR). Cognate interaction between an MHC molecule presenting peptide on an antigen-presenting cell and a specific TCR on a T cell can result in T-cell activation if the appropriate co-stimulatory molecules are present on the antigen-presenting cell. MHC class I molecules consist of three alpha domains and a b2 microglobulin chain, which is not encoded by the MHC gene locus. MHC class II molecules consist of two alpha domains and two beta domains. Peptides that are presented on the class I molecule are usually derived from intracellular proteins, whereas class II molecules present extracellularly derived peptides. The mechanism by which these peptides are transported to the immature MHC molecule is also very different for class I and class II MHC molecules. The MHC is the major identification molecule that triggers allograft rejection, because it determines the difference between self (syngeneic) and non-self (allogeneic). When searching for a suitable organ donor, it is the MHC antigens that are matched between donor and recipient, to give the graft as good a chance as possible of functioning. In defined situations, this potency of the MHC has been exploited to tip the balance of the immune system from immunity to tolerance. The exposure of the recipient of a graft to donor MHC antigens before transplantation to induce tolerance was first investigated in a mouse model by Billingham and colleagues in 1953, when cells from a donor strain were introduced into a recipient mouse *in vitro*.

Following this first attempt, and further studies, pre-transplantation blood transfusions (although not necessarily from graft donors) have been used in the clinic as a means of delivering MHC alloantigens before transplantation, but with limited success. However, the use of blood products also carries inherent risks, such as infections and transfusion reactions; thus, a novel therapy using a more specific approach would eliminate the risks of sensitising transplant recipients to alloantigens that are present in the blood. The delivery of donor genes to cells or tissues in a recipient would offer a highly specific therapy, one that is free from the risks associated with foreign cells and allows transplant recipients to be pre-treated with foreign genes before donor tissue becomes available.

The transfer of MHC genes is also useful in animal models to study the effects of allogeneic MHC antigens on the immune cells of a recipient without the influence of other alloantigens. Such an approach was first carried out by Madsen and colleagues, when a single MHC class I gene from a donor was transfected into a recipient-type mouse cell line and administered to a recipient. Not only was unresponsiveness to a subsequent cardiac allograft achieved in this study, but it showed that the recipient did not need to be exposed to all of the mismatched donor MHC molecules. Although these experiments proved that this strategy could work, transfected recipient cells are not a practical choice in a clinical setting. The next step was made by Wong and colleagues; bone marrow cells from recipient mice were transduced (infected with virus) *ex vivo* with an MHC class I gene using a retroviral gene therapy vector. This approach also resulted in long-term unresponsiveness to a fully allogeneic cardiac allograft, but rejection of a third-party graft, which had MHC class I genes that the recipient had not previously been exposed to.

Another interesting characteristic of MHC molecules is their ability to modify the immune response, depending on whether they are soluble or membrane bound. Observations following human liver transplantation have demonstrated that soluble donor human leukocyte antigens (HLA; human MHC antigens) are present post-transplant at high concentrations. It has been hypothesised that microchimerism (the co-existence of donor cells at very low levels within the recipient) of donor leukocytes alone causes the tolerant state; however, another equally valid explanation is that tolerance is caused by the large quantities of soluble MHC molecules that the liver naturally produces. Soluble donor-specific MHC class I molecules can have immunosuppressive effects, which can be used to promote graft survival in transplantation. Geissler and colleagues used a mouse model in which hepatocytes from a recipient strain were transfected using lipofectin with a plasmid encoding either membrane-bound or soluble allogeneic MHC class I molecules. Hepatocytes that expressed membrane-bound MHC class I molecules were found to prime cytotoxic T-lymphocyte (CTL) precursor cells, whereas exposure to soluble MHC class I molecules reduced the number (frequency) of CTL precursors. This idea has been reinforced by *in vitro* data from Zavazava and colleagues, which showed that soluble HLA class I molecules can induce apoptosis in human alloreactive CTLs.

Immunosuppressive cytokines

The delivery of genes that encode immuno-modulatory molecules to the site of the graft, or to the graft itself, has much scope for reducing the harmful local immune response against foreign tissue that occurs in acute and chronic rejection.

Cytokines are soluble mediators of the immune system, and some of them have immunosuppressive effects. The viral form of interleukin 10 (vIL-10) is a protein that is encoded by the Epstein–Barr virus; it is structurally homologous to mouse and human IL-10 but does not possess the T-cell co-stimulatory properties that IL-10 does. Thus, it is a useful tool in gene transfer to tissue where T-cell activation needs to be switched off or downregulated. DeBruyne and colleagues have demonstrated that gene transfer of vIL-10 to a murine cardiac allograft via vasculature perfusion using DNA–liposome complexes prolonged graft survival (16 days compared with 8 days for untreated grafts). The result was attributed to the vIL-10 gene, because treatment with either an antisense plasmid to vIL-10 or a monoclonal antibody targeted against vIL-10 reversed the graft-prolongation effect. Other cytokine genes, such as transforming growth factor beta (TGF-β), have also been shown to have a significant immunosuppressive effect. This type of approach is not intended to induce immunological tolerance, but might be useful for the delivery of local immunosuppression.

Blockage of co-stimulatory signal

In addition to the production of an intracellular first signal following specific TCR–MHC interaction, full activation of a T cell requires a second co-stimulatory signal, which can be provided by the interaction of CD28 and B7-1 or B7-2 (CD80 or CD86, respectively). Cytotoxic T-lymphocyte antigen 4 (CTLA-4; also called CD152) is an alternative ligand for CD80 and CD86, and is homologous with CD28. CTLA-4 is believed to play a role in the negative regulation of T-cell activation. The blockage of this co-stimulatory signal, for example using a fusion protein, has been shown in many murine and primate studies to inhibit cell-mediated and humoral immune responses *in vivo*. In one such study that used an adenoviral vector to deliver a CTLA-4Ig gene [a fusion protein comprising CTLA-4 and an immunoglobulin (Ig)] intravenously following cardiac allograft transplantation, the median survival time was increased from 6 days in the control group to 23 days in the group treated with the adenoviral vector expressing the CTLA-4Ig transgene. In another investigation by Chahine and colleagues, a transgene for CTLA-4Ig was transfected to both syngeneic and allogeneic mouse muscle precursor cells (myoblasts) and co-transplanted with allogeneic pancreatic islet cells under the kidney capsule of diabetic mice. Syngeneic myoblasts significantly prolonged the survival of the islets from 11 days to 31.7 days; no beneficial effect was observed for the transfected allogeneic myoblasts. The syngeneic myoblasts actively secreted CTLA-4Ig, to create local immunosuppression in the environment of the allogeneic islets, and thus allow them to function. When the myoblasts were allogeneic themselves, the MHC disparity with the recipient was enough to destroy them, thus preventing any CTLA-4Ig from being produced.

Genes associated with chronic rejection

Damage to an allograft can continue for years after transplantation, and despite improvements in immunosuppressive drugs and organ preservation, chronic rejection is still the most important factor in the failure of transplanted grafts. Histologically, during chronic rejection, smooth muscle cells are seen to proliferate in the vasculature of the transplanted organ, sometimes resulting in transplant atherosclerosis; several factors can contribute to this end point. Adhesion molecules such as intercellular adhesion molecule 1 (ICAM-1 or CD54) and growth factors such as vascular endothelial-cell growth factor are upregulated, and inducible nitric oxide synthase is imbalanced.

ICAM-1

ICAM-1 is a member of the Ig superfamily and is very important in both cellular adhesion and T-cell co-stimulation. Strategies to reduce T-cell activation by eliminating the effects of ICAM-1 have been carried out successfully in clinical trials involving renal allograft patients and the use of antibodies targeted against the ICAM-1 molecule. In a study of 18 patients, the anti-ICAM-1 antibody (BIRR1) was given to those patients who had received renal grafts from cadaver donors and were at a high risk for delayed graft function. An adequate level of BIRR1 in the serum (>10 µg/ml) was found to significantly reduce the incidence of both delayed graft function ($p<0.01$) and rejection episodes ($p<0.01$). This therapy has been developed further in murine models to use antisense oligonucleotides that are targeted against the messenger RNA (mRNA) for ICAM-1.

Nitric oxide

The intimal proliferation of vessels in transplanted organs is another indication of chronic rejection. To test the hypothesis that endothelium-derived nitric oxide is an endogenous inhibitor of vascular lesion formation, a Sendai virus virosome (see later section entitled 'Fusigenic virosomes') was used to deliver the endothelial-cell nitric oxide synthase gene *in vivo*. Von der Leyen and colleagues demonstrated that following disruption of the endothelium of rat carotid arteries, using a balloon-injury model, neointimal proliferation was reduced by 70% after transfer of the endothelial-cell nitric oxide synthase gene. A similar approach using a recombinant adenovirus to deliver inducible nitric oxide synthase to sites of arterial injury *in vivo* also demonstrated that the delivery of this gene could increase local levels of nitric oxide and thus prevent intimal hyperplasia.

Oxygen free radicals

Before transplantation, most solid organs are kept cold and without a full blood supply, which results in the effects of cold ischaemia. This, combined with reperfusion of the newly restored blood supply, can induce cell damage mediated by oxygen free radicals, which is thought to increase the chances of chronic rejection. To prevent serious damage, soluble superoxide dismutase (SOD) has been delivered to the graft *ex vivo* in perfusion solutions to 'mop up' free radicals. To date, few studies have been carried out using SOD in gene delivery. One investigation used recombinant adenoviruses encoding SOD (or catalase, which has similar effects) for use in oxidant-injury-related diseases. In this lung-perfusion model of rats, ischaemia-reperfusion injury was evaluated; surprisingly, the overexpression of SOD worsened ischaemia-reperfusion injury. The expression of both SOD and catalase transgenes prevented this increase in ischaemia-reperfusion injury but did not protect from it.

Inhibition of the proto-oncogene c-myb

The proto-oncogene c-*myb* is involved in the mitogen-induced proliferation of vascular smooth muscle cells, which constitutes a major pathway of atherogenesis. An increase in c-*myb* mRNA is seen after smooth muscle cell growth. The inhibition of c-*myb* has been reported to prevent the proliferation and migration of smooth muscle cells that are associated with the intima of blood vessels. Several studies have used antisense strategies or ribozymes to inhibit the proliferation of smooth muscle cells; however, these data are controversial, following reports that the effect might be non-specific and due to four contiguous guanosine nucleotides in the antisense oligonucleotide.

Ribozymes as enzymes to target mRNA

In 1981, Cech and colleagues revealed that proteins were not the only molecules capable of catalysing cellular reactions. Ribozymes are naturally occurring RNA molecules that act as enzymes and catalyse reactions on themselves or on other RNA molecules. Ribozymes were first isolated from the ciliated protozoan *Tetrahymena thermophila,* and subsequently became known as group I introns. The precursor ribosomal RNA (rRNA) molecule that is transcribed from the rRNA gene was found to be self-splicing. To initiate splicing, free guanosine nucleotide is bound to the specific site on the RNA molecule, the intron folds, cuts itself out and splices the exons together. Group II introns (which are found predominantly in fungal mitochondria) use extremely reactive adenine nucleotides within the intron sequence itself to initiate splicing. It soon became apparent that the modification of ribozymes could result in a powerful new molecular tool. Once the original group I introns had been modified to act on an external RNA substrate, ribozymes were then engineered to recognise specific nucleic acid sequences, so that they would cleave in a sequence-specific manner. Many different ribozymes are now in use, the most common being hepatitis delta virus (HDV) ribozyme, hammerhead (HH) ribozyme and hairpin (HP) ribozyme. The HH, HP and HDV ribozymes were all originally isolated in their self-cleaving forms and then modified to act on external substrates. The HP, HH and HDV ribozymes are relatively small, and thus easy to manipulate for gene therapy purposes. Delivery into target cells has been through (1) the direct injection of ribozyme RNA (the ribozymes were modified to increase their half-life), (2) the direct injection of oligodeoxynucleotides encoding ribozymes or (3) the delivery of ribozyme genes using a viral vector such as adenovirus.

The potential disease targets for ribozymes are abundant, including those conditions in which a gene product is overproduced, such as in cardiovascular disease (the genes encoding platelet-activating factor and fibrinogen), arthritis, cancer (oncogene function can be targeted) and many other immunological diseases.

Ribozymes targeting xenoantigens

Ribozymes also have a role to play in targeting genes that are involved in transplantation. In the xenogeneic setting (the transplantation of cells or tissues between species), the sugar epitope galα(1,3)gal is a major xenoantigen, which contributes greatly to the hyperacute rejection response that is characteristic of xenotransplantation. The enzyme that is involved in the production of this sugar, namely α(1,3) galactosyl-transferase, has been targeted using a ribozyme delivered by an adenoviral vector; complement-dependent cytotoxicity was significantly reduced, and would be expected to inhibit hyperacute rejection in subsequent studies, for example *in vivo* xenotransplantation.

Ribozymes targeting Fas ligand and perforin

Du and colleagues have characterised the ability of HH ribozymes to target perforin and Fas ligand (FasL) mRNAs *in vitro* as a therapy to prevent graft versus host disease

(GvHD), which can follow allogeneic bone marrow transplantation. Fas ligand is a cell-surface molecule that belongs to the tumour necrosis factor (TNF) receptor family. When it binds with Fas, a 'death signal' is sent to the Fas-bearing cell, resulting in apoptosis (also known as programmed cell death). Perforin, on the other hand, is a protein that can polymerise to form membrane pores; it is an important part of the killing mechanism used both by CTLs and by natural killer cells. Mice that have been transplanted with bone marrow from genetically engineered 'knockout' mice that do not express perforin or FasL do not develop GvHD. Such a genetic knockout approach cannot be applied to humans, thus the ability to destroy the perforin and FasL mRNAs in donor bone marrow using ribozymes could prove to be a useful method of reducing GvHD.

Ribozymes targeting inflammatory cytokines and chemokines

Another potentially useful target of ribozymes in the transplant setting is inflammatory cytokines, such as interferon gamma (IFN-γ) or tumour necrosis factor alpha (TNF-α), which are upregulated during allograft rejection. Chemokines, which can play a role in directing inflammatory cells to the site of a transplant and amplifying their effects, including RANTES (regulated upon activation, normal T-cell expressed and secreted), interferon-gamma-inducible protein (IP-10), macrophage-inflammatory protein 1 alpha (MIP-1α), macrophage-inflammatory protein 1 beta (MIP-1β) and monocyte chemoattractant protein (MCP-1), are all potent future targets for ribozyme gene therapy in transplantation.

Clinical implications/applications

To date, a total of 3464 patients have been enrolled in 534 human gene transfer protocols worldwide. Of these, 331 are for cancer therapy, and 71 for monogenic diseases, such as cystic fibrosis and ADA deficiency. Most of the gene therapy studies in the transplantation setting are still in the relatively early stages of development. That is to say that they have only been assessed in animal models or have used human tissue *in vitro*. It has been demonstrated, for example, that vIL-10 can be expressed in human islets *in vitro*. This success might lead to *ex vivo* gene transfer of grafted tissues and organs before transplantation, as a means of producing local immunosuppression at the site of the graft.

Research in progress and outstanding research questions

Most of the studies discussed in this chapter have been published recently; moreover, work in many of these areas is still being actively pursued. Although the ultimate aim in transplantation, the induction of donor-specific tolerance in clinical patients, is still a long way off, significant hope for the future is offered by advances in the induction of tolerance in animal models of allogeneic and xenogeneic transplantation. The development of pre-transplant microchimerism in transplant recipients, by the infusion of syngeneic bone marrow transduced with donor MHC class I or MHC class II genes, is an exciting and relevant approach, and advances in HSC isolation should improve the problem of GvHD.

With regards to gene therapy, the main focus of research is the development of new gene transfer vectors that are more efficient and safer, in an effort to overcome the major hurdles that are currently associated with clinical gene therapy. As discussed earlier, these new vectors are hybrids, incorporating the qualities of different vectors, such as targeting capability, strength and duration of transgene expression. The results of these studies are eagerly anticipated so that they can be applied in clinical gene therapy protocols.

Gene-therapy advances offer hope to chemotherapy patients

Recent advances in the design of gene-therapy vectors offer promise for cancer patients who must receive chemotherapy, according to an article published in Nature Reviews Cancer. In the June 2002 issue, Brian Sorrentino, M.D., director of Experimental Hematology at St. Jude Children's Research Hospital, reviews the latest developments in gene therapy and offers a roadmap for future research directions. Scientists and physicians are using gene-therapy approaches to address specific problems involving chemotherapy administration. One challenge is that many tumors are resistant to the chemotherapy drugs used to treat cancer. The drugs can also have a toxic effect on the bone marrow. This toxicity can increase the risk of infection or bleeding and can cause patients to need blood or platelet transfusions.

Researchers have been using gene transfer methods to introduce therapeutic genes into the hematopoietic (blood) stem cells of laboratory animals. These genes are intended to eliminate tumor drug resistance while protecting the normal bone marrow from toxic side effects. Because stem cells can renew themselves and develop into a variety of cell types, they are capable of passing on that protection to other cells in the body. The new cells will also be resistant to the drugs' side effects.

Scientists worldwide and at St. Jude are engineering different kinds of vectors—viruses that have been modified so that they can introduce therapeutic genes into specific cells. Numerous studies have shown that these vectors successfully introduce drug-resistant genes into the cells of laboratory animals. But researchers are still attempting to determine if these strategies can be used successfully in humans.

"We predict that in the next few years, we will be able to use novel chemotherapy regimens that overcome tumor cell resistance mechanisms while protecting patients against bone marrow toxicity through the use of modern gene therapy strategies," Sorrentino said. "This should allow us to expand therapeutic opportunities in cancer treatment by widening the gap between tumor cell killing and toxicity to the patient." Scientists are trying to perfect more efficient gene-transfer techniques so they can genetically modify more stem cells and obtain better responses to chemotherapy. Researchers are also working to create new vector systems and alternate ways of moving genes into stem cells. With additional research, these approaches may soon be widely available to people undergoing chemotherapy.

Gene therapy approaches to the treatment of experimental cancer are usually based on established neoplastic cell lines which are manipulated *in vitro* and subsequently transplanted in host animals. However, the relevance of these artificial models to the biology and therapy of human tumors is uncertain. As recently recognized, to be useful to humans, any new therapeutic approach in experimental animals must be evaluated with regard to its ability to interfere with tumor metastatization, a goal rarely, if ever, investigated at the experimental level. A transgenic mouse model of breast adenocarcinoma (MMTVneu) was developed. The similarities shared by transgenic tumors and their human counterpart include the slow growth in an immunocompetent host, the involvement of the neu (ErbB-2) oncogene in both neoplasias and the ability to metastasize to distant organs (See color plate: Fig 26). In addition, these tumors arise with high reproducibility in 100% of the females within the first three months, allowing an easy assessment of the effects of any novel therapeutic approach.

These mice were used to test several strategies for cancer gene therapy. An approach based on angiogenesis inhibitors was evaluated, alone or in combination with other traditional and non-traditional therapeutic tools. It was demonstrated that mouse model is clearly superior to those based on cultured cell lines and that transgenic models could be used for the study of therapeutic associations involving both conventional and novel strategies such as gene therapy.

Cancer gene therapy is an emerging approach that involves treatment of patients with genetic material to change the way the body deals with the cancer cells. One strategy is to replace or overcome the abnormal genes that cause cancer. VCU scientists have shown that targeting genes in this way facilitates killing cancer cells or making them more sensitive to radiation and/or chemotherapy. These laboratory experiments have served as the foundation for the development of clinical trials that apply this new approach to patients. Physicians in the Department of Radiation Oncology launched the first gene therapy clinical trial at VCU. This study involves the use of an anti-cancer agent called tumor necrosis factor (TNF) that is produced by our own bodies. VCU scientists have designed the agent in a manner that allows them to use radiation to direct the production of TNF only within the tumor. The TNF and radiation then work together to destroy the tumor cells (See color plate: Fig 27). Gene therapy can also be used to target the delivery of toxins or bolster the immune defenses.

Gene Therapy For Parkinson's Disease

A novel gene therapy technique is safe and may be effective at staving off worsening symptoms of Parkinson's disease, according to the first scientific review of a dozen patients who have received the treatment over the last three years. The results were published in Lancet. The patients, half of whom live on Long Island, are in advanced stages of the illness and were no longer responding to medicines when they signed on for the experimental therapy. The study was conducted at The Feinstein Institute for Medical Research and at New York Presbyterian Hospital-Weill Cornell Medical Center in Manhattan.

One woman and 11 men received a surgical infusion of fluid containing a viral vector and genes for a protein called GAD, glutamic acid decarboxylase. This enzyme is critical in controlling a neurotransmitter called GABA. In Parkinson's, GABA is reduced in an area of the brain called the subthalamic nucleus. This region is working on overdrive in the disease process and GABA is an inhibitory transmitter and is important in trying to calm this hyper-reactive circuit.

The gene therapy would be used to reduce symptoms and not alter the underlying disease process. Finding novel therapies are key as many Parkinson's patients stop developing complications after prolonged use of traditional medicines. Brain scans were taken before, during and after the treatment and the scans show that the brain is re-working these abnormal circuits. The patients had about a 27 per cent improvement in symptoms, although the study was an open label design. The scientists are now designing a double-blind placebo controlled trial that would enroll far more patients in an attempt to see whether the gene therapy is effective in reducing symptoms.

The patients' scans showed a quieting of these areas, on the side of the brain where the genes were infused. The study was designed to inject the genes into one side of the brain. Normally, Parkinson's patients have worsening symptoms on one side of the body. The novel strategy included packing genes that make an inhibitory chemical called GABA into pieces of viruses that have been rendered non-infectious. They began studying the experimental treatment in Parkinson's patients in 2003. Some patients continue to show improvement. Parkinson's patients have been willing to step up to the operating table for relief from the tremors, stiffness and rigidity that characterize the disease.

Decades ago, surgeons began to make lesions in parts of the brain involved in the disease, which lessened symptoms. Fetal stem cell surgery was pioneered in Parkinson's patients. And in the past decade, the deep brain stimulation has worked in as many as 70 per cent of patients who have opted for the surgical procedure. If it doesn't work, the electrodes can be removed. (By comparison, it would be impossible to reverse gene therapy.) Gene therapy could be a more natural way to treat the disease. This important study shows that gene therapy can be performed safely and may benefit patients.

Parkinson's is a movement disorder caused by a progressive depletion of the brain chemical dopamine in the substantia nigra. These dopamine-containing cells control movement. When 70 to 80 per cent of these cells are destroyed, a person develops the first symptoms of disease: tremors, slowed movement, muscle rigidity and problems with balance. The main medication used in Parkinson's is L-dopa, which replaces dopamine in cells that are still working normally. Over time, the cell death is so massive that the effects of the medication disappear.

Gene therapy for Parkinson's hitches ride on virus

The latest trial involves pairing an inactive virus known as AAV2 with a gene that causes nerve cells to produce neurturin, a protein that may improve the function of dopamine-producing cells and protect them against damage. Previous studies have shown that laboratory animals given the gene produce neurturin for at least a year, with no decline in expression. It looks like it is permanent once the gene is in there. Probably the neurturin will coax (nerve cells) back into full functioning.

AAV2-NTN is considered harmless in humans because genes that allow it to replicate have been removed. This is a virus that's not known to cause any human diseases explained. It's a virus that almost exclusively goes into neurons, so it targets neurons.

The study drug containing the gene-virus duo, or "gene transfer agent," is called CERE-120. The drug is administered during a surgical procedure in which injections are made directly into the putamen, the part of the brain where dopamine nerve terminals are deteriorating because of Parkinson's disease. Eight injections are made-four on each side of the brain-at different elevations to assure all of the putamen is covered with the agent.

Fifty-one people are being recruited for the study, including four to eight participants at OHSU. They will be assigned to one of two groups: One that receives treatment with the study drug; and one that will have a surgical procedure meant to simulate that used to give the study drug, but without the drug.

Two-thirds of the study group will receive the study drug. If results confirm the continued safety and effectiveness of the study drug following the conclusion of the trial, another study will be opened to allow subjects who received the placebo surgery to be treated with CERE-120. Participation will be made available to all subjects deemed suitable for the surgical procedure by the participant's doctor.

CERE-120 isn't a miracle drug that restores nerve cells destroyed by Parkinson's disease. It's just going to restore the function of dopamine neurons that have been injured by the disease, but not killed. So if the cell has been destroyed and reabsorbed into the body, it's gone and there's nothing which can be done.

Functional restoration could mean less severe symptoms when patients are off their standard Parkinson's disease medications. There will be clinically evaluating the signs of Parkinson's disease symptom changes, if there are any at all, and how the quality of life has improved. The trial also is a test of whether therapeutic genes can be delivered into the nervous system, made to function and beneficial against neurological disease. If this fails, there will be two questions: Did the gene really take, or scientists did not select the right gene?

Gene therapy being used in cardiovascular disease

To date, most gene therapy studies have been done in the laboratory and the earliest experiments seem promising for treatment of cardiovascular disease.

An example of these studies is the use of gene therapy to help increase blood flow to ischemic tissue. Ischemia is a condition in which the flow of blood, and thus oxygen, is restricted to a part of the body. Limb ischemia and myocardial ischemia refer to lack of blood flow and oxygen to the limb and heart muscle, respectively.

The body's first response to less blood flow to the heart is to grow tiny new "collateral" vessels to help blood flow around the blockage (angiogenesis). For unknown reasons, the process eventually switches off. Some proteins in the body can help trigger new blood vessel growth and so increase the oxygen supply to the ischemic tissue. Such angiogenic proteins include the endothelial growth factors, vascular endothelial growth factor (VEGF) and fibroblast growth factor (FGF) and hepatocyte growth factor (HGF).

In gene therapy trials, scientists have used a variety of different ways to deliver the genes for VEGF-1, VEGF-2 and FGF4 into the hearts of patients with advanced myocardial ischemia. After gene therapy, patients had less severe angina (chest pain) and their hearts worked better. Similarly, after gene delivery of VEGF to patients with limb ischemia, the blood supply improved and leg sores healed better. Gene therapy has prevented below-knee amputation in some patients for whom amputation had been recommended.

Gene therapy has also been successful in preventing re-occlusion, or re-blockage, of coronary artery bypass grafts and in keeping arteries open after angioplasty surgery.

Role of Gene Therapy in Heart Failure

Scientists are one step closer to making gene therapy for heart failure a reality.

Molecular biologist Walter J. Koch and his colleagues said in a report prepared for the 71st scientific sessions of the American Heart Association that they have for the first time delivered therapeutic genes throughout a rabbit's heart and have shown that the genes can increase heart function.

The animal experiments are a crucial step in developing a genetic treatment for congestive heart failure, a debilitating and deadly condition in which heart muscle loses its ability to stretch and contract, usually due to clogged arteries caused by coronary artery disease. People with congestive heart failure often experience fatigue, weakness, and an inability to carry out routine daily tasks. Right now, there is no effective means to reverse heart failure, only to treat symptoms.

According to the American Heart Association, about 400,000 new cases are recorded every year in the United States. Death rates from congestive heart failure tripled between 1974 and 1994, making it the leading cause of hospitalization among people 65 and older and costing more than $10 billion a year.

Koch, an associate professor of experimental surgery, and his colleagues have been working for several years to find ways to efficiently deliver genes to the heart to boost heart function.

Early experiments focused on identifying the molecular players that are responsible for efficient pumping action in the heart and showing which ones aren't doing their jobs in failing hearts. Using mouse models and sophisticated genetic techniques, the Duke scientists showed that two key proteins in heart cells work together to regulate heart function.

In diseased hearts, the body releases the hormone norepinephrine, the "fight-or-flight" hormone, directly into the heart, causing it to work up to five times harder than normal. Norepinephrine binds to molecules called beta adrenergic receptors (BARs) present on heart cells. This stimulation initially allows the heart to increase the power of its contractions, but in heart failure it quickly becomes self-defeating: the receptors become desensitized, meaning they no longer are able to respond to hormone stimulation. Desensitization is caused by a second molecule called B-adrenergic receptor kinase (BARK), which in healthy hearts helps restore heart contractions to normal after norepinephrine stimulation. Studies subsequently showed that BARK is elevated in failing human heart tissue.

In 1994, Duke researchers showed that mice genetically altered to produce excess beta adrenergic receptors (bARs) have supercharged hearts that beat faster and stronger than a normal mouse's heart. In 1997, the researchers inserted a gene that encodes the bAR into an adenovirus, the same virus that causes the common cold. When Koch and his colleagues allowed the virus to infect isolated failing rabbit heart cells, the bAR gene made up to 15 times the normal amount found in rabbit cells and restored normal heart signaling.

"Our studies have shown that a malfunctioning beta adrenergic receptor system leads directly to heart failure," Koch said. "By contrast, boosting levels of beta adrenergic receptors or inhibiting BARK can reverse heart failure in our mouse genetic models. Now, we are beginning to see the same result when we deliver these genes to rabbits."

The researchers used the results of the mouse studies to design their latest gene therapy experiments in rabbits. They increased the number of bARs by infusing a common cold virus genetically engineered to contain the bAR gene into a living rabbit's heart and allowing it to spread throughout the coronary arteries. Then they measured the ability of the heart to pump. Animals that received extra copies of bAR, which is equivalent to 10 times the normal number of bARs on heart cells, had significantly increased ability to pump, demonstrated by a 20 per cent increase in the force of the heart's contraction.

The key to the researchers' success, Koch said, was using a new surgical method to ensure that the adenovirus spread throughout the heart. They injected the virus into the left ventricle of live rabbits while the aorta was clamped for a few seconds. This allowed the virus enough time to spread through all the coronary vessels. Clamping the aorta is sometimes used in human heart surgery, Koch said, making this gene delivery method feasible.

In previous experiments, the researchers used a catheter similar to the ones used in opening blocked arteries in people to inject the virus into the coronary arteries, the arteries that feed the heart, in live rabbits. Using this method, Koch and his colleagues demonstrated that they could get genes into heart muscle and that the heart cells made the appropriate protein product, but only in a limited area of heart muscle.

The researchers are continuing experiments with a new generation of gene transfer agents such as the cold virus. But, he said, better gene therapy vectors need to be developed before gene therapy for heart failure in people becomes a practical option.

Gene therapy cures childhood blindness

Researchers at Bangalore for the first time have conducted the first clinical trial to test a revolutionary treatment for blindness in children (May, 2007).

The researchers say that the trial is the first of its kind and could have a significant impact on future treatments for eye disease. The trial involves adults and children who have a condition called Leber's congenital amaurosis (LCA), which is a type of inherited retinal degeneration.

This disease causes progressive deterioration in vision, due to an abnormality in a particular gene called RPE65. This defect prevents normal functioning of the retina, the light-sensitive layer of cells at the back of the eye. This results in severely impaired vision from a very young age and there are currently no effective treatments available.

The new technique that will be used in the trial involves inserting healthy copies of the gene into the cells of the retina to help them to function normally. Restoring the activity in these cells should restore vision. The operation delivers the normal genes to the retina, using a harmless virus or 'vector' to carry the gene into the cells. Previous work using animal models has demonstrated that this gene therapy can improve and preserve vision. During trials, the vision of dogs with the defect was restored to the extent that they were able to walk through a maze without difficulty; something they could not do before the treatment.

As this trial is the first to treat an eye disease using administration of gene therapy vectors to human retinas, the team has carried out extensive pre-clinical testing. The purpose of this trial is to find out how safe and effective the new intervention is in humans.

The team conducting the trial has been developing gene therapy for eye disease for almost 15 years but until now they have been evaluating the technology only in the laboratory. Testing it for the first time in patients is very important and exciting, and represents a huge step towards establishing gene therapy for the treatment of many different eye conditions.

There are many forms of retinal degeneration, meaning the use of gene therapy treatments must be individually developed then tested in a separate clinical trial specifically for that disease. However, the results from this first human trial are likely to provide an important basis for many more gene therapy protocols in the future, as well as potentially leading to an effective treatment for a rare but debilitating disease.

Not all defective genes necessarily produce detrimental effects, since the environment in which the gene operates is also of importance. A classic example of a genetic disease having a beneficial effect on survival is illustrated by the relationship between sickle-cell anemia and malaria. Only individuals having two copies of the sickle-cell gene, which produces a defective blood protein, suffer from the disease. Those with one sickle-cell gene and one normal gene are unaffected and, more importantly, are able to resist infection by malarial parasites. The clear advantage, in this case, of having one defective gene explains why this gene is common in populations in those areas of the world where malaria is endemic.

Gene Therapy for Age-Related Macular Degeneration

In its simplest form, gene therapy involves replacing a defective or absent gene. This approach is being used for the treatment of Leber's congenital amaurosis, which is due to a defect in the RPE65 gene. However, this simple gene replacement approach is not likely to be useful for the treatment of age-related macular degeneration (AMD), since this multifactorial disease is unlikely to be due to simple absence of a protein.

Current and future technologies allow elegant manipulation of the physiology of any micro-environment of the retina including the subretinal space. Gene therapy can be achieved by targeted cellular transfection, manipulation of inducible gene switches, the introduction of ribozyme technology that destroys or inactivates mutant proteins, and/or *ex-vivo* cellular manipulation with subsequent tissue transfer of transfected cells.

BASIC MECHANISMS OF GENE THERAPY

The basic strategies of gene therapy can be separated into gene replacement, gene silencing, or delivery of a gene to produce a protein locally. These genes must be delivered in such a way as to introduce the gene into the cellular genome such that the local effect may be elicited.

Replacing Absent or Defective Gene/Protein. The simplest application of gene therapy occurs when patients develop a disease due to complete absence of a protein product. Typically, this occurs in autosomal recessive diseases in which no protein is made, and absence of the protein leads to disease. The concept here is that a viral vector can be engineered to carry DNA that codes for the missing gene; the DNA is then inserted into the host DNA or maintained as a plasmid within the host nucleus such that transcription of healthy protein will occur. Several factors can be used to ensure the specificity of targeted gene delivery, including targeted vector delivery to a specific anatomic location, such as the vitreous cavity or subretinal space (See color plate:Fig 29a); the use of a cell-specific promoter region on the gene construct to limit gene expression to certain cells types (eg, using a rhodopsin promoter to limit gene expression to photoreceptors as a means of targeting these cells for photoreceptor-based treatments); use of viral vectors that have a high affinity for a particular cell type; or performing gene therapy *ex vivo* in explanted cells in culture, which can then be introduced into the subretinal space or vitreous cavity. In principle, introduction of a specific local source of protein production should be therapeutic if the disease is due to protein absence; the poster child for treatment of autosomal recessive retinal disorders is represented by recent groundbreaking pre-clinical studies of gene therapy in a dog model of Leber's Amaurosis. The role of gene replacement therapy in the treatment of AMD is not known at the current time, but it is unlikely that this disease is due to absence of a single protein.

Gene Silencing. This approach is favored for the treatment of disorders that are due to the presence of a mutant or aberrant protein; normally these disorders are present in diseases that are dominantly inherited. Typically, a normal protein product is present as well due to expression of a normal gene on the other allele, but disease is caused by abnormal protein caused by expression of an abnormal gene. Here the goal of treatment is to destroy or remove the abnormal protein. This can be achieved using RNA inhibition (RNAi and siRNA) or ribozyme technology; where these agents are capable of binding to a cleavage site of the target RNA resulting in disruption of its transcription and expression (See color plate: Fig 29b). Ribozymes or siRNAs can be delivered with viral or liposomal vectors.

Enhanced Production of Beneficial Protein. Another potential application of gene therapy is to introduce a gene that will result in an increase in the local concentration of a beneficial protein. Here, gene therapy would allow continuous expression of a protein that could be used to treat the disease. The best example of this approach in AMD is the recently reported phase 1 study using Adenoviral-vector delivered PEDF to suppress choroidal neovascularization in AMD. Other examples include the production of a soluble VEGF receptor (Flt-1) or delivery of growth factors in animal models of inherited retinal disease.

ETHICAL CONSIDERATIONS

While the body has many billions of cells, only a very small proportion of these cells are involved in reproduction, the process by which our genes are handed on to future generations. In males these cells are located in the testes and in females, in the ovaries. These special reproductive cells are called "germ cells". All other cells in the body, irrespective of whether they are brain, lung, skin or bone cells, are known as "somatic cells".

In somatic gene therapy, *only* somatic cells are targeted for treatment. So any changes to the genes of a person by gene therapy will only impact on the cells of their body and cannot be passed on to their children. Changes to the somatic cells cannot be passed on to future generations (inherited). Somatic Gene therapy treats the individual and has no impact on future generations.

An example of gene therapy

Imagine, for example, a little boy with haemophilia, a condition that is caused by a faulty gene that makes his liver unable to make blood clotting factor VIII. Gene therapy would involve putting a correct copy of the gene which codes for factor VIII into his liver cells so that his liver could then produce adequate levels of factor VIII. While the little boy himself would be cured, the altered genes in his germ cells would remain unchanged and he could still pass the faulty gene on to future generations.

Concerns with gene therapy of the eggs or sperm

The possible genetic manipulation of the egg or sperm cells (germ cells) remains the subject of intense ethical and philosophical discussion. The strong consensus view at present is that the risks of germ-line manipulation far exceed any potential benefit and should not be attempted.

Safety considerations

No therapy, established or experimental, is without some associated risk and the potential benefits of new treatments must always be balanced against such risks. The experience with gene therapy for the immunodeficiency condition SCID as described above illustrates the need for this therapy to be conducted as part of clinical trials. Safety will appropriately remain an important consideration as the field of gene therapy evolves.

Possible Effects of Gene Therapy

Gene therapy is presently conducted only through clinical trials, which often take years to complete. After new drugs or procedures are tested in laboratories, clinical trials are conducted with human patients under strictly controlled circumstances. Such trials usually last 2 to 4 years and go through several phases of research. In the United States, the Food and Drug Administration (FDA) must then approve the new therapy for the marketplace, which can take another 2 years.

The most active research being done in gene therapy for children has been for genetic disorders such as cystic fibrosis. Other gene therapy trials involve children with severe immunodeficiencies, such as adenosine deaminase (ADA) deficiency (a rare genetic disease that makes children prone to serious infection), and those with familial hypercholesterolemia (extremely high levels of serum cholesterol).

Gene therapy does have risks and limitations. The viruses and other agents used to deliver the "good" genes can affect more than the cells for which they're intended. If a gene is added to DNA, it could be put in the wrong place, which could potentially cause cancer or other damage. Genes can also be "overexpressed," meaning they can drive the production of so much of a protein that they can be harmful. Another risk is that a virus introduced into one person could be transmitted to other people or into the environment. Although the National Cancer Institute reports that such problems have not occurred in any of the human gene therapy trials to date, not everyone is excited about this new kind of treatment.

Gene therapy trials with some children present an ethical question (Linda Nicholson, MS, MC, a genetic counselor). Children are not tested to see if they are carriers of genetic diseases, because at this point, there's nothing which can be done about it. If a child knows he has a problem gene, there's no guarantee he will be affected by it, but he has to live with the knowledge for the rest of his life. For example, the gene for some types of breast cancer has been identified. But if a young girl is tested to see if she's a carrier for it, what will she do with that information.

Children may be tested for other disorders if there is a medical treatment or a lifestyle change that could be beneficial-or if knowing one doesn't carry the gene reduces the amount of medical surveillance, Nicholson explains. For example, finding out that a child doesn't carry the gene for a disorder that runs in his/her family might mean that he or she doesn't have to undergo yearly screenings or other regular exams.

Problems of gene therapy

For the safety of gene therapy, the Weismann barrier is fundamental in the current thinking. Soma-to-germline feedback should therefore be impossible. However, there are indications that the Weissman barrier can be breached. One way it might possibly be breached is if the treatment was somehow misapplied and spread to the testes and therefore would infect the germline against the intentions of the therapy.

Some of the problems of gene therapy include:

Short-lived nature of gene therapy–

Before gene therapy can become a permanent cure for any condition, the therapeutic DNA introduced into target cells must remain functional and the cells containing the therapeutic DNA must be long-lived and stable. Problems with integrating therapeutic DNA into the genome and the rapidly dividing nature of many cells prevent gene therapy from achieving any long-term benefits. Patients will have to undergo multiple rounds of gene therapy.

Immune response

Anytime a foreign object is introduced into human tissues, the immune system is designed to attack the invader. The risk of stimulating the immune system in a way that reduces gene therapy effectiveness is always a potential risk. Furthermore, the immune system's enhanced response to invaders it has seen before makes it difficult for gene therapy to be repeated in patients.

Problems with viral vectors

Viruses, while the carrier of choice in most gene therapy studies, present a variety of potential problems to the patient—toxicity, immune and inflammatory responses, and gene control and targeting issues. In addition, there is always the fear that the viral vector, once inside the patient, may recover its ability to cause disease.

Multigene disorders

Conditions or disorders that arise from mutations in a single gene are the best candidates for gene therapy. Unfortunately, some of the most commonly occurring disorders, such as heart disease, high blood pressure, Alzheimer's disease, arthritis, and diabetes, are caused by the combined effects of variations in many genes. Multigene or multifactorial disorders such as these would be especially difficult to treat effectively using gene therapy.

Chance of inducing a tumor (insertional mutagenesis)

If the DNA is integrated in the wrong place in the genome, for example in a tumor suppressor gene, it could induce a tumor (insertional mutagenesis). This has occurred in clinical trials for X-linked severe combined immunodeficiency (X-SCID) patients, in which hematopoietic stem cells were transduced with a corrective transgene using a retrovirus, and this led to the development of T cell leukemia in 3 of 20 patients.

Human Gene Therapy-Harsh Lessons, High Hopes

A 4-year-old girl named Ashanthi DeSilva from the suburbs of Cleveland lay on crisp white hospital sheets with a needle stuck in a vein. She didn't mind; this happened all the time in her chronically sick childhood. At the other end of the intravenous hookup hung a clear plastic bag of very special cells: her own white blood cells, genetically altered to fix a defect she inherited at birth.

A strikingly thin middle-aged doctor stared anxiously at the tiny figure. W. French Anderson, M.D., and his colleagues R. Michael Blaese, M.D., and Kenneth Culver, M.D., all then working at the National Institutes of Health, crossed a symbolic threshold with Ashanthi DeSilva that day, becoming the first group to begin a clinical trial in the new frontier of medical treatment: human gene therapy.

The reason for the excitement was simple: Most diseases have a genetic component and gene therapy holds the hope of curing, not merely treating, a broad range of ailments, including inherited diseases like cystic fibrosis and even chronic conditions like cancer and infectious diseases like AIDS.

At least, that's the theory

In the 10 years since that first genetic treatment on Sept.14, 1990, the hyperbole has exceeded the results. Worldwide, researchers launched more than 400 clinical trials to test gene therapy against a wide array of illnesses. Surprisingly, cancer has dominated the research. Even more surprising, little has worked.

There was initially a great burst of enthusiasm that lasted three, four years where a couple of hundred trials got started all over the world (Anderson, University of Southern California in Los Angeles). Then it was realized that nothing was really working at the clinical level.

And then things got worse

In September 1999, a patient died from a reaction to a gene therapy treatment at the University of Pennsylvania's Institute of Human Gene Therapy in Philadelphia. Jesse Gelsinger, an exuberant 18-year-old from Tucson, Arizona, suffered from a broken gene that causes one of those puzzling metabolic diseases of genetic medicine. An optimistic, altruistic Gelsinger went to Philadelphia to help advance the science that might eventually cure his type of illness. Instead, the experiment killed him.

In the aftermath of his death, there has been a flurry of activity to minimize the chance of future accidental deaths. The Food and Drug Administration, along with the National Institutes of Health, launched several investigations of the University of Pennsylvania studies and others. The inquiries provided disappointing news: Gene therapy researchers were not following all of the federal rules requiring them to report unexpected adverse events associated with the gene therapy trials; worse, some scientists were asking that problems not be made public. And then came the allegations that there were other unreported deaths attributed to genetic treatments, at least six in all.

Probably the clearest evidence of the system [to protect research subjects] not working is that only 35 to 37 of 970 serious adverse events from [a common type of gene therapy trial] were reported to the NIH as required (LeRoy Walters, the recently retired head of the Kennedy Institute of Ethics at Georgetown University and former chairman of NIH's Recombinant DNA Advisory Committee). That is fewer than 5 percent of the serious adverse events.

The news hit the clinical trial community like a thunderclap. The consequences have been immediate and wide-ranging, and may threaten future research. Participation in gene therapy trials was way down because the public was not sure what to make of this (Philip Noguchi, M.D., director of the Cellular and Genetic Therapy Division in FDA's Center for Biologic Evaluation and Research (CBER). Public wanted to know what the government is doing to help restore the confidence in this field.

Responding to the Crisis

The federal government moved quickly to do just that. FDA immediately shut down the trial in which Gelsinger had volunteered, and all clinical gene transfer trials at the University of Pennsylvania in January. The university went on to severely restrict the research of its once-high-flying gene therapy institute director James Wilson, M.D., announcing in May that all his work would be confined to animal and laboratory experiments and that he would be barred from conducting studies in people.

FDA also suspended gene therapy trials at St. Elizabeth's Medical Center in Boston, a major teaching affiliate of Tufts University School of Medicine, which sought to use gene therapy to reverse heart disease, because scientists there failed to follow protocols and may have contributed to at least one patient death. FDA also temporarily suspended two liver cancer studies sponsored by the Schering-Plough Corporation because of technical similarities to the University of Pennsylvania study.

Moreover, as nervousness spread through the field in the months after revelations about Gelsinger's death, some research groups voluntarily suspended gene therapy studies, including two experiments sponsored by the Cystic Fibrosis Foundation and studies at Beth Israel Deaconess Medical Center in Boston aimed at hemophilia. The scientists paused to review their studies and make sure they learned from the mistakes made at the University of Pennsylvania.

In March, the Department of Health and Human Services announced two initiatives by FDA and NIH. The Gene Therapy Clinical Trial Monitoring Plan is designed to ratchet up the level of scrutiny with additional reporting requirements for study sponsors. A series of Gene Transfer Safety Symposia was designed to get researchers to talk to each other, to share their results about unexpected problems and to make sure that everyone knows the rules.

In addition, FDA launched random inspections of 70 clinical trials in more than two dozen gene therapy programs nationwide and instituted new reporting requirements.

So far, the inspections only suggest that one other program appears to be in trouble, he says, but by the fall, We should be able to say accurately [what is] the state of the art of gene therapy and where it needs to improve (Noguchi, FDA).

Meanwhile, President Clinton announced more "new actions designed to ensure that individuals are adequately informed about the potential risks and benefits of participating in research ... and steps designed to address the potential financial conflicts of interest faced by researchers." In addition, the President said in May, "A new legislative proposal is being sent to Congress to authorize civil monetary penalties for researchers and institutions found to be in violation of regulations governing human clinical trials." If the legislation passes, FDA will, for the first time for drugs and biologics, have the power to essentially fine researchers and their institutions, up to $250,000 and $1 million respectively. This is a clear message that we intend to get serious (HHS Secretary Donna E. Shalala, Ph.D., May).

A History of Special Concern

Genetic engineering has always worried the general public.

When scientists first learned to clone genes in the mid-1970s, public reaction ranged from antipathy to hostility. Opponents, fearing that genetically engineered bacteria might escape from a laboratory, shut down the research at Harvard University and the Massachusetts Institute of Technology for months. Twenty-five years ago, in response to public concern, American scientists organized a voluntary moratorium on certain types of gene engineering experiments until safety questions could be resolved.

To help assuage public concern, NIH created its Recombinant DNA Advisory Committee, the RAC—which most simply call the "rack"—to provide a forum for genetic engineering debates to take place in public. As a result, the general opposition subsided.

But the RAC could do little if scientists didn't follow the rules. The promise of gene therapy, the glory of being the first to cure human ills, led at least one very smart scientist to make a very questionable decision. In 1980, an ambitious hematologist at the University of California at Los Angeles tested his gene therapy ideas on patients in Israel and Italy after being denied permission to perform the tests in Los Angeles. The experiments, conducted by Martin Cline, M.D., failed to help his subjects, and they violated federal rules designed to protect research subjects, leading to severe censure of the California scientist.

Ethical issues aside, the bigger problem for gene therapy has been basic biology. It's difficult to get new genes into billions of target cells within the body. Once inserted, the new genes need to function. Frequently, the body suppresses gene expression, essentially turning the new genes off, or destroys the transplanted genes. Although techniques have improved, today's scientists still face these challenges. To solve the problems, independent researchers have sometimes devised their own remedies of unknown safety. FDA began paying careful attention to these laboratory constructs when researchers began to request permission to test them in people under Investigational New Drug applications.

Early investigators were more mom and pop operations (Noguchi). They were individual investigators making their own products ... Almost all of them went on clinical hold because there was a lack of product information. Before FDA could allow them to proceed, technical questions about safety had to be answered, and that took time.

Typically, scientific questions are answered in laboratory and animal studies, but, with gene therapy, clinicians have been anxious to test their ideas in people. Once the NIH physicians treated their tiny patient in 1990, researchers rushed to get into the game with human trials. At the halfway point in the decade, the field was not progressing well. Then-NIH Director Harold Varmus, M.D., himself critical of the gene therapy trials in people, created a committee to review NIH's investment in the field. Varmus wanted to know whether NIH should continue to invest so heavily in the new technology.

The committee's conclusions were bleak:

While the expectations and the promise of gene therapy are great, clinical efficacy has not been definitively demonstrated at this time in any gene therapy protocol, despite anecdotal claims of successful therapy and the initiation of more than 100 ... approved protocols concluded the ad hoc committee co-chairmen Stuart H. Orkin, M.D., of Harvard Medical School and Arno G. Motulsky, M.D., of the University of Washington in Seattle in December 1995. While they saw promise, they also saw challenges. Significant problems remain in all basic aspects of gene therapy. Major difficulties at the basic level include shortcomings in all current gene transfer vectors and an inadequate understanding of the biological interaction of these vectors with the host.

To transfer a repair gene into a patient, the researchers must go through several steps. First, they must isolate the disease-related gene. Then it must be packaged in a vector, usually a disabled virus that cannot reproduce and cause disease, but that can act like a delivery truck to transport the gene inside the patient's cells. Once inside the body's cells, the new gene can begin to function and restore health.

But building an effective delivery truck hasn't been easy. Scientists started by using a type of mouse virus as a vector, engineered so that it cannot replicate itself, that easily infects human cells and integrates the new genes into the cell's chromosomes (structures in the cell that hold the genes). These mouse vectors, however, only infect dividing cells, so researchers switched to adenovirus, a type of human virus that causes the common cold. Because the adenovirus's own genes to reproduce itself have been removed, the remaining viral container is unable to cause an illness.

The Gelsinger Case

When Orkin and Motulsky reported on the technical limitations of gene transfer techniques five years ago, they virtually predicted problems in the clinic. During that same December meeting at which Orkin and Motulsky made their disheartening report, the RAC approved the University of Pennsylvania gene therapy trial for ornithine transcarboxylase deficiency (OTCD). FDA, too, allowed the study to proceed.

The treatment idea was fairly straightforward. OTCD occurs when a baby inherits a broken gene that prevents the liver from making an enzyme needed to break down ammonia. With the OTCD gene isolated, the University of Pennsylvania researchers packaged it in a replication-defective adenovirus. To reach the target cells in the liver, the Philadelphia scientists wanted to inject the adenovirus directly into the hepatic artery that leads to that organ. Some members of the NIH RAC objected, fearing that direct delivery to the liver was dangerous. Nonetheless, after a vigorous public discussion with the University of Pennsylvania researchers, the RAC voted for approval of the study.

At age 18, Jesse Gelsinger was in good health, but was not truly a healthy teenager. He had a rare form of OTCD that appeared not to be linked to his parents, but the genetic defect arose spontaneously in his body after birth. During his youth, he had many episodes of hospitalization, including an incident just a year before the OTCD trial in which he nearly died from a coma induced by liver failure. But a strict diet that allowed only a few grams of protein per day and a pile of pills controlled his disease to the point where he appeared to be a normally active teenager. With the encouragement of his father, Paul Gelsinger, Jesse volunteered for the study, and when he was initially evaluated, his medical condition qualified him to participate.

Gelsinger received the experimental treatment in September 1999. Four days later, he was dead. No one is really sure exactly why the gene therapy treatment caused his death, but it appears that his immune system launched a raging attack on the adenovirus carrier. Then an overwhelming cascade of organ failures occurred, starting with jaundice, and progressing to a blood-clotting disorder, kidney failure, lung failure, and ultimately brain death.

In its investigation, FDA found a series of serious deficiencies in the way that the University of Pennsylvania conducted the OTCD gene therapy trial, some more serious than others. For example, researchers entered Gelsinger into the trial as a substitute for another volunteer who dropped out, but Gelsinger's high ammonia levels at the time of the treatment should have excluded him from the study. Moreover, the university failed to immediately report that two patients had experienced serious side effects from the gene therapy, as required in the study design, and the deaths of monkeys given a similar treatment were never included in the informed consent discussion.

FDA's discussions with the university remain ongoing.

Signs of Progress

Not all the news about gene therapy is bad. It's true that dramatic cures have not been seen to date, but there are tantalizing signs that important advances may be just around the corner.

Ashanthi DeSilva, the girl who received the first credible gene therapy, continues to do well a decade later. She suffered a type of inherited immune disorder called Severe Combined Immune Deficiency, or SCID (pronounced skid), that left her susceptible to every passing microorganism. Without gene therapy, DeSilva would be living like David, the Boy in the Bubble, who had a similar disorder. Instead, the NIH researchers inserted a normal copy of the broken gene into some of her white blood cells, healing them, helping them function normally to restore her immune system. Cynthia Cutshall, the second child to receive gene therapy for the same disorder as DeSilva, also continues to do well.

Scientists, however, have discounted the benefit of the first gene therapies because the girls began receiving a new drug treatment that replaces the missing enzyme just before receiving the genetic therapy. And they continue to receive the drug after the genetic treatment, though gene therapy pioneer Anderson argues that since the drug dose has remained the same while their bodies have grown substantially over the decade, it makes a negligible contribution to their well being.

In April, French scientists reported convincing evidence that they successfully treated a different form of SCID (X-linked severe combined immune deficiency, the type suffered by the boy in the bubble) with gene therapy. Four of the first five babies treated by Alain Fischer, M.D., of the Necker children's hospital in Paris have had a complete or near complete recovery of their immune systems after the treatment.

Meanwhile, researchers at Children's Hospital of Philadelphia, Stanford University and Avigen, Inc., a biotech company in Alameda, Calif., have reported promising results in hemophilia B patients. The team packaged a gene for Factor IX, a blood clotting protein, in a defective adeno-associated virus (AAV). They then used the AAV to insert the gene into patients who suffered abnormal blood clotting because they lack Factor IX. Normally, these hemophilia patients needed to inject Factor IX to prevent uncontrolled bleeding. In June, the researchers reported treating six patients with the Factor IX gene therapy. Even though the dose of the gene therapy was so low that no one expected it to help, it reduced the number of injections of Factor IX that these patients used on an ad hoc basis.

The hemophilia studies are looking promising but will need further study to know whether it is an effective product (Noguchi, FDA).

These two studies suggest the power of genetic treatments.

We do seem to have turned the corner and there are a number of clinical trials that are starting to show success (Anderson).

Even as FDA increases its scrutiny of the field to ensure patient safety, there is a sense of advancement. There is good progress being made, Noguchi says. FDA thinks that gene therapy will work, but don't know for which disease. The recent events in France show that if you the right disease is known, and the right gene is inserted, one can obtain good results.

Gene therapy in the last 18 years

LeRoy Walters provided a valuable perspective on some of the lessons learned by scientists and ethicists over the 18 years since the first human gene therapy protocol was approved. He also offered his predictions for future gene-therapy interventions and discussed some associated ethical dilemmas that society may be facing.

Walters began his talk with two case studies. The first was about David, known as "the boy in the bubble." He was born in 1971 with X-linked severe combined immune deficiency and died 12 years later after receiving a bone marrow transplant that, unknown to doctors, carried a silent Epstein-Barr virus.

In contrast to David's story, Walters continued, is the story of Ashanti, who was born in 1986 with an autosomal recessive form of severe combined immune deficiency. In Ashanti's early years, every environmental microbe attacked her body and made her sick. She was treated with a synthetic enzyme called PEG-ADA, which gradually decreased in efficacy, and in 1990 she became the first patient to receive gene therapy in an approved protocol. She is now almost 13 years old and living a normal life.

In reviewing the history of gene therapy in the United States, Walters referred to a document prepared by an interdisciplinary group in 1984 and 1985. Called "The Points to Consider," it contained 110 questions that investigators were asked to answer as they thought about performing gene therapy on human patients. The questions covered such topics as gene therapy's potential benefits and harms, fairness in selection of recipients, procedures to be followed, recipients' privacy and confidentiality, and possible alternative therapies. The same questions could constitute a checklist for gene therapy today, Walters said.

The review process in the early days was transparent and public, a fact that was important to gene therapy's acceptance. Policymakers knew exactly what was happening, and any member of the public could attend a meeting, see the investigators, hear the questions, and have access to a public list of approved gene therapy protocols.

Walters stated that as of February 1998, 200 therapeutic protocols had been formally reviewed: 23 dealing with HIV infection or AIDS; 33 with single-gene diseases, especially cystic fibrosis; 138 with cancer; and 6 with other diseases. Reviewing what has been learned from the past 18years, he listed the following points:

- Somatic cell gene therapy has been successfully distinguished from more ambitious plans for human genetic engineering.
- The more neutral term "human gene transfer" might have been used, rather than "human gene therapy." "Therapy" seems to promise benefits to the patient; "gene transfer" covers even the PhaseIstudies that test a product's toxicity and are unlikely to be therapeutic to the subjects.
- The success of human gene therapy has been quite modest in the first 8years; unfortunately, some researchers and companies have overstated the early results.

- An optimum location will be needed for a national public review body to examine new biomedical technologies.

Looking to the future, Walters said he thinks we will see prenatal interventions to prevent severe and irreversible damage to fetuses and gene transfer to prevent or treat neurological disease. In studies affecting the brain, the question of what is enhancement and what is cure, treatment, or prevention of disease will arise in an acute form, he said. For example, is it remediation or enhancement to intervene so that a child would have an IQ of 100 instead of 60 or 70?

Walters predicted that, in the next 18years, proposals will emerge for germline genetic intervention, which will require a great deal of preliminary technical work. Instead of the current technologies of adding genes, something analogous to the "search and replace" function on a word processor will be needed to find the malfunctioning gene, splice it out, and replace it with the properly functioning gene.

He pointed out that there are some good moral arguments in favor of germline genetic intervention, whose goal is to prevent or alleviate disease or disability. Such intervention is more efficient than repeating gene therapy generation after generation, and even in utero gene therapy is too late for some diseases. The one case that could justify nuclear transfer in the early embryonic stage, Walters thought, is that in which a woman is likely to pass on a mitochondrial disease to her offspring. In such a situation, he said, after in vitro fertilization it would be justified at perhaps the four-cell stage to remove all the cells' nuclei and fuse them with enucleated egg cells from a donor. Because mitochondria are in the cytoplasm and would be derived from the donor, the resulting embryos would be free from mitochondrial disease. This type of case would involve simultaneous germline intervention and cloning in the technical sense.

Walters ended with a warning against repeating mistakes made in the time of the eugenics movement and the Third Reich. "We can applaud the war on disease that genetic research is waging. It will be a great day when a child is definitively cured of cystic fibrosis or when a particular family line is liberated from the burden of fragileX syndrome. But we will be humane warriors only if, in the midst of the battle, we also show respect for those who courageously cope with disability and for those who cannot yet be cured."

Gene Therapy Patient Tracking System

1. Rationale for a Gene Therapy Patient Tracking System (GTPTS)

FDA has a variety of patient safety programs and procedures in place for assessing and promoting the safety of investigational and approved new drugs and biological agents. For investigational products, FDA requires sponsors to report serious and unexpected adverse events associated with use of a product as soon as possible and within 15 calendar days. Sponsors must report any unexpected fatal or life-threatening experience associated with the use of the product as soon as possible, and within seven days. Prompt reporting allows the agency to analyze reports and take immediate action as indicated. Sponsors summarize other events in annual reports and information amendments that facilitate our periodic comprehensive overview of the effects of a product, including safety. FDA review teams either review related products or meet periodically to share information and discern trends.

Through experience in the review and regulation of gene therapy products, FDA has identified several concerns and issues raised by gene therapy products that differ from those typically raised by more traditional products. Because of the specific issues raised by gene therapy and because continuing and expanding research in gene therapy is creating new demands on oversight systems, FDA has pursued the development of a comprehensive Gene Therapy Patient Tracking System (GTPTS) to help ensure the appropriate oversight and safe development of these therapies.

Objectives

The objectives of the GTPTS are to supplement or replace current systems for assessing and promoting the safety of gene therapy so that the oversight system will be optimized for dealing with relatively specific gene therapy issues. We have identified five areas in which the GTPTS can improve upon pre-existing FDA systems for assessing and promoting product safety.

1. To facilitate collection and analysis of types of information unique to gene therapy, particularly information regarding long-term safety outcome.

Many of the potential adverse effects of concern in gene therapy patients are the same as those of concern for other therapies; however, gene therapy raises some concerns that are relatively unique. Perhaps the greatest area of concern is that of late-occurring toxicities. By permanently altering the genetic makeup of the recipient cells, some forms of gene therapy may cause toxicities that do not manifest themselves until years later. Additionally, some gene therapies use viral vectors with the potential to form latent infections that may emerge clinically years later. Thus, an important objective of the GTPTS is to provide a better system for collection and analysis of such information.

2. To facilitate analysis of safety trends across multiple, related products.

In large part, FDA safety systems are organized around individual products. Most traditional drugs and biologics are either unique or have a small number of closely related products of the same class that are likely to have related safety profiles.

In contrast, large numbers of gene therapy products under development by different sponsors share common features that may affect safety. For example, otherwise unrelated therapies may have closely related vectors, genetic inserts, or concomitant cellular therapies. As the field grows, databases that facilitate safety analyses across products that are related in different ways will lead to enhanced safety assessments.

3. To allow correlation of events by patient.

In current premarketing systems, annual reports and study reports frequently provide the numbers of patients having noteworthy outcomes, whether favorable or adverse. There may be no simple way for the Agency to determine whether or not, for example, one of the four patients experiencing event A was also the patient who experienced event B previously. For standard therapies, most adverse reactions occur over a relatively limited period of time and the need to follow outcomes by patient over time is small. In contrast, many gene therapies are intended to affect the physiology and to persist in the patient for extended periods. Linking early events to later events in the same patient may provide useful insight into the unintended biologic and physiologic properties of the agent.

4. *To facilitate preparation of periodic summaries of gene therapy safety information.*

Periodic review of aggregate gene transfer clinical trial data could help identify trends indicating potential areas of concern. Public discussion of summary safety information would promote awareness among gene therapy study sponsors, research investigators, and the general public of emerging issues in the medical, scientific, and ethical aspects of clinical gene therapy research.

5. *To bridge the gap from pre-market to post-market.*

Typically, the goals of safety oversight differ from the premarket to the postmarket situation and FDA uses different systems. At the time of marketing approval, FDA performs a comprehensive assessment of the premarket safety profile and the sponsor commits to address outstanding issues in post-market studies. Post-marketing surveillance is then designed to capture unanticipated events that may result with more widespread, less controlled use. With gene therapy products, there is a greater potential for delayed serious adverse events, and a need to conduct safety assessment across related products, including investigational and approved products. These factors support a patient tracking system that would coordinate the current premarket and postmarket surveillance systems.

What is the GTPTS?

The GTPTS is a system for the collection and analysis of information pertinent to the safety of gene therapy recipients. Far more than an adverse event database, it represents a comprehensive, integrated collection of procedures, policies, programs, databases and report structures pertinent to the conduct of studies; the collection of short-term and long-term outcome information; the transmission of information to FDA; the storage of information in electronic databases in an accessible and analyzable format; and the analysis and use of the information to make informed regulatory decisions and to increase the understanding of researchers, subjects, and the public. Critical to development of this system is an evaluation of the specific needs of gene therapy oversight with regard to what information should be collected; how best to collect that information; and how best to store, analyze, report, and use the data. One component of the GTPTS involves information technology and the need for adequate databases. No existing database satisfies the needs of the GTPTS. The National Institutes of Health (NIH) and FDA are developing a database application, the Genetic Modification Clinical Research Information

System (GeMCRIS), to facilitate the evaluation and analysis of human gene therapy clinical information. GeMCRIS will be one of the databases used in support of the GTPTS.

FDA has made substantial progress in the development of the GTPTS. To date, FDA has identified data elements regarding product, animal, toxicology, and clinical studies necessary for a patient tracking system database to support the analyses necessary to help ensure safety. We have developed an FDA standardized approach to product characterization and classification. We have initiated efforts to standardize product-testing requirements using appropriate contemporary methods. We have defined clinical outcome information pertinent to a safety assessment of gene therapy products. FDA and NIH have worked together to harmonize adverse event reporting practices. We have identified relevant long-term clinical patient follow-up data information and are developing strategies to collect, abstract, store, analyze, and report the results of these analyses. FDA has redirected educational and training resources to educate sponsors, investigators, and the public about gene therapy safety and product development advances. Abstraction into a temporary archival database of historical IND information from clinical trials has been initiated. FDA is implementing and evaluating procedures and enforcement tools to ensure compliance with data-reporting requirements. Plans are under development for periodic analyses and public discussion of gene therapy information.

This document outlines the GTPTS discussions and activities that have taken place to date including an evaluation of the relevant issues that the Center for Biologics Evaluation and Research (CBER) has identified through internal, interagency, and public discussions (e.g., discussion with the gene therapy industry, FDA advisory committees, NIH/Office of Biotechnology Activities (NIH/OBA), professional societies, and at numerous other public forums). This evaluation provides the basis for current and future development efforts described in the later sections that will help achieve the GTPTS objectives.

Human gene therapy and role of the food and drug administration

The U.S. Food and Drug Administration (FDA) is the primary government agency charged with protecting the health of U.S. citizens by ensuring that drugs, medical devices and biological products are safe and effective before they are used by doctors, nurses, other health care professionals, hospitals and consumers.

The FDA's authority includes any human gene therapy product sold in the United States. Because of FDA's strong involvement in the regulation of human gene therapy products, it is important for consumers to know what FDA is doing to protect them in this area, how FDA is doing it, and how the agency is responding to concerns that human gene therapy, instead of helping patients, may be dangerous and may put patients at risk.

FDA's Center for Biologics Evaluation and Research (CBER) regulates human gene therapies, which fall under the legal definition of a "biologic."

Manufacturers of gene therapy products must test their products extensively and meet FDA requirements for safety, purity and potency before they can be sold in the United States. A manufacturer who is considering selling a gene therapy product in the United States first must tell FDA of its intentions, and then must test the product in a laboratory and then in research animals. When a manufacturer is ready to study the gene therapy product in humans, it must obtain a special permission exemption from FDA before starting. This exemption is called an investigational new drug application or (IND). In the IND, the manufacturer explains how it intends to conduct the study, what possible risks may be involved and what steps it will take to protect patients, and provides data in support of the study. As part of the IND process, the manufacturer also must get approval from a committee of scientific and medical advisors and consumers (called an Institutional Review Board), which focuses on protecting persons who may participate in the study. Researchers also must inform the persons who may be part of the study about the study's potential risks and benefits, and obtain their consent.

When FDA's scientists receive an IND application for gene therapy, they review it carefully before permitting the manufacturer or researcher to begin the study. FDA may ask the study sponsor to do more laboratory tests and include more safeguards to ensure the safety of patients, such as giving patients smaller doses. As discussed in the next section, if unexpected problems arise, FDA may tell the manufacturer to change the study or stop it altogether.

FDA has not yet approved for sale any human gene therapy product. However, gene-related research and development is continuing to grow and FDA is very involved in overseeing this activity. Since 1989, FDA has received about 300 requests from medical researchers and manufacturers to study gene therapy and to develop gene therapy products. Presently, FDA is overseeing approximately 210 active IND gene therapy studies.

FDA functions to protect patients

FDA's number one priority is to have safe human gene therapy studies. FDA does not want to stand in the way of progress that may lead to effective therapies and cures, but the agency knows-as with any medical product or procedure-that new gene therapies may have unknown risks, even as they provide tremendous benefits to patients. Furthermore, FDA is aware of the public's concern about the safety of human gene therapy studies. Therefore, FDA wants to make sure that gene therapy products are as safe as possible, and that people now enrolled in gene therapy studies and those who eventually receive approved gene therapy products will not be harmed.

FDA has dedicated time and resources toward recruiting top-notch scientists, physicians, statisticians and other health professionals, all of whom are working hard to ensure that gene therapy products will be safe. These people are experts in the sciences associated with gene therapy, such as virology (the study of viruses), molecular biology (the study of living cells at the molecular level), pharmacology (the study of drug actions) and toxicology (the study of toxic or adverse reactions from drugs, chemical or other substances). These experts understand how to determine whether gene therapy studies are appropriate, taking into account the safety of the human subjects whom may participate in the studies. FDA staff are trained in how to properly analyze gene therapy study results. They are the people who decide whether or not a manufacturer may begin studying a gene therapy product and, ultimately, will decide if it is safe for human use.

A separate team of FDA professionals with expertise in manufacturing processes travels to a manufacturer's testing and production facilities to make sure that FDA's rules for making gene therapy products are being followed. And yet another team is responsible for reviewing any problems, or adverse events, that happen to patients who are part of a gene therapy study. If a problem is serious and unexpected, a manufacturer or researcher is required by law to notify FDA no later than 15 calendar days from when he learns of the problem. If a patient in the study dies or is at risk of dying, the manufacturer must notify FDA within seven days. At that point, FDA will either demand changes to the human study, require more non-human laboratory studies, put the human study on hold, or stop the study.

FDA works with other scientific experts-in hospitals, medical schools and other medical settings-to better ensure the safety of gene therapy products. FDA consults and receives advice from its Biological Response Modifiers Advisory Committee on scientific issues related to gene therapy products. FDA also works closely with other government agencies, such as the National Institutes for Health (NIH), the major public funding agency for laboratory and clinical research on gene therapy vectors and models, and co-sponsors many meetings with NIH to discuss issues related to gene therapy.

FDA and NIH have complimentary responsibilities with respect to the regulation of human gene therapy. FDA's primary job is to make sure that manufacturers produce high-quality and safe gene therapy products, and that these products are properly studied in human subjects. NIH's primary job is to evaluate the quality of the science involved in human gene therapy research, and to fund the laboratory scientists who invent and refine the tools used for gene transfer clinical studies.

The Recombinant DNA Advisory Committee (RAC), a public advisory committee to the Director of the NIH, holds regular meetings on gene transfer research, and focuses on the scientific, safety, and ethical issues involved. RAC members include physicians, scientists, medical ethicists, consumer activists, and private citizens, as well as *ex officio* members of the federal government, including FDA and the Department of Health and Human Services new Office of Human Research Protection (OHRP).

It's also important to remember that keeping patients safe is not only the government's responsibility. FDA also expects manufacturers and medical researchers to be responsible and to use the highest-quality experimental products, practice good clinical medicine, and accurately communicate information to FDA and to patients.

FDA's plans for the future

Human gene therapy and its safe application to patients will continue to be a high priority for FDA. As advances in our understanding of human gene therapy grow, FDA will use this information to make the best scientific, clinical, and regulatory decisions possible. FDA also will continue to place a high value on the public's concerns and take steps to see that they are addressed appropriately. To accomplish these goals, FDA will strengthen both its educational and enforcement programs for manufacturers, spend more time discussing gene therapy with its own staff, organize gene therapy-based research projects, and continue working with the NIH RAC.

NIH "points to consider" for gene therapy researchers

In anticipation of the first request to perform a human gene therapy experiment, the Human Gene Therapy Subcommittee prepared a document called *Points to Consider in the Design and Submission of Protocols for the Transfer of Recombinant DNA into the Genome of Human Subjects.* This document was approved by the NIH Recombinant DNA Advisory Committee and the Director of the NIH in 1986. The "Points to Consider" document provides guidance to physicians and scientists who are planning to submit proposals to the NIH for gene therapy treatment of patients. It describes the considerations that have been identified in the studies and hearings mentioned previously as the most important in evaluating this new mode of treatment.

In the "Points to Consider," researchers are first asked:

What disease do you intend to treat with gene therapy?

Why do you consider this disease to be an appropriate candidate for treatment with this new method?

In answering these questions, the researcher will discuss the seriousness of the disease, any alternative therapies, and the possible advantages of gene therapy for at least some patients.

Another part of the "Points to Consider" asks:

What laboratory studies have been done, with cells and live animals, that make researchers hopeful that gene therapy will help patients rather than harming them?

Here the researcher will provide the results of studies performed in his/her laboratory or in other laboratories around the world. Especially important will be studies demonstrating that gene therapy does not harm laboratory animals and in fact demonstrates that the desired biological effects occur.

Even if the preceding questions are satisfactorily answered, important questions about the proposed use of gene therapy in patients will remain. The "Points to Consider" ask the following four questions:

What are the probable benefits and harms of the proposed treatment, both to the patient and to others?

If there are several patients who need gene therapy but only one of them can be treated initially, how will selection be made in a way that treats all patients fairly?

How will patients—or, in the case of young children, the parents of patients—be properly informed about the possible benefits and risks of gene therapy?

What steps will be taken to protect the privacy of the patient and the patient's family, while at the same time informing the public about the results of gene therapy? In the Introduction to the "Points to Consider," reference is made to two possible undesirable or unintentional consequences of somatic cell gene therapy: transmission of altered genes to a patient's offspring, and viral infection of persons who come in contact with the patient. The Subcommittee requests that researchers describe what actions will be taken to prevent either event from occurring.

The "Points to Consider" acknowledges the public concern about some aspects of human gene therapy. It reads: "In recognition of the social concern that surrounds the general discussion of human gene therapy, the [Subcommittee] will continue to consider the possible long-range effects of applying knowledge gained from these and related experiments." For the moment, the Subcommittee agrees with the conclusion in the Office of Technology Assessment's report *Human Gene Therapy* that:

"Civic, religious, scientific, and medical groups have all accepted, in principle, the appropriateness of gene therapy of somatic cells in humans for specific genetic diseases. Somatic cell gene therapy is seen as an extension of present methods of therapy that might be preferable to other technologies."

While the RAC and its Subcommittee believe that gene therapy for non-reproductive, or somatic, cells holds promise for patients suffering from certain genetic and other diseases, they will seek to ensure that patients are not subjected to unreasonable risk of harm, excessive discomfort, or unwanted invasion of privacy and that they will receive special care, monitoring, and consideration. The public will be informed about every step that is taken with this new technique.

RECENT BREAKTHROUGHS

Major developments in gene therapy

2007

On 01 May 2007 Moorfields Eye Hospital and University College London's Institute of Opthalmology announced the world's first gene therapy trial for inherited retinal disease. The first operation was carried out on a 23 year-old British male, Robert Johnson, in early 2007, but it is currently too early for results.

2006

Scientists at the National Institutes of Health (Bethesda, Maryland) have successfully treated metastatic melanoma in two patients using killer T cells genetically retargeted to attack the cancer cells. This study constitutes the first demonstration that gene therapy can be effective in treating cancer. The study results have been published in *Science* (October 2006).

In May 2006 a team of scientists led by Dr. Luigi Naldini and Dr. Brian Brown from the San Raffaele Telethon Institute for Gene Therapy (HSR-TIGET) in Milan, Italy reported a breakthrough for gene therapy in which they developed a way to prevent the immune system from rejecting a newly delivered gene. Similar to organ transplantation, gene therapy has been plagued by the problem of immune rejection. So far, delivery of the 'normal' gene has been difficult because the immune system does not recognize the new gene and rejects the cells carrying it. To overcome this problem, the HSR-TIGET group utilized a newly uncovered network of genes regulated by molecules known as microRNAs. Dr. Naldini's group reasoned that they could use this natural function of microRNA to selectively turn off the identity of their therapeutic gene in cells of the immune system and prevent the gene from being found and destroyed. The researchers injected mice with the gene containing an immune-cell microRNA target sequence, and spectacularly, the mice did not reject the gene, as previously occurred when vectors without the microRNA target sequence were used. This work will have important implications for the treatment of hemophilia and other genetic diseases by gene therapy.

In March 2006 an international group of scientists announced the successful use of gene therapy to treat two adult patients for a disease affecting myeloid cells. The study, published in Nature Medicine, is believed to be the first to show that gene therapy can cure diseases of the myeloid system.

2003

In 2003 a University of California, Los Angeles research team inserted genes into the brain using liposomes coated in a polymer called polyethylene glycol (PEG). The transfer of genes into the brain is a significant achievement because viral vectors are too big to get across the "blood-brain barrier." This method has potential for treating Parkinson's disease. See *Undercover genes slip into the brain* at NewScientist.com (March 20, 2003).

RNA interference or gene silencing may be a new way to treat Huntington's. Short pieces of double-stranded RNA (short, interfering RNAs or siRNAs) are used by cells to degrade RNA of a particular sequence. If a siRNA is designed to match the RNA copied from a faulty gene, then the abnormal protein product of that gene will not be produced. See *Gene therapy may switch off Huntington's* at NewScientist.com (March 13, 2003).

2002 and Earlier

New gene therapy approach repairs errors in messenger RNA derived from defective genes. Technique has potential to treat the blood disorder thalassaemia, cystic fibrosis, and some cancers. See *Subtle gene therapy tackles blood disorder* at NewScientist.com (October 11, 2002).

Researchers at Case Western Reserve University and Copernicus Therapeutics are able to create tiny liposomes 25 nanometers across that can carry therapeutic DNA

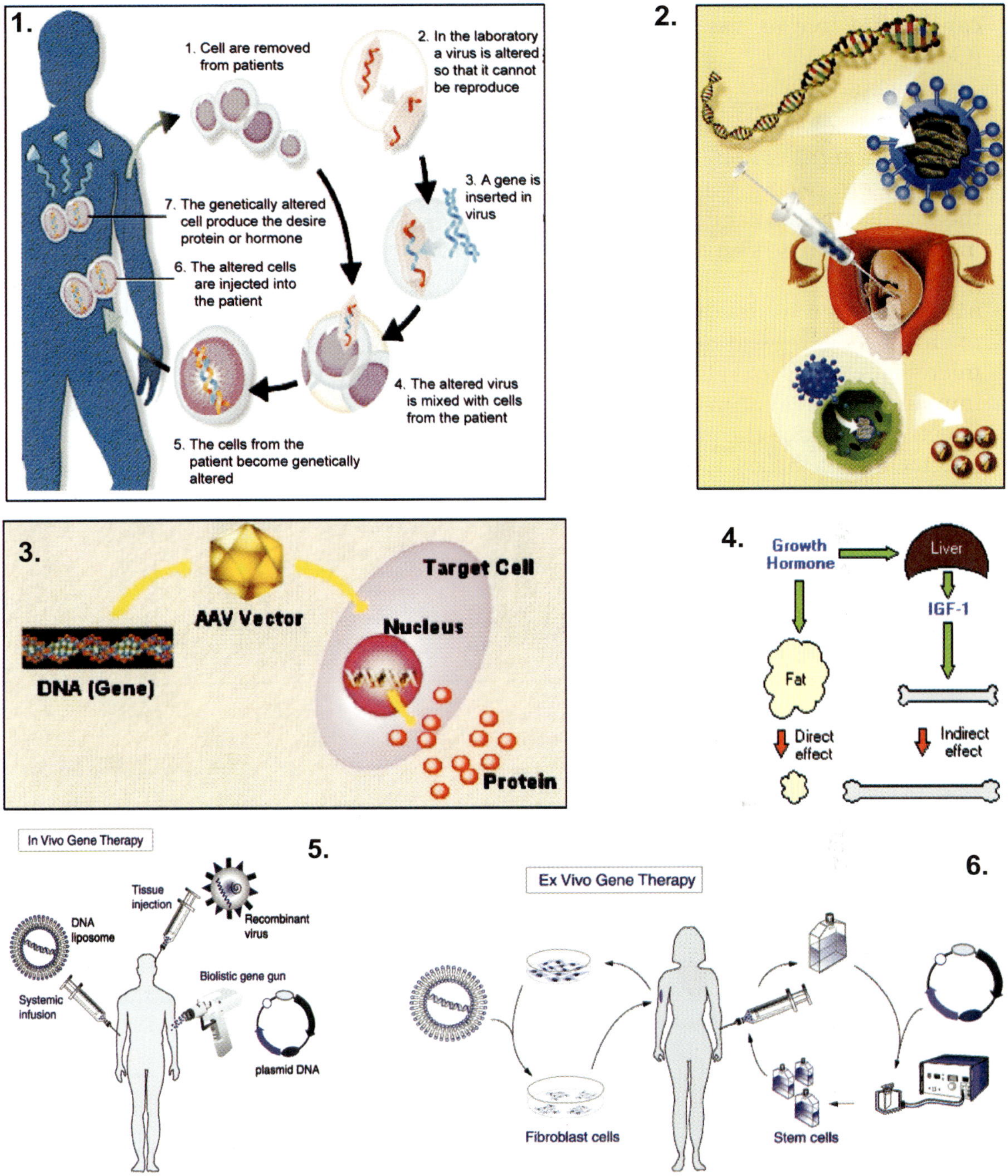

Fig. 1) The various steps in human gene therapy, **2)** Use of gene therapy at embryonic stage to provide healthy copies of gene in order to prevent disease, **3)** Use of Adeno-associated virus as gene therapy vector, **4)** Physiological effects of growth hormone, **5)** Various ways of inserting the healthy copy of gene for *in vivo* gene therapy in humans, **6)** Various ways of inserting the healthy copy of gene for *ex-vivo* gene therapy in humans

7.

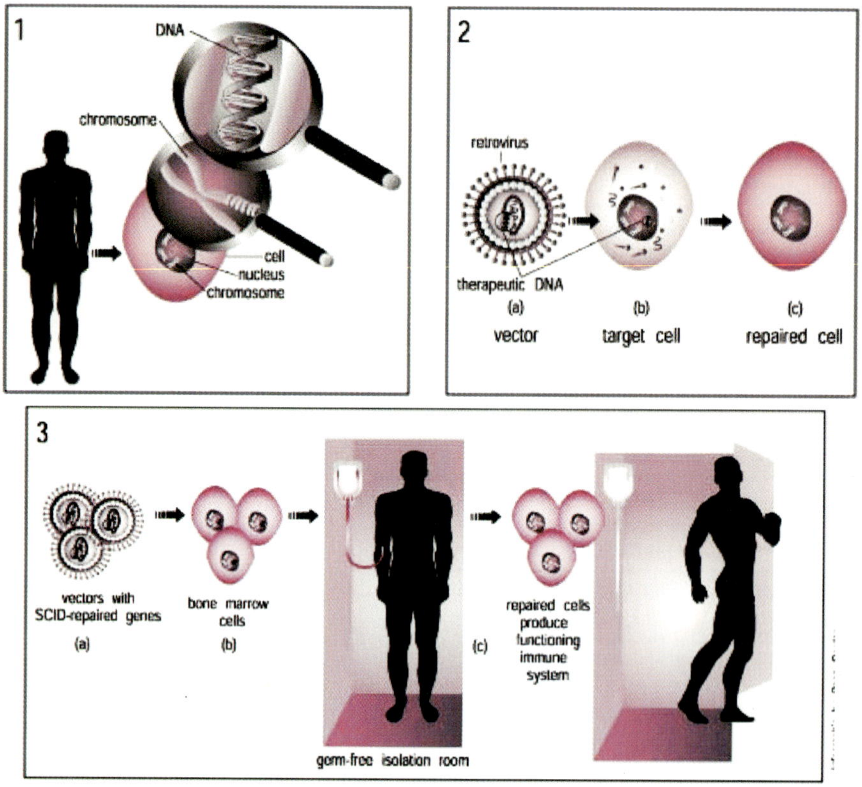

8.

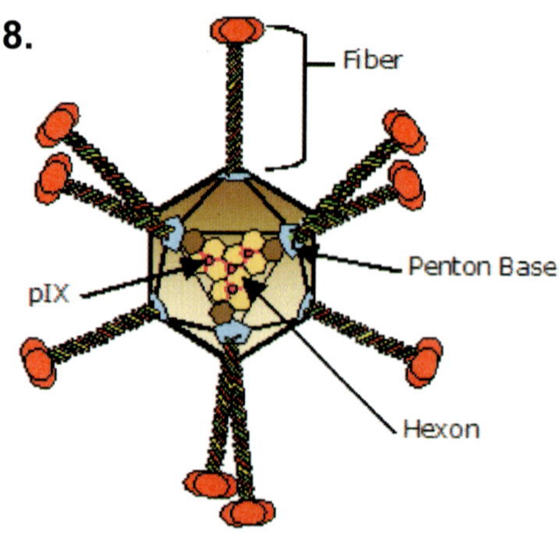

9.

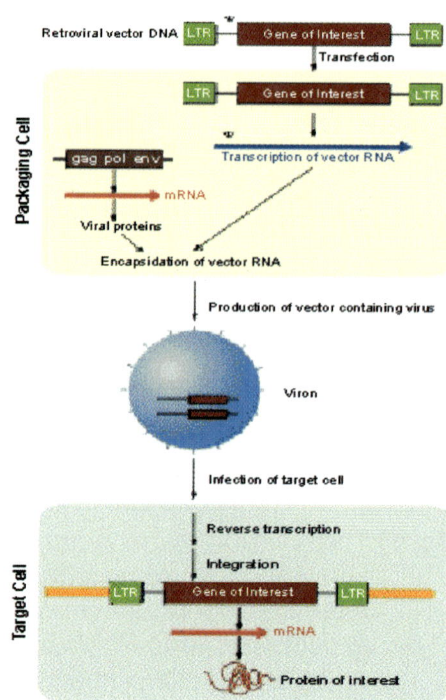

Fig. 7) The Gene therapy procedure being used for treating patient of SCID by using retroviral vectors with SCID repaired genes, **8)** Replication-defective Adenovirus based gene therapy vector, **9)** Replication - Defective Retroviral Gene Transfer Systems

10.

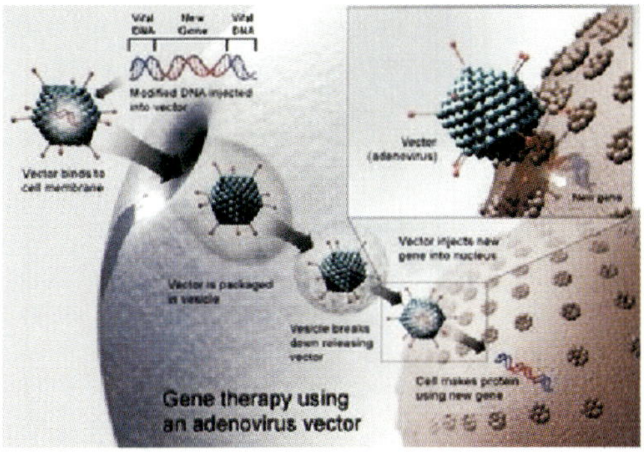

11.

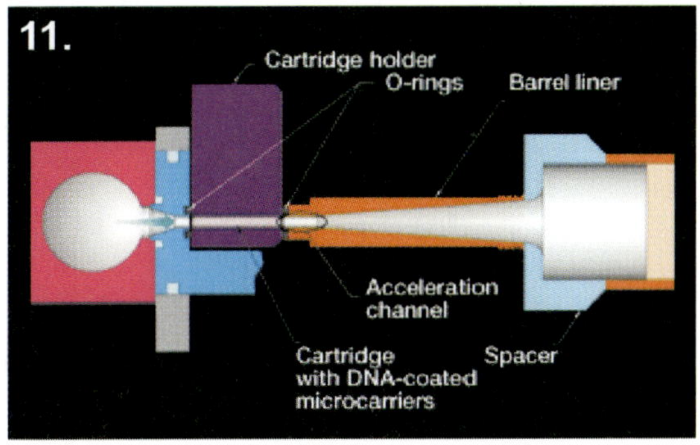

12.

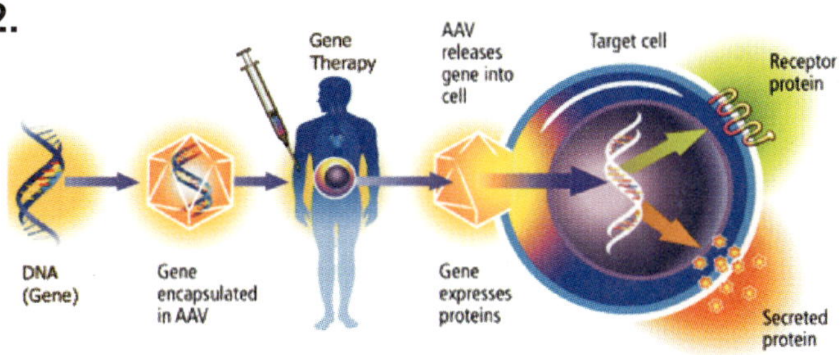

13.

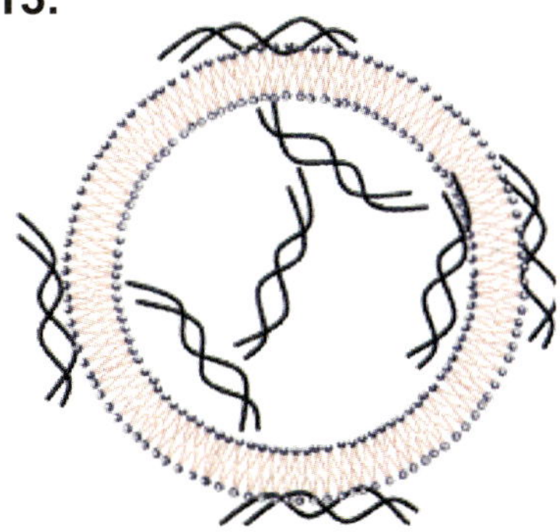

14.

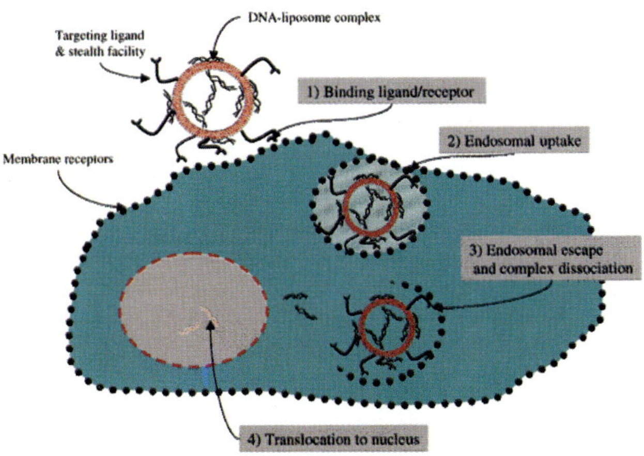

15.

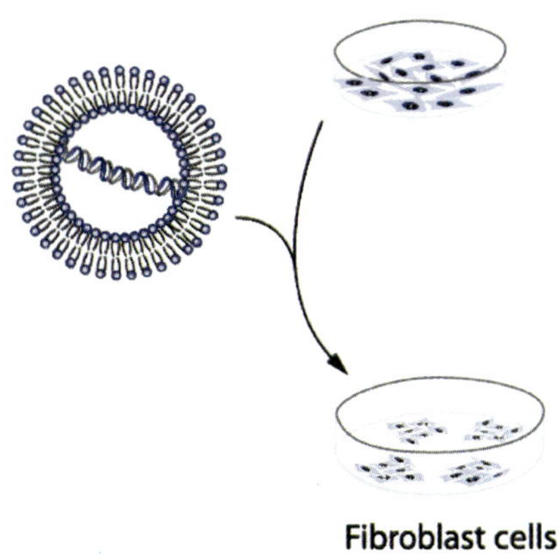

Fig. 10) Gene therapy using an Adenovirus vector, **11)** Use of Adeno Associated Viral Vector in gene therapy, **12)** Gene Gun for ballistic DNA delivery, **13)** DNA-liposome complex, **14)** Liposome mediated DNA transfer, **15)** Pathway used by plasmid-liposome complexes for transfer of gene of interest

16.

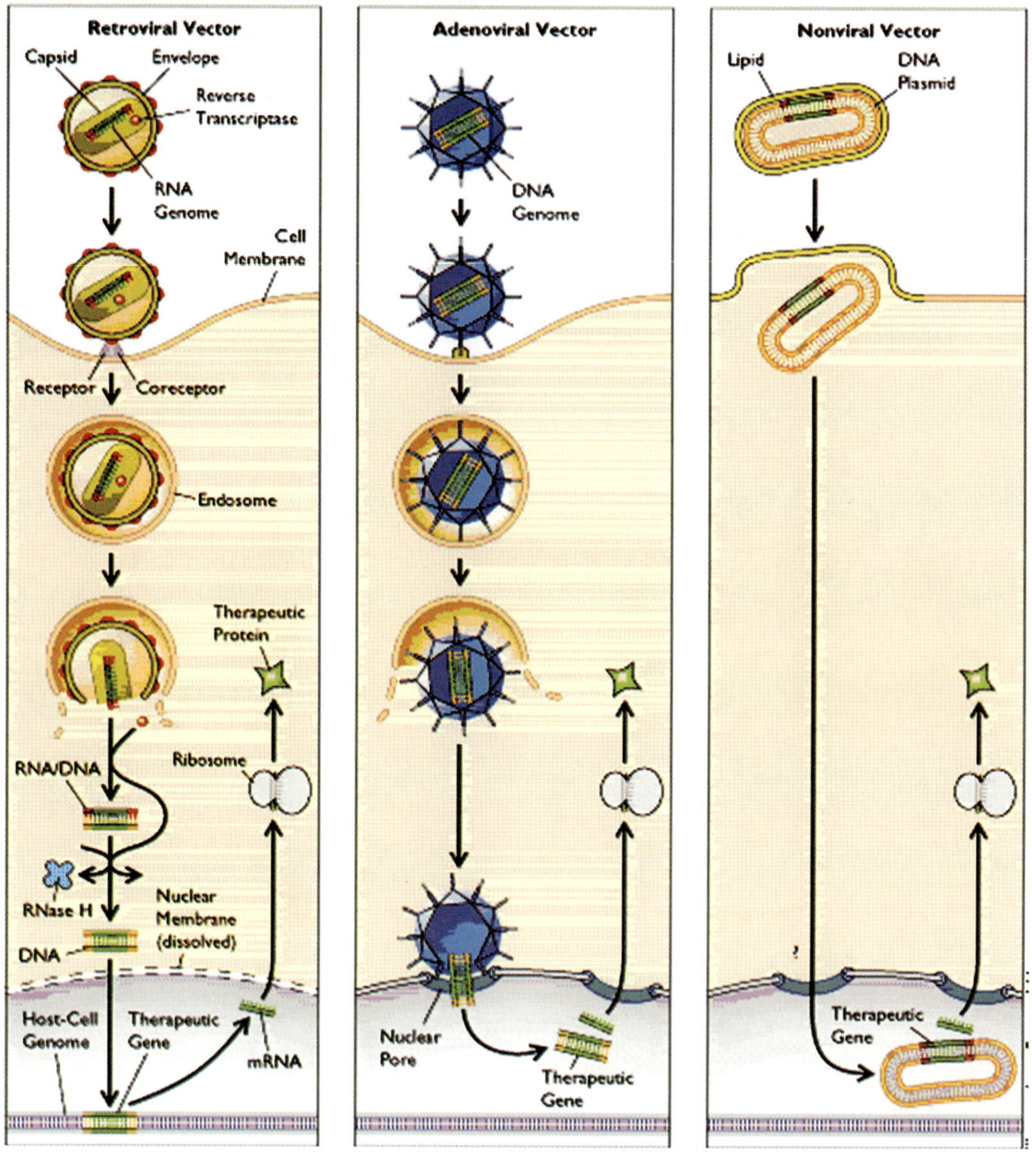

Fig. 16: Idiosyncrasies of a given viral vector arise from differences in viral life cycle. Retroviruses (left) and Adenoviruses (center) are alike in entering cells by binding to specific receptors. The Retroviral genome is, however, a single stranded RNA, requiring conversion into double-stranded DNA for insertion into the host cell genome. Unable to pass through nuclear pores, the genetic material gains access only when the nuclear membrane dissolves during mitosis. The adenoviral genome is itself double-stranded, which enters the nucleus at any time via formation of a capsid-pore complex. To be expressed, it does not require integration into host-cell DNA. Artificial vectors such as lipid coated DNA (right) cannot rely on natural means of entering cells. The vector's extracellular trafficking and even the details of its own structure are not yet well understood.

17.

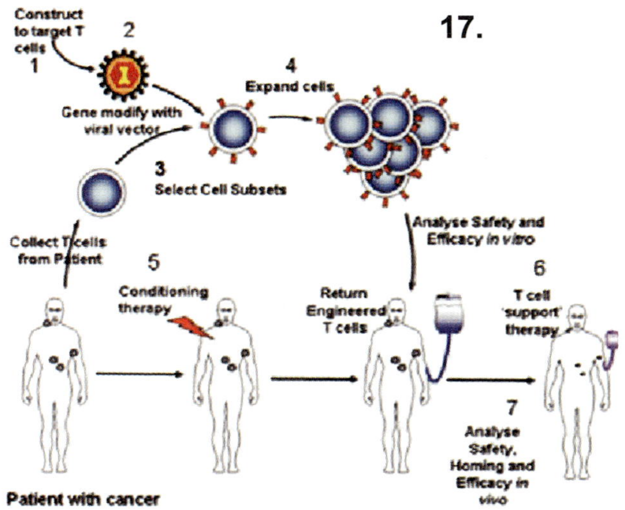

18.

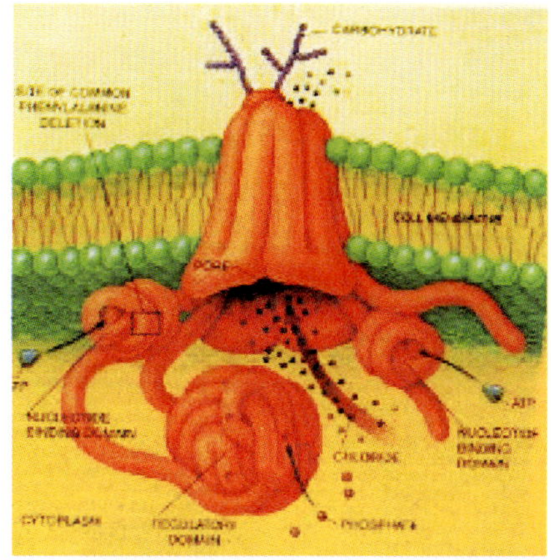

19.

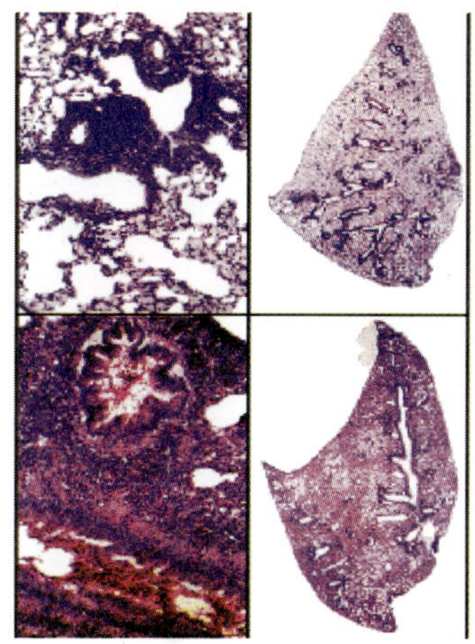

20.

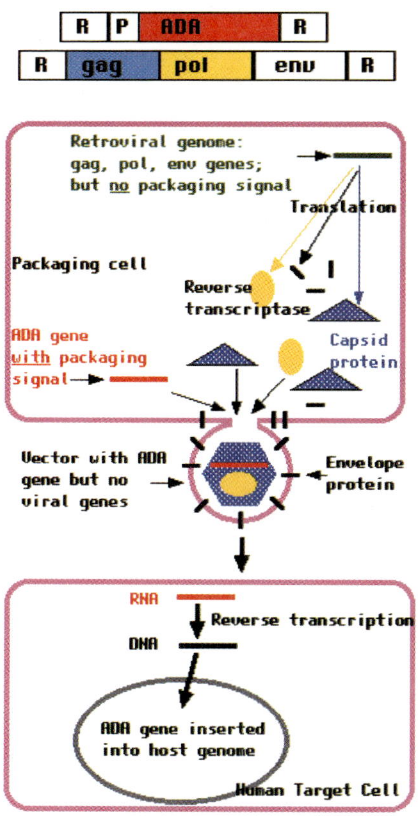

21.

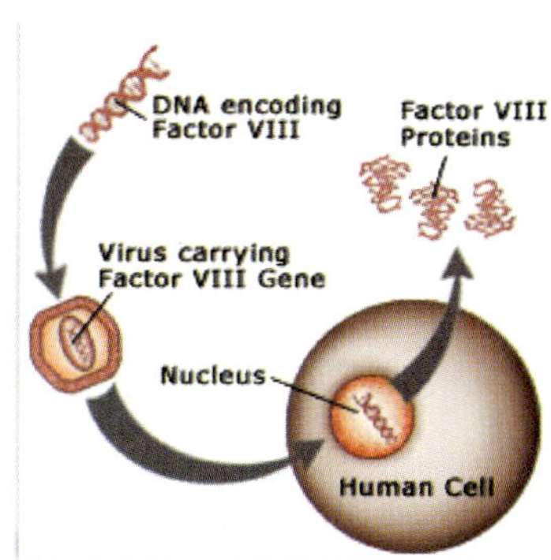

Fig. 17) The processes involved in Engineered T-cell therapy. The attack project will optimize the indicated aspects in model systems to inform future clinical trial design, **18)** Epithelial chloride ion channel functioning, **19)** Dehydration of affected tissues as a result of Cystic fibrosis, **20)** Transfer of ADA gene into human target cell, **21)** Use of viral based gene therapy for Hemophilia.

22.

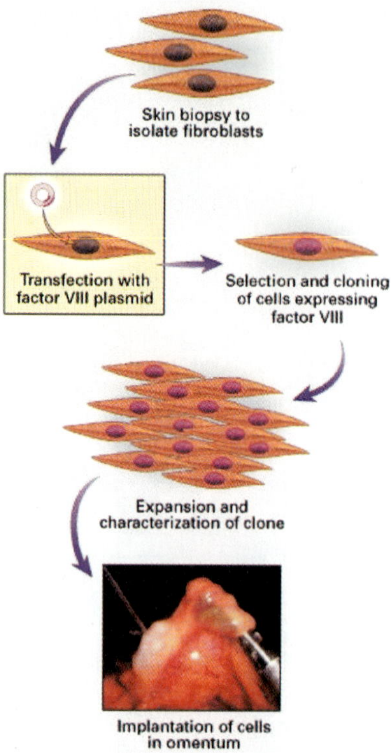

23.

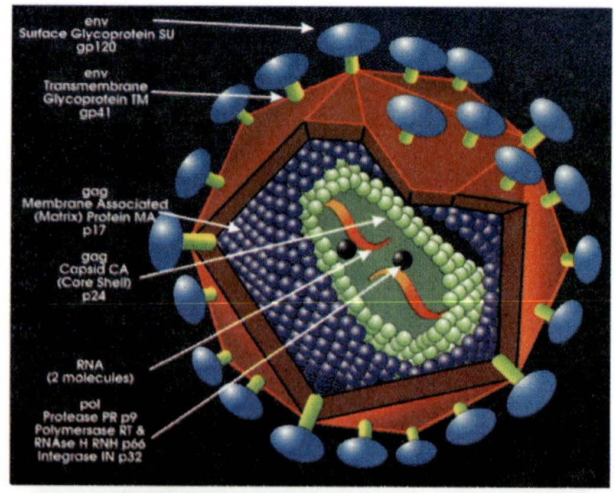

24.

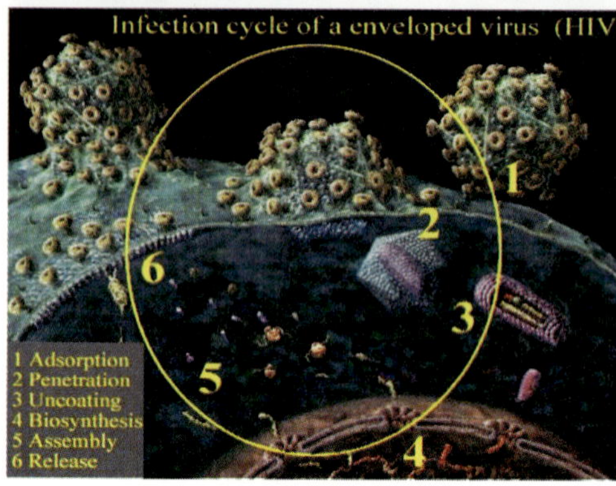

25.

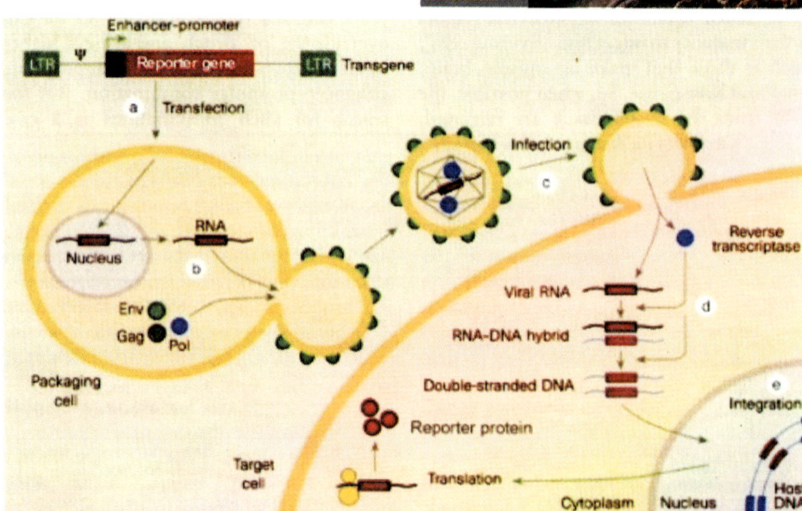

Fig. 22) Non viral procedure of gene therapy for Hemophilia.

23) Structure of HIV virus.

24) The infection cycle of HIV in human cell.

25) HIV based gene therapy vector

26.

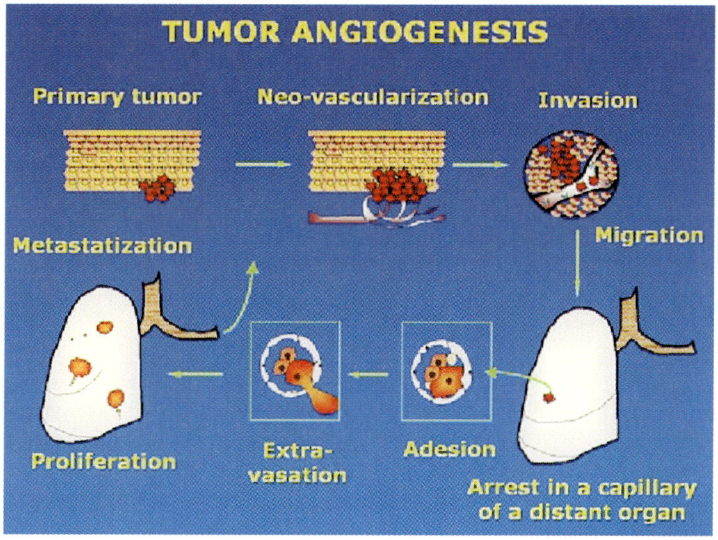

27.

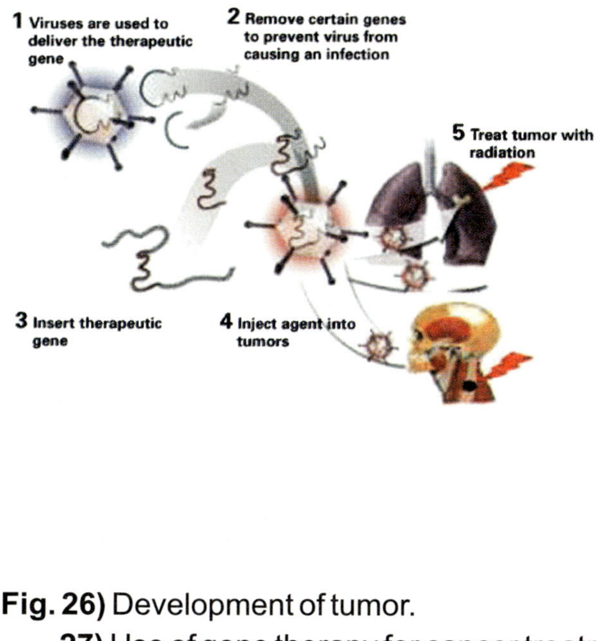

28.

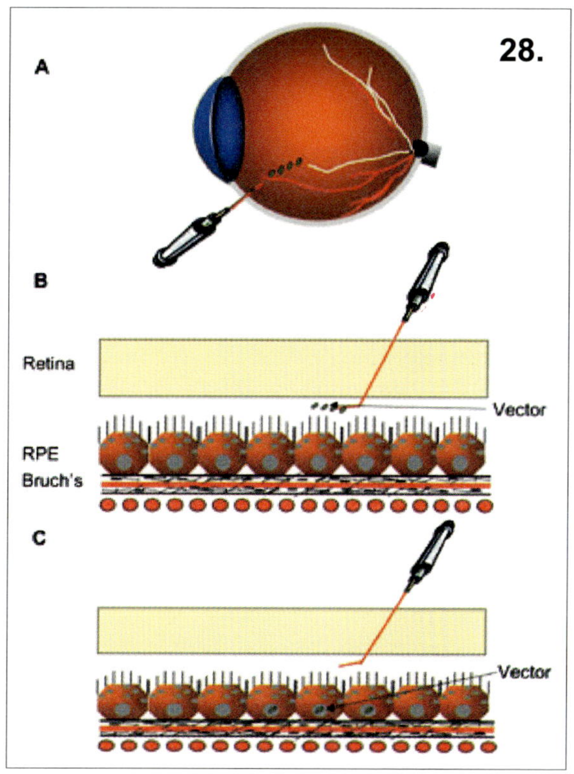

Fig. 26) Development of tumor.

27) Use of gene therapy for cancer treatment.

28) Ocular gene therapy: Intravitreal and subretinal gene-therapy delivery **A).** Intravitreal injection of vector. **B).** Subretinal vector injection. **C).** With time, the virus vector infects the RPE cells and the transmitted gene inserts into the host genome.

through pores in the nuclear membrane. See *DNA nanoballs boost gene therapy* at NewScientist.com (May 12, 2002).

Sickle cell is successfully treated in mice. See *Murine Gene Therapy Corrects Symptoms of Sickle Cell Disease* from March 18, 2002, issue of The Scientist.

The success of a multi-center trial for treating children with SCID (severe combined immune deficiency or "bubble boy" disease) held from 2000 and 2002 was questioned when two of the ten children treated at the trial's Paris center developed a leukemia-like condition. Clinical trials were halted temporarily in 2002, but resumed after regulatory review of the protocol in the United States, the United Kingdom, France, Italy, and Germany. (V. Cavazzana-Calvo, Thrasher and Mavilio 2004; see also *'Miracle' gene therapy trial halted* at NewScientist.com, October 3, 2002).

In 1993 Andrew Gobea was born with a rare, normally fatal genetic disease-severe combined immunodeficiency (SCID). Genetic screening before birth showed that he had SCID. Blood was removed from Andre's placenta and umbilical cord immediately after birth, containing stem cells. The allele that codes for ADA was obtained and was inserted into a retrovirus. Retroviruses and stem cells were mixed, after which they entered and inserted the gene into the stem cells' chromosomes. Stem cells containing the working ADA gene were injected into Andre's blood system via a vein. For four years T-cells (white blood cells), produced by stem cells, made ADA enzymes using the ADA gene. After four years more treatment was needed.

The Future of Gene Therapy

To cure genetic diseases, scientists must first determine which gene or set of genes causes each disease. The Human Genome Project and other international efforts have recently completed the initial work of sequencing and mapping virtually all of the 25,000 to 35,000 genes in the human cell. This research will provide new strategies to diagnose, treat, cure, and possibly prevent human diseases.

Although this information will help scientists determine the genetic basis of many diseases, it will be a long time before diseases actually can be treated through gene therapy. The Human Genome Project has located genes for us, but it's not going to tell us what these genes do (Nicholson). That will be the next step. Once we have that information, we'll be able to take advantage of that knowledge to provide treatment and/or cures.

Gene therapy's potential to revolutionize medicine in the future is exciting, and its expectations for curing and preventing childhood diseases is encouraging. One day it may be possible to treat an unborn child for a genetic disease even before symptoms appear.

Scientists are hoping the mapping of the human genome will lead the way toward cures for many diseases and that the successes of current clinical trials will create new opportunities and challenges. For now, however, it's a wait-and-see situation, calling for cautious optimism.

THE LAST WORD

Researchers React to Gene Therapy's Pitfalls and Promises

The possibility of correcting human genetic disorders by gene therapy caught the imagination of the scientific community as well as that of the public long before the first clinical gene transfer experiment was launched ten years ago. During the ensuing decade, however, the extraordinarily high level of expectation has proven to be grossly optimistic. This prompted Dr. Harold Varmus, then Director of the National Institutes of Health, to appoint two separate committees in the mid-1990s to evaluate the field of gene therapy. The committees independently concluded that clinical applications of gene transfer could not possibly succeed without adequate scientific and technological support, as well as sufficient preclinical studies in relevant animal models of human disease to validate treatment efficacy.

The scientific community has taken these constructive criticisms to heart during the past few years, and dramatic progress has been made in the basic science of viral and non-viral vector development used to transfer genes into patients. As a result, some approaches appear to be working. The Hemophilia B trial currently being conducted at the Children's Hospital of Philadelphia and Stanford University employs intramuscular delivery of the gene expressing human Factor IX, a protein involved in blood clotting. Patients in this trial have exhibited much improved whole blood clotting times for months after the gene treatment. Children in Paris with X-linked SCID, an inherited disorder that destroys the immune system, are able to live at home normally after genetic treatment of bone marrow stem cells. While both trials are in early phases, the encouraging results do provide proof of the scientific principle that human genetic disorders can be corrected by gene transfer.

In September 1999, clinical experiments in gene transfer research suffered their first patient loss as a direct consequence of the gene treatment itself. The patient was a 19-year-old male with a metabolic disorder who received a high dose of a genetic treatment directly into his liver at the University of Pennsylvania. He died from multi-organ failure induced by an adult respiratory distress syndrome secondary to a systemic inflammatory response to the method used to deliver the new gene.

Initial public response to the tragic incident was rather mild, as it is understood that medical research is not without risks. But public perception of clinical gene therapy took a precipitous turn for the worse after the FDA cited the University of Pennsylvania investigators in December 1999 for multiple protocol violations in the trial. In addition, NIH revealed that only 6 percent of all serious adverse events observed in patients during past and current clinical gene transfer studies were reported to the NIH as required. These findings triggered a series of corrective actions by the relevant federal regulatory agencies including the FDA and the NIH, with the full cooperation and enthusiastic support of the American Society of Gene Therapy (ASGT). The Board of Directors of ASGT has accordingly adopted a policy that calls for all of its members to rigorously adhere to federal regulations and institutional guidelines in clinical gene transfer studies. This policy was published in the January issue of Molecular Therapy, the official scientific journal of the Society.

Another revelation that further shook the public's confidence in clinical gene transfer studies at its roots was that some investigators apparently have significant financial interests in the outcome of the clinical research they are conducting. To ensure the public that clinical gene transfer research will be performed with high ethical standards and without financial conflicts in the future, the ASGT Board of Directors adopted a second policy published in the May issue of Molecular Therapy. It prohibits investigators and team members from direct participation in the recruitment and clinical management of trial patients or in gaining participants' informed consent if they have financial interests in companies sponsoring the trial.

Scientists now have the potential through gene therapy to eliminate and prevent hereditary diseases such as cystic fibrosis and hemophilia and possibly heart diseases, AIDS, and cancer. This procedure has not yet been used except for in trial runs with children who have been seriously sick or have been sick with an illness that has been declared incurable by conventional means. Genetic mutations can result in diseases, physical disabilities, and/or and shortened life span. These mutations can also be passed down from generation to generation the same way that hair and eye color does. These too can be cured. The outlook for gene therapy looks very promising. However, gene therapy still needs many improvements before it becomes routine treatment for cardiovascular disease in the clinic. Some areas for improvement include selecting appropriate patient populations for clinical trials, evaluating side effects, selecting and ensuring the safety of the genetic material to deliver, and methods for delivering the gene.

While the recent advances in clinical gene transfer to treat Hemophilia B and X-linked SCID show that gene therapy is indeed an immensely powerful technology that can be developed into novel treatments for a wide variety of human diseases, it is critically important for the scientific community to rise to the occasion and restore full confidence in the public's mind that clinical gene transfer research will be conducted by investigators in a responsible manner and free from financial conflicts. Only then can the full potential of using genes as medicines to treat human disease be realized, which will have a major impact on medicine and health in the 21st century.

REFERENCES

Acland GM Aguirre GD Bennett J *et al* 2005 Long-term restoration of rod and cone vision by single dose rAAV-mediated gene transfer to the retina in a canine model of childhood blindness *Mol Ther* **12** 1072-1082

Acland GM Aguirre GD Ray J *et al* 2001 Gene therapy restores vision in a canine model of childhood blindness *Nat Genet* **28** 92-95

Actions of Anterior Pituitary Hormones Growth Hormone (GH) Medical College of Georgia 2007

Actions of Anterior Pituitary Hormones Physiologic Actions of GH Medical College of Georgia 2007

Adler K J Gifford and Sumner R HIV as a Vector in Gene Therapy [Online] http //wwwpp uwrf edu/%7Ekk00/hivvector/hivvector htm [12-13-99, last date accessed]

Ali RR Sarra GM Stephens C Alwis MD *et al* 2000 Restoration of photoreceptor ultrastructure and function in retinal degeneration slow mice by gene therapy *Nat Genet* **25** 306-310

Alton E W Stern M Farley R *et al* 1999 Cationic lipid-mediated CFTR gene transfer to the lungs and nose of patients with cystic fibrosis a double-blind plcebo-controlled trials *Lancet* **353** 947-954

Alvarez R D Barnes M N Gomez-Navarro J *et al* 2000a A cancer gene therapy approach utilizing an anti-erbB-2 single-chain antibody-encoding adenovirus (AD21) a phase I trial *Clin Cancer Res* **6** 3081-3087

Alvarez R D Gomez-Navarro J Wang M *et al* 2000b Adenovial mediated suicide gene therapy for ovarian cancer *Mol Ther* **2** 524-530

Amado R G and YI S Chen 1999 Lentiviral Vectors — the Promise of Gene Therapy Within Reach? *Science* **285** (5428) 674-76

Ambati J Anand A Fernandez S *et al* 2003 An animal model of age-related macular degeneration in senescent Ccl-2-or Ccr-2-deficient mice *Nat Med* **9** 1390-1397

American Association of Clinical Endocrinologists Medical guidelines for clinical practice for growth hormone use in adults and children 2003 *Endocrine Practice* **9**(1) 64-76

Bagley J *et al* 1998 Long-term expression of the gene encoding green fluorescent protein in murine hematopoietic cells using retroviral gene transfer *Transplantation* **65** 1233-1240, PubMed

Bainbridge JW Mistry A Binley K *et al* 2003 Hypoxia-regulated transgene expression in experimental retinal and choroidal neovascularization *Gene Ther* **10**(12) 1049-54

Bainbridge JW Mistry A De Alwis M *et al* 2002 Inhibition of retinal neovascularisation by gene transfer of soluble VEGF receptor sFlt-1 *Gene Ther* **9** 320-326

Baum C, Dullmann J, Li Z, Fehse B, Meyer J, Williams DA, von Kalle C 2003 *Side effects of retroviral gene transfer into hematopoietic stem cells Blood* **101**(6) 2099-114

Beneveniste P Shu W Cohen A 1995 Interference with thymocyte differentiation by an inhibitor of S-adenosylhomocysteine hydrolase *J Immunol* **155** 536-544

Bengtsson BA 1993 The Consequences of Growth Hormone Deficiency in Adults *Acta Endocrinologica* **128** (Suppl 2) 2-5

Bengtsson BA, Eden S, Lonn L 1993 Treatment of adults with growth hormone (GH) deficiency with recombinant human GH *J Clin Endocrinol Metab* **76**(2) 309-317 [Medline]

Bengtsson BA, Eden S, Lonn L *et al* 1993 Treatment of Adults with Growth Hormone (GH) Deficiency with Recombinant Human GH *Journal of Clinical Endocrinology and Metabolism* **76**;309-317

Bengtsson BA, Eden S, Lonn L *et al* 1993 Treatment of Adults with Growth Hormone (HGH) Deficiency with Recombinant (HGH) *J Clin EndoMetab* **76** 309-17

Bett A J Prevec L Graham F L 1993 Packaging capacity and stability of human adenovirus type 5 vectors *J Virol* **67** 5911-5921

Bett A J Prevec L Graham F L 1993 Packaging capacity and stability of human adenovirus type 5 vectors *J Virol* **67** 5911-5921

Blaese R M Culver K W Miller A D *et al* 1995 T lymphocyte-directed gene therapy for ADA-SCID initial trial results after 4 years *Science* **270** 475-480

Bordignon C Notarangelo L D Nobili N *et al* 1995 Gene therapy in peripheral blood lymphocytes and bone marrow for ADA-immunodeficient patients *Science* **270** 470-475

Boussif O Lezoualc'h Fm Zanta M A *et al* 1995 A versatile vector for gene and oligonucleotide transfer into cells in culture and *in vivo* Polyethylenimine *Proc Natl Acad Sci USA* **92** 7297-7301

Brenner, M K *et al* 1994 Gene transfer and bone marrow transplantation *Cold Spring Harb Symp Quant Biol* **59**, 691-697, PubMed

Brooks G Vile R 1998 Gene therapy for the treatment of disease In Books G ed *Biotechnology in Healthcare An Introduction to Biopharmaceuticals* London Pharmaceutical Press 93-116

Buckley R H Schiff S E Schiff R I *et al* 1999 Hematopoietic stem-cell transplantation for the treatment of severe combined immunodeficiency *N Engl J Med* **340** 508-516

Buckley S M K Waddington S N Jezzard S Lawrence L Schneider H Themis M Coutelle C 2005 Factors influencing adenovirus-mediated airway transduction in fetal mice Mol Ther **12** 484-492

Bunting K D Sangster M Y Ihle J N *et al* 1998 Restoration of lymphocyte function in Janus kinase 3-deficient mice by retroviral-mediated gene transfer *Nat Med* **4** 58-64

Byrnes A P Rusby J E Wood M J *et al* 1995 Adenovirus gene transfer causes inflammation in the brain Neuroscience **66** 1015-1024

Campochiaro PA Nguyen QD Shah SM *et al* 2006 Adenoviral vector-delivered pigment epithelium-derived factor for neovascular age-related macular degeneration results of a phase I clinical trial *Hum Gene Ther* **17** 167-176

Candotti F Oakes S A Johnston J A *et al* 1996 *In vitro* correction of JAK3-deficient severe combined immunodeficiency by retroviral-mediated gene transduction *J Exp Med* **183** 2687-2692

Caplen J J Alton E W Middleton P G *et al* 1995 Liposome-mediated CFTR gene transfer to the nasal epithelium of patients with cystic fibrosis *Nat Med* **1** 39-46

Caplen N J Higginbotham J N Scheel J R *et al* 1999 Adeno-retroviran chimeric viruses as *in vivo* transducting agents *Gene Ther* **6** 454-459

Carpenter D E Stevens J G 1996 Long-term expression of a foreign gene from a unique position in the latent herpes simplex virus genome *Hum Gene Ther* **7** 1447-1454

Carroll PV Christ ER Bengtsson BA 1998 Growth hormone deficiency in adulthood and the effects of growth hormone replacement a review Growth Hormone Research Society Scientific Committee *J Clin Endocrinol Metab* **83**(2) 382-395 [Medline]

Casadevall A 1995 Antibody immunity and invasive fungal infections Infect Immun **63** 4211-8

Casadevall A Scharff MD 1994 "Serum therapy" revisited animal models of infection and the development of passive antibody therapy Antimicrob Agents Chemother **38** 1695-702

Casadevall A Scharff MD 1995 Return to the past the case for antibody-based therapies in infectious diseases Clin Infect Dis **21** 150-61

Cavazzana-Calvo M Hacein-Bey S de Saint Basile G *et al* 2000 Gene therapy of human severe combined immunodeficiency (SCID)-X1 disease *Science* **288** 669-672

Cavazzana-Calvo M Hacein-Bey S de Saint Basile G Gross F Yvon E Nusbaum P Selz F Hue C Certain S Casanova JL Bousso P Deist FL Fischer A 2000 Gene therapy of human severe combined immunodeficiency (SCID)-X1 disease *Science* **288**(5466) 669-72

CFAR Center for AIDS Research at UC San Diego Last update 10-18-99 Lentiviral Vector Core [Online] http //hsrd ucsd edu/Cfar/lenti/lenti html [12-13-99 last date accessed]

Chen CL and others 2002 Hormone replacement therapy in relation to breast cancer JAMA **287** 734-741

Chen J Kubalak S W Chien K R 1998 Ventricular muscle-restricted targeting of the RXRalpha gene reveals a non-cell-autonomous requirement in cardiac chamber morphogenesis Development **125** 1943-1949

Chia S H *et al* (1998) Adenovirus-mediated gene transfer to liver grafts an improved method to maximize infectivity Transplantation 66 1545-1551 PubMed

Chipman JJ Attanasio AF Birkett MA 1997 The safety profile of GH replacement therapy in adults *Clin Endocrinol (Oxf)* **46**(4) 473-81 [Medline]

Chu P *et al* 1998 Retrovirus-mediated gene transfer into human hematopoietic stem cells *J Mol Med* 76 184-192 PubMed

Clemens P R Kochanek S Suanda Y *et al* 1996 *In vivo* muscle gene transfer of full-length dystrophin with an adenoviral vector that lacks all viral genes *Gene Ther* **3** 965-972

Compton S H Mecklenbeck S Mejia J E *et al* 2000 Stable integration of large (>100 kb) PAC constructs in HaCaT keratinocytes using an integrin-targeting peptide delivery system *Gene Ther* 7 1600-1605

Cone R D Mulligan R C 1984 High-efficiency gene transfer into mammalian cells generation of helper-free recombinant retrovirus with broad mammalian host range *Proc Natl Acad Sci USA* **81** 6349-6353

Crystal R G 1995 Transfer of genes to humans early lessons and obstacles to success *Science* **270** 404-410 PubMed

Crystal R G 1995 Transfer to genes to humans early lessons and obstacles to success *Science* **270** 404-410

Cuneo RC Salomon F Wiles CM *et al* 1991 Growth Hormone Treatment in Growth Hormone Deficient Adults II Effects on Exercise Performance *Journal of Applied Physiology* **70** 695-700

Curiel D T *et al* 1992 High-efficiency gene transfer mediated by adenovirus coupled to DNA-polylysine complexes *Hum Gene Ther* **3** 147-154 PubMed

Danel C *et al* 1998 Gene therapy for oxidant injury-related diseases adenovirus-mediated transfer of superoxide dismutase and catalase cDNAs protects against hyperoxia but not against ischemia-reperfusion lung injury *Hum Gene Ther* **9** 1487-1496 PubMed

David A Cook T Waddington S Peebles D Nivsarkar M Knapton H Miah M Dahse T Noakes D Schneider H Rodeck C Coutelle C Themis M 2003 Ultrasound-guided percutaneous delivery of adenoviral vectors carrying ß-galactosidase and human factor IX genes to early gestation fetal sheep in utero *Hum Gene Ther* **14** 353-364

de Boer H Blok GJ Van der Veen EA 1995 Clinical aspects of growth hormone deficiency in adults *Endocr Rev* Feb **16**(1) 63-86 [Medline]

Del Priore LV Sheng Y Johnson E *et al* 2003 Identification of transplanted retinal pigment epithelium with a novel chromosomal marker *Curr Eye Res* **26** 125-31

Demaison C Brouns G Blundell M P *et al* 2000 A defined window for efficient gene marking of severe combined immunodeficient-repopulating cells using a gibbon ape leukemia virus-pseudotyped retroviral vector *Hum Gene Ther* **11** 91-100

Derossi D Calvet S Trembleau A *et al* 1996 Cell internalization of the third helix of the Antennapedia homeodomain in receptor-independent *J Biol Chem* **271** 18188-18193

Derossi D Chassaing G Prochiantz A 1998 Trojan peptides the penetratin system for intracellular delivery *Trends Cell Biol* **8** 84-87

Derossi D Joliot A H Chassaing G *et al* 1994 The third helix of the Antenna-pedia homeodomain translocates through biological membranes *J Biol Chem* **269** 10444-10450

Dong J Y Fan P D Frizzell R A 1996 Quantitative analysis of the packaging capacity of recombinant adeno-associated virus *Hum Gene Ther* **7** 2101-2112

Douglas J T Rogers B E Rosenfeld M E *et al* 1996 Targeted gene delivery by tropism-modified adenoviral vectors *Nature Biotechnol* **14** 1574-1578

Dunbar C E Kohn D B Schiffmann R *et al* 1998 Retroviral transfer of the gluco-cerebrosidase gene into CD34+ cells from patients with Gaucher disease *In vivo* detection of transduced cells without myeloablation *Hum Gene Ther* **9** 2629-2640

Durai Sundar; Mala Mani Karthikeyan Kandavelou Joy Wu Matthew H Porteus and Srinivasan Chandrasegaran 2005 "Zinc finger nucleases custom-designed molecular scissors for genome engineering of plant and mammalian cells" *Nucleic Acids Research* **33** (18) 5978-5990 DOI 10 1093/nar/gki912 PMID 16251401 Retrieved on 2006-05-28

Dzau V J *et al* 1996 Fusigenic viral liposome for gene therapy in cardiovascular diseases *Proc Natl Acad Sci* U S A **93** 11421-11425 PubMed

Edwards AO Ritter R Abel KJ Manning A Panhuysen C Farrer LA 2005 Complement factor H polymorphism and age-related macular degeneration *Science* **308** 421-444

Elliott G O'Hare P 1997 Intercellular trafficking and protein delivery by a herpesvirus structural protein *Cell* **88** 223-233

Ellison K E Bishopric N H Webster K A *et al* 1996 Fusigenic liposome-mediated DNA transfer into cardiac myocytes *J Mol Cell Cardiol* **28** 1385-1399

Felgner J H *et al* 1994 Enhanced gene delivery and mechanism studies with a novel series of cationic lipid formulations *J Biol Chem* **269** 2550-2561 PubMed

Felgner P L Gadek T R Holm M *et al* 1987 Lipofetion a highly efficient lipid-mediated DNA-transfection procedure *Proc Natl Acad Sci USA* **84** 7413-7417

Feng M Jackson W H Jr Goldman C K *et al* 1997 Stable *in vivo* gene transduction via novel adenoviral/ retroviral chimeric vctor *Nat Biotechnol* **15** 866-870

Ferrari G Rossini S Giavazzi R *et al* 1991 An *in vivo* model of somatic cell gene therapy for human severe combined immunodeficiency *Science* **251** 1363-1366

Fischer A 1999 Thirty years of bone marrow transplantation for severe combined immunodeficiency *N Engl J Med* **340** 559-561

Fischer A 2000 T-lymphocyte immunodeficiencies In Roifman C ed *Primary T-Cell Immunodeficiencies Immunology and Allergy Clinics of North America* Philadelphia WB Saunders 113-127

Fischer A Landais P Friedrich W *et al* 1990 European experience of bone-marrow transplantation for severe combined immunodeficiency *Lancet* **336** 850-854

Fisher K J Jooss K Alston J *et al* 1997 Recombinant adeno-associated virus for muscle directed gene therapy *Nat Med* **3** 306-312

Flake A W Roncarolo M G Puck J M *et al* 1996 Treatment of X-linked severe combined immunodeficiency by in utero transplantation of paternal bone marrow *N Engl J Med* **335** 1806-1810

Fletcher SW Colditz GA 2002 Failure of estrogen plus progestin therapy for prevention *JAMA* **288** 366-368

Fowelin J Attvall S Lager I 1993 Effects of treatment with recombinant human growth hormone on insulin sensitivity and glucose metabolism in adults with growth hormone deficiency *Metabolism* **42**(11) 1443-7 [Medline]

Fritz Z 1996 Retroviral vectors for gene transfer http //www-micro msb le ac uk/335/peel/peel2 html

Fujii S Huang S Fong T C *et al* 2000 Induction of melanoma-associated antigen systemic immunity upon intratumoral delivery of interferon-gamma retroviral vector in melanoma patients *Cancer Gene Ther* **7** 1220-1230

Fynan E F Webster R G Fuller D H *et al* 1993 DNA vaccines Protective immunizations by parenteral mucosal and gene-gun inoculations *Proc Natl Acad Sci USA* **90** 11478-11482

Gao X and Huang L (1995) Cationic liposome-mediated gene transfer Gene Ther 2 710-722 PubMed

Gardlik Roman; Roland PÃ¡lffy JÃºlius Hodosy JÃ¡n LukÃ¡cs JÃ¡n TurÅˆa and Peter Celec 2005 "Vectors and delivery systems in gene therapy" *Medical Science Monitor* **11**(4) 110â€"121 PMID 15795707 Retrieved on 2006-05-28

Gilgenkrantz H *et al* 1995 Transient expression of genes transferred *in vivo* into heart using first-generation adenoviral vectors role of the immune response *Hum Gene Ther* **6** 1265-1274 PubMed

Goins W F *et al* 1997 Herpes simplex virus vectors for gene transfer to the nervous system *J Neurovirol* **3** Suppl 1 S80-88 PubMed

Goldman M J *et al* 1997 Lentiviral vectors for gene therapy of cystic fibrosis *Hum Gene Ther* **8** 2261-2268 PubMed

Goldman M J P Lee J Yang and J M Wilson 1997 Lentiviral Vectors for Gene Therapy of Cystic Fibrosis *Human Gene Therapy* **8** 2261-2268

Gorin MB Breitner JC De Jong PT *et al* 1999 The genetics of age-related macular degeneration *Mol Vis* **5** 29

Graham F L Smiley J Russell W C *et al* 1977 Characteristics of a human cell line transformed by DNA from human adenovirus type 5 *J Gen Virol* **36** 59-74

Hacein-Bey H Cavazzana-Calvo M Le Deist F *et al* 1996 Gamma-c gene transfer into SCID X1 patients' B-cell lines restores normal high affinity interleukin-2 receptor expression and function Blood **87** 3108-3116

Hageman GS Anderson DH Johnson LV *et al* 2005 A common haplotype in the complement regulatory gene factor H (HF1/CFH) predisposes individuals to age-related macular degeneration *Proc Natl Acad Sci U S A* **102** 7227-7232

Hageman GS Luthert PJ Victor Chong NH *et al* 2001 An integrated hypothesis that considers drusen as biomarkers of immune-mediated processes at the RPE-Bruch's membrane interface in aging and age-related macular degeneration *Prog Retin Eye Res* **20** 705-32

Haines JL Hauser MA Schmidt S *et al* 2005 Complement factor H variant increases the risk of age-related macular degeneration *Science* **308** 419-421

Halbert C L *et al* 1995 Adeno-associated virus vectors transduce primary cells much less efficiently than immortalized cells *J Virol* **69** 1473-1479 PubMed

Halbert D N Cutt J R and Shenk T 1985 Adenovirus early region 4 encodes functions required for efficient DNA replication late gene expression and host cell shutoff *J Virol* **56** 250-257 PubMed

Harvey B G Leopold P L Hackett N R *et al* 1999 Airway epithelial CFTR mRNA expression in cystic fibrosis patients after repetitive administration of a recombinant adenovirus *J Clin Invest* **104** 1245-1255

He T C Zhou S da Costa L T *et al* 1998 A simplified system for generating recombinant adenoviruses *Proc Natl Acad Sci USA* **95** 2509-2514

Heinzel FP Antibodies In Mandell GL Bennett JE Dolin R editors 1995 Principles and practice of infectious diseases New York Churchill Livingstone 36-57

Hermonat P L Quirk J G Bishop B M *et al* 1997 The packaging capacity of adeno-associated virus (AAV) and the potential for wild-type-plus AAV gene therapy vectors *FEBS Lett* **407** 78-84

Hershfield M S 1995 PEG-ADA an alternative to haploidentical bone marrow transplantation and an adjunct to gene therapy for adenosine deaminase deficiency *Hum Mutat* **5** 107-112

Hershfield M S Mitchell B S 1995 Immunodeficiency diseases caused by adenosine deaminase deficiency and purine nucleotide deficiency In Scriver C R Beaudet A L Sly W S *et al* eds *The Metabolic and Molecular Basis of Inherited Disease* 7th edn New York McGraw-Hill 1725-1768

Hirschhorn R Yang D R Puck J M *et al* 1996 Spontaneous *in vivo* reversion to normal of an inherited mutation in a patient with adenosine deaminase deficiency *Nat Genet* **13** 290-295

HIV as a Vector for Gene Therapy [Online] http //bioinformatik biochemtech uni-halle de/uli/ genetherapy/genehiv htm [12-13-99 last date accessed]

Hoogerbrugge P M van Beusechem V W Fischer A *et al* 1996 Bone marrow gene transfer in three patients with adenosine deaminase deficiency *Gene Ther* **3** 179-183

Horn PA Morris JC Neff T Kiem HP 2004 *Stem cell gene transfer – efficacy and safety in large animal studies Molecular Therapy* **10**(3) 417-31

Humeau LM Binder GK Lu X Slepushkin V Merling R Echeagaray P Pereira M Slepushkina T Barnett S Dropulic LK Carroll R Levine BL June CH Dropulic B 2004 Efficient lentiviral vector-mediated control of HIV-1 replication in CD4 lymphocytes from diverse HIV+ infected patients grouped according to CD4 count and viral load *Mol Ther* **9**(6) 902-13

Hyde S C Southern K W Gileadi U *et al* 2000 Repeat administration of DNA/liposomes to the nasal epithelium of patients with cystic fibrosis *Gene Ther* **7** 1156-1165

Jenkins R G Herrick S E Meng Q H *et al* 2000 An integrin-targeted non-viral vector for pulmonary gene therapy *Gene Ther* **7** 39-400

Jiang S Moriarty-Craige SE Orr M Cai J Sternberg P Jr Jones DP 2005 Oxidant-induced apoptosis in human retinal pigment epithelial cells dependence on extracellular redox state *Invest Ophthalmol Vis Sci* **46** 1054-1061

Jooss K Yang Y Fisher K J *et al* 1998 Transduction of dendritic cells by DNA viral vectors directs the immune response to transgene products in muscle fibers *J Virlol* **72** 4212-4223

Kalpana G V 1999 Retroviral Vectors for Liver-directed Gene Therapy *Seminar in Liver Disease* **19** (1) 27-37

Kawauchi M Suzuki J Morishita R *et al* 2000 Gene therapy for attenuating cardiac allograft arteriopathy using *ex vivo* E2F ecoy transfection by HVJ-AVE-lipsome method in mice and nonhuman primates *Circ Res* **87** 1063-1068

Kim S H Yu S S Park J S *et al* 1998 Construction of retroviral vectors with improved safety gene expression and versatility *J Virol* **72** 994-1004

Kochanek S Clemens P R Mitani K *et al* 1996 A new adenoviral vector replacement of all viral coding sequences with 28 kb of DNA independently expressing both full-length dystrophin and beta-galactosidase *Proc Natl Acad Sci USA* **93** 5731-5736

Koeing M Hoffman E P Bertelson C J *et al* 1987 Complete cloning of the Duchenne muscular dystrophy (DMD) cDNA and preliminary genomic organization of the DMD gene in normal and affected individuals *Cell* **60** 509-517

Kohn D B Hershfield M S Carbonaro D *et al* 1998 T lymphocytes with a normal ADA gene accumulate after transplantation of transduced autologous umbilical cord blood CD34+ cells in ADA-deficient SCID neonates *Nat Med* **4** 775-780

Kohn D B Weinberg K Nolta J A *et al* 1995 Engraftment of gene-modified umbilical cord blood cells in neonates with adenosine deaminase deficiency *Nat Med* **1** 1017-1023

Kopp J Stachel K D Holter W Kandler M Campean V Hertzberg H Carbon R Horch R E Rascher W Schneider H 2005 Haematopoietic stem cell transplantation and subsequent 80 % skin exchange by grafts from the same donor in a patient with Herlitz junctional epidermolysis bullosa Transplantation **79** 255-256

Kotin R M Siniscalco M Samulski R J *et al* 1990 Site-specific integration by adeno-associated virus *Proc Natl Acad Sci USA* **87** 2211-2215

Kukowska-Latallo J F Bielinska A U Johnson J *et al* 1996 Efficient transfer of genetic material into mammalian cells using Starburst polyamidoamine dendrimers *Proc Natl Acad Sci USA* **93** 4897-4902

Kwok W W *et al* 1986 Retroviral transfer of genes into canine hemopoietic progenitor cells in culture a model for human gene therapy *Proc Natl Acad Sci* U S A **83** 4552-4555 PubMed

Lai CC Wu WC Chen SL *et al* 2005 Recombinant adeno-associated virus vector expressing angiostatin inhibits preretinal neovascularization in adult rats *Ophthalmic Res* **37** 50-56

Lang AB Cryz SJ Schurch U Ganss MT Bruderer U 1993 Immunotherapy with human monoclonal antibodies J Immunol **151** 466-72

Lasic D D 1997 Liposomes in Gene Delivery New York CRC Press

Latchman D S 1994 Herpes simplex virus vectors for gene therapy *Mol Biotechnol* **2** 179-195 PubMed

Lee M G 1996 Adenovirus as Vectors for Gene Therapy In Targeting of Drugs 5 Strategies for Oligonucleotide and Gene Delivery in Therapy (Gregoriadis G and McCormack B eds) pp 53-64 Plenum Press New York

Lewis P F Emerman M 1994 Passage through mitosis is required for orncoretro-viruses but not for the human immunodeficiency virus *J Virol* **68** 510-516

Linden R M *et al* 1996 Site-specific integration by adeno-associated virus *Proc Natl Acad Sci* U S A **93** 11288-11294 PubMed

Liu F Song Y Liu D (1999) Hydrodynamics-based transfection in animals by systemic administration of plasmid DNA *Gene Ther* **6** 1258-1266

LoBuglio AF Wheeler RH Trang J Haynes A Rogers K Harvey EB *et al* 1989 Mouse/human chimeric monoclonal antibody in man kinetics and immune response *Proc Natl Acad Sci* USA **86** 4220-4

Lowenstein P R Morrison E E Bain D *et al* 1994 Use of recombinant vectors derived from herpes simplex virus 1 mutant tsK for short-term expression of transgenes encoding cytoplasmic and membrane anchored proteins in post-mitotic polarized cortical neurons and glial cells *in vitro* *Neuroscience* **60** 1059-1077

Lu X Humeau L Slepushkin V Binder G Yu Q Slepushkina T Chen Z Merling R Davis B Chang YN Dropulic B 2004 Safe two-plasmid production for the first clinical lentivirus vector that achieves >99% transduction in primary cells using a one-step protocol J Gene Med 6(9) 963-73

Lu X Yu Q Binder GK Chen Z Slepushkina T Rossi J Dropulic B 2004 Antisense-mediated inhibition of human immunodeficiency virus (HIV) replication by use of an HIV type 1-based vector results in severely attenuated mutants incapable of developing resistance *J Virol* **78**(13) 7079-88

Lund RD Ono SJ Keegan DJ Lawrence JM 2003 Retinal transplantation progress and problems in clinical application *J Leukoc Biol* **74** 151-160

Luskey B D *et al* 1990 Gene transfer into murine hematopoietic stem cells and bone marrow stromal cells *Ann N Y Acad Sci* **612** 398-406 PubMed

Manilla P Rebello T Afable C Lu X Slepushkin V Humeau LM Schonely K Ni Y Binder GK Levine BL MacGregor RR June CH Dropulic B 2005 Regulatory considerations for novel gene therapy products a review of the process leading to the first clinical lentiviral vector *Hum Gene Ther* 2005 Jan;16(1) 17-25 Review Erratum in *Hum Gene Ther* **16**(2) 279

Mann M J Gibbons G H Hutchinson H *et al* 1999a Pressure mediated oligo-nucleotide transfection of rat and human cardiovascular tissues *Proc Natl Acad Sci USA* **96** 6411-6416

Mann M J Whittemore A D Donaldson M C *et al* 1999b *Ex-vivo* gene therapy of human vascular bypass grafts with E2F decoy the PREVENT single-centre randomized controlled trial *Lancet* **354** 1493-1498

Mann R Mulligan R C Baltimore D 1983 Construction of a retrovirus packaging mutant and its use to produce helper-free defective retrovirus *Cell* **33** 153-159

Mazanec MB Kaetzel CS Lamm ME Fletcher D Nedrud JG 1992 Intracellular neutralization of virus by immunoglobulin A antibodies *Proc Natl Acad Sci* **89** 6901-5

McGauley GA Cuneo RC Salomon F *et al* 1990 Psychological Well-Being Before and After Growth Hormone Treatment in Adults with Growth Hormone Deficiency Hormone Research **33** (suppl 4) 52-54

McLaughlin BJ Fan W Zheng JJ *et al* 2003 Novel role for a complement regulatory protein (CD46) in retinal pigment epithelial adhesion *Invest Ophthalmol Vis Sci* **44** 3669-3674

Michael W King 2006 Growth Hormone Indiana State University

Michou A I Santoro L Christ M *et al* 1997 Adenovirus-mediated gene transfer influence of transgene mouse strain and type of immune response on persistence of transgene expression *Gene Ther* **4** 473-482

Miller A D 1992 Human gene therapy comes of age *Nature* **357** 455-460 PubMed

Miller A D *et al* 1993 Use of retroviral vectors for gene transfer and expression *Methods Enzymol* **217** 581-599 PubMed

Morris P J 1997 Transplantation In Encyclopaedia of Immunology (Delves P J and Roitt I M eds) (2nd edn) Academic Press Vol 4 pp 2411-2415

Morshita R Gibbons G H Kaneda Y *et al* 2000 Systemic administration of HVJ viral coat-liposome complex containing human insulin vector decreases glucose level in diabetic mouse a model of gene therapy *Biochem Biophys Res Commun* **273** 666-674

Morsy M A Gu M Motzel S *et al* 1998 an adenoviral vector deleted for all viral coding sequences results in enhanced safety and extended expression of a leptin transgene *Proc Natl Acad Sci USA* **95** 7866-7871

Muhle C Jiang Q J Charlesworth A Bruckner-Tuderman L Meneguzzi G Schneider H 2005 Novel and recurrent mutations in the laminin-5 genes causing lethal junctional epidermolysis bullosa molecular basis and clinical course of Herlitz disease *Hum Genet* **116** 33-42

Mumper R J Wang J Klakmp S L *et al* 1998 Protective interactive noncondensing (PINC) polymers of enhanced plasmid distribution and expression in rat skeletal muscles *J Controlled Release* **52** 191-203

Nabel G J Gordon D Bishop D K *et al* 1996 Immune response in human melanoma after transfer of an allogeneic class I major histocompatibility complex gene with DNA-liposome complexes *Proc Natl Acad Sci USA* **93** 15388-15393

Nabel G J Nabel E G Yang Z Y *et al* 1993 Direct gene transfer with DNA-liposome complexes in melanoma expression biological activity and lack of toxicity in human *Proc Natl Acad Sci USA* **90** 11307-11311

Nagano H and Tilney N L 1997 Chronic allograft failure the clinical problem *Am J Med Sci* **313** 305-309 PubMed

Nakamura M Davila-Zavala P Tokuda H *et al* 1998 Uptake and gene expression of naked plasmid DNA in cultured brain microvessel endothelial cells *Biochem Biphys Res Commun* **245** 235-239

Naldini *et al* 1996 *In vivo* Gene Delivery and Stable Tranduction of Nondividing Cells by a Lentiviral Vector *Science* **272** 263-267

Naldini *et al* 1997 Lentiviral Vectors for in Vivo Gene Delivery The International Symposium on Gene Therapy for Hemophilia [Online] http //www med unc edu/wrkunits/3ctrpgm/thromb/ naldini htm [12-13-99 last date accessed]

Naldini L Blomer U Gallay P *et al* 1996 *In vivo* gene delivery and stable transduction of nondividing cells by a lentiviral vector *Science* **272** 263-267

Nauthton B A Dai Y Sibanda B *et al* 1992 Long-term expression of a retrovirally introduced beta-balactosidase gene in rodent cells implanted *in vivo* using biodegradable polymer meshes *Somat Cell Mol Genet* **18** 451-462

Neumann E Schaefer-Ridder M Wang Y *et al* 1982 Gene transfer into mouse lyoma cells by electroporation in high electric fields *EMBO J* **1** 841-845

Ni Y Sun S Oparaocha I Humeau L Davis B Cohen R Binder G Chang YN Slepushkin V Dropulic B 2005 Generation of a packaging cell line for prolonged large-scale production of high-titer HIV-1-based lentiviral vector *J Gene Med* 7(6) 818-34

Nichols E K 1998 *Human Gene Therapy* Cambridge Massachusettes Harvard University Press

Noone P G Hohneker K W Zhou Z *et al* 2000 Safety and biological efficacy of a lipid-CFTR complex for gene transfer in the nasal epithelium of adult patients with cystic fibrosis *Mol Ther* **1** 105-114

O'Halloran DJ Tsatsoulis A Whitehouse RW 1993 *et al* Increased Bone Density after Recombinant Human Growth Hormone (GH) Therapy in Adults with Isolated GH Deficiency Journal of Clinical Endocrinology and Metabolism **76** 1344-1348

Olsen J C 1998 Gene Transfer Vectors Derived From Equine Infectious Anemia Virus *Gene Therapy* **5** 1481-1487

Onodera M Ariga T Kawamura N *et al* 1998 Successful peripheral T-lymphocyte-directed gene transfer for a patient with severe combined immune deficiency caused by adenosine deaminase deficiency *Blood* **91** 30-36

Orlic D Girard L J Jordan C T *et al* 1996 The level of mRNA encoding the amphotropic retrovirus receptor in mouse and human hematopoietic stem cell sis low and correlates with the efficiency of retrovirus transduction *Proc Natl Acad Sci USA* **93** 11097-11102

Osborne W R Hock R A Kaleko M *et al* 1990 Long-term expression of human adenosine deaminase in mice after transplantation of bone marrow infected with amphotropic retroviral vectors *Hum Gene Ther* **1** 31-41

Parks R J Graham F L 1997 A helper-dependent system for adenovirus vector production helps define a lower limit for efficient DNA packaging *J Virol* **71** 3293-3298

Peel David 1998 Retroviral Vectors and Lentiviral Vectors Department of Microbiology & Immunology University of Leicester [Online] http //science uniserve edu au/mirror/microbiol/335/peel/peel2 html [12-13-99 last date accessed]

Pennington JE 1990 Newer uses of intravenous immunoglobulins as anti-infective agents Antimicrob Agents Chemother **34** 1463-6

Peplinski G R Tsung K and Norton J A 1998 Vaccinia virus for human gene therapy *Surg Oncol Clin N Am* 7 575-588 PubMed

Planelles V 1999 Homepage of Vicente Planelles [Online] http //www urmc rochester edu/gebs/faculty/Vicente_Planelles htm [12-13-99 last date accessed]

Poeschla E M F Wong-Staal and D J Looney 1998 Efficient Transduction of Nondividing Human Cells by Feline Immunodeficiency Virus Lentiviral Vectors *Nature Medicine* **4** (3) 354-357

Porter C D Collins M K Tailor C S *et al* 1996 Comparison of efficiency of infection of human gene therapy target cells iva four different retroviral receptors *Hum Gene Ther* **7** 913-919

Poston R S *et al* 1998 *Ex vivo* gene therapy prevents chronic graft vascular disease in cardiac allografts *J Thorac Cardiovasc Surg* **116** 386-396 PubMed

Puck JM Deschenes SM Porter JC Dutra AS Brown CJ Willard HF Henthorn PS 1993 The interleukin-2 receptor gamma chain maps to Xq13 1 and is mutated in X-linked severe combined immunodeficiency SCIDX1 *Hum Mol Genet* **2**(8) 1099-104

Qin L *et al* (1997) Promoter attenuation in gene therapy interferon-gamma and tumor necrosis factor-alpha inhibit transgene expression *Hum Gene Ther* **8** 2019-2029 PubMed

Rackemann FM 1942 Allergy serum reactions with particular reference to the prevention and treatment of tetanus *JAMA* **226** 726-33

Rapp SR and others 2003 Effect of estrogen plus progestin on global cognitive function in postmenopausal women The Women's Health Initiative Memory Study A randomized controlled trial *JAMA* **289** 2663-2672

Reich SJ Fosnot J Kuroki A *et al* 2003 Small interfering RNA (siRNA) targeting VEGF effectively inhibits ocular neovascularization in a mouse model *Mol Vis* **9** 210-216

Richards J R Brown C E Gogswell J P *et al* 2000 The admid system generation of recombinant adenoviruses by Tn7-mediated transposition in *E coli Biotehniques* **29** 146-154

Riordan J R Rommens J M Kerem B *et al* 1989 Identification of the cystic fibrosis gene cloning and characterization of complementary DNA *Science* **245** 1066-1073

Rizzuto G Cappelletti M Maione D *et al* 1999 Efficient and regulated erythropoietin production by naked DNA injection and muscle electroporation *Proc Natl Acad Sci USA* **96** 6417-6422

Roe T Reynolds T C Yu G *et al* (1993) Integration of murine leukemia virus DNA depends on mitosis *EMBO J* **12** 2099-2108

Rols M P Delteil C Golzio M *et al* 1998 *In vivo* electrically mediated protein and gene transfer in murine melanoma *Nat Biotechnol* **16** 168-171

Rose J A Berns K I Hoggan M D *et al* 1969 Evidence for a single-stranded adenovirus-associated virus genome formation of a DNA density hybrid on release of viral DNA *Proc Natl Acad Sci USA* **64** 863-869

Rosen FS Cooper MD Wedgwood RJ 1995 The primary immunodeficiencies *N Engl J Med* **17**;333(7) 431-40 Review No abstract available

Ross RK and others 2000 Effect of hormone replacement therapy on breast cancer estrogen versus estrogen plus progestin *Journal of the National Cancer Institute* **92** 328-332

Saggese G Ranke MB Saenger P 1998 Diagnosis and treatment of growth hormone deficiency in children and adolescents towards a consensus Ten years after the Availability of Recombinant Human Growth Hormone Workshop held in Pisa Italy 27-28 March 1998 *Horm Res* **50**(6) 320-40 [Medline]

Salmons B Gunzburg WH 1993 "Targeting of retroviral vectors for gene therapy" *Human Gene Therapy* **4**(2) 129-141

Salomon F Cuneo RC Hesp R *et al* 1989 The Effects of Treatment with Recombinant Human Growth Hormone on Body Composition and Metabolism in Adults with Growth Hormone Deficiency *New England Journal of Medicine* **321** 1797-1803

Salomon F Cuneo RC Hesp R *et al* 1989 The Effects of Treatment with Recombinant HGH on Body Composition and Metabolism in Adults with Growth Hormone Deficiency *New Eng JMed* **321** 1797-03

Samulski R J Chang L S Shenk T 1989 Helper-free stocks of recombinant adeno-associated viruses normal integration does not require viral gene expression *J Virol* **63** 3822-3828

Samulski R J Zhu X Xiao X *et al* 1991 Targeted integration of adeno-associated virus (AAV) into human chromosome 19 *EMBO J* **10** 3941-3950

Schairer C and others 2000 Menopausal estrogen and estrogen-progestin replacement therapy and breast cancer risk *JAMA* **283** 485-491

Schiedner G Morral N Parks R J *et al* 1998 Genomic DNA transfer with a high capacity adenovirus vector results in improved *in vivo* gene expression and decreased toxicity *Nat Genet* **18** 180-183

Schmidt Uli 1999 The HIV Page [Online] http //bioinformatik biochemtech uni-halle de/uli/ genetherapy/hiv htm [12-13-99 last date accessed]

Schneider H Adebakin S Themis M Cook T Douar A M Pavirani A Coutelle C 1999 Therapeutic concentrations of human factor IX in mice after gene delivery into the amniotic cavity A model for the prenatal treatment of haemophilia B *J Gene Med* **1** 424-432

Schneider H Coutelle C 1999 In utero gene therapy The case for *Nat Med* **5** 256-257

Schneider H Mühle C Douar A M Waddington S Jiang Q -J von der Mark K Coutelle C Rascher W 2002 Sustained delivery of therapeutic concentrations of human clotting factor IX a comparison of adenoviral and AAV vectors administered in utero *J Gene Med* **4** 46-53

Schuettauf F Vorwerk C Naskar R *et al* 2004 Adeno-associated viruses containing bFGF or BDNF are neuroprotective against excitotoxicity *Curr Eye Res* **29** 379-386

Schultz DW Klein ML Humpert AJ *et al* 2003 Analysis of the ARMD1 locus evidence that a mutation in HEMICENTIN-1 is associated with age-related macular degeneration in a large family *Hum Mol Genet* **12** 3315-3323

Seddon JM Cote J Page WF Aggen SH Neale MC 2005 The US twin study of age-related macular degeneration relative roles of genetic and environmental influences *Arch Ophthalmol* **123** 321-327

Sekhsaria S Fleisher T A Vowells S *et al* **1996** Granulocyte colony-stimulating factor recruitment of CD34+ protenitors ot peripheral blood impaired mobilization in chronic granulomatous disease and adenosine deaminase-deficient severe combined immunodeficiency disease patients *Blood* **88** 1104-1112

Sekul EA Cupler EJ Dalakas MC 1994 Aseptic meningitis associated with high-dose intravenous immunoglobulin therapy frequency and risk factors Ann Intern Med **121** 259-62

Shaked A *et al* 1994 Retroviral-mediated gene transfer into rat experimental liver transplant *Transplantation* **57** 32-34 PubMed

Shalet SM 2004 Growth hormone outgrows growth *Clin Endocrinol (Oxf)* **61**(1) 1-9

Shalet SM Toogood A Rahim A 1998 The diagnosis of growth hormone deficiency in children and adults *Endocr Rev* **19**(2) 203-23 [Medline]

Shears L L *et al* 1998 Efficient inhibition of intimal hyperplasia by adenovirus-mediated inducible nitric oxide synthase gene transfer to rats and pigs *in vivo J Am Coll Surg* **187** 295-306 PubMed

Shenk T 1996 Adenoviridae The Viruses and Their Replication In Fundamental Virology (3rd edn) (Fields B N *et al* eds) pp 979-1016 Lippincott-Raven Publishers Philadelphia

Sheridan P L Bodner M Lynn A *et al* 2000 Generation of retroviral packaging and producer cell lines for large-scale vector production and clinical application improved safety and high titer *Mol Ther* **2** 262-275

Shering A F Bain D Stewart K *et al* 1997 Cell type-specific expression in brain cell culture from a short human cytomegalovirus major immediate early promoter depends on whether it is inserted into herpesvirus of adenovirus vectors *J Gen Virol* **78** 445-459

Shin N H Hartigan-O'Connor D Pfeiffer J K *et al* 2000 Replication of lengthened Moloney murine leukemia virus genomes is impaired at multiple stages *J Virol* **74** 2694-2702

Shinmura K Morishita R Aoki M *et al* 2000 Catheter-delivered *in vivo* gene transfer into rat myocardium using the fusigenic liposomal mediated method *Jpn Heart J* **41** 633-647

Sikorski R and R Peters 1998 Gene Therapy Treating with HIV *Science* **282** (5393) 1438a

Simoes S Slepushkin V Gaspar R *et al* 1998 Gene delivery by negatively charged ternary complexes of DNA cationic liposmes and transferring or fusigenic peptides *Gene Ther* **5** 955-964

Singh M Briones M Ott G *et al* 2000 Cationic microparticles a potent delivery system for DNA vaccines *Proc Natl Acad Sci USA* **97** 811-816

Slade HB 1994 Human immunoglobulins for intravenous use and hepatitis C viral transmission *Clin Diagn Lab Immunol* **1** 613-9

Soudais C Shiho T Sharara L I *et al* 2000 Stable and functional lymphoid reconstitution of common cytokine receptor chain deficient mice by retroviral-mediated gene transfer *Blood* **95** 3071-3077

Stavrou S Kleinberg DL 2001 Diagnosis and management of growth hormone deficiency in adults *Endocrinol Metab Clin North Am* **30**(3) 545-63

Stephan V Wahn V Le Deist F *et al* 1996 Atypical X-linked severe combined immunodeficiency due to possible spontaneous reversion of the genetic defect in T cells *N Engl J Med* **335** 1563-1567

Stewart A K Lassam N J Quirt I C *et al* 1999 Adenovector-mediated gene delivery of interleukin-2 in metastatic breast cancer and melanoma results of a phase 1 clinical trial *Gene Ther* **6** 350-363

Su H Arakawa-Hoyt J Kan YW 2002 Adeno-associated viral vector-mediated hypoxia response element-regulated gene expression in mouse ischemic heart model *Proc Natl Acad Sci U S A* **99** 9480-9485

Tait D L Obermiller P S Hatmaker A R *et al* 1999 Ovarian cancer BRCA1 gene therapy phase I and II trial differences in immune response and vector stability *Clin Cancer Res* **5** 1708-1714

Tanaka T Cohen P Clayton PE 2002 Diagnosis and management of growth hormone deficiency in childhood and adolescence-part 2 growth hormone treatment in growth hormone deficient children *Growth Horm IGF Res* **12**(5) 323-41

Tang M X Szoka F C 1997 The influence of polymer structure on the interactions of catioinic polymers with DNA and morphology of the resulting complexes *Gene Ther* **4** 823-832

Taylor N Uribe L Smith S *et al* 1996 Correction of interleukin-2 receptor functional in X-SCID lymphoblastoid cells by retrovirally mediated transfer to the gamma-c gene *Blood* **87** 3103-3107

Themis M /Schneider H Kiserud T Cook T Adebakin S Jezzard S Hanson M Pavirani A Forbes S Rodeck C Coutelle C 1999 Successful expression of ß-galactosidase and factor IX transgenes in fetal and neonatal sheep after ultrasound-guided percutaneous adenovirus vector administration into the umbilical vein *Gene Ther* **6** 1239-1248

Thoren P E Persson D Karlsson M *et al* 2000 The antennapedia peptide penetratin translocates across lipid bilayers-the first direct observation *FEBS Lett* **482** 265-268

Tighe R and J Fritz 1997 Lentiviral Vectors (HIV-based) [Online] http //www mc vanderbilt edu/ gcrc/gene/hiv htm [12-13-99 last date accessed]

Tomita N Higaki J Ogihara T *et al* 1994 A novel gene-transfer technique mediated by HVJ (Sendai virus) nuclear protein and liposomes *Cancer Detect Prev* **18** 485-491

Triguro D Buciak JB Yang J Pardridge WM 1989 Blood-brain barrier transport of cationized immunoglobulin G enhanced delivery compared to native protein *Proc Natl Acad Sci* **86** 4761-5

Ueno T Matsumura H Tanaka K *et al* 2000 Site-specific integration of a transgene mediated by a hybrid adenovirus/adeno-associated virus vector using the Cre/loxP-expression-switching system *Biochem Biphys Res Commun* **273** 473-478

Vaheri A Pagano J S 1965 Infectious poliovirus RNA a sensitive method of assay *Virology* **27** 434-436

Van De Water TR Lallemend F Eshraghi AA *et al* 2004 Caspases the enemy within and their role in oxidative stress-induced apoptosis of inner ear sensory cells *Otol Neurotol* **25** 627-632

van Ginkel F W McGhee J R Liu GC *et al* 1997 Adenoviral gene elivery elicits distinct pulmonary-associated T helper cell responses to the vector and to its transgene *J Immunol* **159** 685-693

Vance ML Mauras N 1999 Growth hormone therapy in adults and children [see comments] *N Engl J Med* **341**(16) 1206-16 [Medline]

Vives E Brodin P Lebleu B 1997 A truncated HIV-1 Tat protein basic domain rapidly translocates through the plasma membrane and accumulates in the cell nucleus *J Biol Chem* **272** 16010-16017

von der Leyen H E *et al* 1995 Gene therapy inhibiting neointimal vascular lesion *in vivo* transfer of endothelial cell nitric oxide synthase gene *Proc Natl Acad Sci* U S A **92** 1137-1141 PubMed

Wadhwa M S Knoell D L Young A P *et al* 1995 Targeted gene delivery with a low molecular weight glyocpeptide carier *Bioconjug Chem* **6** 283-291

Wahl ML Kenan DJ Gonzalez-Gronow M Pizzo SV 2005 Angiostatin's molecular mechanism aspects of specificity and regulation elucidated *J Cell Biochem* **96** 242-261

Wang Hongjie; Dmitry M Shayakhmetov Tobias Leege Michael Harkey Qiliang Li Thalia Papayannopoulou George Stamatoyannopolous and AndrÃ© Lieber 2005 "A capsid-modified helper-dependent adenovirus vector containing the beta-globin locus control region displays a nonrandom integration pattern and allows stable erythroid-specific gene ex-pression" *Journal of Virology* **79** (17) 10999-11013 Retrieved on 2006-08-15

Wengler G S Lanfranchi A Frusca T *et al* 1996 *In-utero* transplantation of parental CD34 haematopoietic progenitor cells in a patient with X-linked severe combined immunodeficiency (SCIDXI) *Lancet* **348** 1484-1487

Wolff J A Malone R W Williams P *et al* 1990 Direct gene transfer into mouse muscle *in vivo Science* **247** 145-1468

Wong T K Neumann E 1982 Electric field mediated gene transfer *Biochem Biophys Res Commun* **107** 584-587

Wong W *et al* 1997 Retroviral gene transfer of a donor class I MHC gene to recipient bone marrow cells induces tolerance to alloantigens *in vivo Transplant Proc* **29** 1130 PubMed

Wright A Shin S-U Morrison SL 1992 Genetically engineered antibodies progress and prospects *Crit Rev Immunol* 1992;12 125-68

Writing Group for the Women's Health Initiative Investigators 2002 Risks and benefits of estrogen plus progestin in healthy postmenopausal women Principal results from the Women's Health Initiative Randomized Controlled Trial *JAMA* **288** 321-333

Wu WC Lai CC Chen SL *et al* 2005 Long-term safety of GDNF gene delivery in the retina *Curr Eye Res* **30** 715-722

Yang N S Burkholder J Roberts B *et al* 1990 *In vivo* and *in vitro* gene transfer to mammalian somatic cells by particle bombardment *Proc Natl Acad Sci USA* **87** 9568-9572

Yang Y Jooss K U Su Q *et al* 1996a Immune responses to viral antigens versus transgene product in the elimination of recombinant adenovirus-infected hepatocytes *in vivo Gene Ther* **3** 137-144

Yang Y Su Q Wilson J M 1996b Role of viral antigens in destructive cellular immune responses to adenovirus vector-transduced cells in mouse lungs *J Virol* **70** 7209-7212

Yates JR Moore AT 2000 Genetic susceptibility to age related macular degeneration *J Med Genet* **37** 83-87

Yoshida M Hayashi S Sakuma-Mochizuki J *et al* 1999 Introduction of plasmid DNA and oligonucleotides into lung epithelial cells by the hemagglutinating virus of Japan (HVJ)-liposome method *Somat Cell Mol Genet* **25** 49-57

Zarbin M 2004 Current concepts in the pathogenesis of age-related macular degeneration *Arch Ophthalmol* **122** 598-614

Zeitlin P L 2000 Cystic fibrosis gene therapy trials and tribulations *Mol Ther* **1** 5-6

Zuckier LS Rodigues LD Scharff MD 1989 Immunologic and pharmacologic concepts of monoclonal antibodies Semin Nucl Med **19** 166-86

5

RNA INTERFERENCE: SMALL AND SILENT BIOTECHNOLOGICAL TOOL

CHITRA WADHWANI, C P MALIK AND BHAVNEET KAUR

INTRODUCTION

Over the past decade "RNA interference" has emerged as a natural mechanism for silencing gene expression. This ancient cellular antiviral response can be harnessed to allow specific inhibition of the function of any chosen target genes, including those involved in causing diseases such as cancer, AIDS, and hepatitis. RNA interference is already proving to be an invaluable research tool, allowing much more rapid characterisation of the function of known genes. More importantly, the technology considerably helps functional genomics to aid in the identification of novel genes involved in disease processes. But can RNA interference be used as an effective therapeutic strategy? Many people in the biotechnology industry are betting that it can, but first considerable problems relating to delivery to target cells will have to be solved. This problem has proved the undoing of previous wonder technologies such as gene therapy and antisense.

Current understanding of RNA interference

The first hints of the existence of the gene silencing mechanism that is now called RNA interference emerged from work on the genetic modification of plants in the late 1980s. Attempts to deepen the violet hue of petunias by expressing higher levels of an enzyme involved in the synthesis of the pigment unexpectedly resulted in the appearance of many white flowers. While trying to deepen the purple color of these flowers, Rich Jorgensen and colleagues introduced a pigment-producing gene under the control of a powerful promoter. Instead of the expected deep purple color, many of the flowers appeared variegated or even white (Fig B). Jorgensen named the observed phenomenon "cosuppression", since the expression of both the introduced gene and the homologous endogenous gene was suppressed. The introduction of extra copies of the gene had somehow caused a decrease in its expression rather than the anticipated increase. For some time this remained an unexplained oddity. It was soonjoined by similar observations in the filamentous fungus *Neurospora crassa* and then the nematode worm *Caenorhabditis elegans*. Once the large community of developmental biologists working on the worm became involved, the pace quickened.

Thus, RNA interference was observed in petunias, when it was discovered that introduction of a pigment-producing gene under control of a powerful promoter suppressed expression of both the introduced gene and the homologous endogenous gene caused a phenomenon which was called "cosuppression." Cosuppression was subsequently found to occur in many species of plants and fungi also (where it was called "quelling") and to occur at the post-transcriptional level. Such post-transcriptional gene silencing (PTGS) was shown to be mediated by a diffusible, trans-acting molecule in *Neurospora* when Cogoni and colleagues showed that gene silencing could be transferred between nuclei in heterokaryotic strains and in plants when Palauqui and colleagues induced PTGS in a host plant by grafting a silenced, transgene-containing source plant to an unsilenced host. Introduction of transgenes can trigger PTGS, however silencing can also be induced by the introduction of certain viruses (Fig M). The crucial 1998 discovery by Fire *et al.*, in which dsRNA was injected—a mixture of both sense and antisense strands of the target mRNA, rather than either strand alone—into gonads of the nematode *Caenorhabditis elegans* (Fig A). In 1998 the key observation was made that led to the coining of the term RNA interference. It resulted in extremely potent silencing unequivocably identified dsRNA as the inducer of RNA interference i.e double stranded RNA was able to direct the degradation of messenger RNA (mRNA) with sequence complementary to one or other strand.

Short (~25 ntd), antisense RNAs were first implicated in PTGS in plants. Further work in *Drosophila*-derived *in vitro* systems showed that long dsRNA was processed to short interfering RNAs (siRNAs), 21–23 ntd long, that could mediate target mRNA cleavage in the region corresponding to the introduced siRNA.

In *C. elegans*, the demonstration that silencing could be induced by simply feeding nematodes with bacteria that had been engineered to express dsRNA homologous to the target gene, or by soaking worms in such dsRNA, drove rapid adoption of RNAi as a technology for reverse genetics at the organismal level. Likewise, technology was quickly developed for inducing RNAi in *Drosophila* cell culture by bathing or transfection with dsRNA.

Efforts to use RNAi in mammalian cells were hampered at first because of a nonspecific, interferon-mediated response to dsRNA longer than 30 bp that is seen in most mammalian cell lines. With the demonstration that 21–22 ntd siRNAs, either chemically synthesized or expressed from a plasmid vector, could efficiently induce RNAi in mammalian cells without inducing the inteferon response, the door was opened for development of RNAi tools in mammalian systems.

While some researchers were studying the phenomenon of RNAi and working to exploit its use as a tool for reverse genetics, others were trying to understand a distinct but related set of small RNAs—microRNAs (miRNAs). The first miRNAs to be discovered, lin-4 and let-7, were identified through loss-of-function mutations affecting control of postembryonic development in *C. elegans*. Other miRNAs were soon identified in *C. elegans*, *Drosophila*, and mouse by combinations of forward genetics, cDNA cloning, bioinformatics, and reverse genetics. By 2002, miRNAs were firmly established as a large class of conserved regulatory molecules in animals and plants. miRNAs have been shown to play a role in developmental timing, cell death, cell proliferation, and oncogenesis. This class of small RNA may represent 2–3% of the total number of genes in humans, and estimates of miRNA target-binding sites indicate that miRNAs may play a role in regulating as many as 30% of mammalian genes.

A formal connection between the phenomenon of RNA interference and miRNAs was made when the same enzyme, Dicer, was shown to process both long dsRNA into siRNAs and cytoplasmic miRNA precursors (pre-miRNAs) into mature miRNAs. Other components of the RNAi and miRNA silencing pathways have been shown to be closely related, most prominently Argonaute, which is a key component of the RNA-induced Silencing Complex (RISC). It now appears that miRNAs function in a silencing complex that is similar, if not identical, to RISC to regulate expression of target genes either through cleavage of mRNA or translational repression: if the miRNA exhibits perfect complementarity to its target mRNA, the mRNA is cleaved (typically in plants); if there is only partial complementary, translational repression occurs (typically in animals). The further discovery of endogenous small RNAs, distinct from miRNAs, that function in transcriptional silencing and genome stability has driven the adoption of the more general terms "RNA silencing" and "small RNAs" to describe the collection of related silencing pathways and their RNA guides. The term "RNA interference" (RNAi) continues to describe mRNA cleavage that is induced by a dsRNA trigger.

The Nobel Prize

The Nobel Prize in Physiology or Medicine of year 2006 was shared by Professor Andrew Z. Fire at Stanford University, California, USA, and Professor Craig C. Mello at the University of Massachusetts Medical School in Worcester, USA. They received the prize for their discovery that double-stranded RNA triggers suppression of gene activity in a homology-dependent manner, a process named RNA interference (RNAi). Their discovery has revealed a new mechanism for gene regulation, and the biochemical machinery involved plays a key role in many essential cellular processes. Double-stranded RNA synthesized within the cell can reduce or abolish gene activity by RNAi-like mechanisms. This control system for gene expression has proven to be important for both the development of an organism and the physiological functions of cells and tissues. Furthermore, RNAi protects against RNA virus infections, especially in plants and invertebrate animals, and secures genome stability by keeping mobile elements silent. Today, double-stranded RNA is used as a powerful tool to experimentally elucidate the function of essentially any gene in a cell. The discovery of RNAi has already had an immense impact on biomedical research and will most likely lead to novel medical applications in the future.

Antisense RNA

Messenger RNA (mRNA) is single-stranded. Its sequence of nucleotides is called "sense" because it results in a gene product (protein). Normally, its unpaired nucleotides are "read" by transfer RNA anticodons as the ribosome proceeds to translate the message. (See mechanism of translation: Fig C.

However, RNA can form duplexes just as DNA does. All that is needed is a second strand of RNA whose sequence of bases is complementary to the first strand; e.g.,

5′ C A U G 3′ mRNA

3′ G U A C 5′ Antisense RNA

The second strand is called the antisense strand because its sequence of nucleotides is the complement of message sense. When mRNA forms a duplex with a complementary antisense RNA sequence, translation is blocked. This may occur because

- the ribosome cannot gain access to the nucleotides in the mRNA or
- duplex RNA is quickly degraded by ribonucleases in the cell.

With recombinant DNA methods, synthetic genes (DNA) encoding antisense RNA molecules can be introduced into the organism.

Example: The Flavr Savr tomato

Most tomatoes that have to be shipped to market are harvested before they are ripe. Otherwise, ethylene synthesized by the tomato causes them to ripen and spoil before they reach the customer. Transgenic tomatoes have been constructed that carry in their genome an artificial gene (DNA) that is transcribed into an antisense RNA complementary to the mRNA for an enzyme involved in ethylene production. These tomatoes make only 10% of the normal amount of the enzyme. The goal of this work was to provide supermarket tomatoes with something closer to the appearance and taste of tomatoes harvested when ripe.

Another example:

Flower of a tobacco plant carrying a transgene whose transcript is antisense to one of the mRNAs needed for normal flower pigmentation. Flower of another transgenic plant that failed to have its normal pigmentation altered. (van der Krol *et al.*, 1988).

RNA interference (RNAi) or PTGS

In testing the effects of antisense RNA, one should use sense RNA of the same coding region as a control. Surprisingly, preparations of sense RNA often turn out to be as effective an inhibitor as antisense RNA. Why this happens? It seems that the preparations of sense RNA often are contaminated with hybrids: sense and antisense strands that form a double helix of double-stranded RNA (dsRNA). Double-stranded RNA corresponding to a particular gene is a powerful suppressant of that gene. In fact, the suppressive effect of antisense RNA probably also depends on its ability to form dsRNA (using the corresponding mRNA as a template).

The ability of dsRNA to suppress the expression of a gene corresponding to its own sequence is called RNA interference (RNAi). It is also called post-transcriptional gene silencing or PTGS.

RNA interference (RNAi) is a phenomenon leading to post-transcriptional gene silencing (PTGS) after endogenous production or artificial introduction into a cell of small interfering double strand RNA (siRNA) with sequences complementary to the targeted gene (Bosher and Labouesse, 2000). Whereas the transcription of the gene is normal, the translation of the protein is prevented by selective degradation of its encoded mRNA. However, PTGS is not restricted to RNAi and has emerged as a more complex mechanism that involves several different proteins and small RNAs. It is presumed that cells employ RNAi to tightly regulate protein levels in response to various environmental stimuli, although the extent to which this mechanism is employed by specific cell types remains to be discovered. However, the fact that RNAi is operative in cells of organisms ranging from plants, to nematodes and flies, and to mammals attests to its fundamental importance in the selective suppression of protein translation by targeted degradation of the encoding mRNA. Beyond its biological relevance, PTGS is emerging as a powerful tool to study the function of individual proteins or sets of proteins. User-friendly technologies for introducing siRNA into cells, in culture or *in vivo,* to achieve a selective reduction of single or multiple proteins of interest are rapidly evolving. The present article reviews this emerging technology, findings obtained to date using such RNAi methods, and the potential of RNAi based therapeutics for treating human disease.

Principles of RNA Interference

RNA interference most likely evolved as a mechanism for cells to eliminate unwanted foreign genes. Foreign genes are often present in cells at high copy numbers, being present as viral genes, transposable elements, or as plasmids introduced experimentally in cell transfection protocols. It has been known for several decades that the level of expression of transgenes usually decreases as the number of copies present in the cell increases and that endogenous homologous genes can also be suppressed by the presence of the transgene (Napoli *et al.*, 1990). Although such gene silencing can occur at the transcriptional level, it is now recognized that a major mechanism of gene suppression occurs post-transcriptionally, and that a major mechanism for this PTGS is RNAi, the selective degradation of mRNAs targeted by siRNAs (Van Blokland *et al.*, 1994). Such PTGS via RNAi can occur very rapidly with proteins for many genes, being decreased within hours, and completely absent within 24 h (Pruss *et al.*, 1997). Based upon these and other findings initially made in studies of plants (Ratcliff *et al.*, 1997), it seems very likely that RNAi evolved as a mechanism to defend plant cells against viral infections.

HISTORY

The gene expression process is of fundamental importance for all living organisms. Most genes reside in the chromosomes located in the cell nucleus and express themselves via proteins synthesised in the cytoplasm. The genetic material was identified as deoxyribonucleic acid (DNA) in 1944 (Avery *et al.*, 1994) and the double-helical nature of DNA was revealed in 1953 by Francis Crick, James Watson and Maurice Wilkins; who were awarded Nobel Prize in Physiology or Medicine in 1962. At that time, the main problem was how DNA in the nucleus could govern protein synthesis in the cytoplasm. It was proposed that another nucleic acid, single-stranded ribonucleic acid (RNA), acts as an intermediary in the process, and the so-called Central Dogma was formulated, i.e. the idea that the genetic information is transcribed from DNA to RNA and then translated from RNA into protein. The RNA carrying the genetic information was first believed to be the RNA in ribosomes; for several years the hypothesis was formulated as "one gene-one ribosome-one protein". In 1961, Francois Jacob and Jacques Monod presented a visionary gene control model, for which they received the Nobel Prize in Physiology or Medicine in 1965 together with André Lwoff. In their model, they proposed that the gene is transcribed into a specific RNA species, messenger RNA (mRNA). Soon afterwards it was demonstrated that a short-lived, non-ribosomal RNA directs the synthesis of proteins. Subsequently, Marshall Nirenberg and HarGobind Khorana cracked the genetic code and could assign code words (codons; triplets of nucleotides) to the twenty amino acids (they received the Nobel Prize in Physiology or Medicine in 1968 together with Robert Holley). Francis Crick predicted that an RNA molecule could act as an adaptor between mRNA and the amino acid, and a short, stable RNA, transfer RNA (tRNA) was soon identified as the predicted adaptor.

For many years, messenger RNA was believed to correspond to an uninterrupted nucleotide sequence in the DNA. It therefore came as a complete surprise when Phillip Sharp and Richard Roberts showed (1977) that the mRNA sequence could be distributed discontinuously in the genome (the split gene concept; Nobel Prize in 1993). It was known that long RNA molecules (pre-mRNA, heterogeneous nuclear RNA) are trimmed to much shorter mature mRNAs, and Sharp and Roberts therefore suggested that the mRNA sequences, the exons, are likely to be cut out from the primary transcript and spliced, while the intervening sequences, the introns, are degraded. It was immediately realised that the discontinuous arrangement of the mRNA sequences in DNA had important evolutionary implications. In addition, the RNA splicing process could generate different mRNAs, and more than one protein can emanate from the primary transcript (alternative splicing).

The discovery that RNA could act as a catalyst gave a radically new perspective on the roles of RNA (Nobel Prize in Chemistry to Sidney Altman and Thomas Cech in 1989). It was soon revealed that RNA is able to catalyse its own replication and the synthesis of other RNA molecules (the ribozyme concept), which led to the idea that RNA was the first genetic material on earth. An "RNA world" is believed to have existed before DNA took over the role of being the key genetic material, and RNA was relegated to the role of messenger between DNA and protein. Not only did the discovery of catalytic RNA have evolutionary implications, but it also suggested that RNA could play a more active role in gene expression than earlier realised. It is now well established that ribosomal RNA catalyses peptide bond formation during translation.

A large number of small RNA molecules work in conjunction with proteins in ribonucleoprotein (RNP) complexes. There are non-coding RNA molecules that affect transcription (e.g. human 7SK snRNA bound to elongation factors), translation (e.g. SRP RNA in the signal recognition particle), replication (e.g. telomerase RNA) and chromosome structure (e.g. XIST RNA causing X chromosome inactivation). Others regulate RNA processing (e.g. M1 RNA in RNase P, snRNAs and snoRNAs) and RNA editing (guide RNAs). These various RNP particles are now being extensively investigated to understand their specific roles in the cell.

In the early 1980s it was revealed in *Escherichia coli* that small RNA molecules (about 100 nucleotides in length) can bind to a complementary sequence in mRNA and inhibit translation (Mizuno *et al.*, 1984; Nordström and Wagner, 1994). Today, about 25 cases of regulatory trans-acting antisense RNAs are known in *E. coli* (Gottesman, 2004). Regulation of translation by antisense RNA also occurs in eukaryotes as was first demonstrated in 1993 when genes governing the development of the worm *Caenorhabditis elegans* were studied (Lee *et al.*, 1993; Wightman *et al.*, 1993). For many years, this thoroughly documented case of posttranscriptional regulation was regarded as an oddity. The mechanism received more attention when a second example of a small regulatory RNA was found in *C. elegans* (Reinhart *et al.*, 2000), because in this case similar sequences were also present in other species. However, the situation changed drastically when a large number of small RNA molecules, called microRNAs (miRNAs), were revealed in 2001 (Lee and Ambros, 2001; Lau *et al.*, 2001).

Prior to the discovery of RNA interference, a phenomenon called gene (or RNA) silencing was described in plants. It was noted in experiments around 1990 that a cloned gene incorporated into the genome (a transgene) could not only induce or stimulate gene activity but could also inhibit the expression of homologous sequences, a phenomenon called homology-dependent gene silencing. The inhibition of gene activity could take place at the transcriptional level (transcriptional gene silencing, TGS) (Matzke *et al.*, 1989; Wassenegger *et al.*, 1994; Park *et al.*, 1996) or at the posttranscriptional level (posttranscriptional gene silencing, PTGS) (Napoli *et al.*, 1990; van der Krol *et al.*, 1990; Smith *et al.*, 1990; de Carvalho *et al.*, 1992; van Blokland *et al.*, 1994). A PTGS-like process called quelling was also established in the fungus *Neurospora crassa* (Romano and Macino, 1992). Analyses of viral infection in plants gave further insight into the mechanism of PTGS (Lindbo *et al.*, 1993; Dougherty *et al.*, 1995; Fire *et al.*, 1998). However, although it was evident that RNA played a key role in gene silencing, the phenomenon remained enigmatic until the discovery of RNA interference provided a most unexpected explanation with many profound consequences.

C. elegans as a model organism for Discovery of RNA interference

Craig Mello used the nematode worm *C. elegans* as a model organism to investigate how embryonic cells differentiate and communicate during development. In addition, he investigated the mechanism of RNA interference, a form of sequence-specific gene silencing triggered by double-stranded RNA.

Embryogenesis

Embryogenesis in *Caenorhabditis elegans* is a remarkably reproducible process mapped out to the resolution of every single cell division in the wild-type embryo (a total of only 558 cells at hatching). Despite the small number of cells, the newly hatched larvae contain all of the major types of differentiated tissues found in other animals (e.g., intestine, muscles, germline, epidermis, and a nervous system). Because of its relative simplicity, the *C. elegans* embryo provided a valuable opportunity to understand the process of animal development at the level of individual cells.

The *C. elegans* embryo utilizes several mechanisms to produce early blastomeres that are committed to distinct patterns of development. These mechanisms include position-dependent cell-cell interactions as well as the asymmetric expression of maternally provided transcriptional activators and repressors. We are interested in the mechanisms that lead to the initial diversification of cell fates during the first several divisions of the embryo. One key protein involved in this process is the PIE-1 protein, a small Cys3His– zinc finger protein that functions in the specification of certain somatic cell fates and in maintaining the germline fate. After each round of cell division in the early embryo, the PIE-1 protein is localized to one and only one cell until the 28-cell stage of embryogenesis, when it is expressed in the two primordial germ cells. During early embryogenesis, PIE-1 appears to act as a repressor that protects the germline lineage from differentiating in response to transcriptional activators that are present in the germline. Work suggests that PIE-1 not only prevents transcription but also blocks chromatin remodeling in the germline. The mechanism of PIE-1 localization and function is being explored.

In addition to the various mechanisms that lead to the distinct fates of certain early blastomeres, several studies have indicated that cells throughout the embryo share a common mechanism for linking the axis of cell division to cell fate. For example, most of the cell divisions after the eight-cell stage of embryogenesis are oriented along the anterior/ posterior (a/p) axis, and essentially all of these divisions result in a/p daughter cells with different fates. Using a combination of reverse and forward genetics, several components of the signaling pathways that control cell fate and cell polarity in early *C. elegans* embryos have been identified.

Several genes involved in polarity signaling in the embryo encode highly conserved proteins related to Wnt-signaling components. Wnt genes comprise a large family, including numerous vertebrate and invertebrate members, and encode secreted proteins important in cell-cell interactions. In the early *C. elegans* embryo, Wnt signaling acts in parallel with a phosphotyrosine-signaling pathway mediated by SRC-1, a *C. elegans* homolog of tyrosine kinase pp60 Src. Wnt and SRC signaling converge to regulate WRM-1, a homolog of human β-catenin. Mutations that stabilize β-catenin are found in numerous forms of cancer in humans. Genetic tests suggested that signaling overcomes, or down-regulates, a WRM-1 nuclear-export pathway that is mediated by IMB-4, a homolog of the conserved exportin CRM-1. Studies involved a biochemical dissection of signaling as well as extensive genetic screening, with a focus on understanding how these and other signaling pathways function in concert to provide the positional information necessary for patterning the developing embryo.

Reverse Genetics and RNA Interference

During developmental studies, some strange properties of the reverse genetic method were observed. This method was analogous to "antisense", wherein the microinjection of RNA complementary to the messenger RNA is used to block the activity of specific genes. However, although antisense methodology generally causes a partial block of gene activity, a remarkably robust interference effect that could spread from tissue to tissue in the animal was observed. In most experiments, nearly 100 percent of the progeny of an injected animal showed phenotypic effects consistent with blocking the targeted gene's activity. Even more surprising, interference could be transmitted for multiple generations in the germline as an extrachromosomal element.

The remarkable potency and long-lasting effects of RNA interference (RNAi) prompted to investigate the RNAi mechanism. In a collaborative study with Andrew Fire (Carnegie Institution of Washington), it was shown that double-stranded RNA (dsRNA) was 10 to 100 times more efficient than antisense RNA at inducing interference. This exciting discovery soon led to the application of RNAi via dsRNA in other organisms. To date, dsRNA-induced interference has been reported to one degree or another in a variety of animals and plants. In *C. elegans*, dsRNA is so potent at causing interference that simply by soaking worms in RNA, or feeding the worms bacteria expressing a dsRNA, is sufficient to induce interference.

Several mutant strains were identified that are resistant to RNAi. One class of mutants exhibits an activation of the endogenous transposons, selfish DNA elements, in the genome of the animal. This suggests that one *in vivo* function of the silencing mechanism may be to silence the transposons. Molecular and genetic studies of RNAi and of similar gene-silencing mechanisms in other organisms have revealed striking similarities. For example, the *rde-1* gene, required for RNAi in *C. elegans*, has homologs required for gene-silencing mechanisms in *Drosophila*, plants, and fungi. These and other findings indicate that an ancient mechanism underlies the RNAi process, and its now known that related forms of gene regulation are important in development and disease in humans.

Consistent with this latter idea, it was found that *C. elegans* genes related to *rde-1* are required for development and for proper chromosome segregation. The genomes of plants and animals, including humans, contain hundreds of small dsRNA-encoding genes, which have been named micro-RNA genes (miRNAs). The miRNAs are similar in size to the small-interfering RNAs (siRNAs) that mediate RNAi and like siRNAs are processed from dsRNA precursors through the action of an RNase III–related enzyme, Dicer. Understanding the RNAi mechanism will not only lead to better reverse genetic tools for manipulating and studying gene function but will also shed light on an ancient and highly conserved gene regulatory mechanism. Ultimately, it may be possible to harness this natural gene-regulatory mechanism to reduce the expression of disease-related mutant genes.

This work of Fire and Mello was supported in part by a Pew Scholarship and by grants from the National Institutes of Health, the American Cancer Society, and the March of Dimes Birth Defects Foundation.

DISCOVERY OF PTGS AND RNA INTERFERENCE

Andrew Fire and Craig Mello published their break-through study on the mechanism of RNA interference in *Nature* in 1998 (Izant and Weintraub, 1984). It was earlier known that antisense RNA (Guo and Kemphues, 1995), but remarkably also sense RNA (Guo and Kemphues, 1995) could silence genes, but the results were inconsistent and the effects usually modest. However, due to the fact that both sense and antisense RNA could cause silencing Mello argued that the mechanism could not just be a pairing of antisense RNA to mRNA, and he coined the term RNA interference for the unknown mechanism (Rocheleau *et al.*, 1997).

In their *Nature* paper, Fire and Mello tested the phenotypic effect of RNA injected into the worm *C. elegans*. They established that annealed sense/antisense RNA, but neither antisense nor sense RNA alone, caused the predicted phenotype (Fig. E). Furthermore, only injection of double-stranded RNA (dsRNA) led to an efficient loss of the target mRNA (Fig. E). Fire and Mello could present a series of straightforward conclusions in their study. The main results can be summed up as follows: First, silencing was triggered efficiently by injected dsRNA, but weakly or not at all by sense or antisense single-stranded RNAs. Second, silencing was specific for an mRNA homologous to the dsRNA; other mRNAs were unaffected. Third, the dsRNA had to correspond to the mature mRNA sequence; neither intron nor promoter sequences triggered a response. This indicated a posttranscriptional, presumably cytoplasmic mechanism. Fourth, the targeted mRNA disappeared suggesting that it was degraded. Fifth, only a few dsRNA molecules per cell were sufficient to accomplish full silencing. This indicated that the dsRNA was amplified and/or acted catalytically rather than stoichiometrically. Sixth, the dsRNA effect could spread between tissues and even to the progeny, suggesting a transmission of the effect between cells. Furthermore, Fire and Mello made the remark that RNAi could provide an explanation for a phenomenon studied in plants for several years: posttranscriptional gene silencing (PTGS). Finally, they ended their paper by speculating about the possibility that "dsRNA could be used by the organism for physiological gene silencing".

In their *Nature* paper, Fire and Mello did not take a firm stand on the issue of whether dsRNA acts via a transcriptional or posttranscriptional mechanism. However, in a follow-up study published in *PNAS* the same year, Fire provided good evidence for the view that mRNA is the target for dsRNA (recognition via complementary strands), and that the targeted mRNA is degraded prior to translation, *i.e.* dsRNA exerts its effect at the posttranscriptional level (Montgomery *et al.*, 1998). He also presented a specific model showing how *double-stranded* RNA could function in a *catalytic* manner to target homologous mRNAs for degradation. This model differed strikingly from the simple antisense model of that time, which only predicted interaction between an interfering single-stranded RNA molecule and mRNA. It can be added that in this *PNAS* paper, Fire also indicated the possibility that the RNAi mechanism could be a specific "tactical" approach to viral defence in lower organisms (to be compared with the global interferon response in mammals).

Within a year, the presence of RNAi had been documented in many other organisms, including fruit flies, trypanosomes, plants, planaria, hydra and zebrafish (Tuschl *et al.*, 1999). In initial experiments with mammalian cultured cells, it was not possible to elicit a potent and specific RNAi response because of a predominant nonspecific physiological reaction of these cells to long dsRNA. However, when the cells were exposed to a short, 21 nucleotide long, dsRNA, an efficient targeted silencing was also obtained in these cells (Elbashir *et al.*, 2001). Thus, the generality of the RNAi phenomenon among eukaryotes was proven very rapidly; a remarkable exception is the budding yeast, *Saccharomyces cerevisiae*.

The account of the various discoveries in the field of RNA interference is given in Table:

Year	*Discovery*	*Scientists*
1990	Cosuppression of purple colour in Petunia	Napoli *et al.*
1996	Quellingin Neurospora crassa (fungi)	Cogni C. *et al.*
1998	dsRNA injection in *C.elegans*	Fire and Mello
1999	siRNA identified in plants	Hamilton & Baulcombe
2000	siRNA identified into mouse cells	Tuschl and Elbashir
2001	siRNA identified in Drosophila	Elbashir *et al.*
	Functional domains in Dicer identified	Bemstein *et al.*
	Co-purification of Agronaute with RISC	Hammond *et al.*
2003	Incorporation of one strand to RISC	Khorova *et al.*
	Strand with less 5′ stability incorporates to RISC	Schwarz *et al.*
2004	Multiple Dicer genes identified in Drosophila and plants / Evidence for different RISCs	He & Hannon Tijsterman *et al.*
2005	Longer siRNA can be more potent in silencing	Kim *et al.*

Detailed study

PTGS and RNAi were discovered in genetic transformation studies of eukaryotic cells, principally plants and worms, wherein it was shown that mRNAs for the encoded transgene alone, or together with mRNAs for homologous endogenous genes are very low or absent despite high levels of transcription (Fire, 1999; Marathe *et al.*, 2000). The ability to manipulate and monitor gene expression in the plant *Arabidopsis thaliana* and the roundworm *Caenorhabditis elegans* (the genomes of both species are now complete) revealed the process of RNAi and allowed the relatively rapid identification of several genes that regulate the RNAi process.

Transgenes insert into the genomes of plants by recombination in an apparently random manner so that the number of inserted copies, their chromosomal location, and their local arrangement within the chromosome vary among transformants. The observation of an inverse correlation between copy number and the level of gene expression suggested that an increased copy number of a particular gene results in silencing of that gene (Assaad *et al.*, 1993). It was initially thought that such gene silencing was due to reduced gene transcription resulting from interactions between closely linked copies that result in the formation of secondary structures that promote methylation and inhibition of transcription (Ye and Signer, 1996). Further studies showed that transcriptional gene silencing (TGS) could also occur in trans, such that one transgene can be silenced by another transgene introduced either by crossing or transformation. It was then proposed that a silencing RNA is produced by one locus that somehow effects the silencing of the other gene by a mechanism involving RNA-mediated inhibition of transcription (Mette *et al.*, 2000). Although some data were consistent with such mechanisms of transcriptional silencing, additional data suggested the involvement of PTGS. The presence of double-stranded RNAs (dsRNAs) and their cleavage into siRNAs of approximately 23 nucleotides was demonstrated, and it was then shown that expression of dsRNA with sequences corresponding to open reading frames in plants results in PTGS (Hamilton and Baulcombe, 1999). Similarly, expression of dsRNAs with sequences complementary to those of endogenous genes results in the selective silencing of those genes in *C. elegans* (Zamore *et al.*, 2000). Collectively, the studies of *A. thaliana* and *C. elegans* showed that both TGS and PTGS can be initiated by the same RNA degradation pathway—TGS occurs when the dsRNA includes promoter sequences, whereas PTGS occurs when the dsRNA includes coding sequences. Although degradation of dsRNA is common to both mechanisms of gene silencing, the results also indicated that dsRNA-mediated TGS and PTGS involve different specific steps. Although RNAi as a mechanism of PTGS was first discovered in plants and may have evolved as a cellular defense mechanism against foreign DNA and RNA, it is very clear that RNAi is widely employed in most if not all eukaryotic cells as a mechanism to regulate the expression of endogenous genes. In 1998, it was discovered that injection of dsRNA was much more effective for silencing of gene expression in *C. elegans* than was single-stranded antisense RNA (Fire *et al.*, 1998). This experimentally induced PTGS, the first report of the use of

RNAi as a tool in biology, was very potent, and remarkably, the PTGS occurred not only in the worms to which the dsRNA was administered, but also in their progeny. It was then demonstrated that the endogenous mRNA was the target of the injected dsRNA by a posttranscriptional mechanism and involving degradation of the targeted mRNA (Montgomery *et al.*, 1998). Surprisingly, it was further shown that the dsRNA is effective at very low concentrations, such that the copy numbers of the targeted mRNA are far greater than the number of dsRNAs present in the cell (Fire *et al.*, 1998; Kennerdell and Carthew, 1998). In addition, the suppression of the protein encoded by the targeted mRNA was found to persist through many rounds of cell division. The latter two observations strongly suggested that cells possess a mechanism for amplifying the RNAi mechanism. Not only can the RNAi process be maintained within cells of a common lineage, but it can also be transferred between cells, as shown in *C. elegans* where injection of dsRNA into the intestine results in silencing of the targeted gene in all cells of the F1 progeny of that worm (Fire *et al.*, 1998). Indeed, dsRNA can enter cells and induce PTGS when worms are soaked in a solution containing the dsRNA or when the worms are fed bacteria expressing dsRNA (Tabara *et al.*, 1998; Timmons and Fire, 1998). Recently, a transmembrane protein called SID-1 was identified as a possible mediator of intercellular transfer of RNAi (Winston *et al.*, 2002).

Subsequently, other organisms were assayed for their capacity to induce RNAi. Evidence for RNAi in Drosophila was first demonstrated by Kennerdell and coworkers (Kennerdell and Carthew, 1998) who showed the involvement of the *frizzled* and *frizzled2* genes in the wingless pathway after introduction of dsRNA into embryos. Again, several techniques were developed in order to use dsRNA in this organism leading to the establishment of cell-free (Tuschl *et al.*, 1999) and cell culture models (Caplen *et al.*, 2000). A system that employed dsRNA as an extended hairpin-loop RNA was developed to induce heritable gene silencing (Kennerdell and Carthew, 2000). The *Drosophila* system has allowed the identification of several endogenous genes that play key roles in the RNAi process. An RNA nuclease activity called RISC (RNA-induced silencing complex) was discovered that is responsible for the degradation of endogenous mRNAs, as well as small nucleotide fragments (25 nucleotides in length), which could be used as guides by RISC (Hammond *et al.*, 2000). They later characterized RISC as a ribonucleoproteic complex (Hammond *et al.*, 2001). These results were soon extended by showing that RNAi is an ATP-dependent and translation-independent event where the introduced dsRNA is processed into 21–23 nucleotide fragments that guide the cleavage of endogenous transcripts (Zamore *et al.*, 2000). The enzyme responsible for the processing of the dsRNA was later discovered as a RNase III family nuclease named Dicer, a protein with high homology to the *rde-1 C. elegans* gene (Bernstein *et al.*, 2001). To study the functions of RNAi in yeast, Volpe *et al.* (2002) deleted argonaute, dicer, and RNA-dependent RNA polymerase homologs; deletion resulted in the accumulation of complementary transcripts from centromeric heterochromatic repeats and de-repression of transgenes integrated at the centromere and impairment of centromere function. The authors of the latter study proposed that dsRNA arising from centromeric repeats targets the formation and maintenance of heterochromatin through RNAi. In mammalian cells, RNAi was first employed as a tool to induce the silencing of the targeted gene (Wianny and Zernicka-Goetz, 2000). This approach was partially successful in mouse embryos (Wianny and Zernicka-Goetz, 2000; Svoboda *et al.*, 2001) and embryonic cell lines (Billy *et al.*, 2001; Yang *et al.*, 2001; Paddison *et al.*, 2002b) where specific gene silencing was achieved. On the other hand, the introduction of dsRNA into mammalian somatic cells presents a major problem because it can induce (in a manner similar to the silencing observed during viral infection) to the activation of the PKR (protein kinase R) and RNAseL pathway, resulting in the inhibition of protein synthesis and induction of apoptosis (Baglioni and Nilsen, 1983; Clarke and Mathews, 1995; Gil and Esteban, 2000). Interestingly, this shows that, in mammalian cells, the mechanisms for RNA interference are not identical to those in lower organisms although RNAi does operate in at least a subset of mammalian cell types, in a Dicer-dependent manner via post-transcriptional mechanisms (Billy *et al.*, 2001; Paddison *et al.*, 2002b). Elbashir *et al.* (2001a) had the idea of directly introducing 21–23 nucleotide dsRNA (siRNAs) into mouse and human cells to try to avoid the problems associated with the expression of longer dsRNAs. They showed that the siRNA could efficiently trigger silencing in the mammalian cells.

RNAi mechanisms may provide explanations for various biological phenomena that were previously described, but without any understanding of the possible underlying mechanism. For example, it was recently proposed that RNAi mechanisms could explain the controversial process of RNA-mediated memory transfer in planaria (Smalheiser *et al.*, 2001). It will certainly be of interest to investigate possible roles for RNAi in the many different physiological processes that involve modulation of protein levels at a post-transcriptional level.

GENE SILENCING BY RNAi

RNA interference (RNAi) is the process of mRNA degradation that is induced by double-stranded RNA in a sequence-specific manner. RNAi has been observed in all eukaryotes, from yeast to mammals. The power and utility of RNAi for specifically silencing the expression of any gene for which sequence is available has driven its incredibly rapid adoption as a tool for reverse genetics in eukaryotic systems.

The RNAi pathway is thought to be an ancient mechanism for protecting the host and its genome against viruses and rogue genetic elements that use double-stranded RNA (dsRNA) in their life cycles. RNAi is now recognized to be but one of a larger set of sequence-specific cellular responses to RNA, collectively called RNA silencing. These responses have been shown to play a role not only in mRNA and dsRNA stability/ degradation, but also in regulation of translation, transcription, chromatin structure, and genome integrity. In plants and animals, RNA silencing has been adapted to play a critical role in regulation of cell growth and differentiation using a class of small RNAs called microRNAs (miRNAs).

Shortly after the discovery of RNAi, it was shown that PTGS in plants is correlated to the presence of a population of small RNAs (about 25 nucleotides long), and that this RNA contains both sense and antisense RNA sequences (Hamilton and Baulcombe, 1999). It was proposed that this RNA is the determinant of PTGS. Earlier, but subsequent to the discovery of RNAi in animal cells, it had been shown that RNA in double-stranded form also initiates PTGS in plants (Waterhouse *et al.*, 1998).

The biochemistry of RNAi was further elucidated in an *in vitro* system based on *Drosophila* embryo extracts (Tuschl *et al.*, 1999). It could be shown that dsRNA is processed to 21-23 nucleotide long dsRNA fragments (Zamore *et al.*, 2000), which was in good agreement with the data for PTGS in plants (Hamilton *et al.*, and Baulcombe *et al.*, 1999). It was proposed that this short dsRNA, siRNA (small interfering RNA), guided the cleavage of mRNA. Subsequently, Fire and Mello were able to follow the process *in vivo* (Parrish *et al.*, 2000). They established that long dsRNA is cleaved to small RNA (about 25 nucleotides long), and that antisense RNA triggers degradation of mRNA via base-pairing to mRNA. Thus, the trigger, dsRNA, had then been connected to the low molecular weight effector.

The molecular machinery involved in RNAi was subsequently revealed (Fig. F, G, H). In an *in vitro* system, built on *Drosophila* cultured cells, it was demonstrated that a large complex, called RISC (RNA-induced silencing complex) is targeted to the mRNA via a short antisense RNA, and that mRNA is cleaved and subsequently degraded (Hammond *et al.*, 2000). Subsequently, it has been shown that RISC contains at least one member of the argonaute protein family, which is likely to act as an endonuclease and cut the mRNA (nowadays often referred to as the Slicer function). It was also demonstrated that a ribonuclease III-like nuclease, called Dicer, is responsible for the processing of dsRNA to short RNA (Bernstein *et al.*, 2001). In certain systems, in particular plants, worms and fungi, an RNA dependent RNA polymerase (RdRP) plays an important role in generating and/or amplifying siRNA (Cogoni and Macino, 1999).

Thus, in just a few years a vast amount of information accumulated on the specific proteins and protein complexes involved in RNAi and on the molecular details of the particular steps in the process (Hannon, 2002; Mello and Conte Jr; 2004; Meister and Tuschl, 2004; Hammond, 2005).

Small RNAs: siRNAs, rasiRNAs, and miRNAs siRNAs and rasiRNAs

Short interfering RNAs (siRNAs) were originally identified as intermediates in the RNA interference pathway. siRNAs are 21–25 bp dsRNA with dinucleotide 3' overhangs that are formed in the cell from longer dsRNA molecules. The fully assembled RNA-induced silencing complex (RISC) contains only one strand of the siRNA, the guide strand. The guide strand is thought to provide target specificity for RISC-mediated cleavage through perfect base pairing with the mRNA target.

Long dsRNA (several hundred base pairs) can be used to artificially induce the RNAi pathway in fungi, plants, insects, and worms. In mammalian cells, dsRNA >30 bp induces a nonspecific, antiviral interferon response. However, synthetic siRNAs, ~21 bp with dinucleotide 3' overhangs, are potent triggers of RNAi in mammalian cells.

Although siRNAs are typically thought of as originating from exogenous precursors or triggers, it is perhaps not surprising that siRNAs generated from endogenous precursors have been identified. These endogenous siRNAs have been found to function in maintenance of chromatin structure and genome integrity as well as post-transcriptional gene silencing.

Endogenous siRNAs have been identified in plants, fungi, and animals. These siRNAs are derived *in vivo* from perfectly base-paired dsRNA precursors comprised of two distinct RNA strands. In many cases, endogenous siRNAs originate from repetitive elements within the genome, such as heterochromatic regions at centromeres and telomeres, and are therefore known as repeat-associated siRNAs (rasiRNAs). rasiRNAs appear to function, through an RNA-induced transcriptional silencing (RITS) complex, in maintaining the heterochromatic, and hence transcriptionally repressed, state of the region that encodes the rasiRNA. In *Drosophila*, rasiRNAs encoded by the Y-encoded Su (Ste) locus have also been implicated in transregulation of the X-linked Stellate (Ste) locus, but it is not clear if silencing is at the transcriptional or post-transcriptional level. In the protozoan *Tetrahymena*, endogenous siRNAs have also been shown to play a role in specifying sequences that are eliminated during macronucleus formation after sexual conjugation.

Endogenous nonrepetitive siRNAs have also been identified in the model plant *Arabidopsis* as well as in *Caenorhabditis elegans*. In some cases these endogenous siRNAs are complementary to longer, unique, coding RNAs. In *Arabidopsis*, nonrepetitive siRNAs have been implicated in regulation of transcript abundance through mRNA cleavage, indicating an RNAi mechanism similar to that triggered by exogenous siRNAs.

miRNAs

MicroRNAs (miRNAs) are a large class of evolutionarily conserved RNAs found in plants and animals. These small RNAs have been shown to play critical roles in developmental timing, hematopoietic cell differentiation, cell death, cell proliferation, and oncogenesis. miRNAs may represent 2–3% of the total number of genes in humans, and estimates of the number of miRNA target binding sites indicate that miRNAs may play a role in regulating as many as 30% of mammalian genes.

miRNAs are 19–23 nt single-stranded RNAs, originating from single-stranded precursor transcripts that are characterized by imperfectly base-paired hairpins. miRNAs function in a silencing complex that is similar, if not identical, to RISC.

In plants, miRNAs resemble siRNAs in that they exhibit perfect complementarity to binding sites in target mRNA and appear to silence gene expression through mRNA cleavage. Animal miRNAs typically exhibit partial complementarity to target binding sites in the 3' untranslated region (UTR) of target mRNA and repress translation without mRNA cleavage. It originally appeared that the process of miRNA gene silencing in animals involved reduced steady-state protein levels for the targeted gene without a reduction in the corresponding levels of mRNA; however, examples to the contrary have challenged this idea. miRNAs have also been implicated in mRNA turnover pathways in *Drosophila* and humans that use AU-rich elements (ARE), which are found in the 3' UTRs of many short-lived transcripts. miRNAs are also implicated in transcriptional silencing via hypermethylation in *Arabidopsis*.

In *C. elegans*, successful development through its larval stages and on to the adult requires the presence of at least two "microRNAs" ("miRNAs") – single-stranded RNA molecules containing about 22 nucleotides and thus about the same size as siRNAs.

These small single-stranded transcripts are generated by the cleavage of larger precursors using the *C. elegans* version of Dicer.

They act by either destroying or inhibiting translation of several messenger RNAs in the worm (by binding to a region of complementary sequence in the 3' untranslated region [3'-UTR] of the mRNA).

The microRNAs (miRNAs) in *C. elegans* (which were first called "small temporal RNAs") turn out to be representatives of a large class of RNAs that

- are cleaved from larger RNA precursors (by Dicer in animals) encoded by the organism's own genes;
- are found in humans (we have more than 200 miRNA genes), Drosophila, mice, frogs, fish, and plants (Arabidopsis) as well as in *C. elegans;*
- contain 19–25 nucleotides;
- may be expressed in only certain cell types and at only certain times in the differentiation of a particular cell type.

While direct evidence of the function of many of these newly-discovered gene products remains to be discovered, they probably regulate gene expression by regulating messenger RNA (mRNA), either

- destroying the mRNA when the sequences match exactly (the usual situation in plants) or
- repressing its translation when the sequences are only a partial match. In this latter case, it probably requires several miRNAs to bind simultaneously in the 3'-UTR.

These "riboregulators" have two traits ideally suited for this:

- Being so small, they can be rapidly transcribed from their genes.
- They do not need to be translated into a protein product to act (in contrast, e.g., to transcription factors).

A study reported in *Cell* (Rhoades, *et al.*, 110: 513, 23 Aug 2002) found that most of the miRNAs examined in *Arabidopsis* matched the messenger RNAs (mRNAs) of transcription factors. Presumably these miRNAs repress their target mRNA. If so, this could provide a mechanism for daughter cells to escape from the transcription factors of their parent cell and thus be able to enter a different path of differentiation.

Of the 46 miRNAs expressed in the Drosophila embryo, 25 have been shown to be essential to normal development. (See Leaman *et al., Cell*, 1 July 2005.)

MicroRNAs regulate (repress) expression of genes in mammals as well. Genome analysis has revealed thousands of human genes whose transcripts (mRNAs) contain sequences to which one or more of our miRNAs might bind. *RAS*, for example, has binding sites for a miRNA, and a defect in the production of that miRNA is associated with especially aggressive lung cancer.

And a study reported in *Nature* (Lim *et al.*, 433: 769, 17 Feb 2005) used DNA chip analysis to show that when a particular miRNA was expressed in HeLa cells,

- a miRNA normally expressed in the brain repressed mRNA production by 174 different genes while
- a miRNA normally expressed in cardiac and skeletal muscle repressed mRNA production by 96 genes – all but 8 of them different from the those repressed by the brain miRNA.

As work proceeds rapidly in this field, the pattern that begins to emerge is that:

- Many genes – especially those involved in such housekeeping activities (e.g., cellular respiration) common to all cells – do not have 3'-UTRs that can be blocked by any of the miRNAs encoded in the genome (humans have about 230 miRNA genes).
- The genes that must be expressed in a particular type of differentiated cell and/or at a particular time in the life of that cell
- do not express any of the miRNA genes that could block their expression but
- do express miRNA genes that block the expression of other genes for specialized functions that would not be appropriate in that cell at that time.

Thus repression of gene expression by miRNAs appears to be a mechanism to ensure proper, coordinated gene expression as cells differentiate along particular paths. For example, when zygote genes begin to be turned on in the zebrafish blastula, one of them encodes a miRNA that triggers the destruction of the maternal mRNAs that have been running things up to then.

So miRNAs may play as important role as transcription factors in coordinating the expression of multiple genes in a particular type of cell at particular times.

Biogenesis of miRNAs and Endogenous siRNAs

miRNAs are synthesized in the nucleus as long (up to 1000 nt) RNA polymerase II transcripts, called pri-miRNAs, that are characterized by imperfect hairpin structures.

Many pri-miRNAs arise from intergenic regions, or are in antisense orientation to known genes, indicating independent transcription units. Other genes for miRNAs are found in intronic regions and could be transcribed as part of the primary transcript for the corresponding gene. A dsRNA-specific endonuclease, Drosha, in conjunction with a dsRNA-binding protein, called Pasha in *Drosophila* and DGCR8 in humans, processes the pri-miRNA into hairpin RNAs 70–100 nt in length, called pre-miRNAs. Pre-miRNAs are transported to the cytoplasm via an Exportin-5 dependent mechanism. There Dicer works with a dsRNA-binding partner, Loqs in *Drosophila* and TRBP in humans, to process the pre-miRNA into mature, single-stranded miRNA and load it into an RNA silencing complex that is similar to RISC. Loqs and TRBP are functionally analogous to the dsRNA-binding protein R2D2 which partners with Dicer to process and assemble siRNAs into RISC.

Endogenous dsRNA precursors for siRNAs could, in principle, arise from bidirectional transcription of genomic regions encoding siRNAs. However, in *C. elegans*, *Arabidopsis*, and fungi, silencing requires an RNA-dependent RNA polymerase (RdRP), raising the possibility that RdRP activity generates dsRNA from single-stranded RNA transcripts. Synthesis of dsRNA using the target mRNA itself as a template (using siRNA as a primer, or via de novo synthesis) could explain the phenomenon of "spreading" of RNAi, the production of siRNAs encoded by the target gene but not by the trigger RNA. Spreading is associated with "systemic silencing" seen in *C. elegans* and plants, in which silencing is inherited or is spread to distant parts of the organism. RdRP enzymes have not been found in *Drosophila* and mammals, and the spreading and systemic silencing phenomena are not seen in these organisms.

Many questions remain about endogenous siRNA production and function. How the cell distinguishes between normal mRNA and aberrant single-stranded RNA that is processed to dsRNA triggers for RNA silencing is unknown. Endogenous siRNA precursors are presumably transported to the cytoplasm before Dicer processing; the mechanism by which this happens is unknown, but is presumed to be similar if not identical to Exportin-5 mediated transport of pre-miRNAs. The biogenesis of endogenous siRNAs remains to be more clearly defined in all systems.

Although small RNAs differ in their origins, some of the proteins involved in their production and function in silencing complexes are closely related, and in some cases are identical. These include the most well characterized components of the silencing machinery, Dicers and Argonautes. Dicer produces small RNAs from longer precursors and plays a key role in assembling the guide strand of the small RNA with a pathway-specific complex of silencing proteins. Argonaute is a core component of all silencing complexes characterized so far and associates with the small RNA in the silencing complex. In the case of RNAi, Argonaute has been shown to be the catalytic engine of mRNA cleavage.

Mechanism of RNAi: An ancient antiviral mechanism

The only RNA molecules normally found in the cytoplasm of a cell are molecules of single-stranded RNA. If the cell finds molecules of double-stranded RNA (dsRNA), it uses an enzyme Dicer to cut them into fragments containing 19 base pairs (~2 turns of a double helix) with two additional nucleotides at the opposite end of each strand. The two strands of each fragment then separate — releasing the antisense strand. With the aid of a protein, it binds to a complementary sense sequence on a molecule of mRNA. If the base-pairing is exact, the mRNA is destroyed. Because of their action, these fragments of RNA have been named "short (or small) interfering RNA" (siRNA). The complex of siRNA and protein is called the "RNA-induced silencing complex" (RISC).

siRNAs can also interfere with transcription

There is growing evidence that siRNAs can also inhibit the transcription of genes

- perhaps by binding to complementary sequences on DNA or
- perhaps by binding to the nascent RNA transcript as it is being formed.

In fission yeast, at least, the siRNA is complexed with one molecule of each of three different proteins. The entire complex is called the RITS complex ("RNA-induced initiation of transcriptional gene silencing"). How these siRNAs are synthesized in the cytosol and gain access to the DNA in the nucleus is unknown.

Synthetic siRNA molecules that bind to gene promoters can — in the laboratory — repress transcription of that gene. The repression is mediated by methylation of the DNA in the promoter and, perhaps, methylation of histones in the vicinity.

There is a strain of rice (LGC-1) that produces abnormally low levels of proteins called glutelins. It turns out that of several glutelin genes found in rice

- two closely-similar glutelin genes are located back to back on the same chromosome.
- In LGC-1, a deletion has occurred between the two genes which removes the signal that would normally stop transcription after the first gene.
- Thus RNA polymerase II transcribes right past the first gene and on into the second.
- The result is a messenger RNA with almost-identical sequences running in opposite directions.
- This causes the mRNA to fold up into a molecule of double-stranded RNA (dsRNA).
- A Dicer-like enzyme cuts up the dsRNA into small interfering RNAs (siRNAs) that suppress further transcription of those genes as well as other glutelin genes.

A clearer picture of PTGS emerged from several different basic observations, including the necessity of transcriptionally active genes and the ability of RNA viruses to silence a homologous endogenous gene (English *et al.*, 1997). Within the last 3 years, studies have identified several of the molecules that mediate RNAi, and the mechanism whereby these molecules effect the selective degradation of targeted mRNAs. It is now clear that the production of dsRNA with sequence complementary to the mRNA being targeted is fundamental to the process of PTGS; single-stranded RNA is not sufficient to induce PTGS. The importance of dsRNA is supported by a wealth of data. Transgenes engineered to synthesize dsRNA require only a few copies of the dsRNA to achieve PTGS and can induce cosuppression. There are several ways such transgenes produce dsRNA including the synthesis of long hairpin mRNAs by transcription of an inverted repeat (Kennerdell and Carthew, 2000; Tavernarakis *et al.*, 2000), and transcription of complementary sense and antisense strands by opposing promoters (Wang *et al.*, 2000). Other studies have shown that although cells may initially produce very long dsRNAs, they are cleaved into smaller dsRNAs, 21–25 nucleotides in length, that actually mediate RNAi (Hamilton and Baulcombe, 1999). How does a small dsRNA with a sequence complementary to a specific mRNA effect PTGS? The proteins that mediate the RNAi process have been identified using several approaches, most notably genetic screens for mutants resistant to RNAi in *C. elegans* and resistant to PTGS in *Neurospora* and *Arabidopsis*. These studies have identified homologous genes in each species, and subsequent identification of mammalian homologs, that encode proteins required for RNAi. Three homologous genes identified in the initial screens are *rde-1 in C. elegans* (Tabara *et al.*, 1999), *qde-2 in Neurospora* (Cogoni *et al.*, 1996) and *ago-1* in *Arabidopsis* (Dalmay *et al.*, 2000; Fagard *et al.*, 2000). The proteins encoded by each of these genes share homologies with the eIF2C translation factors. In *C. elegans, rde-1* is required for germline transmission of RNAi, but not for transmission of RNAi among cells of the worm (Grishok *et al.*, 2000 suggesting that it plays a role in the production of the RNAi signal. Another set of homologous genes involved in RNAi, identified in the screens for PTGS/RNAi mutants, include *ego-1* in *C. elegans* (Tabara *et al.*, 1999; Smardon *et al.*, 2000), *qde-1* in *Neurospora* (Cogoni and Macino, 1999) and *sgs2/sde1* in *Arabidopsis* (Mourrain *et al.*, 2000). The latter proteins appear to be required for PTGS and may act by catalyzing the production of dsRNA in cells because they contain motifs similar to those of RNA-directed RNA polymerases that convert a single-stranded RNA template into a dsRNA. Further studies in *C. elegans* have *identified the smg-2, smg-5, and smg-6 genes as being* involved in PTGS. However, the products of these genes are not required for the initial silencing by dsRNA, but are required for long-term maintenance of the gene suppression (Domeier *et al.*, 2000). Smg-2 apparently amplifies the RNAi signaling such that it persists for the lifetime of the worm, whereas SMG-5 and SMG-6 are phosphatases that may facilitate Smg-2's actions by dephoshorylating it. Interestingly, Smg-2 shares a high degree of homology with yeast UPF1, a protein known to have RNA binding, ATPase and helicase activities (Page *et al.*, 1999).

A working model for RNAi is shown in Figs. I and J. The first step is the production of dsRNA directed against an mRNA. The second step involves the recognition of dsRNA and its processing to produce 21–23 nucleotide siRNAs. The "effector step" is the recognition of the target mRNA by the siRNAs and the selective degradation of that mRNA. The introduction of dsRNA into cells, whether produced endogenously from exogenous plasmids or viral vectors, results in its recognition by an enzyme that cleaves the dsRNA into 21- to 23-nucleotide double-stranded fragments in an ATP-dependent, processive manner with a 2-nucleotide 3'-overhang and a 5'-phosphorylated end (Zamore *et al.*, 2000; Elbashir *et al.*, 2001b). This nuclease was identified as an enzyme called Dicer that is highly conserved among plants, fungi, worms, flies, and mammals; it is a member of the RNase III family of dsRNA-specific ribonucleases (Bernstein *et al.*, 2001). Dicer enzymes recognize and process dsRNA (Bernstein *et al.*, 2001; Ketting *et al.*, 2001) and are essential for RNAi (Bernstein *et al.*, 2001; Grishok *et al.*, 2001; Ketting *et al.*, 2001). Dicer is thought to function as a dimer based upon knowledge of bacterial RNase III and structural evidence; crystallographic and modeling studies of RNase III suggest a mechanism for double-stranded RNA cleavage (Blaszczyk *et al.*, 2001). Dicer not only processes dsRNA into siRNAs, but also processes endogenous regulatory RNAs called micro-RNAs. The *C. elegans* RNAi pathway gene *rde-4* encodes a dsRNA binding protein that interacts during RNAi with RNA identical to the trigger dsRNA; RDE-4 protein also interacts with Dicer and a conserved DExH-box helicase (Tabara *et al.*, 2002). These and additional data obtained by the authors in the latter study suggest that RDE-4 and RDE-1 function together to detect, retain, and present dsRNA to Dicer for processing. Different domains of Dicer have been identified including a dsRNA binding domain, an RNase III activity domain, a helicase activity domain and a PAZ domain (Piwi-Argonaut-Zwille domain, a region of a hundred amino acids, which could mediate interaction with argonaute proteins) (Bernstein *et al.*, 2001). Mouse Dicer is very similar to human Dicer with a predicted size of 1906 amino acids and molecular mass of 215 kDa, and contains a tandem repeat of RNase III catalytic domains, dsRNA binding region, a DExH/DEAH helicase motif and a PAZ domain (Nicholson and Nicholson, 2002). The mouse Dicer gene is located in chromosome 12 and the gene is widely expressed in cells throughout the body in embryonic and adult life.

Once generated, the small 21–23 nucleotide dsRNA fragments called siRNA are then recognized by a multiprotein complex called RISC and used as a guide for the recognition and degradation of the target mRNA (Tuschl *et al.*, 1999; Hammond *et al.*, 2000; Zamore *et al.*, 2000; Nykanen *et al.*, 2001). Experiments in *Drosophila* showed that RISC is present as a precursor complex that can be activated by ATP to form a complex with endonuclease activity that can cleave endogenous mRNAs (Hammond *et al.*, 2000, 2001; Nykanen *et al.*, 2001). The specific components of the RISC are not known, but do include members of the Argonaute family (Hammond *et al.*, 2001) that have been implicated in many processes previously linked to post-transcriptional silencing. (Also see Fig. K & L)

Moreover, RISC should include protein responsible for *endo*-and *exo*-nuclease activity. Recently, RISC activity was studied in a human model. Two proteins of the Argonaute family, eIF2C1 and eIF2C2, were identified in the affinity-purified human RISC; the authors further showed that RISC uses single-stranded siRNAs as a guide to cleave the endogenous mRNA. In their studies of the mechanism of RNAi in human cells, Chiu *et al.* (2002) provided evidence that the status of the 5′-hydroxyl terminus of the antisense strand of a siRNA determines RNAi activity, whereas blocking the 3′ terminus does not prevent RNAi. They found that an Aform helix structure was required for the formation of antisense-target RNA duplexes. Surprisingly, RNAi still occurred when the siRNA duplex was cross-linked by psoralen, suggesting that complete unwinding of the siRNA helix is not necessary for RNAi activity. Thus, it appears that amplification of RNA by RNA-dependent RNA polymerase is not essential for RNAi in human cells.

It is likely that additional proteins modify the different steps in the RNAi process. For example, recent experiments have shown that the *Drosophila* homolog of the fragile X mental retardation protein interacts with Dicer and RISC suggesting a possible role in the RNAi machinery (Caudy *et al.*, 2002; Ishizuka *et al.*, 2002). The latter results also raise the possibility of a role of abnormalities in RNAi in various human diseases.

Technical Considerations in the Use of RNA Interference

In several respects the approaches for silencing gene expression using RNAi methods are similar to those used for antisense DNA-mediated suppression of gene expression. In principle, any cloned gene can be targeted by designing RNA oligonucleotides or RNA-expressing viral vectors with sequences complementary to the mRNA transcribed from the target gene. The present section of this review article is intended to provide readers who are planning to use, or have just begun to use, RNAi technology into their experimental tool kit with practical information on designing and performing experiments using RNAi methods. In addition, we provide descriptions of emerging technical approaches for selective gene silencing *in vitro* and *in vivo* using RNAi. Examples of studies that employed methods described in the section can be found in Table 1.

A. Design and Synthesis of Small Interfering RNAs

Double-stranded RNAs of 20–23 nucleotides (siRNAs), which can be synthesized in large quantities and transfected into cells, are the most commonly used reagents for RNAi in cultured cells. All that is needed to implement siRNA-mediated silencing of expression of a gene of interest is the cDNA sequence of that gene, and commercially available reagents with which to perform the synthesis. Although targeting of siRNAs to any region of an mRNA would be expected to induce degradation of the mRNA and therefore abolish production of the encoded protein, empirical data suggest that the probability of achieving selective silencing can be increased by targeting the siRNAs to specific regions of the mRNA. Ambion (Houston, TX) recommends the following approach for designing an siRNA. 1) Beginning with the AUG start codon of the target gene transcript, downstream scanning is done for AA dinucleotide sequences; each AA and the 3′ adjacent 19 nucleotides are potential siRNA targets. 2) The sequences of the potential target sequences are compared to sequences in the species-appropriate genome database (www.ncbi.nlm.nih.gov/BLAST/) and any target sequences that are homologous to

Table 1 : Applications of RNA interference

Gene	Organism	In vitro/In vivo	Cell Line/Tissue	RNA Species	Delivery Method	Reference
Acyl-CoA binding protein	Human	in vitro	HeLa HopG2 Chang Cells	siRNA	Oligofectamine	Faergenian and Knudsen, 2002
Androgen receptor ipolyCAG	Human	in vitro	HEK293T cells	siRNA	Cell Fectin	Caplen et al., 2002
β-galactosidasa	Mouse	in vivo	Embryo	siRNA	Microinjection/ electroporalion	Calegari et al. 2002
β-tubulin	Mouse	in vitro	P19 cells	siRNA, shRNA (transcribe) shRNA (expressed)	Lipofectamine 2000	Yu et al. 2002
CD4, p24	Human, Virus	in vitro	Magl.CCR5, HoLa-CD4 cells	siRNA	Oligofectamine	Novins et al. 2002
CD4/CD8a	Mouse	in vitro	T lymphocytes primary T cells and bone marrow derived dentritic cells, hematopoietic stem cells, ES cells, zygotes, adult mice	siRNA	Electroporation/	McManus et al. 2002
CD8, CD25, Bim, p53	Mouse	in vitro/in vivo		shRNA (U6 promoter-driven)	Lentivirus	Rubinson et al. 2003
Cdc14a	Human	in vitro	HeLa cells	siRNA	Oligofectamine	Mailand et al. 2002
Cdc14A	Human	in vitro	HeLa cells	siRNA	Oligofectamine	Mailand et al. 2002
CDH1, p53	Human	in vitro	MCF-7	shRNA (H1 promoter)	Retrovirus	Brummalkamp et al. 2002
Centrin-1	Human	in vitro	HeLa cells	siRNA	Oligofectamine	Salisbury et al. 2002
Cep 135	hamster	in vitro	CHO cells	siRNA	Lipofectamine, Fugene	Ohta et al. 2002
Dlcer	Human	in vitro	DEK293T cells	shRNA plasmid	Calcium phosphate	Paddison et al. 2002[b]
DNA-PK	Human	in vitro	Fibroblesis	siRNA	Liposomes	Peng et al. 2002
Drep1	Human	in vitro	HEK293T cells	shRNA plasmid		Calcium phosphate
Paddison et al. 2002[b]						
Fortilin, myelold cell leukemia 1 protein	Human	in vitro	U2OS	siRNA	Transit solution	Zhang et al. 2002
GalTIIA, GalTIIB	Human	in vitro	HeLa	siRNA	Oligofectamine	Bal et al. 2001
GFP	Human	in vitro	293T	shRNA construct	Lentivirus	Abbas-Terki et al. 2002
GFP	Mouse	in vivo	Brain/striatal cells	shRNA	Adenovirus	Xia et al. 2002
GFP	Mouse	in vivo (mouse)	Whole organism	shRNA (H1 promoter)	Lentivirus	Tiscomia et al. 2003
GFP, luciferase, β-catenin	Human	in vitro	HeLa S3/EBNA-1	shRNA (U6 promoter-generated)	Lipofectamine, calcium	Donze and Picard, 2002
Hepatitis C-NS5B	Virus	in vitro	Huh-7.5	siRNA	Electroporation	Randall et al. 2003
HIV-1 rev	Virus	in vitro	293/EcR cells	shRNA (U6 promoter-driven)	Lipofectamine Plus	Lee et al. 2002
HIV-gag	Virus	in vitro	HEK293/T lymphocytes	siRNA/flourine derivatized siRNA	Lipofectin or naked addition (derivatized)	Capodici et al. 2002
Human papillomavirus genes E6 and E7	Virus	in vitro	Cervical carcinoma cells	siRNA	Oligofectamine	Jiang and Milner, 2002
Karyopherin α2, α3	Porcine	in vitro	Embryo	siRNA	Microinjection	Cabol and Prether, 2003
Kinase interacting stathmin	Human	in vitro	HEK 293 cells	siRNA	Lipofectamine 200	Boehm et al. 2002
Kinesin E15	Human	in vitro	HeLa	siRNA	Calcium phosphate, oligofectamine electroporation	Weil et al. 2002
Lamin A/C	Human	in vitro	HeLa cells	shRNA (U6 promoter-	Lipofectin Plus	Paul et al. 2002
Luciferase		in vitro	293T/Cos1/NIH3T3/HeLe	shRNA plasmid	Calcium phosphate	Paddison et al. 2002a
Luciferase		in vivo (mouse)	Liver, kidney, spleen, lung, pancreas	siRNA	Tall vein injection	Lewis et al. 2002
Luciferase		in vivo (mouse)	Liver	siRNA	Hydrodynamic transfection naked RNA	McCaffrey et al. 2002

Table 1 continued

Gene	Organism	In vitro/In vivo	Cell Line/Tissue	RNA Species	Delivery Method	Reference
Luciferase, femin A/C	Human (lamin A/C)	in vitro	HeLa, Cos-7, HEK293, HEK 293	siRNA	Oliogofectamine	Elbashir et al. 2002a
mAPH-la and b, nicastrin, presenilin	Human	in vitro	HeLa cells	siRNA	Oligofectamine	Lee et al. 2002
Mina53	Human	in vitro	HeLa, rat3Y1MyeB cells	siRNA	Oligofectamine	Tsuneoka et al. 2002
Mps1 kinase	Human	in vitro	HeLa cells	siRNA	Oligofectamine	Stucke et al. 2002
Myocyte enhancer factor 2A	Mouse	in vitro	Primary cerebellar granulo neurons	shRNA	Calcium phosphate	Gaudilliere et al. 2002
Nuclear, Obf2-related kinase, p75	Human	in vitro	HeLa cells	shRNA	Retrovirus	Devroe and Silver, 2002
Nuclear, cytoskeletal, cell cycle	Human	in vitro	HeLa	siRNA	Oligofectamine	Harborth et al. 2001
Oct 3/4, c-mos	Mouse	in vitro	Oocytes	siRNA	Microinjection	Kim et al. 2002
Oml/Htre2	Human	in vitro	U2Os	siRNA	Effectene	Martins et al. 2002
Orc6	Human	in vitro	HeLa cells	siRNA	Liposomes	Presanth et al. 2002
p110 subunit of P13 kinase, akt1, akt2	Human	in vitro	HeLa	shRNA plasmid, siRNA	Effectene, oligofectamine	Czaudema et al. 2003
p21	Mouse	in vitro	Prostate cancer cell line	siRNA	Liposomes	Ukomadu and Dutta, 2003
p53	Human	in vitro	HEK293T cells	shRNA	Lentivirus	Barton and Medzhhitov, 2002
p53	Human	in vitro	H1299 cells	siRNA	Lipofectamine	Martinez et al. 2002
p53	Human	in vitro	SK-N-SH cells	siRNA	Oligofectamine	Kartasheva et al. 2002
p65 subunit of NF-kB	Human	in vitro	MAGI/Jurkat cells	siRNA	Oligofectamine	Surabhi and Gaynor, 2002
PKCα	Human	in vitro	HEK293T cells	siRNA and shRNA	Cationic lipid	Leirdal and Sioud, 2002
Polio-specific capsid, polymerase	Virus	in vitro	HeLa S3, mouse embryonic bibroflasts	siRNA	Lipofectamine 2000	Gitlin et al. 2002
Polio-like kinase	Human	in vitro	HeLa cells	siRNA	Oligofectamine	Liu and Erikson, 2002
Poly (ADP-ribose) polymerase	Mouse, Rat	in vitro	Neuro 2A-derived cell line	siRNA	Lipofectamine	Gan et al. 2002
PRC1	Human	in vitro	HeLa cells	siRNA	Oligofectamine	Mollinari et al. 2002
ront1/hUpf1 and rent2/hUpf2	Human	in vitro	HeLa cells	siRNA	Liposomes	Mendell et al. 2002
RNA-specific adenosine deaminase-1	Human	in vitro	Huh-7	siRNA	Lipofectamine 2002	Wong and Lazinski, 2002
Shc	Human	in vitro	HeLa cells	siRNA	Oligofectamine	Kisielow et al. 2002
Speedy (Spy)	Human	in vitro	HEK 293T cells	siRNA	Calcium phosphate	Porter et al. 2002
Sphingosine kinase-1	Human	in vitro	HUVE cells	siRNA	Oligofectamine	Ancellin et al. 2002
Various PKC isoforms	Human, Rat	in vitro	HeLa, Cos-7, HEK293, rat-1	siRNA	Oliogofectamine	Irle et al. 2002

Pls kinase, phosphatidylin ositol 3-kinase

other coding sequences are eliminated from consideration. 3) 3–4 target sequences are selected along the length of the gene for production of siRNAs. Of course it is important for all siRNA experiments to include negative control siRNAs with the same nucleotide composition but a scrambled sequence.

Chemical synthesis was the first method used to produce siRNAs, but now they can be produced in any laboratory using *in vitro* transcription methods. One protocol involves the synthesis of DNA oligonucleotides that include an 8-base sequence complementary to the 5′ end of a T7 promotor primer. Each gene specific oligonucleotide is annealed to the T7 promoter primer, and a fill-in reaction using Klenow fragment produces a double-stranded template for use in an *in vitro* transcription reaction. The two RNA products of the *in vitro* transcription reactions are hybridized to each other, treated with DNase (to remove the DNA template) and RNase (to even the ends of the dsRNA), and the RNA is column purified. Another protocol for the production of siRNAs takes advantage of the availability of recombinant human Dicer. Large *in vitro* transcribed RNA templates are cleaved by Dicer to produce multiple species of 22 base pair siRNAs (Dicer siRNA generation kit; Gene Therapy Systems Inc., Dan Diego, CA). An advantage of the latter method is that, because it produces a mixture of different siRNAs directed against the same mRNA target, the probability of obtaining gene silencing is increased.

B. Construction of Plasmids and Viral Vectors for RNA Interference

There are several reasons why expression plasmids and viral vectors are being used in basic and applied RNAi research. One major reason is that expression vectors allow continuous production of siRNAs in cells and, therefore, sustained depletion of the protein encoded by the targeted mRNA. A second reason is that, particularly with viral vectors, the transfection efficiency of certain types of cells, particularly postmitotic cells can be greatly increased. A third advantage of viral vectors is that they are typically more effective in obtaining sustained expression (and gene silencing, in the case of RNAi) *in vivo*. For example, adenoviral vectors have been extensively used to express genes in postmitotic neurons *in vivo* (Smith and Romero, 1999). Short hairpin RNAs (shRNAs) can be transcribed from RNA polymerase III promoters in cells in culture or *in vivo* allowing continuous suppression of expression of the targeted mRNA (Paddison *et al.*, 2002a). The latter authors proposed the use of this technology in the generation of transgenic mice as an alternative approach to gene knockout mice. Brummelkamp and colleagues (Brummelkamp *et al.*, 2002a) developed a novel vector system for the stable expression of siRNAs in mammalian cells. They used the polymerase-III H1-RNA gene promoter, which produces a small RNA transcript lacking a poly-adenosine tail and has a well defined transcription start and termination signals. The construct also allows cleavage of the transcript at the second uridine after the termination resulting in a transcript that resembles the ends of synthetic siRNAs. They designed a gene-specific insert that specified a 19-nucleotide sequence derived from the target transcript, separated by a short spacer from the reverse complement of the same 19-nucleotide sequence resulting in the production of a 19-base pair stem-loop structure. This vector system was shown to be effective in sustained suppression of target gene expression in several different types of cultured cells.

Several laboratories have constructed plasmids that contain DNA templates for the synthesis of siRNAs under the control of the U6 promoter. For example, Sui *et al.* (2002) inserted DNA fragments that acted as templates for the synthesis of small RNAs under the control of the mouse U6 promoter that directs the synthesis of a Pol III-specific RNA transcript to generate an RNA composed of two identical 21-nucleotide sequence motifs in an inverted orientation, separated by a 6-base pair spacer of nonhomologous sequences. Five thymidines that function as a termination signal for Pol III were added at the 3′ end of the repeat; the resulting RNA is predicted to fold back to form a hairpin dsRNA with a 3′ overhang of several thymidines. Using this plasmid, they were able to demonstrate the efficient inhibition of expression of three different endogenous genes (lamin A/C, CDK-2, and DNA methyltransferase) in cultured human cells. A similar approach that employed U6 promoter-driven siRNAs with four uridine 3′ overhangs was used to effectively suppress expression of ectopically expressed genes as well as the endogenous β-catenin gene (Miyagishi and Taira, 2002). Retroviral delivery systems have been developed based upon several commercially available vectors. For example, a retroviral siRNA vector was developed in which the U6 promoter and anti-target gene hairpin was subcloned into pMSCVpuro (BD Biosciences Clontech, Palo Alto, CA) at the unique NsiI site just upstream from the 3′ long terminal repeat (Devroe and Silver, 2002). Using this retroviral siRNA delivery system, they demonstrated the efficient and sustained depletion of the NDR kinase and the transcriptional coactivator p75 in cultured cells. Lentiviral systems for shRNA delivery have also been developed. Lentiviruses can infect noncycling and postmitotic cells, and also provide the advantage of not being silenced during development allowing generation of transgenic animals through infection of embryonic stem cells or embryos (Naldini, 1998; Lois *et al.*, 2002; Pfeifer *et al.*, 2002). Using this approach, silencing of green fluorescent protein (GFP) in GFP-positive transgenic mice has been shown after transduction with lentiviruses expressing shRNA directed against the GFP protein (Tiscornia *et al.*, 2003). More recently, Rubinson *et al.* (2003) used lentivirus-delivered shRNA to induce silencing of CD8 and CD25 in cycling primary T cells and the pro-apoptotic molecule Bim in primary bone marrow-derived dentritic cells. Lentiviral-mediated silencing of CD8 in hematopoietic stem cells was still present after injection of the cells in lethally irradiated congenic mice. Moreover, *in vivo* silencing for CD8 or p53 was also observed after infection of ES cells or zygotes leading to stable and functional silencing in adult RNAi transgenic mice.

C. Transfection Methods

Several different transfection methods previously used to introduce oligodeoxynucleotides and DNA plasmids into cells have been used to successfully introduce siRNAs into cells. However, it has become clear there is no single transfection method that can be successfully applied to all cell types under all experimental conditions. It is therefore important to optimize transfection conditions so that maximum gene silencing is acheived. The following transfection parameters have been shown to affect transfection and gene silencing efficacy: cell culture conditions, including cell density and medium composition; the type and amount of transfection agent; the quality and amount of siRNA; and the length of time that the cells are exposed to the siRNA. For proliferating cells, a subconfluent cell density is preferable. For postmitotic cells such as neurons, cell densities in the range of 200 to 500 cells per mm2 of culture surface work well (O. Milavet and M. P. Mattson, unpublished data). Because proteins in serum can bind to and/or degrade siRNAs, the transfection should be performed in serumfree medium. Differences have been reported in the ability to transfect and silence gene expression between adherent and nonadherent cells. For example, the ErbB3 gene was readily silenced in adherent carcinoma cells using liposome-mediated siRNA transfection, whereas the same transfection method was ineffective in nonadherent myeloma cells (Walters and Jelinek, 2002). Postmitotic cells such as neurons and muscle cells tend to be more difficult to transfect using liposomes compared to mitotic cells such as stem cells, fibroblasts, and tumor cells.

Calcium phosphate-mediated transfection has been used successfully by several laboratories (Donza and Picard, 2002; Weil *et al.*, 2002). The most commonly used and effective transfection method for short-term suppression of gene expression using RNAi is to incorporate siRNAs into liposomes. There are an increasing variety of such transfection reagents including: Oligofectamine, Lipofectamine-2000 and CellFectin from Invitrogen (Carlsbad, CA) (Caplen *et al.*, 2002; Gan *et al.*, 2002; Gitlin *et al.*, 2002; Irie *et al.*, 2002; Wong and Lazinski, 2002; Mise-Omata *et al.*, 2003); Effectene from Qiagen (Valencia, CA) (Martins *et al.*, 2002); and si-PORT-Amine and siPORT-Lipid from Ambion (Austin, TX). Other methods that have proven effective for transfecting siRNAs into cultured cells include electroporation (Calegari *et al.*, 2002; McManus *et al.*, 2002; Randall *et al.*, 2003), microinjection (Calegari *et al.*, 2002; Kim *et al.*, 2002), and hydrodynamic shock (McCaffrey *et al.*, 2002). Similar transfection methods have been used to introduce RNA-expressing plasmids into cultured cells (Iratni *et al.*, 2002; Czauderna *et al.*, 2003). An example of the kind of results obtained with an optimized transfection protocol that employed Oligofectamine is where levels of the cellular prion protein are markedly decreased in mouse neural precursor cells using two different siRNAs. Xia and coworkers (Xia *et al.*, 2002) described the development of methods for using a viral-mediated delivery system for gene silencing in mice *in vivo*. They constructed a 21-base pair hairpin representing sequences directed against enhanced green fluorescent protein (eGFP), and tested its ability to reduce target gene expression in mammalian culture cells. Two different constructs were used, one in which the siRNA hairpin targeted against eGFP was placed under the control of the cytomegalovirus promoter and contained a fulllength simian virus-40 (SV40 polyadenylation (polyA) cassette, and the second in which the hairpin was juxtaposed almost immediate to the cytomegalovirus transcription start site (within 6 base pairs) and was followed by a synthetic, minimal polyA cassette. Cotransfection of these constructs with an eGFP expression plasmid showed that the second construct was effective in silencing eGFP expression (silencing was correlated with the generation of a 63-base pair RNA specific It should be recognized that as RNAi technology advances it will likely be possible to produce RNAi "knockout" mice (or other mammals) in which the expression of a protein of interest is repressed by the expression of its corresponding RNAi related molecule (shRNA or miRNA, for example). The resulting animal could be seen as an equivalent of its knockout generated by targeted gene disruption but with much more flexibility and efficiency. For example, cell type-specific promoters could be used to effect PTGS only in cells of interest; in many cases this may circumvent embryonic lethality resulting from gene deletion from all cells. It would also be quicker and less costly to produce RNAi transgenic animals compared with conventional knockouts.

RNAi: Some possibilities

RNAi has been found to operate in such diverse organisms as plants, fungi, and animals such as *Drosophila melanogaster, Caenorhabditis elegans,* and even mice and the zebrafish. Such a universal cell response must have an important function. What could it be?

- Some viruses of both plants and animals have a genome of dsRNA. And many other viruses of both plants and animals have an RNA genome that in the host cell is briefly converted into dsRNA. So RNAi may be a weapon to counter infections by these viruses by destroying their mRNAs and thus blocking the synthesis of essential viral proteins.
- Transposons (one type of "junk" DNA) may be transcribed into RNA molecules with regions that are double-stranded. RNAi could then destroy these.
- RNA interference may be the unexpected dividend of another basic process of controlling gene expression discussed below.

RNAi as a tool

In any case, the discovery of RNAi adds a promising tool to the toolbox of molecular biologists. Introducing dsRNA corresponding to a particular gene will knock out the cell's own expression of that gene. (Feeding *C. elegans* on *E. coli* manufacturing the dsRNA will even do the trick.

In the 24 March 2005 issue of Nature, Sönnichsen *et al* reported that they have injected dsRNAs corresponding to 20,326 of *C. elegans's* genes (98% of the total!) and monitored the effect of each on embryonic development from the completion of meiosis (following fertilization) through the second mitotic division that produces the 4-cell embryo.

They found that at least 661 different genes altered some process during this period:

- about half of them involved in cell division and
- half in general cell metabolism.

(Another thousand genes produced phenotypic effects that were seen at later stages of development).

Because RNAi can be done in particular tissues at a chosen time, it often provides an advantage over conventional gene "knockouts" where the missing gene is carried in the germline and thus whose absence may kill the embryo before it can be studied.

In mammalian cells, introducing dsRNA fragments only reduces gene expression temporarily. However, Brummelkamp *et al.* reported in the 19 April 2002 issue of Science that they have succeeded in introducing into (mammalian) cells a DNA vector that can continuously synthesize a siRNA corresponding to the gene that they want to suppress. Two months later the cells still failed to manufacture the protein whose gene had been turned off by RNAi.

Two papers in the 25 July 2002 issue of Nature report that human cells in culture can be protected from

- poliovirus
- HIV-1

by siRNAs with sequences that match RNA molecules encoded by the viruses.

And the 19 June 2003 issue of Nature reports on coffee plants that were engineered to express a transgene that makes siRNA that interferes – by RNAi – with the expression of a gene needed to make caffeine. So perhaps "decaf" coffee will one day no longer require the chemical removal of caffeine from coffee beans.

Amplification of RNAi

In *C. elegans*, plants, and *Neurospora*, the introduction of a few molecules of dsRNA has a potent and long-lasting effect. In plants, the gene silencing spreads to adjacent cells (through plasmodesmata) and even to other parts of the plant (through the phloem). RNAi within a cell can continue after mitosis in the progeny of that cell. Triggering of RNAi in *C. elegans* can even pass through the germline into its descendants.

Such amplification of an initial trigger signal suggests a catalytic effect. It turns out that these organisms have RNA-dependent RNA polymerases (RdRPs) that uses the mRNA targeted by the initial antisense siRNA as a template for the synthesis of more siRNAs. Synthesis of these "secondary siRNAs even occurs in adjacent regions of the mRNA. So not only can these secondary siRNAs target additional areas of the original mRNA, but they are potentially able to silence mRNAs of other genes that may carry the same sequence of nucleotides.

This phenomenon, called "transitive RNAi",

- may complicate the interpretation of gene suppression experiments as the expression of other genes may be suppressed in addition to the target gene;
- raises a warning flag for the use of RNAi to suppress single genes in human therapy (although RdRPs and amplification have not been observed in mammalian cells).

Significance of the discovery of RNAi

It was evident from the very beginning that the significance of the discovery of RNAi would be exceptional. The far-reaching consequences of the discovery can be summed up as follows (Fig. 4):

1. **RNAi protects against viral infections:** The finding of Fire and Mello that cells can process injected dsRNA and eliminate homologous single-stranded RNA suggested that RNAi could constitute a defence mechanism against viral attacks. It had earlier been shown that plant cells have an efficient defence against viruses based on the PTGS phenomenon (Covey *et al.*, 1997; Ratcliff *et al.*, 1997). When it became apparent that PTGS is the plant equivalent to RNAi, this early work in plants supported the proposition that RNAi is involved in protecting cells from viral attacks. Today, we know that this anti-viral mechanism is at work in plants, worms and flies, whereas it is still unclear how relevant it is for vertebrates, including man.

It is believed that endogenous RNAi mechanisms evolved, at least in part, to protect cells against infectious pathogens such as viruses. For example, by producing siRNAs directed against genes required for viral replication, the infected cells prevented propagation of the virus (Lindenbach and Rice, 2002). The development of RNAi technology has verified the importance of RNAi mechanisms in viral invasion and is revealing the underlying molecular interactions. Treatment of various types of cultured cells with siRNAs directed against both viral and cellular targets has revealed specific roles for the targeted proteins in the processes of viral invasion or replication. When cells were treated with siRNAs directed against the HIV-1 tat or reverse transcriptase genes, or against the NF-β p65 subunit of the host cells, the expression of these viral and cellular proteins were decreased and HIV-1 replication was inhibited, demonstrating the ability of RNA interference to elucidate the biological roles of cellular and viral regulatory factors involved in the control of HIV-1 gene expression (Surabhi and Gaynor, 2002). When human and mouse cells were pretreated with siRNAs to the poliovirus genome, the replication of the virus in those cells was dramatically reduced (Gitlin *et al.*, 2002). In another study it was shown that siRNAs can efficiently silence both cellular lamin AC and hepatitis C virus RNAs in Huh-7 hepatoma cell lines resulting in an 80-fold decrease in HCV RNA within 4 days (Randall *et al.*, 2003).

The latter study further showed that the same siRNAs could be used to almost completely eliminate the virus from cells with an established infection. The expression of the human papilloma virus E6 and E7 genes was silenced resulting in the accumulation of cellular p53 protein, transactivation of the cell cycle control p21 gene and reduced cell growth (Jiang and Milner, 2002). A final example of the use of RNAi in the discovery of mechanisms of viral infection is a study of hepatitis delta virus (HDV) which uses a host-encoded RNA-editing machinery to express two essential proteins from the same coding sequence (Wong and Lazinski, 2002). The authors employed siRNAs to deplete either a small or long form, or both forms, of an adenosine deaminase that acts on RNA (ADAR)1.

They found that editing during viral replication was only inhibited when both forms of the enzyme were depleted, demonstrating a cooperative interaction between the short and long forms of ADAR1 in viral replication.

2. **RNAi secures genome stability by keeping mobile elements silent:** It was proposed early on that RNAi/PTGS in *C. elegans* and plants could block the action of transposons (mobile elements in the genome). Subsequently, it could be shown that when components of the RNAi machinery are mutated in *C. elegans,* transposons are activated and the mobile elements cause disturbances in the function of the genome (Ketting *et al.*, 1999; Dabara *et al.*, 1999). It has been proposed that in transposon-containing regions of the genome both DNA strands are transcribed, dsRNA is formed, and the RNAi process eliminates these undesirable products. As short dsRNAs can also operate directly on chromatin and suppress transcription (see below), this would be another mode to keep transposons inactive (see below under 4). Even if the mechanisms are not yet fully revealed, it is clear that if the RNAi machinery is not efficient, the transposons are not kept under control and can start to jump and cause deleterious effects in the genome.

It has been argued that RNA silencing could represent an "immune defence" of the genome (Plasterk, 2002). Close to 50% of our genome consists of viral and transposon elements that have invaded the genome in the course of evolution. The RNAi machinery can recognize invading double-stranded viral RNA (or the double-stranded replicative form of the viral RNA) and suppress the infection by degradation of the RNA. The RNAi system thus shares important features with the vertebrate immune system: it recognizes the invading parasite (dsRNA), raises an initial response and subsequently amplifies the response to eliminate the foreign element.

3. **RNAi-like mechanisms repress protein synthesis and regulate the development of organisms:** Soon after the discovery that short RNA is the effector of RNAi, it was shown that there is a class of endogenous RNA molecules of the same size in worms, flies, mice and humans; this small RNA was called microRNA (miRNA) (Lee and Ambros, 2001; Lau *et al.*, 2001; Lagos-Quintana *et al.*, 2001). Plants also contain this class of endogenous RNA (Reinhart *et al.*, 2002). The revelation of miRNA led to intense research on the nature of these RNA molecules. The *C. elegans lin-4* (Lee *et al.*, 1993) and *let-7* (Reinhaert *et al.*, 2000 RNAs were regarded as prototypes, and examples of similar cases were soon revealed in several organisms (Sharp, 2006). The small miRNAs are processed from larger hairpin-like precursors by an RNAi-like machinery (Murchison *et al.*, 2004; Zamore and Haley, 2005) (Fig. J). The miRNAs can regulate gene expression by base-pairing to mRNA, which results in either degradation of the mRNA or suppression of translation. Today, it is estimated that there are about 500 miRNAs in mammalian cells, and that about 30% of all genes are regulated by miRNAs. It is known that miRNAs play an important role during development in plants, *C. elegans* and mammals. Thus, the miRNA-dependent control of gene expression represents a new major principle of gene regulation. However, the full significance of small regulatory RNAs is probably still not apparent.

4. **RNAi-like mechanisms keep chromatin condensed and suppress transcription:** It was known from work in plants that gene silencing could take place at the transcriptional level (TGS). After the discovery of RNAi, it was soon shown that TGS in plants operates via RNAi-like mechanisms (Mette, 2000; Sijen *et al.*, 2001). In the fission yeast *Schizosaccharomyces pombe* (Hall *et al.*, 2002; Volpe *et al.*, 2002), and later on in *Drosophila* and vertebrates, it was found that similar processes keep heterochromatic regions condensed and transcriptionally suppressed. In addition, the RNAi-like machinery regulates the activity of genes in the immediate vicinity of the condensed blocks of chromatin. The phenomenon is still not understood at the molecular level although histone modifications, binding of specific chromatin condensing proteins (HP1), and DNA methylation all play important roles (Sharp, 2006). It is, however, evident that this action on chromatin is most important for proper functioning of the genome and for maintenance of genome integrity.

5. **RNAi offers a new experimental tool to repress genes specifically :** The targeted action of RNAi immediately suggested that this phenomenon could be utilized as a general method to suppress specific genes and look for the resulting phenotypic effect. It was also soon evident that this could be accomplished in such an efficient manner that essentially any gene in an organism could be studied functionally. This has proved particularly easy in *C. elegans,* where simply feeding the worms with bacteria expressing the double stranded RNA has been found to cause RNA interference throughout the tissues of the worm. After the initial work in *C. elegans,* this technical approach could be applied to cells from almost all organisms, including mammalian cells. This targeted gene silencing by RNAi has already had a tremendous impact in studies of the function of individual genes. Researchers working on mammalian systems have had more difficulties. Most of the work has concentrated on introducing small interfering double stranded RNAs into cells in tissue culture. A popular method has been to make these synthetically *in vitro*. However, as mammalian cells will not readily take up naked nucleic acids, the RNAs have to be complexed with agents such as cationic lipids to allow them to enter the cells. Synthetic siRNAs can cause efficient inhibition of expression of homologous genes, although only for a few days. As the effect is rarely complete, it is generally termed a "knock down" to distinguish it from the "knock out" achieved by deletion of the gene.

It is now exploited not only in cultured cells but also in transgenic organisms. Another way of introducing siRNAs into cells is to use expression vectors such as engineered viruses to direct expression of short RNA sequences that will form hairpins owing to the presence of complementary sequences of about 20 nucleotide pairs. These short hairpin RNAs are then processed within the cell to remove the loop and form siRNA duplexes. Viral vector mediated RNA interference can result in long term inhibition of target gene expression. Retroviruses, adenoviruses, and lentiviruses have all been used as vehicles for RNA interference constructs.

6. **RNAi might be a useful approach in future gene therapy :** The possibility to achieve RNAi-governed gene regulation in transgenic organisms has stimulated many explorations of whether this would be a useful option for medical therapy (Dorsett and Tuschl, 2004; Hannon, 2004). Promising results have been reported in several animal models (Soutschek *et al.*, 2004; Morrisey *et al.*, 2005; Palliser, 2006; Zimmermann TS *et al.*, 2006) and even in recent clinical trials, but it is too early to predict the outcome of these challenging efforts.

Because its target is so specific, the possibility of using RNAi to shut down the expression of a single gene has created great excitement that a new class of therapeutic agents is on the horizon.

Antisense RNA that is complementary to the proto-oncogene BCL-2 is being examined as a possible therapy for certain B-cell lymphomas and leukemias.

Other promising advances:

- In mice and monkeys, an intravenous injection of a chemically-modified siRNA specific for the mRNA of apolipoprotein B
- Reduced the levels of the apolipoprotein B mRNA
- With an accompanying reduction in the level of the protein, which resulted in lowering their cholesterol and LDL levels.
- Introducing siRNAs targeting genes of herpes simplex virus-2 (HSV-2) protected mice from an otherwise lethal infection by the virus.

Other applications of RNAi

Establishing Gene Function

The most widely used RNAi technology has been in cell culture and *in vivo* studies aimed at understanding the function of an individual (or multiple) proteins. The kinds of studies demonstrate the power and flexibility of RNAi for unambiguously establishing (or excluding) a function of individual proteins in various cellular processes. It should also be noted that cells can be transfected with different combinations of multiple siRNAs, each directed against a specific mRNA of interest, to elucidate the specific contributions of the proteins to a biological process involving a multiprotein complex. For example, Wojcik and DeMartino (2002) recently took advantage of the latter feature of RNAi methods to elucidate roles for different subunits of the proteasome in its assembly and function. The power of *C. elegans* and *Drosophila* molecular genetics is providing the opportunity to use genome-wide RNAi to rapidly establish functions of genes in a specific process. For example, Ruvkun and colleagues (Lee *et al.*, 2002c) systematically inactivated 5690 genes in *C. elegans* using RNAi to identify genes that limit lifespan. They identified a mitochondrial leucyl-tRNA synthetase gene and showed that mutations of this gene that impair mitochondrial function increase lifespan, which was associated with decreased ATP levels and oxygen consumption.

Signal Transduction

Intercellular messenger molecules play vital roles in the development and proper functioning of all organisms. Among the most prominent of such signals in mammals are growth factors, cytokines, cell adhesion molecules, neurotransmitters, steroids, and gases such as nitric oxide. Specific receptors located on the cell surface or within the cell transduce responses to the ligand via signaling cascades that are often complex, involving kinases and transcription factors, for example. RNAi methods provide powerful tools for establishing the roles of individual proteins in the signal transduction pathway employed by a specific ligand. Several recent studies have demonstrated the efficacy of siRNAmediated knockdown of signal transduction proteins and have elucidated roles for those proteins in biological responses of cells. Adaptor proteins of the Shc family transduce signals from a diverse group of growth factors that signal through receptor tyrosine kinases. Liposomemediated introduction of siRNA against a single isoform of ShcA into HeLa cells was used to selectively reduce levels of that Shc revealing its role in the regulation of cell proliferation (Kisielow *et al.*, 2002). Neurotrophin receptors and integrins (receptors for extracellular matrix molecules such as laminin) are often coupled to a signaling pathway involving phosphatidylinositol 3-kinase and Akt kinase (Gary and Mattson, 2001; Cheng *et al.*, 2003). Decreasing the amount of the 110 subunit of phosphatidylinositol 3-kinase using siRNAs resulted in a marked decrease in the growth and tissue invasiveness of tumor cells (Czauderna *et al.*, 2003). Small molecule inhibitors have been widely employed to study the functions of various protein kinases in cells. However, most such inhibitors are not specific and affect multiple kinases. RNAi has been successfully employed to unequivocally establish the roles of specific kinases in signal transduction processes. For example, Irie and coworkers (Irie *et al.*, 2002) demonstrated the ability to knockdown levels of specific subtypes of protein kinase C in cultured human and rat cells in a species-specific manner. The existence of an extracellular system for the production of sphingosine-1-phosphate was demonstrated in a study in which siRNAs directed against the sphingosine kinase-1 enzyme were used to inhibit its production and export in cultured endothelial cells (Ancellin *et al.*, 2002).

Calcium plays important roles as an intracellular signal that mediates a variety of responses of cells to environmental stimuli. Mechanisms for regulating levels of calcium in the cytoplasm are complex and involve movements of calcium ions across the plasma membrane, as well as into and out of endoplasmic reticulum and mitochondria.

A key role for inositol 1,4,5-trisphosphate (IP3)-mediated release of intracellular calcium in the maturation of mouse oocytes was demonstrated in which siRNAs against the IP3 receptor-1 were injected into germinal vesicle-intact oocytes (Xu *et al.*, 2003). The siRNAs reduced IP3 receptor-1 levels by 90% and, following insemination, blocked the intracellular calcium oscillations that play a critical role for the first steps in development.

In another study, RNAi-mediated depletion of the endoplasmic reticulum calcium-ATPases resulted in lethality in *C. elegans* (Cho *et al.*, 2000), demonstrating a pivotal role for calcium uptake by this organelle in cell functions and survival.

Cell Cycle Regulation

Because of their fundamental importance for development, tissue homeostasis, and stem cell biology and their centrality to the field of cancer research, genes that regulate the cell cycle have been a prominent focus of studies that employ RNAi. Examples of the kinds of findings obtained using RNAi technologies to knockdown the expression of cell cycle genes are as follows. A requirement of the Mps1 kinase for the spindle assembly checkpoint was established, whereas a role for this kinase in centrosome duplication was ruled out (Stucke *et al.*, 2002). A key role for a novel centrosomal protein called Cep135 in microtubule organization was revealed (Ohta *et al.*, 2002), and a role for centrin-2 in centriole duplication was established (Salisbury *et al.*, 2002). RNAi was used to show that the regulation of cell cycle progression in response to mitogens is controlled by the cyclin-dependent kinase inhibitor p27 (Kip1) (Boehm *et al.*, 2002). Depletion of Plk1 using siRNAs results in activation of cyclin B and inhibits centrosome amplification in hydroxyurea-treated U2OS cells (Liu and Erikson, 2002). RNAi was used to show that the protein PRC1 is a microtubule-associated protein that facilitates bundling of microtubules (Mollinari *et al.*, 2002) and that the processes of centrosome separation and chromosome segregation require the phosphatase Cdc14A in culture human cells (Mailand *et al.*, 2002). Depletion of the origin recognition complex subunit ORC6 using RNAi resulted in cells with decreased DNA replication, multipolar spindles and aberrant mitosis (Prasanth *et al.*, 2002). A novel human protein called Speedy, expressed only during the G1/S phase of the cell cycle, was shown to enhance cell proliferation by enhancing the activity of Cdk2 (Porter *et al.*, 2002).

Development

The complex and remarkably rapid events that occur during development of the fertilized egg into an adult organism remain largely a mystery. There would appear to be a great potential for RNAi technology to unravel the cellular and molecular events that regulate developmental processes. Methods for silencing single or multiple selected genes in developing embryos *in vivo* and stem cells in culture are beginning to reveal the functions of specific proteins in developmental processes. Nodal is a secreted factor that plays a key role in the formation and patterning of the mesoderm during gastrulation. RNAi was used to demonstrate that the transcriptional corepressor DRAP1 inhibits the transcription factor FoxH1 and thereby regulates signaling by Nodal during mouse embryogenesis (Iratni *et al.*, 2002). Depletion of karyopherins and in cleavage stage porcine embryos revealed different requirements of these two proteins in embryogenesis

(Cabot and Prather, 2003). In another study of embryogenesis, siRNAs directed against the mRNAs encoding Oct-3/4 and c-mos resulted in depletion of the encoded proteins, and phenotypes similar to those observed in Oct-3/4 and c-mos knockout mice (Kim *et al.*, 2002). A key role for microtubule-associated protein-2 in the regulation of dendrite outgrowth in developing brain neurons was demonstrated using siRNAs (Krichevsky and Kosik, 2002). The transcription factor myc is known to play a fundamental role in the regulation of cell proliferation. A key role for the novel myc target gene mina53 in the regulation of cell proliferation by myc was demonstrated using RNAi technology (Tsuneoka *et al.*, 2002).

Macromolecular Synthesis and Degradation

Regulated synthesis of nucleic acids, proteins, and lipids, and the turnover of those macromolecules is essential for cell survival and functions. Although biochemical technologies have allowed the identification of various biosynthetic pathways and mechanisms for the degradation of proteins and other macromolecules, the functions of specific proteins in such processes are not well understood. Recent studies have employed RNAi to advance the understanding of the regulation of macromolecular synthesis and degradation. For example, depletion of galactosyltransferase II using siRNAs revealed a critical role for this enzyme in the biosynthesis of the linkage region of glycosaminoglycans (Bai *et al.*, 2001). The regulation of the processing of RNAs transcribed from coding and noncoding regions of the genome is an active area of investigation because of the recent realization of its importance in the regulation of gene expression. Mendell and coworkers (Mendell *et al.*, 2002) used RNAi to show that rent1/Upf1 plays distinct roles in the regulation of splicing and decay of nonsense transcripts. Transfection of cultured HeLa and HepG2 cells with siRNAs directed against the mRNA encoding the acyl-CoA binding protein resulted in cessation of cell proliferation, cell detachment, and death, demonstrating that acyl-CoA binding protein is essential for cell survival (Faergeman and Knudsen, 2002).

Cell Motility

The migration of cells within and among tissues, and extensions of cells such as the axons and dendrites of neurons, is central to the structural and functional organization of all tissues. Although there are a few pharmacological agents that have proven useful in studying cell motility (the actin-depolymerizing agent cytochalasin D and the microtubule-stabilizing agent taxol, for example), a detailed understanding of the molecular regulation of cell motility is lacking. Recent studies have taken advantage of RNAi methods to elucidate roles for specific proteins in regulating cell motility. Depletion of the cytoskeletal linker protein trypanin using siRNAs resulted in a loss of directional cell motility in African trypanosomes that is caused by impaired coordination of the flagellar beating (Hutchings *et al.*, 2002). The growth and guidance of axons in developing neurons is regulated by substrate-associated and soluble ligands encountered by the axonal growth cone. RNAi-mediated depletion of the integrin-interacting protein MIG-15 resulted in a dysregulation of commissural axon navigation in *C. elegans* (Poinat *et al.*, 2002). RNAi has also shown that although actin binding protein 1 is essential for the processes of endocytosis, it is not necessary for lamellipodia formation in human embryonic kidney cells (Mise-Omata *et al.*, 2003). Membrane microdomains called lipid rafts and clathrin-associated domains are increasingly recognized as sites of signal transduction and endocytosis in eukaryotic cells. Proteins critical for the function of the signaling and endocytic functions of these membrane domains are being identified using RNAi approaches. For example, RNAi methods were used to establish an essential role for a J-domain protein called auxilin in clathrin-mediated endocytosis in *C. elegans* (Greener *et al.*, 2001).

Cell Death

In many tissues throughout the body, cells have a finite life span and then undergo apoptosis, a form of programmed cell death in which the cell dies in a well controlled manner and is removed from the tissue without adversely affecting adjacent healthy cells. Apoptosis also plays a key role in sculpting the cellular structure of tissues during embryonic and postnatal development (Baehrecke, 2002). Of course, abnormal cell death is a major problem in a variety of diseases, including neurodegenerative disorders such as Alzheimer's and Parkinson's diseases, and ischemic vascular conditions (Mattson, 2000). Considerable progress is being made in understanding the molecular mechanisms that regulate cell death, with the goal of identifying key targets for therapeutic intervention. Drug development efforts have resulted in several exciting classes of agents that either prevent the death of cells such as neurons, or induce selective death of cancer cells. For example, inhibitors of the pro-apoptotic protein p53 (Duan *et al.*, 2002; Zhu *et al.*, 2002) and caspases (Eldadah and Faden, 2000 are being developed for use in neurodegenerative disorders. RNAi might be used in lieu or in combination with such drugs.

RNAi has recently been employed to establish roles for specific genes in apoptotic and anti-apoptotic pathways. For example, a critical role for p73 in p53-mediated apoptosis was demonstrated by showing that depletion of p73 using siRNA prevents cell death (Kartasheva *et al.*, 2002). Depletion of the catalytic subunit of DNA dependent protein kinase using siRNAs increased the sensitivity of human fibroblasts to radiation-induced death because of an impaired ability to sense and repair the DNA damaged by the radiation (Peng *et al.*, 2002). RNAi was used to establish a role for the calcium-binding protein calreticulin in necrotic cell death in *C. elegans* (Xu *et al.*, 2001). Many cells express one or more inhibitor-of-apoptosis proteins (IAPs) that can prevent apoptosis by directly binding to and inhibiting caspases. RNAi was used to identify the serine protease Omi/ HtrA2 as a mammalian XIAP-binding protein that sensitizes cells to apoptosis (Martins *et al.*, 2002). In a study of cell death in neurons, RNAi was used to show that myocyte enhancer factor 2A is required for activity-dependent cell survival (Gaudilliere *et al.*, 2002). In another study, RNAi was used to deplete cells of apoptosis inducing factor (AIF), thereby preventing apoptosis (Wang *et al.*, 2002).

Associations between expression of a gene and a particular biological process are often taken as evidence for a major role for the protein encoded by that gene in that biological process. However, correlations do not establish cause-effect relationships. A well known example of this fact is found in the record of studies of the transcription factor NF-β. Because NF-β was shown to be activated in several different types of cells during the process of apoptosis, it was assumed that NF-β functioned in the cell death process. However, subsequent studies in which NF-β activity was selectively blocked revealed that this transcription factor actually induced the expression of anti-apoptotic proteins and that cells died more readily when NF-β function was blocked (Mattson and Camandola, 2001). Thus, in addition to revealing the function of a specific protein, RNAi can also be used to establish that a protein is not involved in a particular process.

Therapeutic Applications of RNA Interference

The most obvious clinical uses of RNAi are for diseases in which selective depletion of one or a few specific proteins would be expected to slow or halt the disease process in the affected cells. Ideally this would be accomplished with no or tolerable side effects. Although there are candidate gene targets for many different diseases, we will focus on four different types of diseases that are very common and for which RNAi approaches are currently being tested in preclinical studies.

A. Cancer

There are two general abnormalities in cancer cells—they exhibit dysregulation of the cell cycle resulting in uncontrolled growth and they are resistant to death as a result of abnormalities in one or more proteins that mediate apoptosis (Nam and Parang, 2003). The goals for RNAi approaches for cancer therapy are therefore to knock out the expression of a cell cycle gene and/or an anti-apoptotic gene in the cancer cells thereby stopping tumor growth and killing the cancer cells. To selectively eliminate cancer cells without damaging normal cells, the RNAi would be targeted to a gene specifically involved in the growth or survival of the cancer cell, or the siRNAs would be selectively delivered into the cancer cells.

For many years antisense oligodeoxynucleotide technology was pursued in preclinical studies of cancer therapies but with discouraging results overall (Jansen and

Zangemeister-Wittke, 2002). Recent studies have clearly demonstrated advantages of RNAi methods for the growth suppression and killing of cancer cells. In one study, siRNA was shown to be greater than an order of magnitude more potent than antisense DNA in suppressing gene expression in human hepatoma and pancreatic cancer cell lines (Aoki *et al.*, 2003). In another study four different myeloid leukemia cell lines (HL-60, U937, THP-1, and K562) were transfected with dsRNA duplexes corresponding to the endogenous c-raf and bcl-2 genes (Cioca *et al.*, 2003). Levels of Raf-1 and Bcl-2 proteins were markedly decreased in each of the transfected cell lines; combined RNAi for c-raf and bcl-2 induced apoptosis in HL-60, U937, and THP-1 cells and increased their sensitivity to the DNA-damaging agent etoposide. Activation of tumor necrosis factor (TNF) receptors and related death receptors can induce death of some cancer cells, but may simultaneously activate pathways that promote cell survival; one protein that inhibits the TNF cell death pathway is called FLIP (FLICE-like inhibitory protein). When FLIP expression was suppressed in cancer cells using siRNAs, the cells were more sensitive to being killed when death receptors were activated (Siegmund *et al.*, 2002).

Viral vectors have also been used to express siRNAs and inhibit cancer cell growth and tumorogenicity. For example, a retroviral vector was used to specifically and stably inhibit expression of the oncogenic K-RAS(V12) allele in human tumor cells (Brummelkamp *et al.*, 2002b). Depletion of K-RAS(V12) resulted in loss of anchorage-independent growth and tumorigenicity. In addition to blocking the expression of normal genes that are required for cancer cell growth and survival, RNAi can be used to target specific cancer-causing mutations. For example, dsRNA was employed to target the MBCR/ ABL fusion site to kill leukemic cells with such a rearrangement (Wilda *et al.*, 2002). Leukemic cells without BCR/ABL rearrangement were not killed by M-BCR/ ABL-dsRNA. Several other studies have demonstrated efficacy of liposome-mediated or viral vector-mediated transfection of cancer cells in suppressing their growth and/or inducing their death (Zhang *et al.*, 2002a). The next step in the development of RNAi technology for cancer therapy will be to establish methods for targeting tumor cells *in vivo*. Another approach might be to target genes that promote angiogenesis. Tumor cells require a rich supply of blood and achieve this by stimulating the process of angiogenesis; it may therefore be possible to inhibit tumor growth by targeting the vascular endothelial cells involved in angiogenesis. As evidence, it was shown that depletion of the crk adaptor protein using RNAi inhibited the migration of cultured vascular endothelial cells (Nagashima *et al.*, 2002).

B. Infectious Diseases

Diseases caused by viruses and bacteria continue to be major causes of death worldwide and are an increasing concern because of the emergence of resistant strains and the potential use of infectious pathogens by terrorists (Tan *et al.*, 2000; Franz and

Zajtchuk, 2002). Currently, HIV infection has reached epidemic proportions in many African countries and also continues to be a major cause of morbidity and death among homosexuals and intravenous drug users. Other prominent infectious diseases include influenza, hepatitis, Lyme disease, and West Nile virus. Many deaths also result from bacterial infections, with pneumonia and sepsis being prominent examples. The ability of RNAi to inhibit the replication or cellular uptake of viruses and other infectious agents has been clearly demonstrated in cell culture studies and, therefore, holds promise for the treatment of human patients. The ability of HIV-1 to infect cells and replicate can be severely compromised by targeting of viral genes using siRNAs. Examples include the suppression of HIV-1 replication in human cells transfected with siRNA directed against tat and the rev gene (Capodici *et al.*, 2002; Jacque *et al.*, 2002; Lee *et al.*, 2002a; Novina *et al.*, 2002). Transfection of human cells with siRNAs directed against different genes in the poliovirus genome resulted in resistance of the cells to infection with poliovirus (Gitlin *et al.*, 2002). The ability of siRNAs targeting the gene encoding the death receptor Fas to protect mice from liver failure and fibrosis in two models of autoimmune hepatitis was tested by Song and colleagues (Song *et al.*, 2003). Intravenous injection of Fas siRNA specifically reduced Fas protein levels in the livers of mice during a 10-day period. Fas siRNA treatment abrogated hepatocyte necrosis and inflammatory infiltration and markedly reduced serum concentrations of transaminases demonstrating a clear hepatoprotective effect of the siRNA therapy.

C. Cardiovascular and Cerebrovascular Diseases

Cardiovascular disease is the leading cause of death in the United States and many other industrialized countries. It most commonly results from the progressive occlusion of arteries in a process called atherosclerosis, which can ultimately culminate in a myocardial infarction or stroke. Atherosclerosis involves damage to vascular endothelial cells, local production of inflammatory cytokines, and the recruitment of macrophages to the site forming foam cells; in addition, apoptosis of foam cells and vascular smooth muscle cells occurs (Geng and Libby, 2002). The severe ischemia that occurs in heart or brain cells during a myocardial infarction or stroke results in the death of cardiac muscle cells or neurons. Although some of the cells die rapidly by necrosis, many other cells die more slowly by apoptosis; data from animal studies suggest that such cardiac myocytes and brain neurons that die by apoptosis can be saved (Mattson *et al.*, 2000; Zhao and Vinten-Johansen, 2002). It may be possible to use RNAi technology to intervene in the process of atherosclerosis or to reduce the damage to heart tissue and brain cells that patients suffer following a myocardial infarction or stroke. A key step in the process of atherosclerosis is the up-regulation of cell adhesion molecules in vascular endothelial cells, which play an essential role in the recruitment of macrophages to the site of endothelial damage. The production of cell adhesion molecules can be selectively suppressed in cultured cells (Jarad *et al.*, 2002). In another study relevant to the pathogenesis of atherosclerosis, it was shown that mevastatin, an inhibitor of cholesterol synthesis, suppresses cell proliferation by inhibiting cyclin-dependent kinase-2 (Ukomadu and Dutta, 2003).

D. Neurodegenerative Disorders

Alzheimer's disease, Parkinson's disease, Huntington's disease, and amyotrophic lateral sclerosis are examples of relatively common age-related neurodegenerative disorders that are increasing as average life expectancy increases. Each disorder is characterized by the dysfunction and death of specific populations of neurons: hippocampal and cortical neurons involved in learning and memory processes in Alzheimer's disease, dopamine-producing neurons in the substantia nigra that control body movements in Parkinson's disease, and spinal cord motor neurons in amyotrophic lateral sclerosis. Specific genetic mutations are responsible for a small percentage of cases of Alzheimer's and Parkinson's disease and amyotrophic lateral sclerosis (Hardy, 2001), whereas all cases of Huntington's disease result from mutations (polyglutamine expansions) in the huntingtin protein (Rubinsztein, 2002). Studies of patients, and of animal and cell culture models of each disease, have revealed shared biochemical cascades that result in neuronal death. Those cascades include increased oxidative stress, dysregulation of cellular calcium homeostasis and apoptosis (Mattson, 2000). There have therefore been two different strategies for preventative and therapeutic interventions in neurodegenerative disorders. One strategy is to block the disease-specific events that are believed to initiate the neurodegenerative process, whereas the second strategy targets downstream events in the neurodegenerative cascade. For example, an abnormality in the proteolytic processing of the amyloid precursor protein is believed to be a key early event in Alzheimer's disease pathogenesis, and two enzymes called secretases are responsible for cleaving of amyloid precursor protein to generate the neurotoxic amyloid peptide are being targeted for drug development. RNAi has recently been used to identify additional proteins, such as APH-1, that are critical for production of amyloid peptide (Lee *et al.*, 2002b). Downstream targets include proteins involved in the production of reactive oxygen species, in the regulation of calcium homeostasis, and in the process of apoptosis (Mattson, 2000, 2003).

Recent studies have shown that cultured neurons can be efficiently transfected with siRNAs and that the targeted genes are effectively silenced. In one study it was shown that cultured neurons can be depleted of the p75 neurotrophin receptor, a protein in the TNF receptor family that has been implicated in neuronal apoptosis in certain settings (Higuchi *et al.*, 2003). Pro-apoptotic members of the Bcl-2 family (Colussi *et al.*, 2000) and caspases (Quinn *et al.*, 2000) have been effectively targeted and neuronal death prevented, using RNAi methods. Caplen and colleagues (Caplen *et al.*, 2002) performed studies aimed at determining whether RNAi could be used to target the pathogenic process in inherited neurodegenerative disorders caused by polyglutamine expansions. They used *Drosophila* and human tissue culture models of spinobulbar muscular atrophy, a disease caused by CAG expansions in the gene encoding the androgen receptor. They assessed the abilities of different siRNAs to selectively inhibit expression of transcripts that included a truncated human androgen receptor gene containing different CAG repeat lengths (16–112 repeats). They found that RNA duplexes containing CAG repeat tracts only induced gene-specific inhibition when flanking androgen receptor sequences were included. Sequence-specific small dsRNAs of 22 nucleotides rescued the toxicity and caspase-3 activation induced by plasmids expressing a transcript encoding an expanded polyglutamine tract. Thus, it is possible, at least in cell culture, to selectively silence a transcript associated with an important group of genetic diseases by RNAi.

Several aspects of neuronal cell biology provide opportunities for novel uses of RNA. Neurons possess complex morphologies with long axons and dendrites, and synapses that are often located at considerable distances from the cell body (for example, the presynaptic terminals of the axons of the lower motor neurons that innervate muscles in the foot in humans can be more than a meter from their cell bodies in the spinal cord). Recent findings suggest that the neurodegenerative cascades that occur in different neurodegenerative disorders may be activated first in synapses. Indeed, it has been shown that apoptotic biochemical cascades can be activated in synapses causing their degeneration (Mattson *et al.*, 1998). Accordingly, pharmacological agents (p53 and caspase inhibitors; Glazner *et al.*, 2000; Gilman *et al.*, 2003) and antisense treatments (Par-4 antisense; Duan *et al.*, 1999) that target apoptotic cascades have been shown to protect synapses in cell culture models of neurodegenerative disorders. RNAi technology would seem to be an ideal approach to target synaptic proteins involved in the pathogenesis of neurodegenerative disorders.

RNA interference in mammals

Initial attempts to induce RNA interference responses in human cells were unsuccessful. Introduction of double stranded RNA into mammalian cells induces a powerful set of quite different antiviral responses characterised by production of interferons, resulting in inhibition of all gene expression and rapid cell death, limiting the ability of a virus to replicate and spread throughout the organism. It seemed that RNA interference might have been lost all together, replaced by the more recently evolved interferon system that is not found in invertebrates.

However, there were hints that RNA interference might still exist in mammals. The breakthrough came when short, double stranded RNA molecules of less than about 30 nucleotide pairs long were shown to be unable to induce the interferon response. As the global shut down of gene expression no longer occurred with these "small interfering RNAs" (siRNAs), they could be seen to be capable of directing a sequence specific degradation of homologous mRNA in a manner very similar to that in plants, worms, and flies. siRNAs of about 20 nucleotide pairs in length, when introduced into mammalian cells, directly engage RISC and promote silencing of the expression of genes with the same sequence. Parsimonious nature had kept RNA interference as a back-up system even after the evolution of the interferon system.

RNA interference as a research tool

Long before RNA interference had been established as operating in mammalian cells, researchers working on worms had recognised the great power that it promised as a research tool. The sequencing of the genomes of humans and most commonly studied model organisms has led to a situation in which the identities of very large numbers of genes are known but little is understood about their function. A cheap and easy way of ablating gene function holds out massive hope for improving our ability to untangle the complex regulatory pathways that control cellular behaviour in health and disease. RNA interference allows analysis in a matter of days of the effect of loss of gene function at the cellular level that would have taken several months or even years by previous methods such as homologous recombination.

RNA interference as a functional genomics tool

DNA microarray technology has now enabled the level of expression of every gene in the genome to be determined under any condition. This has led to a vast accumulation of information about genes whose expression is significantly altered in various disease states. For example, huge databases have been established of genes that are aberrantly regulated in cancers. In a few cases this has resulted in the identification of key genes involved in the formation of the tumour and provided important new therapeutic targets. However, most of the time the pattern of gene expression is far too complex to allow identification of the relatively small number of misexpressed genes that are involved in causing or maintaining the disease rather than the much larger number that are innocent bystanders.

The ability of RNA interference to provide relatively easy ablation of gene expression has opened up the possibility of using collections of siRNAs to analyse the significance of hundreds or thousands of different genes whose expression is known to be up-regulated in a disease, given an appropriate tissue culture model of that disease. Perhaps more important still is the possibility of using genome-wide collections of siRNAs, whether synthetic or in viral vectors, as screening tools. This has attracted much attention recently from both academic and industrial researchers. The libraries of RNA interference reagents can be used in one of two ways. One is in a high throughput manner, in which each gene

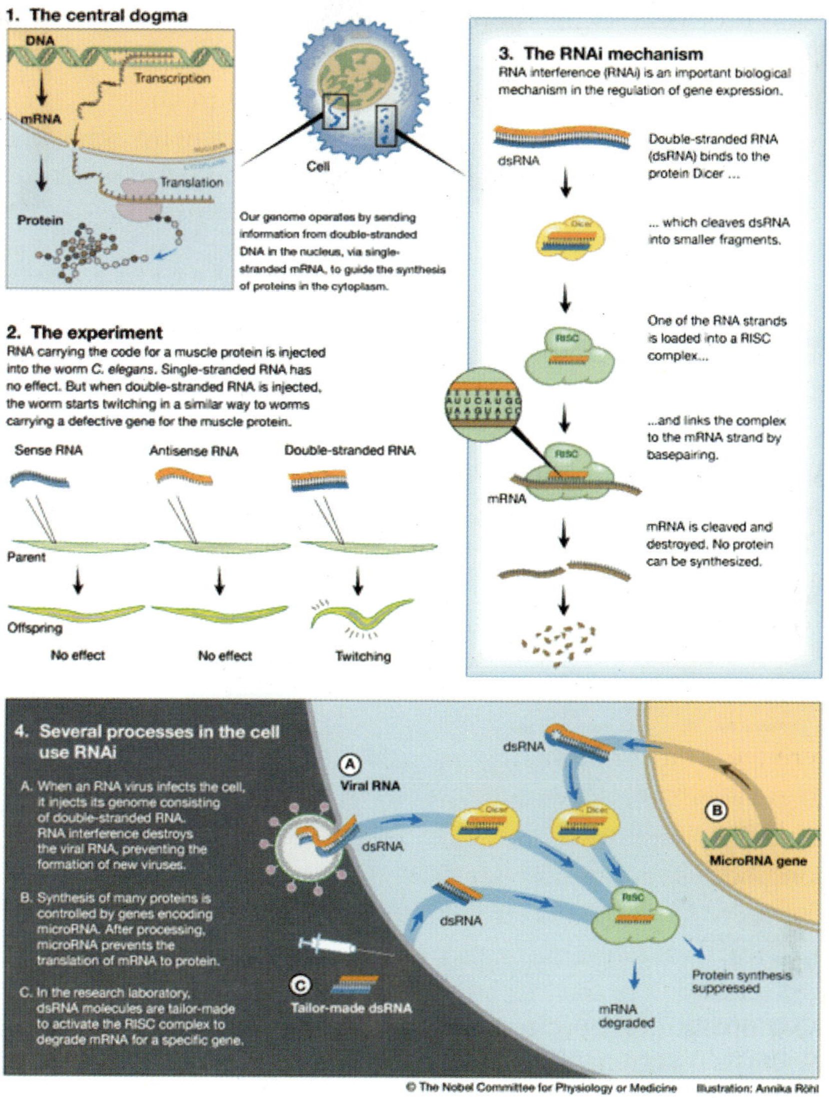

Fig. A: 1) DNA in the cell's nucleus codes for mRNA which gives results into protein synthesis, **2)** Phenotypic effect after injection of single-stranded or double-stranded unc-22 RNA into the gonad of C. elegans. The unc-22 gene encodes a myofilament protein. Decrease in unc-22 activity is known to produce severe twitching movements. Injected double-stranded RNA, but not single-stranded RNA, induced the twitching phenotype in the progeny, **3)** The RNA interference process and the biochemical machinery involved. Double-stranded RNA is cut into short pieces (siRNA) by the endonuclease Dicer. The antisense strand is loaded into the RISC complex and links the complex to the mRNA strand by base-pairing. The RISC complex cuts the mRNA strand, and the mRNA is subsequently degraded, **4)** Use of RNAi in cell.

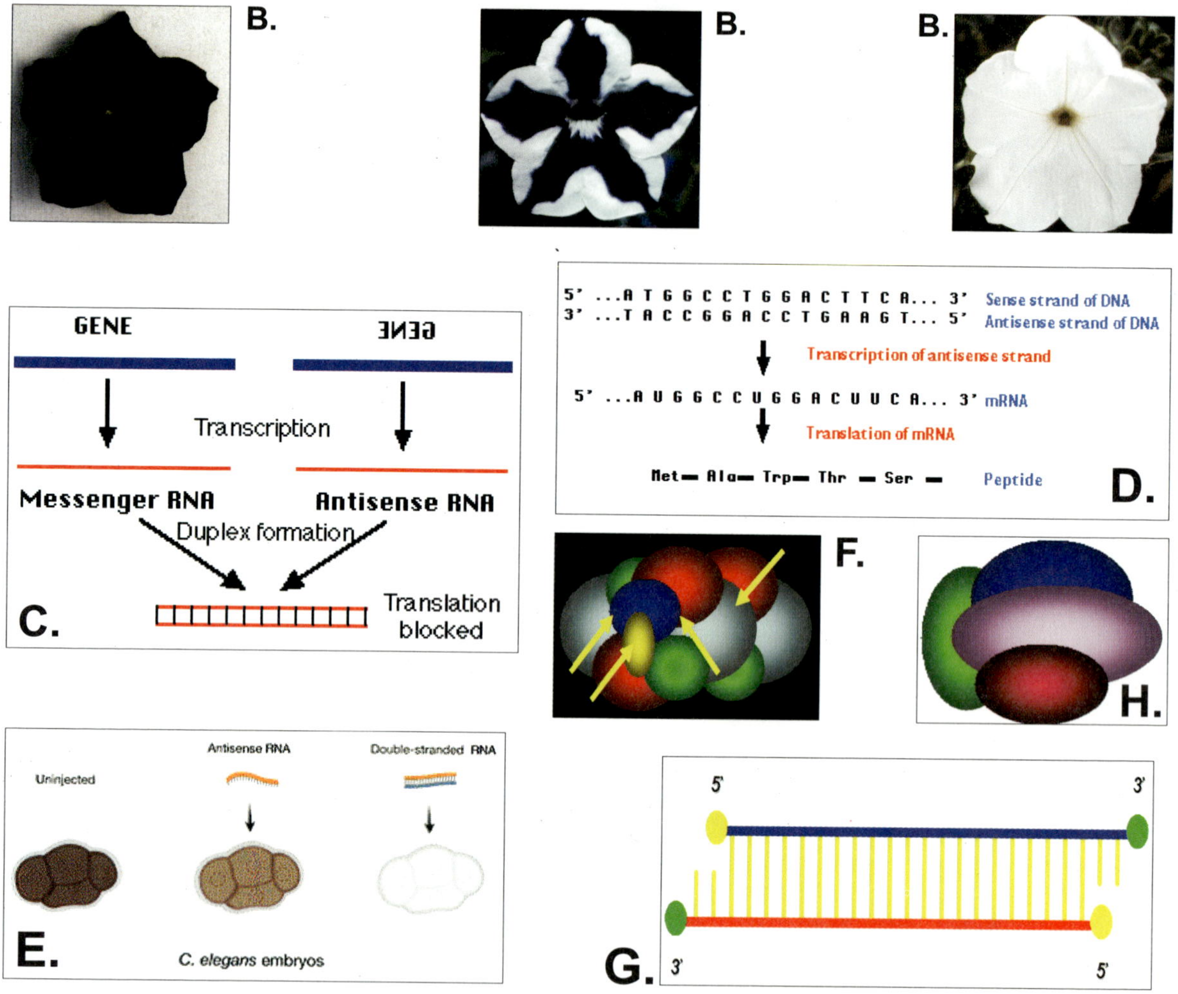

Fig. B). Cosuppression of purple colour in Petunia flowers, **C)** Transcription of antisense DNA leads to sense mRNA which in turn leads to protein via translation, **D)** If an antisense copy of mRNA is inserted it leads to blocking of sense mRNA leading to blockage of translation, **E)** The effect on mex-3 mRNA content in embryos after injection of single-stranded or double-stranded mex-3 RNA into the gonad of C. elegans. mex-3 mRNA is abundant in the gonads and early embryos. The mRNA was lost after injection of double-stranded RNA, while injection of antisense RNA only reduced the content of mRNA to some extent. The extent of brown colour reflects the amount of mRNA present, **F)** Dicer: RNAse III family members show specificity for ds RNAs and cleave them with 3' overhangs of 2 to 3 nucleotides and 5' hydroxyl termini (Bernstein et al., 2001), **G)** siRNA: Input dsRNA is digested into 21-23 nucleotides small interfering RNAs (siRNA) also called as "Guide RNA" with 2 nucleotide 3' overhangs (Hutvagner and Zamore, 2002; Bernstein et al., 2001), **H)** RISC- RNA induced Gene silencing complex : It has 2 RNA binding proteins, RNA/DNA Helicase, Translation initiation factor, RNA-dependent RNA Polymerase (RdRP) and Transmembrane protein.

Post-transcriptional gene silencing by ds RNA

Initiation Step

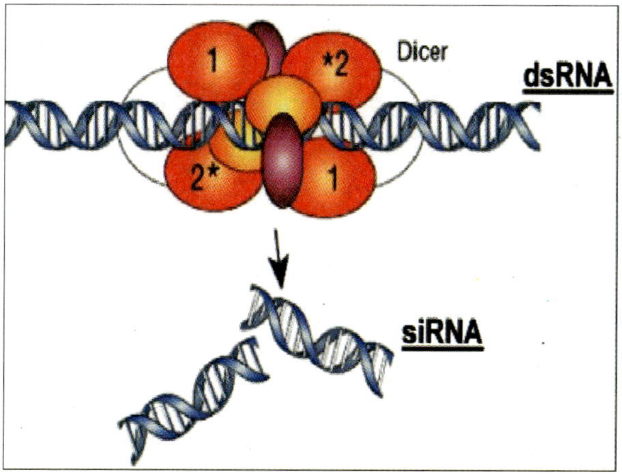

Effector Step

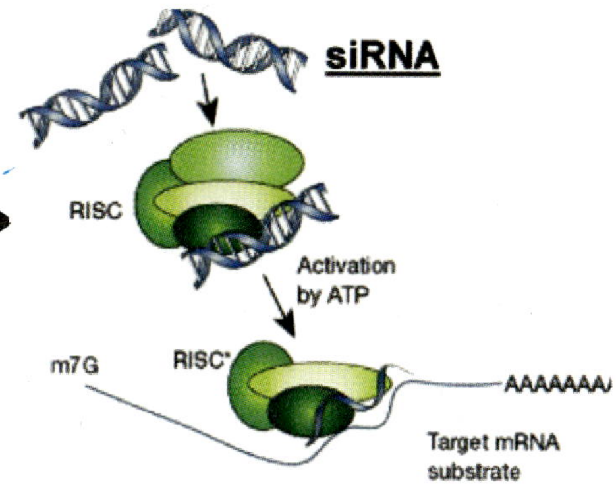

Fig. I: 1) THE INITIATION STEP:

- dsRNA is cleaved into approx. 22 nucleotide fragments
- such fragments have been shown to be necessary for silencing
- RNAse III displays specificity for dsRNA
- RNAse III is divided into three classes, one of which contains two catalytic domains and helicase and PAZ (Piwi/Arganaute/Zwille) domains named dicer enzymes (dicer enzymes are evolutionary conserved between Drosophila, Arabidopsis, tobacco, C.elegans, mouse and Neurospora and have been shown to recognize and process dsRNA into smaller fragments of species specific size).

2) THE EFFECTOR STEP:

- Approx. 22 nucleotides long fragments are double stranded duplexes with 2 nucleotide 3' overhangs and 5' phosphate termini (short interfering siRNA)
- siRNA bind to a multi component nuclease that recognizesand destroys mRNAs (RISC).

3) THE AMPLIFICATION STEP:

- Dicer produces siRNAs
- siRNAs bind to RNA dependent RNA polymerase (RdRP)
- RdRP produces new dsRNA

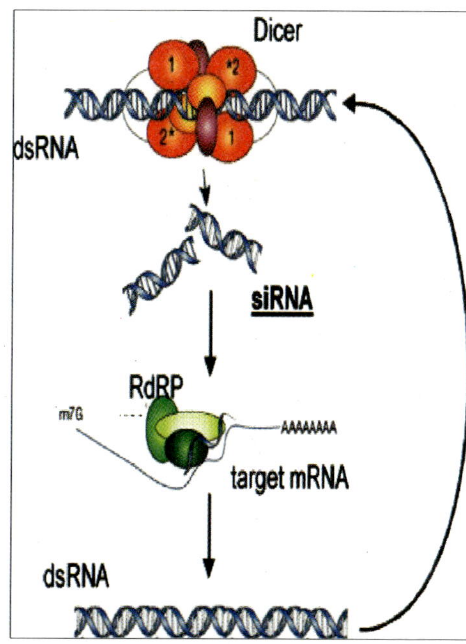

Amplification Step

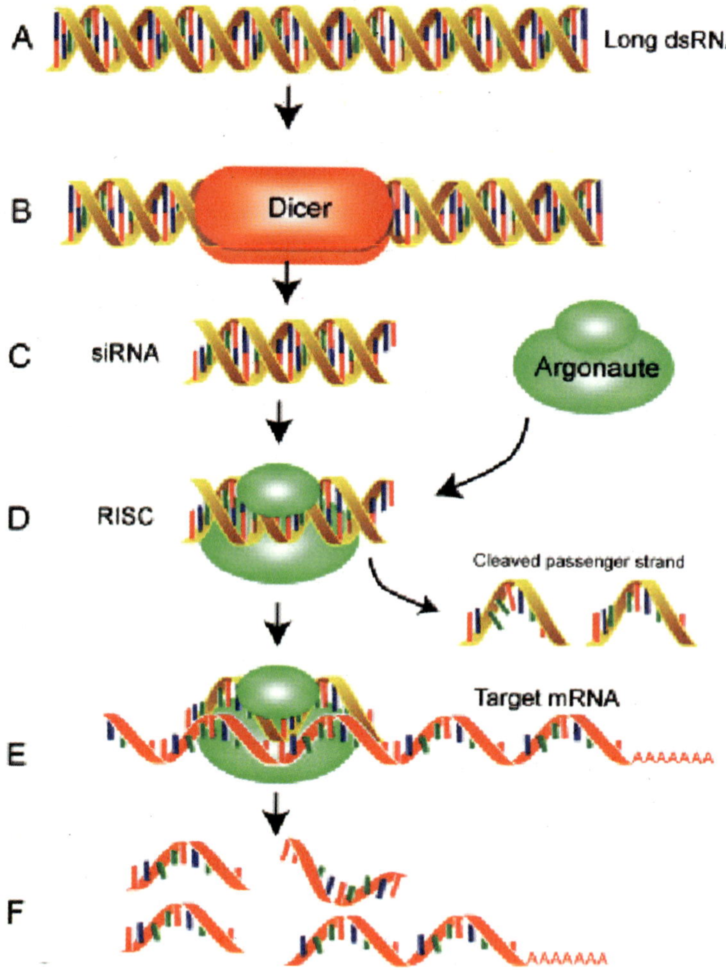

Fig. J: Mechanism of RNA interference

- **A).** On entering the cell, long dsRNAs act as a trigger of RNAi process.
- **B).** It is first processed by the RNAse III enzyme Dicer in an ATP-dependent reaction.
- **C).** Dicer processes dsRNAs into 21-23 nt short interfering RNA (siRNA) with 2-nt 3' overhangs. siRNA can also be synthesized outside the cell and then be introduced into a cell.
- **D).** The siRNAs are incorporated into the RNA-inducing silencing complex (RISC) which consists of an Argonaute (Ago) protein as one of its main components. Ago cleaves and discards the passenger (sense) strand of the siRNA duplex leading to activation of the RISC.
- **E & F).** The remaining guide (antisense) strand of the siRNA guides RISC to its homologous mRNA, resulting in the endonucleolytic cleavage of the target mRNA.

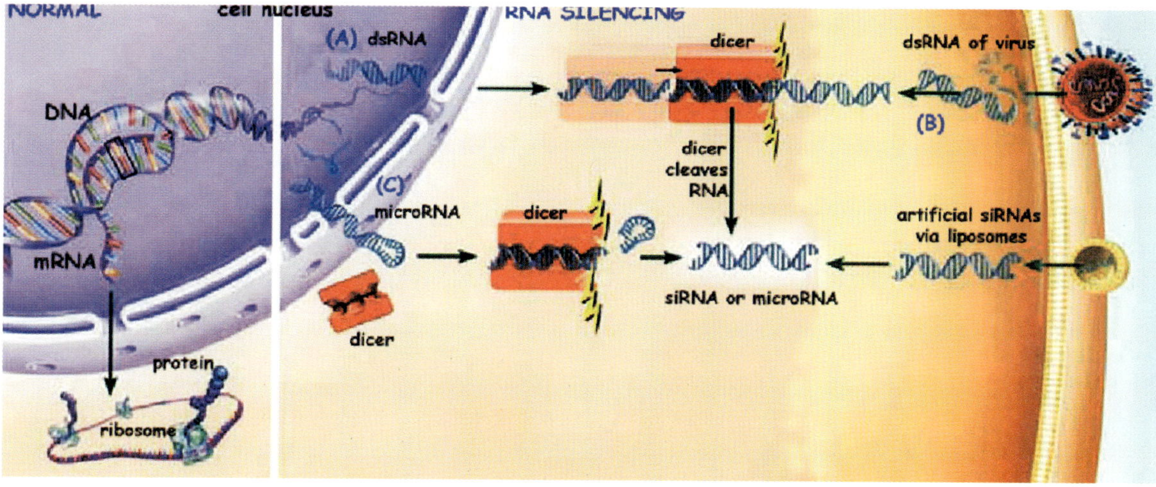

Fig. K

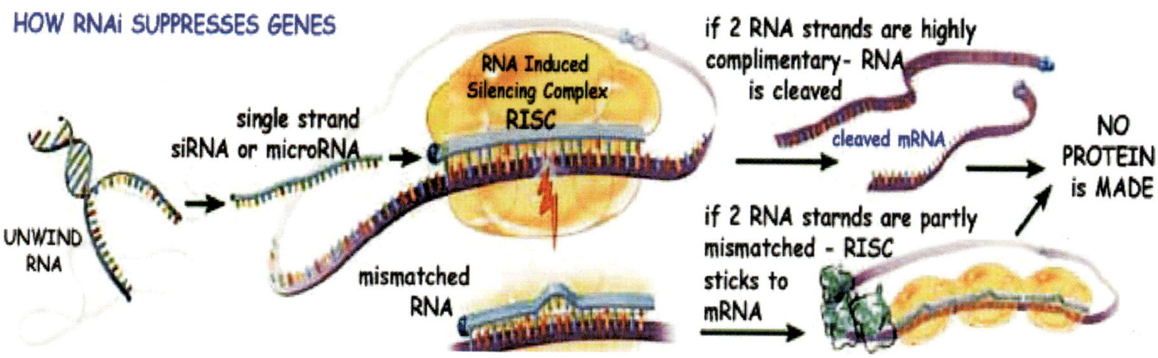

Fig. L

Fig. K: Cells can inhibit the expression of individual genes (stop proteins from being made) by interfering with a mRNA being transcribed. This is done via small double-stranded RNA. An enzyme named DICER snips short interfering RNAs (siRNA) from longer double stranded RNAs made by (A) self-copying gene sequences, (B) by replicating viruses, or (C) regulatory RNA sequences known as microRNAs. All the RNAs (A, B, & C) are cleaved by DICER enzyme into short siRNA pieces that can suppress gene expression.

Fig. L: The short siRNA pieces (A, B, & C above) unwind into single strand RNAs, which then combine with proteins to form a complex called RISC [RNA-Induced Silencing Complex]. The RISC then captures a native mRNA molecule that complements the short siRNA sequence. If the pairing [native mRNA and siRNA piece] is essentially perfect, the native mRNA is cut into useless RNA fragments that aren't translated. If however, the pairing is less than perfect then the RISC complex binds to the mRNA and blocks ribosome movement along the native mRNA also halting translation. The net effect is no protein is made.

in the genome is knocked down one at a time and the cells or organism scored for a desired outcome—for example, death of a cultured cancer cell but not a normal cell. Owing to the very large numbers of assays needed to look at the involvement of all 35 000 or so genes in the human genome, this approach is very labour intensive. The approach has been used successfully on a relatively small scale to investigate cell death signalling by TRAIL (tumour necrosis factor related apoptosis inducing ligand), an agent that might have therapeutic potential against various cancers. In addition, the approach was used to identify the familial cylindromatosis tumour suppressor gene (CYLD) as a de-ubiquitinating enzyme in the nuclear factor B pathway. As aspirin is known to target this pathway, this work suggested a novel therapeutic approach to this rare inherited cancer.

The other approach is to use large pools of RNA interference viral vectors and apply a selective pressure that only cells with the desired change in behaviour can survive. The identity of the genes knocked down in the surviving cells can then be identified by sequencing the RNA interference vectors that they carry. This method is being used to investigate genes involved in neurodegenerative diseases, diabetes, and cancer. It has recently been used successfully to identify several novel components of the p53 tumour suppressor gene signalling pathway. Both approaches show considerable promise in identifying novel genes that may make important therapeutic targets for inhibition either by conventional drug discovery methods or, more controversially, by RNA interference itself.

RNA interference as a novel therapeutic agent

RNA interference clearly has much promise in the laboratory, but how about in the clinic? In principle, RNA interference might be used to treat any disease that is linked to elevated expression of an identified gene. This might make it suitable for combating viral diseases, cancers, and inflammatory diseases, to name but three areas. Indeed, in tissue culture models, impressive results have been achieved against various cancer cells by using RNA interference to target oncogenes and against HIV, influenza, and polio viruses by targeting viral genes. However, a huge gap exists between achieving such results *in vitro* and in a whole animal or patient.

RNA interference *in vivo*

Despite the problems of delivery, RNA interference has been used effectively in the mouse to block expression of a hepatitis C virus protein in the liver. In addition, the same group has used specific RNA interference to block hepatitis B virus infection in mice. They achieved delivery by injecting large amounts of synthetic double stranded RNA or DNA encoding a short hairpin RNA into the portal vein. A similar approach was taken to target the Fas protein, an important inducer of programmed cell death, resulting in protection of mice from fulminant hepatitis caused by injection with agonistic Fas-specific antibodies.

The problems seen with the use of viral vectors in gene therapy mean that many researchers in RNA interference are favouring the use of synthetic siRNA duplexes rather than gene expression vectors that will direct the production of such molecules within the target cell. A large number of biotechnology companies have programmes to develop synthetic RNA interference therapies for various diseases. These include Sirna Therapeutics (Boulder, Colorado) for macular degeneration; Avocel (Sunnyvale, California) for hepatitis C; Alnylam Pharmaceuticals (Cambridge, Massachusetts) for Parkinson's disease; CytRx (Los Angeles, California) for obesity, type 2 diabetes, and ALS; Acuity Pharmaceuticals (Philadelphia, Pennsylvania) for macular degeneration and diabetic retinopathy; and Sequitur (Natick, Massachusetts) for hepatic insufficiency, respiratory syncytial virus, asthma, and cancer.

Given sufficient research into delivery methods, some of these diseases will probably eventually be treated effectively by RNA interference based therapeutics. Success is more likely in those diseases with a simple genetic basis rather than in complex multigene disorders such as cancer. Diseases involving sites where delivery of synthetic RNA is more straightforward will also be more likely to be effectively treated. The bitter experiences with antisense and gene therapy mean that the likely problems should not be underestimated, but perhaps this time the reality may—eventually—live up to the hype.

Although a big improvement on previous methods, RNA interference has its limitations. Not every sequence works—most researchers get a success rate of about one in three. In addition, although the effects are generally thought to be highly sequence specific, some question marks remain as to whether or not some of the effects seen are "off target." Some residual activation of the interferon system has been reported, as well as degradation of closely related, but non-identical, mRNAs.

Delivery problems

The major challenge in turning RNA interference into an effective therapeutic strategy is the delivery of the RNA interference agents, whether they are synthetic short double stranded RNAs or viral vectors directing production of double stranded RNA, to the target cells within the body. Lessons can be learnt from two earlier technologies that held out much initial therapeutic promise but have ultimately failed to deliver effective treatments.

One of these is antisense. This uses short pieces of single stranded DNA complementary to the mRNA that was to be targeted. The resulting RNA-DNA hybrids forming in the cell can block translation of the mRNA by the protein synthesis machinery and also promote its degradation. Despite nearly two decades of work, antisense has failed to prove its efficacy in the clinic, although several clinical trials have been done. In part this reflects the fact that antisense in general provides a much less robust inhibition of gene expression than RNA interference, but also major difficulties arose in getting the antisense oligonucleotides to their target cells without them being degraded elsewhere in the body.

Summary

First discovered in plants the nematode *Caenorhabditis elegans*, the production of small interfering RNAs (siRNAs) that bind to and induce the degradation of specific endogenous mRNAs is now recognized as a mechanism that is widely employed by eukaryotic cells to inhibit protein production at a posttranscriptional level. The endogenous siRNAs are typically 19- to 23-base double-stranded RNA oligonucleotides, produced from much larger RNAs that upon binding to target mRNAs recruit RNases to a protein complex that degrades the targeted mRNA. Methods for expressing siRNAs in cells in culture and *in vivo* using viral vectors, and for transfecting cells with synthetic siRNAs, have been developed and are being used to establish the functions of specific proteins in various cell types and organisms. RNA interference methods provide several major advantages over prior methods (antisense DNA or antibody-based techniques) for suppressing gene expression. Recent preclinical studies suggest that RNA interference technology holds promise for the treatment of various diseases. Pharmacologists have long dreamed of the ability to selectively antagonize or eliminate the function of individual proteins—RNAi technology may eventually make that dream a reality.

- RNA interference is an ancient natural antiviral mechanism that directs silencing of gene expression in a sequence specific manner
- RNA interference can be exploited artificially to inhibit the expression of any gene of interest
- The principal systems for achieving RNA interference are short synthetic double stranded RNA molecules and gene expression vectors that direct their production in the cell
- Libraries of RNA interference molecules have been constructed that allow the analysis of gene function on a genome-wide scale
- RNA interference systems could be used clinically to suppress gene expression as a therapeutic strategy in many diseases characterised by elevated gene function

The Future of RNA Interference in Biology and Medicine

Even at this early stage of understanding the molecular mechanisms of RNAi and in the development of methods for the use of RNAi technology for selective gene silencing, it is clear that RNAi will be a widely used tool for establishing the functions of genes. The ability to selectively deplete a single protein of interest in cultured cells using siRNAs, and plasmids and viral vectors, is now established. Improvements on the currently available protocols for RNAi are being made, and the methods are being applied by thousands of investigators in diverse fields. With the advent of these methods has come an explosion of studies that have employed RNAi. Indeed, although there were only nine publications listed on Medline in the years 1987 through 1998 inclusive, there are 631 listed from 1999 to the time of writing of this article. The current status of RNAi as an experimental tool is such that many investigators are now aware of the technology, but most have not yet implemented it in their own studies. The development of RNAi kits by several companies will facilitate the implementation of RNAi methods by essentially any investigator, regardless of their knowledge of RNAi mechanisms. However, major hurdles remain to be crossed, including the application of RNAi methods to *in vivo* studies. Once it becomes possible to reliably target a specific gene and deplete the protein encoded by that gene from one or all cell types in an organism, a wealth of information will flow from studies of processes ranging from embryogenesis to the function of organ systems. As for uses of RNAi in medicine, its potential remains to be established. The application of gene therapy approaches for the treatment of specific diseases has progressed much more slowly than initially anticipated.

There are, of course, many potential gene targets for therapeutic intervention using RNAi. Studies that employ RNAi to counteract a disease process *in vivo* are emerging. For example, RNAi that targeted the Fas gene (Song *et al.*, 2003) or the hepatitis C virus genome.

Potential Therapeutic Targets of RNAi

Cell Cycle Proteins

Anti-Apoptotic Proteins

Apoptotic Proteins

Ligands/receptors/signaling

Oxidative Stress-Related Proteins

Inflammation

Pathogen-Specific Genes

Gain-of-Function Mutations

1. **Cancer**
2. **Cardiovascular disease**
3. **Neurodegenerative disorders**

4. Infectious diseases

5. Autoimmune disorders

6. Obesity/diabetes

The ability to target a specific gene or genes using siRNAs or vector-mediated RNA expression methods, suggests the potential of RNAi to block the disease process or relieve symptoms of the disease. Depletion of proteins critical for the cell cycle, such as cyclins, cyclin-dependent kinases, or telomerase, might be effective in treating cancers and some neurodegenerative disorders. Blocking the production of anti-apoptotic proteins such as Bcl-2, inhibitor of apoptosis proteins and antioxidant enzymes may be used to kill cancer cells. Conversely, RNAi-mediated suppression of expression of apoptotic proteins (Bax, Par-4, p53, AIF, and caspases, for example) may slow or stop the degenerative processes in degenerative myocardial, neurological, and autoimmune disorders. Ligands, receptors, and downstream signal transduction proteins critical for a specific disease process might also be targeted. For example, it might be possible to suppress the appetite of obese patients by targeting neuropeptide Y- or ghrelin-producing cells using shRNA-expressing viral vectors. Genes that encode proteins involved in oxidative stress and inflammation (nitricoxide synthase, cyclooxygenases, and tumor necrosis factor, for example) might be targeted in autoimmune and infectious or inflammatory diseases. Viral and bacterial genes are obvious targets for RNAi-based therapeutic intervention in infectious diseases.

(Wilson *et al.*, 2003) protected mice from hepatitis. One might expect that the next demonstrations of successful treatment of disease in mice will come from models of cancer and neurodegenerative disorders. Because of the potential of RNAi for therapeutic intervention, major efforts should be placed on preclinical studies using this technology.

Conclusions

The discovery that cells have a special mechanism for suppressing the expression of homologous genes by recognizing and processing double-stranded RNA was totally unexpected and has dramatically expanded our knowledge of gene control. Remarkably, the RNAi machinery can handle double-stranded RNA entering the cell as well as double-stranded RNA generated within the cell. The development of an organism and proper function of its cells and tissues are dependent on an intact RNAi machinery. Infection by RNA viruses can be blocked by RNAi, especially in plants and lower animals, and foreign elements in the genome (viruses and transposons) can be kept silent. Finally, the discovery of RNAi has not only provided us with a powerful new experimental tool to study the function of genes but also raises expectations about future applications of RNAi in medicine.

GLOSSARY

Argonaute

Argonaute proteins are found in all RISC and miRNA-containing ribonucleoprotein complexes and have been shown to be the catalytic site of mRNA cleavage in RISC.

Argonautes are ~100 kDa proteins characterized by conserved domains called PAZ and PIWI. Argonautes can be organized by sequence into two subfamilies, Ago and Piwi, based on a higher degree of homology to either *Arabidopsis* AGO1 or to *Drosophila* Piwi. The Ago group functions in RNAi and miRNA silencing pathways.

Cosuppression

Silencing of an endogenous gene caused by the introduction of a trans-gene or infection by a virus. This term, which can refer to silencing at the post-transcriptional (PTGS; post-transcriptional gene silencing) or transcriptional (TGS; transcriptional gene silencing) level, is used primarily by researchers working with plants. In *Neurospora crassa*, this phenomenon is known as quelling.

Dicer

A ~200 kDa multidomain, RNase III family enzyme that functions in processing dsRNA to siRNA and assembly of the guide strand into RISC. Dicer progressively cleaves dsRNA at 21–25 bp intervals to generate siRNAs with 2-nt 3' overhangs and phosphorylated 5' ends. The predicted structure of Dicer includes an ATPase/RNA helicase domain, a conserved PAZ domain that is shared with Argonaute, two catalytic RNase III domains, and a C-terminal dsRNA binding domain (dsRBD).

dsRNA

Double-stranded RNA

Guide Strand

The strand of the siRNA that is assembled with RISC and provides sequence specificity for target mRNA cleavage. The guide strand is in the anti-sense orientation with respect to the mRNA.

miRNA

MicroRNA, a large class of evolutionarily conserved, noncoding, RNA originating from longer transcripts characterized by imperfect hairpin structures. miRNAs are 19–23 nt RNAs processed from pre-miRNA precursors by Dicer, the same enzyme that processes siRNAs.

Passenger Strand

The strand of the siRNA that dissociates from the siRNA during assembly with RISC.

PAZ Domain

Conserved domain found in both Dicer and Argonaute. It is thought that the 3' end of the guide strand of an siRNA is in contact with the PAZ in RISC.

PIWI Domain

Conserved domain found in Argonaute and thought to be the catalytic site for mRNA cleavage. It is thought that the 5' end of the guide strand of the siRNA contacts the PIWI domain in RISC.

Post-transcriptional Gene Silencing (PTGS)

Silencing of an endogenous gene caused by the introduction of a homologous dsRNA, trans-gene or virus. In PTGS, the transcript of the silenced gene is synthesized but does not accumulate because it is rapidly degraded. This is a more general term than RNAi, since it can be triggered by several different means.

R2D2

A small, dsRNA binding protein that works with Dicer in assembly of siRNA with RISC.

rasiRNA

Repeat-associated siRNA, an endogenous form of siRNA that originates from repetitive elements within the genome.

RdRPs

RNA-dependent RNA Polymerases. RdRPs may play a role in amplifying RNA triggers for silencing. Although they are not found in insects and mammals, they are present in other eukaryotes that have RNA silencing pathways.

RISC

RNA-induced silencing complex. A nuclease complex, composed of proteins and siRNA, that targets and cleaves endogenous mRNAs complementary to the siRNA within the complex.

RITS

RNA-induced transcriptional silencing complex. A complex of proteins and rasiRNA that inhibits transcription of target DNA through heterochromatin formation. RITS contains at least one protein in common with RISC, Argonaute.

RNAi, RNA Interference

A process of mRNA cleavage and degradation that is induced by double-stranded RNA in a sequence-specific manner.

RNA Silencing

The collective term for the pathways that use small RNAs as guides to specifically modify expression of targeted genes or genomic regions.

RNase III Family

A group of double-stranded RNA-specific endonucleases characterized by production of dsRNA fragments with 2–3-nt 3' overhangs and 5' phosphorylated ends, similar to those produced by Dicer.

Small RNA

A collective term for siRNA, rasiRNA, and miRNA

siRNA

Short interfering RNA, siRNAs are 21–25 bp dsRNA with dinucleotide 3' overhangs that are processed from longer dsRNA by Dicer in the RNA interference pathway. Introduction of synthetic siRNAs can induce RNA interference in mammalian cells. siRNAs can also originate from endogenous dsRNA precursors.

shRNA

Short hairpin RNA. shRNAs are used in plasmid- or vector-based approaches for supplying siRNAs to cells to produce stable gene silencing. A strong promoter is used to drive transcription of a target sequence designed to form hairpins and loops of variable length, which are then processed to siRNAs by the cellular RNAi machinery.

siRISC

A term for RISC assembled with the guide strand of an siRNA.

REFERENCES

Abbas-Terki T Blanco-Bose W Deglon N Pralong W and Aebischer P 2002 Lentiviral-mediated RNA interference *Hum Gene Ther* **13** 2197–2201

Ancellin N Colmont C Su J Li Q Mittereder N Chae SS Stefansson S Liau G and Hla T 2002 Extracellular export of sphingosine kinase-1 enzyme Sphingosine 1-phosphate generation and the induction of angiogenic vascular maturation *J Biol Chem* **277** *6667–6675*

Aoki Y Cioca D Oidaira H Kamiya J and Kiyosawa K 2003 RNA interference may be more potent than antisense RNA in human cancer cell lines *Clin Exp Pharmacol Physiol* **30** 96–102

Assaad FF Tucker KL and Signer ER 1993 Epigenetic repeat-induced gene silencing (RIGS) in Arabidopsis *Plant Mol Biol* **22** 1067–1085

Avery OT MacLeod CM and McCarty M 1944 Studies on the chemical nature of the substance inducing transformation of *Pneumococcal* types *J Exp Med* **79** 137-158

Aza-Blanc P Cooper CL Wagner K Batalov S Deveraux QL Cooke MP 2003 Identification of modulators of TRAIL-induced apoptosis via RNAi-based phenotypic screening *Mol Cell* **12** 627-37

Baehrecke EH 2002 How death shapes life during development *Nat Rev Mol Cell Biol* **3** 779–787

Baglioni C and Nilsen TW 1983 Mechanisms of antiviral action of interferon *Interferon* **5** 23–42

Bai X Zhou D Brown JR Crawford BE Hennet T and Esko JD 2001 Biosynthesis of the linkage region of glycosaminoglycans cloning and activity of galactosyltransferase II the sixth member of the beta 1 3-galactosyltransferase family (beta 3GalT6 *J Biol Chem* **276** 48189–48195

Banerjee D and Slack F 2002 Control of developmental timing by small temporal RNAs a paradigm for RNA-mediated regulation of gene expression *Bioessays* **24** 119–129

Barton GM and Medzhitov R 2002 Delivery of small interfering RNA into primary cells *Proc Natl Acad Sci USA* **99** 14943–14945

Berns K Hijmans EM Mullenders J Brummelkamp TR Velds A Heinerikx M *et al* 2004 A large-scale RNAi screen in human cells identifies new components of the p53 pathways *Nature* **428** 431-7

Bernstein E Caudy AA Hammond SM and Hannon GJ 2001 Role for a bidentate ribonuclease in the initiation step of RNA interference *Nature (Lond)* **409** 363–366

Bernstein E Caudy AA Hammond SM and Hannon GJ 2001 Role for a bidentate ribonuclease in the initiation step of RNA interference *Nature* **409** 363-366

Billy E Brondani V Zhang H Muller U and Filipowicz W 2001 Specific interference with gene expression induced by long double-stranded RNA in mouse embryonal teratocarcinoma cell lines *Proc Natl Acad Sci USA* **98** 14428–14433

Blaszczyk J Tropea JE Bubunenko M Routzahn KM Waugh DS Court DL and Ji X 2001 Crystallographic and modeling studies of RNase III suggest a mechanism for double-stranded RNA cleavage *Structure (Camb)* **9** 1225–1236

Boehm M Yoshimoto T Crook MF Nallamshetty S True A Nabel GJ and Nabel EG 2002 A growth factor-dependent nuclear kinase phosphorylates p27(Kip1 and regulates cell cycle progression *EMBO (Eur Mol Biol Organ) J* **21** 3390–3401

Bosher JM and Labouesse M 2000 RNA interference genetic wand and genetic watchdog *Nat Cell Biol* **2** E31–E36

Brummelkamp TR Bernards R and Agami R (2002a) A system for stable expression of short interfering RNAs in mammalian cells *Science (Wash DC)* **296** 550–553

Brummelkamp TR Bernards R and Agami R (2002b) Stable suppression of tumorigenicity by virus-mediated RNA interference *Cancer Cell* **2** 243–247

Brummelkamp TR Nijman SM Dirac AM Bernards R 2003 Loss of the cylindromatosis tumour suppressor inhibits apoptosis by activating NF-kappaB *Nature* **424** 797-801

Cabot RA and Prather RS 2003 Cleavage stage porcine embryos may have differing developmental requirements for karyopherins alpha2 and alpha3 *Mol Reprod Dev* **64** 292–301

Calegari F Haubensak W Yang D Huttner WB and Buchholz F 2002 Tissue specific RNA interference in postimplantation mouse embryos with endoribonuclease-prepared short interfering RNA *Proc Natl Acad Sci USA* **99** 14236–14240

Caplen NJ Fleenor J Fire A and Morgan RA 2000 dsRNA-mediated gene silencing in cultured Drosophila cells a tissue culture model for the analysis of RNA interference *Gene* (Amst) **252** 95–105

Caplen NJ Taylor JP Statham VS Tanaka F Fire A and Morgan RA 2002 Rescue of polyglutamine-mediated cytotoxicity by double-stranded RNA-mediated RNA interference *Hum Mol Genet* **11** 175–184

Capodici J Kariko K and Weissman D 2002 Inhibition of HIV-1 infection by small interfering RNA-mediated RNA interference *J Immunol* **169** 5196–5201

Carrington JC Ambros V 2002 Role of microRNAs in plant and animal development *Science* **301** 336-8

Caudy AA Myers M Hannon GJ and Hammond SM 2002 Fragile X-related protein and VIG associate with the RNA interference machinery *Genes Dev* **16** 2491–2496

Cheng A Wang S Yang D Xiao R and Mattson MP 2003 Calmodulin mediates brain-derived neurotrophic factor cell survival signaling upstream of Akt kinase in embryonic neocortical neurons *J Biol Chem* **278** 7591–7599

Chiu YL and Rana TM 2002 RNAi in human cells basic structural and functional features of small interfering RNA *Mol Cell* **10** 549–561

Cho JH Bandyopadhyay J Lee J Park CS and Ahnn J 2000 Two isoforms of sarco/endoplasmic reticulum calcium ATPase (SERCA) are essential in Caenorhabditis elegans Gene (Amst) **261** 211–219

Cioca DP Aoki Y and Kiyosawa K 2003 RNA interference is a functional pathway with therapeutic potential in human myeloid leukemia cell lines *Cancer Gene Ther* **10** 125–133

Clarke PA and Mathews MB 1995 Interactions between the double-stranded RNA binding motif and RNA definition of the binding site for the interferon-induced protein kinase DAI (PKR) on adenovirus VA RNA *RNA* **1** 7–20

Coburn GA Cullen BR 2002 Potent and specific inhibition of human immunodeficiency virus type 1 replication by RNA interference *J Virol* **76** 9225-31

Cogoni C and Macino G 1999 Gene silencing in Neurospora crassa requires a protein homologous to RNA-dependent RNA polymerase *Nature (Lond)* **399** 166–169

Cogoni C and Macino G 1999 Gene silencing in *Neurospora crassa* requires a protein homologus to RNA-depednet RNA polymerase *Nature* **399** 166-169

Cogoni C Irelan JT Schumacher M Schmidhauser T Selker EU and Macino G 1996 Transgene silencing of the al-1 gene in vegetative cells of *Neurospora* is mediated by a cytoplasmic effector and does not depend on DNA-DNA interactions or DNA methylation *EMBO (Eur Mol Biol Organ) J* **15** 3153 3163

Colussi PA Quinn LM Huang DC Coombe M Read SH Richardson H and Kumar S 2000 Debcl a proapoptotic Bcl-2 homologue is a component of the Drosophila melanogaster cell death machinery *J Cell Biol* **148** 703–714

Covey S Al-Kaff N Lángara A and Turner D 1997 Plants combat infection by gene silencing *Nature* **385** 781-782

Czauderna F Fechtner M Aygun H Arnold W Klippel A Giese K and Kaufmann J 2003 Functional studies of the PI(3)-kinase signalling pathway employing synthetic and expressed siRNA *Nucleic Acids Res* **31** 670–682

Dalmay T Hamilton A Rudd S Angell S and Baulcombe DC 2000 An RNAdependent RNA polymerase gene in Arabidopsis is required for posttranscriptional gene silencing mediated by a transgene but not by a virus *Cell* **101** 543–553

Damm-Welk C Fuchs U Wossmann W Borkhardt A 2003 Targeting oncogenic fusion genes in leukemias and lymphomas by RNA interference *Semin Cancer Biol* **13** 283-92

de Carvalho F Gheyson G Kushnir S van Montagu M Inzé D *et al* 1992 Suppression of beta-1 3-glucanase transgene expression in homozygous plants *EMBO J* **11** 2595-2602.

Devroe E and Silver PA 2002 Retrovirus-delivered siRNA *BMC Biotechnol* **2** 15

Domeier ME Morse DP Knight SW Portereiko M Bass BL and Mango SE (2000) A link between RNA interference and nonsense-mediated decay in *Caenorhabditis elegans Science* (Wash DC) **289** 1928–1931

Donze O and Picard D 2002 RNA interference in mammalian cells using siRNAs synthesized with T7 RNA polymerase *Nucleic Acids Res* **30** e46

Dorsett Y and Tuschl T 2004 siRNAs Applications in functional genomics and potential as therapeutics *Nature Reviews* **3** 318-329

Dougherty WG and Parks TD 1995 Transgenes and suppression telling us something new? *Curr Opin Cell Biol* **7** 399-405

Duan W Rangnekar V and Mattson MP 1999 Par-4 production in synaptic compartments following apoptotic and excitotoxic insults evidence for a pivotal role in mitochondrial dysfunction and neuronal degeneration *J Neurochem* **72** 2312–2322

Duan W Zhu X Ladenheim B Yu QS Guo Z Oyler J Cutler RG Cadet JL Greig NH and Mattson MP 2002 p53 inhibitors preserve dopamine neurons and motor function in experimental parkinsonism *Ann Neurol* **52** 597–606

Eddy SR 2001 Non-coding RNA genes and the modern RNA world *Nat Rev Genet* **2** 919–921

Elbashir SM Harborth J Lendeckel W Yalcin A Weber K and Tuschl T 2001a Duplexes of 21-nucleotide RNAs mediate RNA interference in cultured mammalian cells *Nature (Lond)* **411** 494–498

Elbashir SM Harborth J Lendeckel W Yalcin A Weber K and Tuschl T 2001 Duplexes of 21-nucleotide RNAs mediate RNA interference in cultured mammalian cells *Nature* **411** 494-498

Elbashir SM Harborth J Lendeckel W Yalcin A Weber K Tuschl T 2001 Duplexes of 21-nucleotide RNAs mediate RNA interference in cultured mammalian cells *Nature* **411** 494-8

Elbashir SM Martinez J Patkaniowska A Lendeckel W and Tuschl T 2001b Functional anatomy of siRNAs for mediating efficient RNAi in Drosophila melanogaster embryo lysate *EMBO (Eur Mol Biol Organ) J* **20** 6877–6888

Eldadah BA and Faden AI 2000 Caspase pathways neuronal apoptosis and CNS injury *J Neurotrauma* **17** 811–829

English JJ Davenport GF Elmayan T Vaucheret H and Baulcombe DC 1997 Requirement of sense transcription of homology-dependent virus resistance and trans-inactivation *Plant J* **12** 597–603

Faergeman NJ and Knudsen J 2002 Acyl-CoA binding protein is an essential protein in mammalian cell lines *Biochem J* **368** 679–682

Fagard M Boutet S Morel JB Bellini C and Vaucheret H 2000 AGO1 QDE-2 and RDE-1 are related proteins required for post-transcriptional gene silencing in plants quelling in fungi and RNA interference in animals *Proc Natl Acad Sci USA* **97** 11650–11654

Fire A 1999 RNA-triggered gene silencing *Trends Genet* **15** 358–363

Fire A Xu S Montgomery MK Kostas SA Driver SE and Mello CC 1998 Potent and specific genetic interference by double-stranded RNA in Caenorhabditis Elegans *Nature (Lond) 391 806–811*

Fire A Xu S Montgomery MK Kostas SA Driver SE and Mello CC 1998 Potent and specific genetic interference by double-stranded RNA in *Caenorhabditis elegans Nature* **391** 806-811

Fire A Xu S Montgomery MK Kostas SA Driver SE Mello CC 1998 Potent and specific genetic interference by double-stranded RNA in *Caenorhabditis elegans Nature* **391** 806-11

Fire A Xu S Q Montgomery M K Kostas S A Driver S E Mello C C 1998 Potent and specific genetic interference by double-stranded RNA in *Caenorhabditis elegans Nature* **391** 806-811

Franz DR and Zajtchuk R 2002 Biological terrorism understanding the threat preparation and medical response *Dis Mon* **48** 493–564

Frischmeyer PA and Dietz HC 1999 Nonsense-mediated mRNA decay in health and disease *Hum Mol Genet* **8** 1893–1900

Gañ L Anton KE Masterson BA Vincent VA Ye S and Gonzalez-Zulueta M 2002 Specific interference with gene expression and gene function mediated by long dsRNA in neural cells *J Neurosci Methods* **121** 151–157

Gary DS and Mattson MP 2001 Integrin signaling via the PI3-kinase-Akt pathway increases neuronal resistance to glutamate-induced apoptosis *J Neurochem* **76** 1485–1496

Gaudilliere B Shi Y and Bonni A 2002 RNA interference reveals a requirement for myocyte enhancer factor 2A in activity-dependent neuronal survival *J Biol Chem* **277** 46442–46446

Ge Q McManus MT Nguyen T Shen CH Sharp PA Eisen HN *et al* 2003 RNA interference of influenza virus production by directly targeting mRNA for degradation and indirectly inhibiting all viral RNA transcription *Proc Natl Acad Sci U S A* **100** 2718-23

Geng YJ and Libby P 2002 Progression of atheroma a struggle between death and procreation *Arterioscler Thromb Vasc Biol* **22** 1370–1380

Gil J and Esteban M 2000 Induction of apoptosis by the dsRNA-dependent protein kinase (PKR) mechanism of action *Apoptosis* **5** 107–114

Gilman CP Chan SL Guo Z Zhu X Greig NH and Mattson MP 2003 p53 is present in synapses where it mediates mitochondrial dysfunction and synaptic degeneration in response to DNA damage and oxidative and excitotoxic insults *Neuromolecular Med* **3** 159–172

Gitlin L Karelsky S and Andino R 2002 Short interfering RNA confers intracellular antiviral immunity in human cells *Nature (Lond)* **418** 430–434

Gitlin L Karelsky S Andino R 2002 Short interfering RNA confers intracellular antiviral immunity in human cells *Nature* **418** 430-4

Glazner GW Chan SL Lu C and Mattson MP 2000 Caspase-mediated degradation of AMPA receptor subunits a mechanism for preventing excitotoxic necrosis and ensuring apoptosis *J Neurosci* **20** 3641–3649

Gottesman S 2004 The small RNA regulators of *Escherichia coli* Roles and mechanisms *Ann Rev Microbiol* 58 303-328

Greener T Grant B Zhang Y Wu X Greene LE Hirsh D and Eisenberg E 2001 *Caenorhabditis elegans* auxilin a J-domain protein essential for clathrinmediated endocytosis *in vivo Nat Cell Biol* **3** 215–219

Grishok A Pasquinelli AE Conte D Li N Parrish S Ha I Baillie DL Fire A Ruvkun G and Mello CC 2001 Genes and mechanisms related to RNA interference regulate expression of the small temporal RNAs that control C elegans developmental timing *Cell* **106** 23–34

Grishok A Tabara H and Mello CC 2000 Genetic requirements for inheritance of RNAi in *C Elegans Science (Wash DC)* **287** 2494–2497

Guo S and Kemphues KJ 1995 *Par-1* a gene required for establishing polarity in C *elegans* embryos encodes a putative *Sex*/Thr kinase that is asymmetrically distributed *Cell* **81** 611-620

Hall IM Shankaranarayana GD Noma K Ayoub N Cohen A and Grewal SI 2002 Establishment and maintenance of a heterochromatic domain *Science* **297** 2232-2237

Hamilton AJ and Baulcombe DC 1999 A species of small antisense RNA in posttranscriptional gene silencing in plants *Science (Wash DC)* **286** 950–952

Hamilton AJ and Baulcombe DC 1999 A species of small antisense RNA in posttranscriptional gene silencing in plants *Science* **286** 950-952

Hammond SM 2005 Dicing and slicing The core machinery of the RNA interference pathway *FEBS Lett* **579** 5822-5829

Hammond SM Bernstein E Beach D and Hannon GJ 2000 An RNA-directed nuclease mediates post-transcriptional gene silencing in Drosophila cells *Nature (Lond)* **404** *293–296*

Hammond SM Bernstein E Beach D and Hannon GJ 2000 An RNA-directed nuclease mediates post-transcriptional gene silencing in *Drosophila* cells *Nature* **404** 293-296

Hammond SM Boettcher S Caudy AA Kobayashi R and Hannon GJ 2001 Argonaute2 a link between genetic and biochemical analyses of RNAi *Science (Wash DC)* **293** 1146–1150

Hannon GJ 2002 RNA interference *Nature* **418** 244-251

Hannon GJ 2002 RNA interference *Nature* **418** 244-51

Hannon GJ and Rose JJ 2004 Unlocking the potential of the human genome with RNA interference *Nature* **431** 371-378

Harborth J Elbashir SM Bechert K Tuschl T and Weber K 2001 Identification of essential genes in cultured mammalian cells using small interfering RNAs *J Cell Sci* **114** 4557–4565

Hardy J 2001 The genetic causes of neurodegenerative diseases *J Alzheimers Dis* **3** 109–116

Higuchi H Yamashita T Yoshikawa H and Tohyama M 2003 Functional inhibition of the p75 receptor using a small interfering RNA *Biochem Biophys Res Commun* **301** 804–809

Holtorf H Schob H Kunz C Waldvogel R and Meins F Jr 1999 Stochastic and nonstochastic post-transcriptional silencing of chitinase and beta-1 3-glucanase genes involves increased RNA turnover-possible role for ribosome-independent RNA degradation *Plant Cell* **11** 471–484

Hutchings NR Donelson JE and Hill KL 2002 Trypanin is a cytoskeletal linker protein and is required for cell motility in African trypanosomes *J Cell Biol* **156** 867–877

Hutvagner G and Zamore PD 2002 A microRNA in a multiple-turnover RNAi enzyme complex *Science (Wash DC)* **297** 2056–2060

Iratni R Yan YT Chen C Ding J Zhang Y Price SM Reinberg D and Shen MM 2002 Inhibition of excess nodal signaling during mouse gastrulation by the transcriptional corepressor DRAP1 *Science (Wash DC)* **298** 1996–1999

Irie N Sakai N Ueyama T Kajimoto T Shirai Y and Saito N 2002 Subtype and species-specific knockdown of PKC using short interfering RNA *Biochem Biophys Res Commun* **298** 738–743

Ishizuka A Siomi MC and Siomi H 2002 A Drosophila fragile X protein interacts with components of RNAi and ribosomal proteins *Genes Dev* **16** 2497–2508

Izant JG and Weintraub H 1984 Inhibition of thymidine kinase gene expression by anti-sense RNA A molecular approach to genetic analysis *Cell* **36** 1007-1015

Jacque JM Triques K and Stevenson M 2002 Modulation of HIV-1 replication by RNA interference *Nature (Lond)* **418** 435–438

Jansen B and Zangemeister-Wittke U 2002 Antisense therapy for cancer–the time of truth *Lancet Oncol* **3** 672–683

Jarad G Wang B Khan S DeVore J Miao H Wu K Nishimura SL Wible BA Konieczkowski M Sedor JR and Schelling JR 2002 Fas activation induces renal tubular epithelial cell beta 8 integrin expression and function in the absence of apoptosis *J Biol Chem* **277** 47826–47833

Jiang M and Milner J 2002 Selective silencing of viral gene expression in HPVpositive human cervical carcinoma cells treated with siRNA a primer of RNA interference *Oncogene* **21** 6041–6048

Kartasheva NN Contente A Lenz-Stoppler C Roth J and Dobbelstein M 2002 p53 induces the expression of its antagonist p73 Delta N establishing an autoregulatory feedback loop *Oncogene* **21** 4715–4727

Kennerdell JR and Carthew RW 1998 Use of dsRNA-mediated genetic interference to demonstrate that frizzled and frizzled 2 Act in the wingless pathway *Cell* **95** 1017–1026

Kennerdell JR and Carthew RW 2000 Heritable gene silencing in Drosophila using double-stranded RNA *Nat Biotechnol* **18** 896–898

Ketting RF Fischer SE Bernstein E Sijen T Hannon GJ and Plasterk RH 2001 Dicer functions in RNA interference and in synthesis of small RNA involved in developmental timing in *C Elegans Genes Dev* **15** 2654–2659

Ketting RF Haverkamp TH van Luenen HG and Plasterk RH 1999 Mut-7 of *C elegans* required for transposon silencing and RNA interference is a homolog of Werner syndrome helicase and RNase D *Cell* **99** 133-141

Kim MH Yuan X Okumura S and Ishikawa F 2002 Successful inactivation of endogenous Oct-3/4 and c-mos genes in mouse preimplantation embryos and oocytes using short interfering RNAs *Biochem Biophys Res Commun* **296** 1372–1377

Kisielow M Kleiner S Nagasawa M Faisal A and Nagamine Y 2002 Isoformspecific knockdown and expression of adaptor protein ShcA using small interfering RNA *Biochem J* **363** 1–5

Krichevsky AM and Kosik KS 2002 RNAi functions in cultured mammalian neurons *Proc Natl Acad Sci USA* **99** 11926–11929

Lagos-Quintana M Rauhut R Lendeckel W and Tuschl T 2001 Identification of novel genes coding for small expressed RNAs *Science* **294** 853-858

Lau NC Lim LP Weinstein EG and Bartel DP 2001 An abundant class of tiny RNAs with probable regulatory roles in *Caenorhabditis elegans Science* **294** 858-862

Lee NS Dohjima T Bauer G Li H Li MJ Ehsani A Salvaterra P and Rossi J 2002a Expression of small interfering RNAs targeted against HIV-1 rev transcripts in human cells *Nat Biotechnol* **20** 500–505

Lee RC and Ambros V 2001 An extensive class of small RNAs in *Caenorhabditis elegans Science* **294** 862-864

Lee RC Feinbaum RL and Ambros V 1993 The *C elegans* heterochronic gene *lin-4* encodes small RNAs with antisense complimentarity to *lin-14 Cell* **75** 843-854

Lee SF Shah S Li H Yu C Han W and Yu G 2002b Mammalian APH-1 interacts with presenilin and nicastrin and is required for intramembrane proteolysis of amyloid-beta precursor protein and Notch *J Biol Chem* **277** 45013–45019

Lee SS Lee RY Fraser AG Kamath RS Ahringer J and Ruvkun G 2002c A systematic RNAi screen identifies a critical role for mitochondria in *C elegans* longevity *Nat Genet* **33** 40–48

Leirdal M and Sioud M 2002 Gene silencing in mammalian cells by preformed small RNA duplexes *Biochem Biophys Res Commun* **295** 744–748

Lewis DL Hagstrom JE Loomis AG Wolff JA and Herweijer H 2002 Efficient delivery of siRNA for inhibition of gene expression in postnatal mice *Nat Genet* **32** 107–108

Lindbo J Silva-Rosales L Proebsting W and Dougherty W 1993 Induction of a highly specific antiviral state in transgenic plants Implications for regulation of gene expression and virus resistance *Plant Cell* **5** 1749-1759

Lindenbach BD and Rice CM 2002 RNAi targeting an animal virus news from the front *Mol Cell* **9** 925–957

Liu X and Erikson RL 2002 Activation of Cdc2/cyclin B and inhibition of centrosome amplification in cells depleted of Plk1 by siRNA *Proc Natl Acad Sci USA* **99** 8672-8676

Lois C Hong EJ Pease S Brown EJ and Baltimore D 2002 Germline transmission and tissue-specific expression of transgenes delivered by lentiviral vectors *Science (Wash DC)* **295** 868–872

Mailand N Lukas C Kaiser BK Jackson PK Bartek J and Lukas J 2002 Deregulated human Cdc14A phosphatase disrupts centrosome separation and chromosome segregation *Nat Cell Biol* **4** 317–322

Makimura H Mizuno TM Mastaitis JW Agami R and Mobbs CV 2002 Reducing hypothalamic AGRP by RNA interference increases metabolic rate and decreases body weight without influencing food intake *BMC Neurosci* **3** 18

Marathe R Anandalakshmi R Smith TH Pruss GJ and Vance VB 2000 RNA viruses as inducers suppressors and targets of post-transcriptional gene silencing *Plant Mol Biol* **43** *295–306*

Martinez J Patkaniowska A Urlaub H Luhrmann R and Tuschl T 2002 Singlestranded antisense siRNAs guide target RNA cleavage in RNAi *Cell* **110** 563–574

Martins LM Iaccarino I Tenev T Gschmeissner S Totty NF Lemoine NR Savopoulos J Gray CW Creasy CL Dingwall C and Downward J 2002 The serine protease Omi/HtrA2 regulates apoptosis by binding XIAP through a reaper-like motif *J Biol Chem* **277** 439–444

Mattson MP 2000 Apoptosis in neurodegenerative disorders *Nat Rev Mol Cell Biol* **1** 120–129

Mattson MP 2003 Excitotoxic and excitoprotective mechanisms abundant targets for the prevention and treatment of neurodegenerative disorders *Neuromolecular Med* **3** 65–94

Mattson MP and Camandola S 2001 NF-kappaB in neuronal plasticity and neurodegenerative disorders *J Clin Investig* **107** 247–254

Mattson MP Culmsee C and Yu ZF 2000 Apoptotic and antiapoptotic mechanisms in stroke *Cell Tissue Res* **301** 173–187

Mattson MP Keller JN and Begley JG 1998 Evidence for synaptic apoptosis *Exp Neurol* **153** 35–48

Matzke M Primig M Trnovsky J and Matzke A 1989 Reversible methylation and inactivation of marker genes in sequentially transformed plants *EMBO J* **8** 643-649

McCaffrey AP Meuse L Pham TT Conklin DS Hannon GJ and Kay MA 2002 RNA interference in adult mice *Nature (Lond)* **418** 38–39

McCaffrey AP Meuse L Pham TT Conklin DS Hannon GJ Kay MA 2002 RNA interference in adult mice *Nature* **418** 38-9

McCaffrey AP Nakai H Pandey K Huang Z Salazar FH Xu H *et al* 2003 Inhibition of hepatitis B virus in mice by RNA interference *Nat Biotechnol* **21** 639-44

McManus MT Petersen CP Haines BB Chen J and Sharp PA 2002 Gene silencing using micro-RNA designed hairpins *RNA* **8** 842–850

Meister G and Tuschl T 2004 Mechanisms of gene silencing by double-stranded RNA *Nature* **431** 343-349

Mello CC and Conte Jr D 2004 Revealing the world of RNA interference *Nature* **431** 338-342

Mendell JT Rhys CM and Dietz HC 2002 Separable roles for rent1/hUpf1 in altered splicing and decay of nonsense transcripts *Science (Wash DC)* **298** 419–422

Mette MF Aufsatz W van der Winden J Matzke M and Matzke A 2000 Transcriptional gene silencing and promoter methylation triggered by double stranded RNA *EMBO J* **19** 5194-5201

Mette MF Aufsatz W van der Winden J Matzke MA and Matzke AJ 2000 Transcriptional silencing and promoter methylation triggered by double-stranded *RNA EMBO (Eur Mol Biol Organ) J* **19** *5194–5201*

Mise-Omata S Montagne B Deckert M Wienands J and Acuto O 2003 Mammalian actin binding protein 1 is essential for endocytosis but not lamellipodia formation functional analysis by RNA interference *Biochem Biophys Res Commun* **301** 704–710

Miyagishi M and Taira K 2002 U6 promoter-driven siRNAs with four uridine 3′ overhangs efficiently suppress targeted gene expression in mammalian cells *Nat Biotechnol* **20** 497–500

Mizuno T Chou M-Y and Inouye M 1984 A unique mechanism regulating gene expression Translational inhibition by a complementary RNA transcript (micRNA) *Proc Natl Acad Sci* **81** 1966-1970

Mollinari C Kleman JP Jiang W Schoehn G Hunter T and Margolis RL 2002 PRC1 is a microtubule binding and bundling protein essential to maintain the mitotic spindle midzone *J Cell Biol* **157** 1175–1186

Montgomery MK Xu S and Fire A 1998 RNA as a target of double-stranded RNA-mediated genetic interference in Caenorhabditis Elegans *Proc Natl Acad Sci USA* **95** 15502–15507

Montgomery MK Xu S and Fire A 1998 RNA as target of double-stranded RNA-mediated genetic interference in *Caenorhabditis elegans Proc Natl Acad Sci* **95** 15502-15507

Morrisey DV *et al* 2005 Potent and persistent *in vivo* anti-HBV activity of chemically modified siRNAs *Nat Biotechnol* **23** 1002-1007

Mourrain P Beclin C Elmayan T Feuerbach F Godon C Morel JB Jouette D Lacombe AM Nikic S Picault N Remoue K Sanial M Vo TA Vaucheret H 2000 Arabidopsis SGS2 and SGS3 genes are required for posttranscriptional gene silencing and natural virus resistance *Cell* **101** 533–542

Murchison EP and Hannon GJ 2004 miRNAs on the move miRNA biogenesis and the RNAi machinery *Curr Opin Cell Biol* **16** 223-229

Nagashima K Endo A Ogita H Kawana A Yamagishi A Kitabatake A Matsuda M and Mochizuki N 2002 Adaptor protein Crk is required for ephrin-B1-induced membrane ruffling and focal complex assembly of human aortic endothelial cells *Mol Biol Cell* **13** 4231–4242

Naldini L 1998 Lentiviruses as gene transfer agents for delivery to non-dividing cells *Curr Opin Biotechnol* **9** 457–463

Nam NH and Parang K 2003 Current targets for anticancer drug discovery *Curr Drug Targets* **4** 159–179

Napoli C Lemieux C and Jorgensen R 1990 Introduction of a chimeric chalcone synthase gene into petunia results in reversible co-suppression of homologous genes in trans *Plant Cell* **2** 279–289

Napoli C Lemieux C and Jorgensen R 1990 Introduction of a chimeric chalcone synthase gene into petunia results in reversible co-suppression of homologous genes *in trans Plant Cell* **2** 279-289

Napoli C Lemieux C Jorgensen R 1990 Introduction of a chimeric chalcone synthase gene into petunia results in reversible co-suppression of homologous genes in trans *Plant Cell* **2** 279-89

Nicholson RH and Nicholson AW 2002 Molecular characterization of a mouse cDNA encoding Dicer a ribonuclease III ortholog involved in RNA interference *Mamm Genome* **13** 67–73

Nordström K and Wagner EG 1994 Kinetic aspects of control of plasmid replication by antisense RNA *Trends Biochem Sci* **19** 294-300

Novina CD Murray MF Dykxhoorn DM Beresford PJ Riess J Lee SK Collman RG Lieberman J Shankar P and Sharp PA 2002 siRNA-directed inhibition of HIV-1 infection *Nat Med* **8** 681–686

Nykanen A Haley B and Zamore PD 2001 ATP requirements and small interfering RNA structure in the RNA interference pathway *Cell* **107** 309–321

Ohta T Essner R Ryu JH Palazzo RE Uetake Y and Kuriyama R 2002 Characterization of Cep135 a novel coiled-coil centrosomal protein involved in microtubule organization in mammalian cells *J Cell Biol* **156** 87–99

Paddison PJ Caudy AA and Hannon GJ 2002b Stable suppression of gene expression by RNAi in mammalian cells *Proc Natl Acad Sci USA* **99** 1443–1448

Paddison PJ Caudy AA Bernstein E Hannon GJ and Conklin DS 2002a Short hairpin RNAs (shRNAs) induce sequence-specific silencing in mammalian cells *Genes Dev* **16** *948–958*

Page MF Carr B Anders KR Grimson A and Anderson P 1999 SMG-2 is a phosphorylated protein required for mRNA surveillance in Caenorhabditis elegans and related to Upf1p of yeast *Mol Cell Biol* **19** 5943–5951

Palliser D 2006 An siRNA-based microbicide protects mice from lethal herpes simplex virus infection *Nature* **439** 89-94

Park Y-D Moscone EA Iglesis VA Vaucheret H Matzke AJM and Matzke MA 1996 Gene silencing mediated by promoter homology occurs at the level of transcription and results in meiotically heritable alterations in methylation and gene activity *Plant J* **9** 183-194

Parrish S Fleenor J Xu S Mello C and Fire A 2000 Functional anatomy of a dsRNA trigger Differential requirements for the two trigger strands in RNA interference Molec *Cell* **6** 1077-1087

Paul CP Good PD Winer I and Engelke DR 2002 Effective expression of small interfering RNA in human cells *Nat Biotechnol* **20** 505–508

Peng Y Zhang Q Nagasawa H Okayasu R Liber HL and Bedford JS 2002 Silencing expression of the catalytic subunit of DNA-dependent protein kinase by small interfering RNA sensitizes human cells for radiation-induced chromosome damage cell killing and mutation *Cancer Res* **62** 6400–6404

Pfeifer A Ikawa M Dayn Y and Verma IM 2002 Transgenesis by lentiviral vectors lack of gene silencing in mammalian embryonic stem cells and preimplantation embryos *Proc Natl Acad Sci USA* **99** 140–145

Plasterk RHA 2002 RNA Silencing The Genome's Immune System *Science* **296** 1263-1265

Poinat P De Arcangelis A Sookhareea S Zhu X Hedgecock EM Labouesse M and Georges-Labouesse E 2002 A conserved interaction between beta1 integrin/ PAT-3 and Nck-interacting kinase/ MIG-15 that mediates commissural axon navigation in *C elegans Curr Biol* **12** 622–631

Porter LA Dellinger RW Tynan JA Barnes EA Kong M Lenormand JL and Donoghue DJ 2002 Human Speedy a novel cell cycle regulator that enhances proliferation through activation of Cdk2 *J Cell Biol* **157** 357–366

Prasanth SG Prasanth KV and Stillman B 2002 Orc6 involved in DNA replication chromosome segregation and cytokinesis *Science (Wash DC)* **297** 1026–1031

Pruss G Ge X Shi XM Carrnington JC and Vance VB 1997 Plant viral synergism the potyviral genome encodes a broad-range pathogenicity enhancer that transactivates replication of heterologous viruses *Plant Cell* **9** 1749–1759

Quinn LM Dorstyn L Mills K Colussi PA Chen P Coombe M Abrams J Kumar S and Richardson H 2000 An essential role for the caspase dronc in developmentally programmed cell death in Drosophila *J Biol Chem* **275** 40416–40424

Randall G Grakoui A and Rice CM 2003 Clearance of replicating hepatitis C virus replicon RNAs in cell culture by small interfering RNAs *Proc Natl Acad Sci USA* **100** 235–240

Ratcliff F Harrison B and Baulcombe D 1997 A similarity between viral defence and gene silencing in plants *Science* **276** 1558-1560

Ratcliff F Harrison BD and Baulcombe DC 1997 A similarity between viral defense and gene silencing in plants *Science (Wash DC)* **276** 1558–1560

Reinhart BJ Slack FJ Basson M Pasquinelli AE Bettinger JC Rougvie HR Horvitz HR and Ruvkun G 2000 The 21-nucleotide *let-7* RNA regulates developmental timing in *Caenorhabditis elegans Nature* **403** 901-906

Reinhart BJ Weinstein EG Rhodes MW Bartel B and Bartel DP 2002 MicroRNAs in plants *Genes Dev* **16** 1616-1626

Rocheleau CE Downs WD Lin R Wittman C Bei Y Cha YH Ali M Priess JR and Mello CC 1997 Wnt signaling and an APC-relatred gene specify endoderm in early *C elegans* embryos *Cell* **90** 707-716

Romano N and Macino G 1992 Quelling Transient inactivation of gene expression in *Neurospora crassa* by transformation with homologous sequences *Mol Microbiol* **6** 3343-3353

Rubinson DA Dillon CP Kwiatkowski AV Sievers C Yang L Kopinja J Zhang M McManus MT Gertler FB Scott ML and Van Parijs L 2003 A lentivirus-based system to functionally silence genes in primary mammalian cells stem cells and transgenic mice by RNA interference *Nat Genet* **33** 401–406

Rubinsztein DC 2002 Lessons from animal models of Huntington's disease *Trends Genet* **18** 202–209

Ruiz-Echevarria MJ Czaplinski K and Peltz SW 1996 Making sense of nonsense in yeast *Trends Biochem Sci* **21** 433–438

Salisbury JL Suino KM Busby R and Springett M 2002 Centrin-2 is required for centriole duplication in mammalian cells *Curr Biol* **12** 1287–1292

Sharp PA 2006 The Biology of Short RNAs In "RNA World" Cold Spring Harbor Laboratory Press Cold Spring Harbor NY

Siegmund D Hadwiger P Pfizenmaier K Vornlocher HP and Wajant H 2002 Selective inhibition of FLICE-like inhibitory protein expression with small interfering RNA oligonucleotides is sufficient to sensitize tumor cells for TRAILinduced apoptosis *Mol Med* **8** 725–732

Sijen T Vijn I Rebocho A van Blokland R Roelofs D *et al* 2001 Transcriptional and posttranscriptional gene silencing are mechanistically related *Curr Biol* **11** 436-440

Smalheiser NR Manev H and Costa E 2001 RNAi and brain function was McConnell on the right track? *Trends Neurosci* **24** 216–218

Smardon A Spoerke JM Stacey SC Klein ME Mackin N and Maine EM 2000 EGO-1 is related to RNA-directed RNA polymerase and functions in germ-line development and RNA interference in C *elegans Curr Biol* **10** 169–178

Smith CJS Watson CF Bird CR Ray J Schuch W and Grierson D 1990 Expression of a truncated tomato polygalacturonase gene inhibits expression of the endogenous gene in transgenic plants *Mol Gen Genet* **224** 447-481

Smith GM and Romero MI 1999 Adenoviral-mediated gene transfer to enhance neuronal survival growth and regeneration *J Neurosci Res* **55** 147–157

Song E Lee SK Wang J Ince N Ouyang N Min J Chen J Shankar P and Lieberman J 2003 RNA interference targeting Fas protects mice from fulminant hepatitis *Nat Med* **9** 347–351

Song E Lee SK Wang J Ince N Ouyang N Min J *et al* 2003 RNA interference targeting Fas protects mice from fulminant hepatitis *Nat Med* **9** 347-51

Soutschek J *et al* 2004 Therapeutic silencing of an endogenous gene by systematic administration of modified siRNAs *Nature* **432** 173-178

Stucke VM Sillje HH Arnaud L and Nigg EA 2002 Human Mps1 kinase is required for the spindle assembly checkpoint but not for centrosome duplication *EMBO (Eur Mol Biol Organ) J* **21** *1723–1732*

Sui G Soohoo C Affar el B Gay F Shi Y Forrester WC Shi Y 2002 A DNA vector-based RNAi technology to suppress gene expression in mammalian cells *Proc Natl Acad Sci USA 99 5515–5520*

Surabhi RM and Gaynor RB 2002 RNA interference directed against viral and cellular targets inhibits human immunodeficiency virus type 1 replication *J Virol* **76** 12963–12973

Svoboda P Stein P and Schultz RM 2001 RNAi in mouse oocytes and preimplantation embryos effectiveness of hairpin dsRNA *Biochem Biophys Res Commun* **287** 1099–1104

Tabara H Grishok A and Mello CC 1998 RNAi in *C elegans* soaking in the genome sequence *Science (Wash DC)* **282** 430–431

Tabara H Sarkissian M Kelly WG Fleenor J Grishok A Timmons L Fire A and Mello CC 1999 The Rde-1 gene RNA interference and transposon silencing in *C elegans Cell* **99** 123–132

Tabara H Sarkissian M Kelly WG Fleenor J Grishok A Timmons L Fire A and Mello CC 1999 The rde-1 gene RNA interference and transposon silencing in *C elegans Cell* **99** 123-132

Tabara H Yigit E Siomi H and Mello CC 2002 The dsRNA binding protein RDE-4 interacts with RDE-1 DCR-1 and a DExH-box helicase to direct RNAi in *C elegans Cell* **109** 861–871

Tan YT Tillett DJ and McKay IA 2000 Molecular strategies for overcoming antibiotic resistance in bacteria *Mol Med Today* **6** 309–314

Tavernarakis N Wang SL Dorovkov M Ryazanov A and Driscoll M 2000 Heritable and inducible genetic interference by double-stranded RNA encoded by transgenes *Nat Genet* **24** *180–183*

Tewari M Vidal M 2003 RNAi on the apoptosis TRAIL the mammalian cell genetic screen comes of age *Dev Cell* **5** 534-5

Timmons L and Fire A 1998 Specific interference by ingested dsRNA *Nature (Lond)* **395** 854

Tiscornia G Singer O Ikawa M and Verma IM 2003 A general method for gene knockdown in mice by using lentiviral vectors expressing small interfering RNA *Proc Natl Acad Sci USA* **100** 1844–1848

Tsuneoka M Koda Y Soejima M Teye K and Kimura H 2002 A novel myc target gene mina53 that is involved in cell proliferation *J Biol Chem* **277** 35450–35459

Tuschl T Zamore PD Lehmann R Bartel DP and Sharp PA 1999 Targeted mRNA degradation by double-stranded RNA *in vitro Genes Dev* **13** 3191–3197

Tuschl T Zamore PD Lehmann R Bartel DP and Sharp PA 1999 Targeted mRNA degradation by double-stranded RNA *in vitro Genes Devel* **13** 3191-3197

Ukomadu C and Dutta A 2003 Inhibition of cdk2 activating phosphorylation by mevastatin *J Biol Chem* **278** 4840–4846

van Blokland K van der Geest N Mol J and Kooter J 1994 Transgene-mediated suppression of chalcone synthase expression in *Petunia hybrida* results from an increase in RNA turnover *Plant J* **6** 861-877

Van Blokland R Van Der Geest N Mol JMN and Kooter JM 1994 Transgenemediated expression of chalcone synthase expression in Petunia hybrida results from an increase in RNA turnover *Plant J* **6** 861–877

van der Krol A Mur L Beld M Mol JNM and Stuitje AR 1990 Flavonoid genes in petunia Addition of a limited number of gene copies may lead to a suppression of gene expression *Plant Cell* **2** 291-299

Van der Krol AR Mur LA Beld M Mol JN Stuitje AR 1990 Flavonoid genes in petunia addition of a limited number of gene copies may lead to a suppression of gene expression *Plant Cell* **2** 291-9

Volpe TA Kidner C Hall IM Teng G Grewal SI and Martienssen RA 2002 Regulation of heterochromatic silencing and histone H3 lysine-9 methylation by RNAi *Science (Wash DC)* **297** 1833–1837

Volpe TA Kidner C Hall IM Teng G Grewal SI and Martienssen RA 2002 Regulation of heterochromatic silencing and histone H3 lysine-9 methylation by RNAi *Science* **297** 1833-1837

Walters DK and Jelinek DF 2002 The effectiveness of double-stranded short inhibitory RNAs (siRNAs) may depend on the method of transfection *Antisense Nucleic Acid Drug Dev* **12** 411–418

Wang X Yang C Chai J Shi Y and Xue D 2002 Mechanisms of AIF-mediated apoptotic DNA degradation in Caenorhabditis elegans *Science (Wash DC)* **298** 1587–1592

Wang Z Morris JC Drew ME and Englund PT 2000 Inhibition of Trypanosoma brucei gene expression by RNA interference using an integratable vector with opposing T7 promoters *J Biol Chem* **275** 40174–40179

Wassenegger M Heimes S Riedel L and Sänger H 1994 RNA-directed de novo methylation of genomic sequences in plants *Cell* **76** 567-576

Waterhouse PM Graham MW and Wang M-B 1998 Virusa resistance and gene silencing in plats can be induced by simultaneous expression of sense and antisense RNA *Proc Natl Acad Sci* **95** 13959-13964

Weil D Garcon L Harper M Dumenil D Dautry F and Kress M 2002 Targeting the kinesin Eg5 to monitor siRNA transfection in mammalian cells *Biotechniques* **33** 1244–1248

Wianny F and Zernicka-Goetz M 2000 Specific interference with gene function by double-stranded RNA in early mouse development *Nat Cell Biol* **2** 70–75

Wightman B Ha I and Ruvkun G 1993 Posttranscriptional regulation of the heterochronic gene *lin-14* by *lin-4* mediates temporal pattern formation in *C elegans Cell* 75 855-862

Wilda M Fuchs U Wossmann W and Borkgardt A 2002 Killing of leukemic cells with a BCR/ABL fusion gene by RNA interference (RNAi) *Oncogene* **21** 5716–5724

Wilson JA Jayasena S Khvorova A Sabatinos S Rodrigue-Gervais IG Arya S Sarangi F Harris-Brandts M Beaulieu S and Richardson CD 2003 RNA interference blocks gene expression and RNA synthesis from hepatitis C replicons propagated in human liver cells *Proc Natl Acad Sci USA* **100** 2783–2788

Winston WM Molodowitch C and Hunter CP 2002 Systemic RNAi in *C elegans* requires the putative transmembrane protein SID-1 *Science (Wash DC)* **295** 2456–2459

Wojcik C and DeMartino GN 2002 Analysis of Drosophila 26 S proteasome using RNA interference *J Biol Chem* **277** 6188–6197

Wong SK and Lazinski DW 2002 Replicating hepatitis delta virus RNA is edited in the nucleus by the small form of ADAR1 *Proc Natl Acad Sci USA* **99** 15118–15123

Xia H Mao Q Paulson HL and Davidson BL 2002 siRNA-mediated gene silencing *in vitro* and *in vivo Nat Biotechnol* **20** 1006–1010

Xu K Tavernarakis N and Driscoll M 2001 Necrotic cell death in *C elegans* requires the function of calreticulin and regulators of Ca2 release from the endoplasmic reticulum *Neuron* **31** 957–971

Xu Z Williams CJ Kopf GS and Schultz RM 2003 Maturation-associated increase in IP(3)receptor type 1 role in conferring increased IP (3) sensitivity and Ca (2) oscillatory behavior in mouse eggs *Dev Biol* **254** 163–171

Yang S Tutton S Pierce E and Yoon K 2001 Specific double-stranded RNA interference in undifferentiated mouse embryonic stem cells *Mol Cell Biol* **21** 7807–7816

Ye F and Signer ER 1996 RIGS (repeat-induced gene silencing) in *Arabidopsis* is transcriptional and alters chromatin configuration *Proc Natl Acad Sci USA* **93** 10881–10886

Yu JY DeRuiter SL and Turner DL 2002 RNA interference by expression of short-interfering RNAs and hairpin RNAs in mammalian cells *Proc Natl Acad Sci USA* **99** 6047–6052

Zamore PD and Haley B 2005 Ribo-gnome The Big World of Small RNAs *Science* **309** 1519-1524

Zamore PD Tuschl T Sharp PA and Bartel DP 2000 RNAi Double-stranded RNA directs the ATP-dependent cleavage of mRNA at 21 to 23 nucleotide intervals *Cell* **101** 25–33

Zamore PD Tuschl T Sharp PA and Bartel DP 2000 RNAi Double-stranded RNA directs the ATP-dependent cleavage of mRNA at 21 to 23 nucleotide intervals *Cell* **101** 25-33

Zhang D Li F Weidner D Mnjoyan ZH and Fujise K 2002a Physical and functional interaction between myeloid cell leukemia 1 protein (MCL1) and Fortilin The potential role of MCL1 as a fortilin chaperone *J Biol Chem* **277** 37430–37438

Zhang H Kolb FA Brondani V Billy E and Filipowicz W 2002b) Human Dicer preferentially cleaves dsRNAs at their termini without a requirement for ATP *EMBO (Eur Mol Biol Organ) J* **21** 5875–5885

Zhao ZQ and Vinten-Johansen J 2002 Myocardial apoptosis and ischemic preconditioning *Cardiovasc Res* **55** 438–4

Zhu X Yu QS Cutler RG Culmsee CW Holloway HW Lahiri DK Mattson MP and Greig NH 2002 Novel p53 inactivators with neuroprotective action syntheses and pharmacological evaluation of 2-imino-2 3 4 5 6 7-hexahydrobenzothiazole and 2-imino-2 3 4 5 6 7-hexahydrobenzoxazole derivatives *J Med Chem* **45** 5090–5097

Zimmermann TS *et al* 2006 RNAi-mediated gene silencing in non-human primates *Nature* **441** 111-114

6

CYANOBACTERIAL FLAVODOXIN PROVIDES MULTIPLE STRESS TOLERANCE

MATÍAS ZURBRIGGEN, VANESA B. TOGNETTI, ESTELA M. VALLE AND NÉSTOR CARRILLO

THE PROBLEM

Plants growing under natural conditions unavoidably face episodic situations of environmental stress in the course of their life times. They have developed numerous strategies to survive in such adverse conditions. Crops, in contrast, are selected by humans for their high productivity in agriculture, but this is usually not accompanied by increasing resistance to hostile environments. Diseases, unfavorable climates, or inappropriate soils are responsible for most agricultural losses. Analysis of major crops with economically valuable reproductive or vegetative structures (corn, soybeans, barley, potatoes, among others) shows potential record yields 3- to 7-fold greater than average yields (Boyer, 1982).

One approach to obtain plants adapted to unfavorable environments may be to improve essential nutrient (nitrogen, phosphorus, or iron) acquisition. Although iron is abundant, plants need to solubilize it from insoluble oxides in alkaline, calcareous soils, which cover approximately one-third of the earth's surface and represent a major deterrent for agriculture (Mori, 1999). Plants deprived of iron develop interveinal chlorotic symptoms in young leaves and a general decrease in photosynthetic activity that can lead to death. Chlorosis has been attributed to inhibition of chlorophyll synthesis, which requires the function of Fe-containing enzymes, but chlorophyll-binding proteins and other photosynthetic components are down-regulated with relative independence of pigment levels. Chloroplasts are therefore primary targets of iron deficiency.

On the other hand, plants sustain different types of environmental hardships, such as drought, flood, chilling, salinity, and radiation. Being motionless organisms, a plant's defense to adverse conditions is limited to physiological and biochemical responses, including stomatal closure, osmotic adjustment, ion pumping, etc. In response to regulated changes in gene expression, central and secondary metabolisms are redirected to cope with the undesired effects of the hostile situation. This is achieved by up-regulating the synthesis of proteins and metabolites involved in protection (i.e., compatible solutes, antioxidant enzymes and compounds, heat-shock proteins, etc.). Whenever these defensive mechanisms are overcome by the intensity of the adverse condition, the plant is under environmental stress.

Although, different types of nutritional and environmental adversities have their own features and display idiosyncratic responses, they all have in common a significant perturbation of the electron transfer network of the stressed cells, leading to electron derivation to non-productive routes, breakage of redox homeostasis, and eventually to the generation of reactive oxygen species (ROS) such as singlet oxygen, the superoxide and hydroxyl radicals, and hydrogen peroxide.

The Cyanobacterial/Algal Solution

Photosynthetic microorganisms respond to adverse environments by eliciting a replacement strategy, which involves substitution of sensitive enzymes and proteins by resistant isofunctional versions. Most conspicuous among them is the induction of flavodoxin expression to take over the functions of the ubiquitous iron-sulfur protein ferredoxin, whose levels are down-regulated under conditions of environmental stress or iron starvation (Tognetti *et al.*, 2006; 2007). Flavodoxins are small soluble electron carrier proteins containing flavin mononucleotide that shuttle electrons among a plethora of donors and acceptors. Their properties as redox carriers largely match those of ferredoxin. In heterotrophic bacteria, flavodoxins (and ferredoxins) are reduced enzymatically (largely but not exclusively by ferredoxin-NADP$^+$ reductases) to provide low potential reducing equivalents to many different metabolic pathways such as dinitrogen fixation, sulfate and nitrate assimilation, nucleotide synthesis, and C_2 carbon metabolism (Zurbriggen *et al.*, 2007). In cyanobacteria, the prokaryotic ancestors of modern-day chloroplasts, flavodoxin and ferredoxin are reduced photosynthetically at the level of photosystem I in vegetative cells. Ferredoxin is normally the preferred electron carrier, whereas flavodoxin acts as a backup whose expression is induced under adverse conditions.

While ferredoxins are present in virtually all types of organisms, the flavodoxin gene displays a rather limited distribution. Within photosynthetic species, flavodoxin is common, but not universal, among cyanobacteria. The gene entered the algal world during the endosymbiotic event that gave origin to photosynthetic eukaryotes and spread into all major algal lineages, with the corresponding product being targeted to chloroplasts. However, flavodoxin presence is less frequent in algae than in cyanobacteria and generally confined to species living in the open oceans, where nutritional and environmental conditions are usually extreme.

The Plant Solution

Substitutive strategies as those displayed by photosynthetic microorganisms are absent in higher plants. Indeed, the flavodoxin gene is not found in the plant genome and the considerable adaptive advantages derived from its expression were irreversibly lost somewhere during the evolution from green algae to vascular plants. Instead, plants have developed complex and multi-step responses to both adverse environments and iron deficit. In the first case, the response operates at various levels, including avoidance of the hostile conditions, scavenging of toxic compounds, and damage repair. In the second case, the strategy focuses on optimization of iron intake from scarcely available sources.

Noteworthy, cyanobacterial flavodoxin is still able to productively interact *in vitro* with plant enzymes whose prokaryotic ancestors used this flavoprotein as usual or occasional substrate. These observations prompted us to evaluate if reintroduction of this flavoprotein in plants could improve tolerance to abiotic stress and iron deficit as it occurs in microorganisms (Tognetti *et al.*, 2006; 2007).

DESIGN AND CHARACTERIZATION OF TRANSGENIC TOBACCO PLANTS EXPRESSING FLAVODOXIN IN CHLOROPLASTS

Preparation of the transformants

The flavodoxin gene from the cyanobacterium *Anabaena* PCC7119 was cloned between the promoter and terminator regions of the cauliflower mosaic virus 35S, which directs constitutive expression in tobacco. A presequence, encoding a chloroplast transit peptide, was fused in-frame to the N-terminal end of the gene to direct the product to plastids. Several independent lines were obtained via *Agrobacterium* transformation and made homozygous by self-pollination followed by selection of the segregants (Tognetti *et al.*, 2006). Lines expressing ≥ 60 μM flavodoxin in chloroplasts were selected because levels of the foreign protein were similar to that of endogenous ferredoxin (~80 μM). When grown under normal conditions with water and iron *ad libitum*, transgenic plants displayed the same phenotypic features as wild-type (WT) non-transformed siblings with respect to growth rate, flowering time, and seed production, indicating that integration and expression of the foreign gene was inconsequential in phenotypic terms whenever plants were not stressed.

Flavodoxin-expressing plants could grow in iron-limited soils

Transgenic plants differed significantly from WT siblings when cultured in different iron-restricted media (Fig. 1), including soil watered with an alkaline solution that led to progressive iron depletion. Non-transformed plants were bleached, and underwent growth arrest and inhibition of key metabolic routes, including photosynthesis and amino acid synthesis. In contrast, the transformants gained weight and height, maintained photosynthetic rates at 40% of the levels found in untreated controls, and could complete their life cycles and produce viable seeds[4]. Metabolic profiling indicated that steady-state levels of storage sugars, intermediate metabolites, and most amino acids were dramatically decreased in stressed WT plants, but not in the transgenic lines.

Flavodoxin expression in chloroplasts confers tolerance to abiotic stress

Plants expressing flavodoxin in plastids were able to withstand a remarkable range of environmental adversities, including high temperature, chilling, drought, high light intensities, salinity, and ultraviolet radiation (Tognetti *et al.*, 2006). They were also tolerant to oxidative stress conditions generated in chloroplasts by poisoning with the herbicide paraquat (Tognetti *et al.*, 2006). As in the case of iron deprivation, WT plants suffered different types of damage, including chlorosis, necrosis, inhibition of photosynthesis and other metabolic pathways, and eventually death. Flavodoxin expression in chloroplasts prevented all these symptoms to a great extent or completely (Fig. 2). ROS accumulation, which was prominent in stressed WT plants, was entirely avoided in the transformants.

MECHANISMS UNDERLYING FLAVODOXIN PROTECTION

As anticipated, both iron deficit and abiotic stress caused an imbalance in the distribution of electrons originated in the photosynthetic electron transport chain, due to ferredoxin decrease and augmented delivery of reducing equivalents to adventitious acceptors such as oxygen. Flavodoxin restores the redox homeostasis of the plastid by productively interacting with endogenous routes and partners, including ferredoxin-$NADP^+$ reductase and thioredoxin reductase, that are crucial for optimal rates of CO_2 assimilation, and ferredoxin-dependent reactions of amino acid synthesis. Flavodoxin also contributes significantly to antioxidant protection by providing the reducing power needed to regenerate the active form of peroxiredoxin, the most abundant peroxidase of chloroplasts. The collected results indicate that flavodoxin maintains the welfare of plants

1.

2.

Fig. 1) Flavodoxin expression confers tolerance to iron starvation. Wild-type (WT) and transgenic tobacco plants expressing 63 μM flavodoxin in chloroplasts were cultured for 4 weeks in Murashige-Skoog (MS) agar containin 30 μM $FeSO_4$-EDTA, and then transferred to iron-free MS agar supplemented with 10 mM $CaCO_3$ (pH 8.0). Plants were photographed 20 days after transfer.

Fig. 2) Increased tolerance to water deficit in transgenic tobacco plants expressing Anabaena flavodoxin in chloroplasts. Wild-type (WT) and transgenic tobacco plants expressing 63 M flavodoxin in chloroplasts were grown in soil for 8 weeks and then subjected to a 3-day water deprivation regime.

under stress by avoiding electron misrouting and ROS formation, rather than by combating the deleterious consequences of these processes.

CONCLUSIONS AND PERSPECTIVES

Manipulation of tolerance to iron deficit and hostile environments in crop plants represents a particularly demanding challenge for breeders and geneticists, due to its agronomical importance and the complexities of the plant responses. So far, strategies pursued to accomplish this task have been based on boosting endogenous protective systems such as those involved in ROS scavenging and iron uptake and storage. The success of this type of approach has been variable, as could be expected from the complexities and multigenic nature of the responses involved. The strategy described here is based on a different conceptual background. It relies on the assumption that most of the damage suffered by stressed or iron-starved plants is caused by faulty electron distribution within plastids and cells, resulting in impairment of key metabolic, regulatory, and dissipative pathways and/or ROS accumulation. Introduction of an alien protein that restores electron transfer to productive routes was expected to significantly ameliorate the undesired consequences of growth under sub-optimal conditions. The results obtained (Tognetti *et al.*, 2006; 2007) largely confirmed the working hypothesis, leading to transgenic plants that displayed a remarkable tolerance to iron starvation and to various sources of abiotic stress of agronomical importance. Since flavodoxin-expressing plants took up the same amounts of iron as their WT siblings (Tognetti *et al.*, 2007) the incorporated metal could be reallocated to other demanding sources in these lines.

Besides increased tolerance, the transformants displayed other features of potential agronomical relevance:

i) tolerant lines could be obtained by transformation with a single gene;

ii) the bacterial origin of the transgene rules out unwanted complications such as silencing, co-suppression, and/or endogenous regulatory loops; and

iii) flavodoxin expression remains unnoticed, in phenotypic terms, when plants are grown under normal conditions. Then, (re)introduction of flavodoxin into chloroplasts restored some of the selective advantages that allowed photosynthetic microorganisms to thrive in hostile media. Engineering of this ancient trait, either alone or in combination with other protective devices, opens entirely novel possibilities for the design of valuable crops with greater plasticity to face the challenges of environmental hardships.

Transformation of related Solanaceae (tomato, potato) yielded essentially the same tolerant phenotypes, and protocols to incorporate flavodoxin into several other crops, including cereals, are currently underway. The use of available experimental resources such as custom-ready promoters, tissue-specific expression, transplantomics, etc., could make the flavodoxin gene-based strategy a most useful biotechnological tool to improve crop yields and to gain at least some wastelands for agriculture.

REFERENCES

1. Boyer JS 1982 Plant Productivity and Environment *Science* **218** 443-448

2. Mori S 1999 Iron acquisition by plants *Current Opinions in Plant Biology* **2** 250-253

3. Tognetti VB, Palatnik JF, Fillat MF, Melzer M, Hajirezaei MR, Valle EM, Carrillo N 2006 Functional Replacement of Ferredoxin by a Cyanobacterial Flavodoxin in Tobacco Confers Broad-Range Stress Tolerance *Plant Cell* **18** 2035-2050

4. Tognetti VB, Zurbriggen MD, Morandi EN, Fillat MF, Valle EM, Hajirezaei MR, Carrillo N 2007 Enhanced plant tolerance to iron starvation by functional substitution of chloroplast ferredoxin with a bacterial flavodoxin *Proc Natl Acad Sci USA* **104** 11495-11500

5. Zurbriggen MD, Tognetti VB, Carrillo N 2007 *IUBMB Life* **59**, 355-360

7

RECENT ADVANCES IN BIOTECHNOLOGY

BHAVNEET KAUR, C.P. MALIK AND CHITRA WADHWANI

Recent developments in the field of what we call a **'marvel technology'** i.e. **'biotechnology'** has left humankind with awe and wonder. The revolutionary changes brought about by molecular biology and biomedical research in the last 21 years is overwhelming and incredible. This has led to the proposition of expecting the most unexpected. What may be an idea or a thought beyond conception today might be a reality five years from now. Our forefathers could be termed as 'pioneer biotechnologists' for they used to make curd from milk. Thus 'biotechnology' can be defined in very simple terms as 'the use of living organisms or products derived from them for the benefit of humankind.'

This youngest branch of 'life sciences' was born in the year 1976 when a 27-year old venture capitalist, Robert Swanson and a University of California professor, Herb Boyer founded a company known as 'Genentech'. Two years later in 1978, Biogen Idec was founded. Since then thousands of biotech companies have sprung up all over the globe marking the birth of a huge industry which is going to rule the world for at least another 200 years. The 'biotech industry' which harbours immense potential in the fields of medicine, agriculture, pharmaceuticals, biomedical research, etc. and has the answer to all problems that the world is facing today.

'Recombinant DNA technology or genetic engineering' can be labelled as a **'Technology of Future'**. Name any aspect and it has the potential to solve it. It can:

- remove all black genes from humans.
- cure all kinds of illnesses and ailments.
- provide food to millions of hungry people.
- can produce replacement for fossil fuels (non-renewable source of energy).
- preserve/protect endangered germplasm.

Think of any process and product and biotechnology has the answer. Let us take a look at a few research findings/ breakthroughs which this amazing technology has achieved in the past one and half year. Modern biotechnology includes recombinant DNA technology, hybridoma technology, gene transfer and cell-fusion techniques, among others. Main advances have been made in four segments- healthcare biotechnology, agriculture and veterinary biotechnology, Industrial biotechnology and R & D product and services. A few breakthroughs in biotechnology are discussed below :

MEDICAL BIOTECHNOLOGY/HEALTHCARE

1. **New York scientists** have achieved a scientific breakthrough by creating embryonic stem cells. This has been done by stimulating unfertilised eggs. This enormous advance is a significant step toward producing transplant tissue that is genetically matched to women. Such cells can develop into any tissue of the body, and scientists hope to harness them for producing specialized tissues like nerve cells or pancreas cells to treat a range of illnesses.

2. Scientists at the **University of Florida, College of Medicine**, have found that injecting patients (children aged 2-7 years) suffering from type 1 diabetes with their own umbilical cord blood may slow down the destruction of their insulin producing cells. The therapy has been found to reduce disease severity as well as rejuvenate the immune system. Cord blood, which is rich in immune regulatory cells, restarted the non-functional immune systems.

   Seven children, aged 2 to 7 years with type 1 diabetes, who had their cord blood stored, were recruited for the study and were infused with the blood. Thirteen randomly selected youngsters of similar age and diabetes duration, who had been intensively treated with insulin, were also put in place as a control group. The team believes that cord blood, which contains stem cells capable of differentiating into insulin producing cells, and infused cells, could have stimulated islets to regenerate. A1C tests (to measure blood glucose control) and total daily insulin use from diagnosis to six months after infusion were compared. The children who received cord blood transfusions had lower average A1Cs (7% as against 8.04% in the control group receiving insulin therapy alone). They found that children who had transfusions also required much lower average total daily insulin than the control group, 0.45 as against 0.69 units of insulin per Kg per day.

3. Lead researcher, **Amalia Sapag and coworkers at the University of Chile** in Santiago have engineered adenoviruses using antisense technology. They have blocked the gene for the enzyme, aldehyde dehydrogenase involved in alcohol metabolism. The team claims to have reduced the enzyme's activity in rat's livers by 80% and it was found that rats bred to crave for alcohol drank 50% less for more than a month after being injected with engineered adenoviruses.

Therefore, engineered adenoviruses may help alcoholics quit the habit of drinking.

4. Research to develop anti-venom from camel serum has reached a crucial stage at the **National Research Centre on Camel (NRCC), Bikaner.** Anti-venin or anti-venom is a biological product used in the treatment of venomous bites or stings. It is created by injecting a small quantity of the targeted venom into the body of an animal such as a horse or camel. The subject animal then suffers from an immune response to the venom, producing antibodies against the venom's active molecule. This molecule can be harvested from the animal's blood and used to treat those suffering from bites. The NRCC in association with Sardar Patel Medical College, Bikaner found that anti-venom prepared from camel serum was more effective than the one derived from horses. The possible reason is that anti-venom from horses has a bigger molecular size due to which it sometimes fails to reach the infected part of the body. Black cobra and Bandi (a viper) are found in large numbers in western Rajasthan and are responsible for a lot of deaths. This new anti-venom vaccine would be a potential life-saviour.

5. **Edible rice vaccine** has been developed for the deadly water-borne disease, cholera at the University of Tokyo. This vaccine will not need refrigeration or injections for delivery. It has successfully passed the test in mice. According to the team leader, **Hiroshi Kiyono**, the rice vaccine has provided a two tier immunity in mice who consumed the transgenic rice. The vaccine does not dissolve in the stomach and lasts till it reaches the intestine. Here it instigates the immune response against the cholera bacteria. The vaccine has been created by inserting part of the cholera bacterium, *Vibrio cholerae* into the Kitaak rice plant. About 30 μg of the genes have been introduced into each grain of rice by the Japanese researchers. The team plans to initiate primate trials of the rice. The vaccine has demonstrated a superior level of immunity in the mice by producing antibodies in the nose, mouth and urinary tracts mucosal surfaces. It holds great promise for a country like India where rice is the staple food of 85% of the country's population.

6. **Danish scientists at the University of Copenhagen** have developed an inexpensive and efficient technique to remove sugar molecules from RBCs. The enzymes employed for the task were bacterial glycosidases isolated from *Elizabethkingia meningosepticum* and *Bacterioides fragilis*. Therefore, the blood group barrier has been broken and all human beings have now become universal donors. The scientists led by cellular biologist, Henrik Clausen have designed a conversion machine which can simultaneously treat eight units of blood with

the enzymes in 90 minutes. The machine would be commercially available in another five years.

7. Scientists at the University of Nevada have created the world's first human-sheep chimera – which has the body of a sheep and half human organs. **Professor Esmail Zanjani** has spent seven years and £ 5 million for perfecting the technique which involves injecting adult human cells into a sheep's foetus. This has brought the prospect of animal organs being transplanted into human beings one step closer. Then stem cells were isolated and injected into the peritoneum of a sheep's foetus. These cells got distributed throughout the metabolic system into the circulatory system of all the organs in the body. When the lamb is born two months later, it would have a liver, heart, lungs and brain that are partly human and available for transplant. This technology would benefit patients who are waiting for an organ transplant.

8. **British scientists at the Newcastle University** have grown a **miniature human liver** from stem cells. They expect to produce entire organs for life saving transplants in another 20 years time. The liver tissue has been created from stem cells which were removed from the umbilical cord minutes after birth. These were subsequently placed in a 'bioreactor' developed by NASA to mimic weightlessness. This property allowed the cells to multiply rapidly than usual rate of multiplication. Also various hormones and chemicals were added to condition the stem cells to differentiate into liver cells.

9. Molecular biologist and pioneer in protein and antibody engineering, **Gregory Winter** has conceived the idea of making an individual into a self healing hospital. He has invented three technologies: (a) To make almost human antibodies from mice – 95% human antibodies and 5% mouse – what in scientific jargon are called humanised antibodies; (b) Made pure human antibodies to combat rheumatoid arthritis, Crohn's disease and cirrhosis; (c) Has made miniantibodies called dAbs – for dual use that can bind two molecules at a time.

   Presently, work is going on recombinant technology so as to put proteins into people through injection. Therefore, a time may come when body cells could be coaxed to make antibodies to attack tumours, so no antibody injections would be required. One's own body could be the factory to make one's own cells and antibodies.

10. Genetically engineered mice carrying a copy of the gene cyclin A2, in actively expressing condition even after birth were found to make new heart cells. According to **Dr. Hina Chaudhry**, a researcher at Columbia University Medical

Center in New York, these genes enabled the mice to have a 77% better heart function, post heart attack, than the other mice which did not carry the gene.

11. US Researchers have manipulated **"Skinny gene"** in mice which keeps them thin. The gene (also called as adp) was discovered 50 years ago in fruit flies by **Winifred Doane**. This gene when upregulated made the mice skinny and when downregulated made them fat. The gene probably has a regulatory function. It might offer a way to prevent obesity and diabetes in people.

12. New Gene Therapy tools for inherited blindness have been identified. Researchers have identified DNA elements that control when and where genes linked to blindness are turned on. **Joseph Corbo and colleagues** of Washington University School of Medicine, St. Louis used a computational analysis of DNA to look for elements that can turn on genes in the eye's light-sensing photoreceptor cells. Their search identified hundreds of potential cis-regulatory elements, which are segments of DNA involved in turning genes on and off. They confirmed 19 of these elements, more than doubling the number known to scientists. The newly discovered elements can be used as switches to activate blindness therapies. Corbo and others are currently using the results of this study to design gene therapy vectors for a form of Leber's congenital amaurosis, an inherited disorder that leaves patients blind from birth.

13. **Doctors in Rome** have created a vagina using stem cells for two patients suffering from rare malformations. The vaginal tissue was grown using stem cells from the patients' own bodies. In the first case, a 28-year-old woman received a tiny square of mucous membrane and has slowly grown a vagina. A 17-year-old girl underwent the same procedure Tuesday in Rome's Umberto I hospital. According to lead researcher, **Cinzia Marchese**, who grew the tissue, the procedure is a first step toward possible breakthroughs in creating intestinal, oral and eye tissues.

14. **University of Florida** (UF) researchers are developing devices that can interpret signals in the brain and stimulate neurons to perform correctly, advances that might someday make it possible for a tiny computer to fix diseases or even allow a paralyzed person to control a prosthetic device with his thoughts. This research is supported by $2.5 million grant from the National Institutes of Health. UF researchers from the College of Medicine, the College of Engineering and the McKnight Brain Institute have teamed up to create a "neuroprosthetic" chip designed to be implanted in the brain. They are currently (July 2007) studying the concept in rats but are aiming to develop a prototype of the device within the next four years that could be tested in people. The project aims at correcting conditions such as paralysis or epilepsy and a variety of other diseases.

The chip UF researchers are seeking to develop would be implanted directly into the brain tissue, where it could gather data from signals, decode them and stimulate the brain in a self-contained package without wires. In the interim, UF researchers are studying implantable devices in rats and are evaluating an intermediate form of the technology - placing electrodes on the surface of the brain - in people.

They have also developed new techniques using surface electrodes to access signals almost as precisely as they could with sensors implanted in the brain. Developing these techniques is a big step forward in understanding how to best decode a patient's intent from their brain waves and should have broad implications for delivering therapy.

15. **Brigham and Women's Hospital in Boston** has granted a small surgical team permission to perform partial face transplants. The move makes Brigham and Women's the second hospital in the United States to give permission to perform the controversial procedure. The world's first face transplant was done a year and a half ago on a woman in France who had been mauled by her dog. Since then, the procedure has been fiercely debated because patients who undergo the surgery must take powerful drugs that were once reserved for patients with life-threatening conditions. Those who argue against face transplants say if the body rejects the tissue, another operation to remove the tissue could leave the patient more scarred than before. Some doctors argue that face transplants can dramatically transform the lives of people who are disfigured to the point where they do not work or go out.

16. A herd of goats has been engineered to produce a therapeutic human protein called **antithrombin** in their milk in central Massachusetts (July 2007). The animals are owned by a company called **GTC Biotherapeutics**, which extracts the protein for use in treating a human hereditary condition in which people lack antithrombin, which helps keep blood from clotting. GTC's drug has already been approved in Europe, and the company is now conducting clinical trials in hopes of filing for United States approval early next year.

    Other companies, too, are at work on medicines that would be extracted from a transgenic animal's milk or blood, saying the approach might be less expensive than other ways of making protein drugs. Pharming Group, a Dutch company, has applied for European approval of a drug made in the milk of transgenic rabbits. Pharmathene, based in Annapolis, Md., is developing transgenic goats whose milk produces an antidote to nerve gas. **Hematech, a South Dakota company** owned by the pharmaceutical division of Krin, the Japanese beer company, has developed cows with human immune systems to produce human antibodies to treat diseases. Other companies and academic scientists are developing chickens that produce drugs in their eggs and pigs with organs that would not be acutely rejected if transplanted into people.

Medical therapies produced in genetically engineered animals would generally be regulated like other drugs. But the new federal regulatory guidelines being developed for transgenic animals might also cover at least the creation and testing of the drug-bearing creatures. And those rules would apply if the companies tried to sell their animals for food after their drug-producing days were over.

17. Scientists with the **Institute of Stem Cell Biology and Medicine at University of California-Los Angeles** (UCLA) have produced from human embryonic stem cells a highly pure, large quantity of functioning neurons that will allow them to create models of and study diseases such as Alzheimer's, Parkinson's, prefrontal dementia and schizophrenia. Researchers previously had been able to produce neurons from human embryonic stem cells. However, the percentage of neurons in the cell culture was not high and the neurons were difficult to isolate from the other cells. UCLA's **Yi Sun**, an associate professor of psychiatry and biobehavioral sciences, and Howard Hughes Medical Institute investigator **Thomas Südhof** at the University of Texas Southwestern Medical Center were able to produce 70 to 80 per cent of neurons in cell culture.

Sun and Südhof also were able to isolate the neurons and determine that they had a functional synaptic network, which the neurons use to communicate. Because they were functional, the neurons can be used to create a variety of human neurological disease models. Previously, the system to grow and isolate neurons was very messy and it was unknown whether those neurons were functioning. Now the scientists have been able to purify so many more neurons out of the cell culture and they were, surprisingly, healthy enough to form synapses. These cells will be excellent for doing gene expression studies and biochemical and protein analyses.

To isolate the cells, the team added an enzyme that digests a sort of protein matrix that holds cells in culture together. The neurons could then be separated from the neural stem cells that had not yet differentiated a sort of chemical round-up that isolated the neurons. The cells were then put into a cell strainer that allowed passage through of the isolated neurons. The large number of pure neurons produced will allow Sun and her team to study their biological form and structure, the genes they express, the development of synapses and the electric and chemical communication activities within the synapse network. Thus it may now be possible to study the cellular properties of neurons in a very defined way that will tell us what goes wrong in diseases such as

Alzheimer's and Parkinson's. By studying the chemical and electrical transmissions, a way may be found to treat these debilitating diseases.

18. **Insulin grown in plants** has been shown to relieve diabetes in mice, according to a new study. Diabetic mice showed signs of recovery after they were treated with insulin grown in plants, showed the study, conducted by a research team at the **University of Central Florida**. In the study, the researchers genetically engineered tobacco plants with the insulin gene and then administered freeze-dried plant cells to five-week-old diabetic mice as a powder for eight weeks. By the end of the study, the diabetic mice had normal blood and urine sugar levels, and their cells were producing normal levels of insulin. These results and prior research indicate that insulin capsules could someday be used to prevent diabetes before symptoms appear and treat the disease in its later stages according to lead researcher biochemist Henry Daniell. Daniell has also proposed using lettuce instead of tobacco to produce the insulin because the crop can be produced cheaply and avoids the negative stigma associated with tobacco.

    Capsules of insulin produced in genetically modified lettuce could hold the key to restoring the body's ability to produce insulin and help millions of Americans who suffer from insulin-dependent diabetes, said the study. Insulin-dependent, or Type 1, diabetes is an autoimmune disease in which the body's immune system attacks and destroys insulin and insulin-producing beta cells in the pancreas. Insulin is a hormone that is needed to convert sugar, starches and other food into energy. Insulin typically is given through shots and not pills so the hormone can go straight into the bloodstream. In Daniell's method, plant cell walls made of cellulose initially prevent insulin from degrading. When the plant cells containing insulin reach the intestine, bacteria living there begin to slowly break down the cell walls and gradually release insulin into the bloodstream. Though produced in lettuce, the insulin would be delivered to human patients as a powder in capsules because the dosage would need to be controlled carefully.

    The National Institutes of Health provided 2 million U.S. dollars to fund the study. If human trials are successful, the impact of the research could affect millions of diabetics worldwide and dramatically reduce the costs of fighting a disease that can lead to heart and kidney diseases and blindness.

19. **DNA vaccine against multiple sclerosis (MS)** has been developed which appears safe and produces beneficial changes in the brains and immune systems of individuals with multiple sclerosis. In patients with MS, the immune system attacks the myelin sheaths that protect nerve cells in the brain and spinal cord. The nerve cell's axon, which transmits messages to other neurons, is eventually

destroyed. The cause of MS is unknown, but evidence points to the involvement of immune cells and antibodies that recognize and attack specific substances in the myelin, such as myelin basic protein. Certain cytokines, small proteins produced by cells that trigger inflammation, also may play a role.

Amit Bar, M.D., of the Montreal Neurological Institute and colleagues tested a DNA vaccine, BHT-3009, that encodes a full-length human myelin basic protein. Between 2004 and 2006, the researchers administered the vaccine to 30 patients with relapsing-remitting MS [characterized by symptomatic periods and periods of remission] or secondary progressive MS [when symptoms progressively worsen, but there still may be periods of remission]. After one, three, five and nine weeks, participants received intramuscular injections of placebo or BHT-3009 (in doses of 0.5 milligrams, 1.5 milligrams or 3 milligrams), with or without 80 milligram pills of atorvastatin calcium, a lipid-lowering drug previously shown to be effective in autoimmune conditions. After 13 weeks, participants who initially received placebo received four injections of BHT-3009. Magnetic resonance imaging (MRI) and other safety evaluations were performed at the beginning of the study, and again after five, nine, 13, 26, 38 and 50 weeks. BHT-3009 is safe and well tolerated, provided favorable trends on brain MRI and produced beneficial antigen-specific immune changes. These changes included a reduction in the number of cytokine-producing $CD4^+$ T cells (a type of white blood cell) specifically targeting myelin proteins. This reduction was found in the blood as well as in the cerebrospinal fluid of three patients who voluntarily underwent lumbar puncture after completing the course of BHT-3009. Atorvastatin did not appear to provide additional benefit.

There were no increases in clinical relapses, disability, drug-associated laboratory abnormalities, adverse events or the number and volume of contrast-enhancing [visible on MRI] lesions on brain MRI with BHT-3009 treatment compared with placebo. In fact, there was a trend toward a decrease in the number and volume of contrast-enhancing lesions in the brain in patients treated with BHT-3009 compared with placebo.

Based on these results, a phase 2b trial—a randomized clinical trial in approximately 290 patients—of BHT-3009 is already under way. If successful in MS, antigen-specific DNA vaccines can be developed for prevention or treatment of related diseases, such as type 1 diabetes mellitus, systemic lupus erythematosus, rheumatoid arthritis and myasthenia gravis.

20. In a landmark study, researchers at the BC Cancer Agency have discovered that all blood stem cells are not created equal. The discovery adds another layer of understanding about the basic biology of blood stem cells, which may lead to

improved treatments for leukemia patients. Dr. Connie Eaves and her team at the **BC Cancer Agency's Terry Fox Laboratory** identified distinct subtypes of blood stem cells in the adult mouse. Each stem cell subtype behaves uniquely and produces different types of mature blood cells in a transplant setting. For the first time it has been shown that in mice there are different types of blood stem cells that can perpetuate themselves for a lifetime. This could have important implications for understanding human leukemia since many leukemias are thought to arise from a single blood stem cell. If this discovery holds true in humans, it could explain why certain leukemias are harder to treat. Transplanting specific stem cell subtypes that produce different amounts of mature blood cells might also make bone marrow transplants more effective and improve outcomes for patients. Blood stem cells are highly versatile, self-sustaining cells that can generate multiple types of other specialized cells that make up our blood, including immune cells, platelet-producing cells, red blood cells and other infection-fighting white cells.

The discovery was made by isolating a tiny sub-population of adult mouse bone marrow that is highly enriched in blood stem cells and then transplanting them as single cells into irradiated mice whose own blood cells could be distinguished from the progeny of the injected cell. The power of a single cell transplant is that all of the cells it subsequently produces can be tracked throughout the entire post-transplant period. The next step is to isolate the different subtypes of blood stem cells and study the genes and proteins that dictate their behaviour.

21. Biorefining of corn brings gelatin production into the 21st century. Corn plants have been engineered into natural factories for producing gelatin to replace animal-sourced gelatin widely used by the pharmaceutical industry for manufacturing capsules and tablets. This may lead to a safe, inexpensive source of this protein for manufacturers who now rely on material obtained as a by-product of meat production. Today, production of gelatin, a jelly-like substance, relies on the same fundamental methodology employed since commercial production began in the 17th century.

Gelatin is derived from the breakdown of collagen, which is a component of skin, tendon, bone, cartilage and connective tissue of animals. While there are no naturally occurring plant sources of gelatin, scientists have successfully modified plants, such as corn, to have a gene that results in the production of **"recombinant" gelatin**. About 55,000 tons of animal-sourced gelatin are used every year to produce capsules and tablets for medicinal purposes. Plant-derived recombinant gelatin would address concerns about the possible presence of infectious agents in animal by-products and the lack of traceability of the source of the raw materials currently used to make gelatin. However, finding ways to recover and purify recombinant gelatin from plants has remained a challenge because only very low levels accumulate at the early stages of the development process.

Now, scientists at Iowa State University in Ames and FibroGen, Inc., in South San Francisco have developed a purification process to recover these small quantities of recombinant gelatin present in the early generations of transgenic corn. The method uses a four-step recovery system to separate the recombinant protein from other corn proteins with sufficient purity. Protein production from transgenic plants is a challenging process, with potential pitfalls all along the way. It is important to develop methods in the early stages of the development program to purify gelatin to demonstrate that it can be produced properly. The study establishes transgenic corn as a viable way to produce gelatin and potentially other products. Reasearchers may also be able to develop a variety of "designer" gelatins, with specific molecular weights and properties tailored to suit various needs of products containing gelatin.

Corn is an ideal production unit, because it can handle high volumes at a low cost. In addition the recombinant gelatin is free from the safety concerns of using meat by products. The purification process relies on chromatographic and filtration techniques, building upon methods developed by Fibro Gen to recover recombinant gelatin produced in yeast. Ultra filtration takes advantage of the size difference between the recombinant protein and other corn proteins. This step greatly reduces the process volume for later chromatographic steps and is crucial to achieving a high purification factor.

22. New research from **Rockefeller University** shows that neurons that cannot produce microRNAs, tiny single strands of RNA that regulate the expression of genes, slowly die in a manner similar to what is seen in such human neurodegenerative disorders as Alzheimer's and Parkinson's diseases. This study in mice shows that these tiny snippets of RNA are essential for survival of mature neurons. MicroRNAs are needed if certain neurons are to function and survive, and that means they are likely involved in survival of other neurons as well according to senior investigator, Paul Greengard, head of the Laboratory of Molecular and Cellular Neuroscience. The researchers have hypothesized that abnormalities in microRNA expression might be causing or modifying disease progression. The researchers specifically found that mice engineered to stop expression of microRNAs in cerebellar cortex neurons after birth experienced a slow decline in function, resulting in death of the neurons, known as Purkinje

cells. Because this brain area helps control motor function, mice without functioning Purkinje neurons could no longer work correctly. Since the use of these particular cells was a model system testing deletion of microRNAs, the results can likely be extended to other types of neurons, such as those involved in memory and higher thinking. Since their discovery in 1993, microRNAs have been found to be powerful regulators of gene expression, but mainly in cells that are developing. The research team cross-bred three different kinds of mice. One, created by co-author Dónal O'Carroll, in the Rockefeller Laboratory of Lymphocyte Signaling, is known as a "Dicer conditional" mouse. It gives researchers the ability to delete a gene known as Dicer, whose protein is required to produce microRNAs. They cross-bred these mice with another line, produced by researchers at the University of Hong Kong, that expresses a protein, Cre-recombinase, that inactivates Dicer in postnatal Purkinje cells. The offspring of these mice were then bred with a mouse engineered to express green fluorescent proteins when Dicer is deleted. In this way, the researchers were able to follow Dicer deletion and could then test for the presence of different microRNAs known to be expressed in the adult brain. They found that some microRNAs were deleted right away but that others took longer, and during this time, the cells were basically stable although slowly degenerating. Eventually the "mice showed symptoms reminiscent of those seen in humans with neurodegenerative disorders, and by 18 weeks almost all of the Purkinje cells had died. Conclusively, a roadmap for identification of genes which might be involved in neurodegeneration has been created.

23. The effectiveness of **vitamin D therapy** is partly dependent on how quickly it will be broken down according to Biochemistry professor Glenville Jones, an expert in the field of vitamin D metabolism. Prof. Jones and colleagues have studied the enzyme responsible for breaking down the vitamin and hope to develop a way to prevent this from happening by blocking that response. Their studies reveal that changing a single amino acid in the hydroxylase enzyme will cause it to take a completely different pathway. Although scientists have known for 25 years that the enzyme is capable of taking two different pathways, until now they could not explain why this occurs. Earlier study of the enzyme had shown that its pathway pattern is species specific. Some species, including humans and rats, favour one pathway, while others – most notably the Opossum – favour the other pathway. The researchers studied cells from animals in both categories using liquid chromatography mass spectrometry. Altering a single amino acid completely changes the enzyme from a human pattern to an Opossum pattern. The Queen's researchers believe the hydroxylase enzyme plays an important role in human cell functions. When vitamin D drugs are

used in an attempt to arrest certain types of cancer, for example, the tumour responds by making more of this enzyme. If the tumour response is blocked it will be possible to treat some tumours with vitamin D compounds.Vitamin D deficiency has also been correlated with other diseases, including multiple sclerosis, muscle weakness, and bone-related disorders.

24. Working with an organic semiconductor, electrical-engineering researchers at the **University of Arkansas** have fabricated and tested two similar but slightly different biosensors that can measure important physiological signs. Integrated into "smart" fabrics - garments with wireless technology - the sensors will be able to monitor a patient's respiration rate and body temperature in real time and thus provide point-of-care diagnostics to health-care professionals and greater freedom for patients. Although there has been some success at this effort over the past decade, traditional materials are not suitable for manufacturing low-cost, large-area sensor devices. The advantages of organic semiconductors will allow manufacturers to produce devices that are light, flexible and easily integrated into biomedical applications such as smart vests and fabrics. The researchers - Ji and Soyoun Jung, a graduate student in electrical engineering, under the direction of Vijay Varadan, Distinguished Professor of electrical engineering - worked with pentacene, a hydrocarbon molecule, and carbon nanotubes to develop the two types of sensors - a temperature sensor and a strain sensor. The addition of carbon nanotubes with pentacene increases sensor sensitivity. As an organic semiconductor, pentacene is efficient and easy to control. Both sensors were fabricated directly on flexible polymeric substrates.

    The strain sensor, which would monitor respiration rate, consisted of a Wheatstone bridge, an instrument that measures unknown electrical resistance, and a thin pentacene film that acted as a sensing layer. The system would work when a physiological strain, such as breathing, creates a mechanical deformation of the sensor, which then affects the electrical current's resistance. The researchers found that the smaller the sensor, the more sensitive it was to current variations. For the temperature sensor, the researchers used what is known as a thin-film transistor, which is a special kind of transistor that deposits thin film semiconductors on substrates. The thin-film transistor helped the researchers observe electrical current in linear response to temperature change. Most importantly, in low voltage areas, the current displayed the highest sensitivity to temperature changes. The success of the research is promising for patients whose vital signs must be continuously monitored. According to Varadan the sensors and wireless networks can fit on garments such as undershirts. With this technology, the smart fabric can monitor vital signs and collect and send data to an information hub in real time. The information can provide immediate detection of physiological abnormalities, which will allow physicians to begin treatment or prevent illness before problems reach an acute stage.

25. A 30-year scientific debate over how specialized cells in the inner ear amplify sound in mammals has been settled more in favor of bouncing cell bodies rather than vibrating, hair-like cilia, according to investigators at **St. Jude Children's Research Hospital**. The finding could explain why dogs, cats, humans and other mammals have such sensitive hearing and the ability to discriminate among frequencies. The work also highlights the importance of basic hearing research in studies into the causes of deafness. This discovery helps explain the mechanics of hearing and what might be going wrong in some forms of deafness. There are a variety of causes for hearing loss, including side effects of chemotherapy for cancer. The long-standing argument centers around outer hair cells, which are rod-shaped cells that respond to sound waves.

Located in the fluid-filled part of the inner ear called the cochlea, these outer hair cells sport tufts of hair-like cilia that project into the fluid. The presence of outer hair cells makes mammalian hearing more than 100 times better than it would be if the cells were absent. As sound waves race into the inner ear at hundreds of miles per hour, their energy—although dissipated by the cochlear fluid—generates waves in the fluid, somewhat like the tiny waves made by a pebble thrown into a pond. This energy causes the hair cell cilia in both mammals and non-mammals to swing back and forth quickly in a steady rhythm. In mammals, the rod-shaped body of the outer hair cell contracts and then vibrates in response to the sound waves, amplifying the sound. A protein called prestin has been described that is the motor in mammalian outer hair cells triggers this contraction. While both mammals and non-mammals have cilia on their outer hair cells, only mammalian outer hair cells have prestin, which drives this cellular contraction, or somatic motility. The contraction pulls the tufts of cilia downward, which maximizes the force of their vibration. In mammals, both the cilia and the cell itself vibrate. One group of scientists believes that somatic motility in mammalian outer hair cells is simply a way to change the height of the cilia in the fluid to maximize the force with which the cilia oscillate. That, in turn, would amplify the sound. An opposing group of scientists maintains that although the vibration of the outer hair cell body itself—somatic motility—does maximize the vibration of the cilia, the cell body works independently of its cilia. That is, vibration of the mammalian cell dominates the work of amplifying sound in mammals. According to St. Jude researchers altering the position of the cilia in the fluid changes the ability of the cilia to amplify sound, and hearing should be affected when the mutant prestin made the cell extend itself. Therefore, the team developed a line of genetically modified mice that carried only mutant prestin in their outer hair cells. The researchers then tested the animals' responses to sound. Results of the studies showed no alteration in hearing, which suggested that it did not matter whether the outer hair cells contracted or extended itself, that is, raised or lowered the cilia. There was no effect on amplification. The researchers concluded that somatic motility was not simply a way to make cilia do their job better; rather, there is no connection between the hair cell contractions and how the cilia do their job. Instead, somatic motility, generated by prestin, is the key to the superior hearing of mammals.

26. An international group of scientists, led by a team from the **Wellcome Trust Centre for Human Genetics at Oxford University**, have discovered a gene that increases an individual's chances of being left-handed. The research, which involved over 40 scientists from 20 research centers around the world, revealed a gene called LRRTM1; the first to be discovered which has an effect on handedness. Although little is known about LRRTM1, the Oxford team suspects that it modifies the development of asymmetry in the human brain. The researchers also discovered that LRRTM1 might slightly increase the risk of developing schizophrenia. People with schizophrenia often have unusual patterns of brain asymmetry and handedness, LRRTM1 also showed a possible effect on the risk of developing schizophrenia. Schizophrenia is a disorder of the brain, which results in impaired perception and thought. It affects roughly one per cent of adults worldwide. Future study on the roles of LRRTM1 in the developing brain, and other genes with which LRRTM1 interacts is being conceived.

27. The researchers of Massachusetts General Hospital and the University of North Carolina School of Medicine report in the journal Nature that **LKB1** is not only a **"tumor-suppressor"** gene for non-small cell lung cancer in mice, it also may be more powerful than other, better-known suppressors. If further research shows LKB1 has a similar effect in human lung cells, it could influence the way non-small cell lung cancer is diagnosed and treated. If tumors with LKB1 mutations are found to be especially fast-growing, for example, patients with such tumors might be candidates for more aggressive therapy. People born with defective versions of LKB1 often develop Peutz-Jeghers syndrome, which is marked by intestinal growths and an increased risk for certain cancers. Non-inherited mutations of the gene have been found in some lung cancers. This suggested that LKB1 normally thwarts tumors from forming. Mutated versions may be unable to act as a brake on cancer. To find out, the investigators ran a series of experiments in mice with a defective form of a gene called Kras, which

drives the formation and growth of lung cancer. They tracked the development of lung cancer in animals with mutated LKB1 and compared it to the experience of animals with abnormalities in either of two well-known tumor-suppressor genes. They found that while Kras "cooperated" with the mutated tumor-suppressor genes to produce lung cancer, it cooperated even more strongly with mutated LKB1. The LKB1-deficient tumors grew more rapidly and spread more frequently than the others, and comprised all three types of non-small cell lung cancer — squamous cell carcinoma, large-cell carcinoma, and adenocarcinoma — rather than just one or two. "This suggests that LKB1 plays a role at major stages of the tumors' development: initiation, differentiation of normal lung cells into cancer cells, and metastasis." An examination of human non-small-cell lung tissue suggests LKB1 mutations play a role there as well. Of 144 samples analyzed, 34 per cent of the lung adenocarcinomas and 19 per cent of the squamous cell carcinomas contained abnormal versions of the gene, researchers report. This suggests there may be additional lung tumor-suppressor genes yet to be discovered.

28. Researchers at the **Louisiana State University**, School of Veterinary Medicine, have used the herpes virus to potentially fight **breast cancer**, which, according to the American Cancer Society, is the most common cancer among women. The tumor emits signals to down-regulate the T-cells. Herpes virus can be engineered to awaken those cells and modulate the immune system so that it recognizes the tumor cells and destroys them. The **herpes virus was engineered** to selectively replicate in cancer cells; it does not affect normal cells. Besides destroying cancer cells, this herpes virus also has the potential to work as a vaccine. A German company, MediGene, has a similar herpes virus, which is in Phase 3 clinical trials for liver cancer.

29. A significant breakthrough has been achieved at **Medicago Inc.**, by using its proprietary expression system to produce a **vaccine** candidate for **H5N1 Avian Influenza** in highly immunogenic particles called Virus-Like Particles ("VLPs"). Medicago is a biotechnology company focused on the development, production and commercialization of vaccines and therapeutic protein-based biopharmaceuticals using a proprietary manufacturing system developed from its expertise in the genetic engineering of plants. VLPs have significant advantages over conventional vaccines as they are known to enhance immunity and therefore increase protection against disease. These particles are similar to the virus from which they were derived from; however they lack viral nucleic acid, which results in the best compromise between safety (not infectious) and efficacy (highly immunogenic). Medicago has initiated preclinical studies with its

proprietary VLP against H5N1 Avian Influenza virus and is now accumulating data.

30. The first research to show the involvement of a gene known as **Dmp1 in human lung cancer** will hopefully lead to an increased understanding on what goes wrong at the cellular level to cause the disease, according to researchers from **Wake Forest University School of Medicine** reporting in *Cancer Cell*. The research team found that the Dmp1 gene - which normally works to suppress tumor formation - is non-functional in about 35 per cent of human lung cancers. "Lung cancer is one of the most lethal types of cancer and understanding more about its cause could be a first step to developing new treatments," said senior scientist Kazushi Inoue, an assistant professor of pathology. Earlier studies in mice found that the gene is involved in activating tumor suppressors known as p53 and Arf. The researchers believe that when the Dmp1 gene isn't functional, these tumor suppressors are not available to stop tumor growth by killing cancer cells.

    The researchers found that deletion of one copy of Dmp1 was found in 30-40 per cent of mice lung tumors as well as in human non-small-cell lung cancers. As with all human genes, there are two copies of Dmp1 - one from each parent. Even with one deleted copy, the effects of the gene are silenced. The Dmp1 gene is located on human chromosome 7q21, a region that is often deleted in human malignancies. In addition, abnormality of the p53 gene is one of the most common events in human lung cancers. This work provides evidence that Dmp1 is a physiological regulator of the Arf-p53 pathway in humans and is primarily involved in lung cancer. The gene may also be a target for future drug development since high expression of Dmp1 significantly inhibited the growth of some lung cancer cells.

31. New research reported at **UC, San Francisco** in the journal "Nature Genetics" identifies naturally occurring processes that allow many **genes to both slow aging and protect against cancer** in the much-studied ***Caenorhabditis elegans*** roundworm. Many of the worm genes have counterparts in humans, suggesting that new drugs may some day ensure a long, cancer-free life. The new research and a related study the scientists reported in "Science" last year indicate that cellular changes leading to longevity antagonize tumor cell growth. This finding led to the understanding that lifespan is regulated by genes (daf-2 and daf-16) and is therefore changeable, rather than the inevitable result of the body's breakdown. The discovery in worms has been confirmed in other animals including mice. The gene daf-2 codes for a receptor for insulin and also for an insulin-like protein that promotes growth. It influences daf-16, which makes a

so-called transcription factor—a protein that determines when and where hundreds of other genes are turned on.

The scientists used an established tumor model in the worms. Then, starting with a list of 734 genes known to be targets of daf-16, they identified 29 genes that either promote or suppress tumor cell growth. They did this using several techniques, including RNA interference or RNAi, a powerful tool that allows scientists to control the expression of just one kind of gene at a time. The genes that stimulated tumor growth also accelerated aging itself, and the genes that prevented tumor growth slowed down the aging process and extended lifespan. These findings greatly strengthen the view that the controls of lifespan and cancer have deep, common roots.

CREATION OF NOVELTY

1. A study at the **University of California**, partly funded by the US Army Research Centre unravelled that spider silk from Black Widow is stronger than steel of the same density while also being elastic and flexible. The spiders are difficult to be kept together because they eat each other. Instead scientists will put silk genes into bacteria and make ingredients for the threads. Cheryl Hayashi, associate professor of biology at the University said that spider silk could be used to make body armour for troops once the liquid silk would be spun into solid fibre which takes the shape of a thread or a fabric.
2. US Scientists developed modified mosquitoes that are malaria resistant. This has raised the possibility of replacing regular mosquitoes with the resistant ones that do not spread the disease. Jason Ragson, Department of Molecular Microbiology and Immunology at John Hopkins University, and his coworkers have performed experiments with *Plasmodium berghei* (infects mice) and remarked that modified (genetically engineered) mosquitoes are proof of principle. They have yet to engineer *P. falciparum* which causes malaria in humans.
3. **Jeff Leid and Tim Vail at Center** for Microbial Genetics and Genomics at Northern Arizona University have recently (August 2007) developed the first Lateral Flow Assay device to detect microorganisms. Leid invented the device in collaboration with Tim Vail, an associate professor of Biochemistry at NAU, Jennifer Kofonow, an NAU biology graduate student, Mark Shirtliff, an assistant professor in biomedical sciences at the University of Maryland, and Rebecca Brady, a biomedical sciences graduate student at the University of Maryland.

The microorganisms remain bound together in a protective wall, known as biofilms, which attach to inanimate substances, such as rocks in streams or catheters in humans, or to living tissue such as human heart valves and bone. Certain bacteria produce biofilms, which act as a defense against drugs and the immune system, making it difficult to treat an infection. Biofilms cause more than 70 per cent of community and hospital-acquired infections, such as staph infections, according to the Centers for Disease Control and Prevention. Biofilms are an important problem with implanted medical devices as well as in the establishment of chronic infections. They are notoriously less susceptible to antibiotics than their single-cell, non-community orientated bacteria and are less susceptible to attack and killing from the human immune system. Leid explained that because of this, patients often suffer chronic and recurring infections. Without a specific way for physicians and health-care workers to diagnose these infections, patient treatment may be delayed.

Developed by a team of faculty and students from NAU and the University of Maryland, the Lateral Flow Assay works in similar ways to the test used to diagnose strep throat. The device identifies the presence of biofilm-specific antibodies in patients by allowing the antibodies to bind to biofilm-specific proteins on the device. If physicians and other health-care workers can diagnose these infections early, there will be a much greater chance for clinical treatments to work and for patient health to improve dramatically. This new device will allow doctors to test drugs and materials simultaneously and will save time and money. The researchers are working to fine-tune the treatment protocols for using the new device.

4. A Japanese geneticist, **Hiroshi Nagashima** and his research team created the world's first fourth-generation cloned pig, an achievement that could help scientists in medical and other research. The male pig was born at Tokyo's Meiji University in July 2007. Earlier attempts to clone animals for several generations were problematic. Scientists had thought that was because the genetic material in the nucleus of the donor cell degraded with each successive generation. But the team's findings show that a large mammal can be cloned for multiple generations - in this case, the clone of a clone of a clone of a clone - without degradation. Mice have already been successively cloned for multiple generations. The cloning success rate has been rather low until now, regardless of species. Researchers have been trying to improve efficiency, but there hasn't been any real progress. The results of this study may prove important to breeders of other large animals, such as racehorses or bulls, looking for ways to keep a prized animal in the gene pool.

5. **Eclyse** the **zorse** is striking proof of how an offspring inherits genes from both parents - which in her case was a male zebra and a female horse. The result is shown in her amazing coat which looks like a zebra's that has been partly covered in white paint. While most **zebra-horse crossbreeds** have stripes across their entire body, Eclyse has only two such patches, on her face and rump. The 1-year-old was born after her mother, Eclipse, was taken from her German safari park home to visit a ranch in Italy. There she was left to roam freely with other horses and a number of zebras. One zebra called Ulysses took a shine to her and there was some horseplay. When she arrived home to Germany, Eclyse's mother surprised her keepers by giving birth to a baby zorse.

   Eclyse has become a major attraction at her home safari park at Schloss Holte-Stukenbrock, near the German border with Holland. In Africa, horses and zebras are often crossbred and used as trekking animals. Hybrids are an interesting curiosity. The mule is perhaps the most famous cross - a combination between a horse and a donkey - and an animal of economic importance because it is a hard worker. Hybrids are not easy to create, however. The mating pair's different number of chromosomes - the "packets" of DNA in each cell - makes a pregnancy hard to achieve. A horse has 64 chromosomes; the zebra has 44. The zorse that results from cross-breeding will have a number of chromosomes that is somewhere in between. The zorse can only result where the sire is the zebra. If it had been the other way around there would have been no pregnancy. It's how nature works.

6. **Thomas Fischer**, an associate professor of chemistry and biochemistry at **Florida State University**, is designing a "smart" traffic system similar to those in major metropolises. A major difference, though, is its size: Fischer's traffic grid is small enough to fit on a tiny microchip. Working with an FSU postdoctoral associate, Pietro Tierno, and another colleague, Professor Tom H. Johansen of the University of Oslo in Norway, Fischer has designed a **"lab on a chip"** – a small device that, when exposed to very low magnetic fields, might one day be used as a portable tool for quickly diagnosing a variety of human illnesses. The Current method of treatment is a doctor takes a blood sample and sends it out to a laboratory. In three or four days, the lab results come back and the doctor has a better idea of what ails the patient.

   With the 'lab on a chip,' however, it might be possible to take a single drop of the patient's blood, place it on a small chip, and then be able to provide a very quick, inexpensive and—most important—accurate diagnosis. Fischer explained that the device would work by exposing the blood sample to very low magnetic field oscillations. In so doing, certain microscopic particles within the sample would be manipulated into "commuting" through an array of magnetic bubbles on the surface of the chip. Observing where various particles align themselves then would enable medical professionals to determine the nature of the patient's illness. Single molecules marking the presence or absence of a disease will be attached to magnetic particles a billion times smaller than a marble. The magnetic traffic system then will guide these particles to different positions on the chip depending on their molecular marking.

7. Peanut allergy is one of the most common causes of deaths from food allergies, accounting for more than 100 deaths in the US alone per year. It is characterized by more severe symptoms and at a higher rate on minimal contact as compared to other food allergies. Up to now there is no available cure for peanut allergy. Avoidance of peanuts is very difficult because it is commonly used as an ingredient in several food preparations. With the advent of genetic engineering, novel strategies are now being tested to solve the problem of food allergies from the source. A group of **US scientists** constructed transgenic peanut lines expressing significantly lower amount of Ara h 2, the immunodominant allergen found in peanuts, using RNAi. Enzyme-Linked ImmunoSorbent Assay (ELISA) using sera of patients allergic to peanut was used to monitor Ara h 2 expression. Several constructs showed complete absence of the Ara h 2 protein. Other phenotypic features and characteristics of the transgenic peanuts are the same as that of the wild type. Similar studies are now being made to silence other food allergens using RNAi.

8. Researchers from **MIT, Alnylam Pharmaceuticals** and other institutions have demonstrated the safety of a promising type of genetic therapy that could lead to treatments for a wide range of diseases such as cancer. The research focuses on **RNAi**, a key part of the body's genetic machinery. RNAi works by using short strands of RNA to block the expression of specific genes. RNAi has huge potential as a therapeutic agent.

GENOMICS

1. The entire **genetic code of breast and colon cancers** has been cracked by the scientists at the **John Hopkins Kimmel Cancer Centre**. The genetic map that has been deciphered shows that nearly 200 mutated genes, most previously unknown, help tumours emerge, grow and spread. The research paper has appeared in the esteemed journal 'Science.'

2. A team co-led by UCSF's Brian Shoichet, Steven Almo of the Albert Einstein College of Medicine, and Frank Raushel of **Texas A&M University** described

the first success **"decrypting" an enzyme's function from its structure**. Subject to working of this new strategy with other enzymes, it may become a potent tool to determine how key enzymes work in the body. And since enzyme action is crucial to disease, the technique opens an efficient route to drug discovery. The team used molecular docking, a computer-aided modeling strategy by virtue of which atom-by-atom structure of an enzyme was determined. This was followed by screening of many thousands of molecules for one that fits into the empty "active site" of the enzyme.

For validation of the technique, the team extracted an enzyme from a bacterium known as *Thermotoga maritima*, which normally lives at very high temperatures and pressures near volcanic ocean vents. Its structure was determined. With the nature of the substrate in hand, the scientists have discovered that the enzyme works in a previously uncharacterized metabolic pathway in bacteria.

3. Sandip Ghosal, Associate Professor of mechanical engineering in **Northwestern's McCormick School of Engineering and Applied Science** devised an inexpensive and accurate way of sequencing DNA i.e. **nanosequencing**. This technique can detect individual DNA molecules with accuracy, but the speed of the DNA past the nanopore is so fast that it is not possible to read the bases. The method is based on classical hydrodynamics in which DNA is pulled through a nanopore's channel by electric force. The individual bases can be easily read as the strand passes if the resistive force which is created is exploited. According to the scientist, DNA pulls some of the fluid surrounding the molecule through the channel with it. The lubrication forces arising in this fluid layer create the resistance that opposes the electrical pulling force. The theoretical model is consistent with the experimental results.

   Understanding the mechanics of DNA translocation will allow scientists to make alterations, to figure out how to apply more friction to slow down the flow of the DNA.

4. **Craig Venter** and his team created **synthetic genome** of a bacterial cell - constructed chemically from the building blocks of DNA. Giovanni Murtas of the Enrico Fermi Centre at the University of Rome, Italy, reported successful synthesis proteins in cell-like compartments. Murtas and his team initiated the process of protein synthesis in cell-like self-assembling spheres bounded by lipid membranes, known as "liposomes". A similar feat was achieved in 2004 by Vincent Noireaux and Albert Libchaber of Rockefeller University in New York, but while they seeded their lipid vesicles with an extract of *Escherichia coli* bacterial cells, Murtas and his colleagues used a recipe of 37 enzymes and a

range of smaller molecules to enable protein synthesis. Murtas and his colleagues are trying to incorporate into their vesicles genes for enzymes that can form new lipids, which they hope will make the liposomes grow to the point that they split apart to form smaller daughter vesicles. This artificial organism can become a workhorse for biotechnology that could be adapted to do useful tasks such as making complex biochemicals.

The novel "genome transplant" that Venter's team announced has proved in principle that they can do just that. The researchers, led by John Glass of the J. Craig Venter Institute in Rockville, Maryland, managed to transfer the genome of *Mycoplasma mycoides* to a related parasite called *Mycoplasma capricolum*. Both species infect goats, sheep and cows. Judging from the proteins they produced, the resulting cells seem to have been completely transformed into *M. mycoides* The team took a strain of *M. mycoides* resistant to the antibiotic tetracycline, broke open the cells and then used enzymes to digest away their proteins, leaving its circular chromosomes intact. Next, they incubated these chromosomes with *M. capricolum* cells in a culture medium containing a polymer called polyethylene glycol (PEG). PEG makes cell membranes fuse, and the researchers speculate that some *M. capricolum* cells fuse together, encapsulating an *M. mycoides* chromosome as they do so. The cells containing multiple genomes soon divide, putting one genome into each daughter cell. The researchers then treated their cultures with tetracycline, which wipes out those containing the host *M. capricolum* genome while cells containing the *M. mycoides* genome survive. Though the transplantation worked in only about 1 in every 150,000 cells, that was enough to give healthy colonies of transformed bacteria containing no *M. capricolum* DNA.

Venter says that efforts to synthesise his minimal genome from scratch are still in progress, but once it is ready, the transplant method should allow the first bacterium with a synthetic genome to be created with little delay.

5. Scientists in **France** have deciphered the entire genome of the plant producing wine grapes. The findings could pave the way for gene-based manipulations to boost flavour of wine and improve resistance against disease. Dozens of researchers analyzing the Pinot Noir varietal of *Vitis vinifera*, the core species from which virtually all grape wine is made, have found twice as many genes contributing to aroma than in other sequenced plants, suggesting that wine flavours could be traced to the genome level. *V. vinifera* is only the fourth complete genetic sequence ever produced for a flowering plant, and the first for a fruit crop. The other three are rice, the poplar, and Thale Cress, a species of wild plant related to mustard and cabbage.

The global wine industry generates some 200 billion dollars (150 billion euros) a year in revenue. The researchers discovered that *V. vinifera* had large "families" of genes "directly correlated with the aromatic features of wine," especially related to tannins and terpenes. The study also identified the genetic source in the plant of resveratrol, the anti-oxidant in red wine that has been widely associated with health benefits ranging from anti-aging to boosting anti-viral treatments. Research is already underway to isolate a gene that could increase resistance to oidium, a common form of mildew to which Pinot Noir is especially vulnerable. This would also help reduce the use of chemicals in grape growing.

6. The **genome sequence** of the fungal plant pathogen, *Fusarium graminearum* has been deciphered. The sequencing has provided scientists a road map to someday combat a fungus that infects wheat and barley crops, rendering them unusable. Frances Trail, Michigan State University associate professor of plant biology and plant pathology, and Jonathan Walton, professor in the MSU-Department of Energy (DOE) Plant Research Laboratory, joined scientists around the world in picking over the inner workings of the fungus. The real estate in some parts of the chromosomes, where many switches of disease and toxins reside, is unstable. Other areas of the chromosomes, where basic metabolism and other vital functions dwell, are stable. Those unstable areas are places where the organism is ready to evolve. In those genes there's a lot of mutation. They can change a lot without killing the fungus. The genes that are involved in basic metabolism can't change without killing the fungus. This gives insight to what makes a pathogen a pathogen. Understanding the layout of the genome is a high-stakes proposition. This fungus is a serious pathogen of wheat and barley in Michigan and throughout the Midwest. It causes *Fusarium* head blight, which reduces grain yields, and taints grain with mycotoxins that have been found to be detrimental to human and animal health.

   Fusarium begins its blighting ways as pinprick-sized pods that spit spores into the air. The spores float over grain fields, landing on flowering wheat and barley. The spores colonize the wheat flowers. The often cool, wet weather of the Midwest provides an ideal environment for the fungus to take hold. The result: fields of blight, identified by withered, bleached heads of grain. At harvest, many of the grains are shrunken and white, and harbor the mycotoxins.

   The fungal plant pathogen has some 14,000 genes sequenced. Trail said the roles of some of them are understood, including which ones help form the spores or help produce toxins. Trail's team figures that there are 2,000 genes dedicated to making the spores. Those spores have to get out to cause the new disease cycle. It is likely that new strategies can now be deployed to control the growth of the pathogen.

Understanding the sequence is the first step in the process. From there, the task is understanding the makeup of the genes - where they're strong and organized, where they're unstable and ready to change strategy. Flexibility in the pathogenic-holding parts of the chromosome is the reason this fungus can produce so many different mycotoxins – including zearalenone, which can mimic sex hormones in mammals, including possibly people, and potentially cause developmental and reproductive problems. The research was funded by a joint program between the U.S. Department of Agriculture, the National Science Foundation and the DOE as well as supported by the Michigan Agricultural Experiment Station. The sequencing was performed at the Broad Institute at MIT.

7. Researchers at the **University Of Pennsylvania School Of Medicine** have developed a new method to screen drug resistant forms of HIV. A growing number of drug-resistant strains of HIV are a threat to the effectiveness of current treatments despite anti-HIV drug cocktails decreasing the number of HIV-related deaths and improving the quality of life for HIV patients. Existing methods of detecting drug-resistant forms of HIV are expensive, time consuming, and often fail to identify small populations of drug-resistant HIV. The new screening method analyzes multiple HIV variants at the same time, while also saving time and money. By combining two genetic tests, Frederic D. Bushman, Professor of Microbiology, and colleagues, rapidly obtained gene sequences from multiple drug-resistant HIV samples at once.

   To overcome drug resistance, patients must be treated with drugs to which the HIV virus is still susceptible. To tailor the most effective anti-HIV treatment for a patient, antiretroviral resistance screenings are conducted before a patient begins or changes drug therapy. In developing countries where HIV/AIDS is most prevalent, resistance testing is rare due to the high costs of screening. In cases where resistance testing is available, most screening techniques are not sensitive enough to analyze small populations of drug-resistant strains of HIV.

   To increase the sensitivity of HIV screening techniques and decrease the time and cost of each test, Bushman and others examined seven samples of mutated strains of HIV, including three HIV samples from patients who had experienced antiretroviral multi-drug resistance. DNA bar coding, in which different DNA molecules are indexed using DNA sequence tags, allowed Bushman and others to map multiple sequences of HIV mutants simultaneously. Drug-resistant mutations were identified using a new DNA sequencing technique called

pyrosequencing, which allows researchers to determine millions of bases of a DNA sequence in a single one-day experiment. By combining the two methods, researchers were able to quantify and characterize hundreds of thousands of HIV variants for drug resistance in a single test.

8. A research team led by scientists at the **Broad Institute of Harvard and MIT** and the **Massachusetts General Hospital** has created genome-wide chromatin maps for embryonic stem (ES) cells and two cell types derived from them, by applying a powerful new technology for sequencing DNA. The work provides a framework for mapping the complete chromatin landscape of almost any kind of cell. One of the most surprising findings suggests that cells contain an explicit chromatin-based code that reveals the developmental choices they have already made as well as those decisions that lie ahead. Unraveling the mysteries of chromatin holds great promise for understanding how cells in the body – with nearly identical DNA – assume such different forms and functions. By applying this new technology for sequencing DNA, it is now possible to look across the genome at chromatin, with greater resolution and efficiency than ever before.

   To decipher this "epigenetic" code requires ways of determining precisely which chromatin proteins sit at which locations along a cell's DNA. In principle, scientists could infer the locations by using specialized DNA chips. In practice, though, the technique has proven slow and expensive to construct genome-wide maps of mammalian chromatin. But now, a new method of massively parallel DNA sequencing has given rise to a powerful approach for readily churning out whole-genome maps of chromatin structure. The technology–based on single-molecule sequencing–makes it possible to read billions of DNA letters simultaneously. Single molecule-based methods for decoding DNA have thrown open the doors to a plethora of unexplored questions in chromatin, epigenetics and many other areas of biology.

   Empowered by this new technology, the researchers set out to study chromatin in cells with drastically different behaviors. They analyzed an assortment of chromatin proteins, each with a distinct chemical tag that switches genes either on or off. The scientists examined these proteins in mouse ES cells – known for their unusual ability to form nearly any tissue – as well as two other types of descendant cells that are more limited in the developmental paths they can choose.

9. One of the most remarkable findings involves a way of using chromatin to look into a cell's past to determine the **developmental decisions** it has already made, and to peer into the future to read its potential choices. The fortuneteller lies in

a unique form of modified chromatin known as a "bivalent domain", which marks the control regions of important genes. Such domains merge both activating and repressive chemical tags, keeping genes quiet yet poised for later activity. Bivalent domains had been noted for their role in ES cells, helping keep these cells' developmental options wide open. But with the new genome-wide chromatin data, the scientists discovered that these domains also function in more specialized kinds of stem cells. In neural stem cells, for example, bivalent domains sit near genes important to various types of brain cells, but are notably absent from genes that would be active only in, say, skin cells or blood cells.

10. Scientists thought that most new genes were formed from existing genes, but Cornell researchers have discovered a gene in some fruit flies that appears to be unrelated to other genes in any known genome. The **new gene, called hydra**, exists in only a small number of species of ***Drosophila*** fruit flies, which suggests it was created about 13 million years ago, when these *melanogaster* subgroup species diverged from a common ancestor. And early evidence indicates that the new gene is functional (as opposed to being nonfunctional "junk" DNA) and is likely to express a protein involved in late stages of sperm cell development (spermatogenesis). This finding is consistent with work of other scientists who are discovering that many of the most recently formed functional genes in any species also are expressed in male testes and appear related to spermatogenesis. This is a de novo – 'out of nowhere' – gene, according to Hsiao-Pei Yang, a senior research associate in Cornell's Department of Molecular Biology and Genetics. The researchers do not yet know how the hydra gene was created, but they speculate that the gene may have developed from a piece of DNA junk called a transposable element (also known as a "jumping gene"), which may have been inserted into the genome by a virus. These transposons are known to copy and insert themselves into DNA sequences. For example, one theory is that when a transposon sits next to a gene and then jumps to a new location, it carries part of the gene sequence it was next to and inserts it in the new location. Often, transposable elements appear to have no function or may be harmful and are eliminated by natural selection, but researchers are beginning to think transposons may be a source for creating new functional genes as well.

11. In two major traits that set humans apart from chimps and other primates – those involving brains and diet – **gene regulation**, the complex cross-talk that governs when genes are turned on and off, appears to be significantly different. Positive selection, the process by which genetic changes that aid survival and reproduction spread throughout a species, has targeted the regulation of many genes known to be involved in the brain and nervous system and in nutrition,

according to **Ralph Haygood**, a post-doctoral fellow in the laboratory of **Duke biology professor Gregory Wray**. His group looked at the regulatory sequences immediately adjacent to 6,280 genes on the DNA of chimps, humans and the rhesus macaque, a more distant primate relative that has 88 per cent the same genes as humans. These regulatory stretches of DNA are where proteins bind to the genome to initiate a gene's function. And it is here that evolution has apparently fine-tuned the performance of genes resulting in the dramatic differences in the human brain. The mouse genome had been used as a reference point for comparing the coding sequences of humans and chimps, but the non-coding sequences generally evolved much faster. While the biochemistry that cells use to turn food into energy is essentially the same across most animal species, the fine-tuning of how an organism deals with the different sorts of sugars and complex carbohydrates in its diet lies in the regulatory sequences.

Chimps are fruit-eaters, for the most part, and would not last long away from their fruit-rich forest. The sugars in their diet are relatively simple to break down and convert to cellular fuel. Humans, on the other hand, eat a wider array of foods, including many the chimps would simply not be able to digest like starchy root crops. The researchers found dramatic differences in the regulatory regions of their genes for breaking down more complex carbohydrates. It may be that parts of the human metabolism are cranked up to digest carbohydrates down to simpler sugars. Regulatory changes have adapted to changing circumstances without changing the essential chemistry of metabolism. This may set the stage for a more focused analysis of the human diet. To do a genome-wide analysis of regulatory regions, Haygood and post-doctoral fellow Olivier Fedrigo had to adapt some of the statistical tools used for genome-wide analysis of coding regions. To be sure their results would be robust, they focused on just the most reliably accurate published DNA sequences in common between the three animals, discarding two-thirds of the genome to ensure accuracy.

12. Botanists at **Oregon State University** have discovered that a **single plant gene can cause resistance to one disease at the same time it produces susceptibility to a different disease** – the first time this unusual phenomenon has ever been observed in plants. The finding may help scientists better understand the pathways that genetic disease resistance can take. Plant diseases are a multi-billion dollar problem in agriculture, and scientists for decades have been trying to develop new varieties of plants with resistance to one disease or another. The research also explains why an epidemic of "Victoria blight," a fungal disease, occurred in the United States in the 1940s. The Pc-2 gene in a

widely-planted, imported variety of oats provided good resistance to oat rust, which is a costly crop disease - but the same gene also caused susceptibility to Victoria blight, and its use had to be discontinued as a result. "The blight fungus makes a toxin that causes disease in susceptible plants - that is, only plants that carry this gene," said Jennifer Lorang, an OSU research associate. "But it also turned out that the same gene can provide disease protection. This is very unusual, and should provide insight into genetic influences on disease resistance and susceptibility."

Most work that has been done on plant diseases is focused on disease resistance, the researchers said, and less has been done on the genetic basis for disease susceptibility. Among other things, the study suggests that plants bred for resistance to one disease may inadvertently be changed in ways that make them susceptible to a different disease. It also indicates that the physiological basis for disease resistance and susceptibility may have some similarities. The actual plant used to identify these genetic pathways was *Arabidopsis*, a small plant in the mustard family, which is frequently used for genetic research. The scientists put the Pc-2-like gene in *Arabidopsis*, which has a similar function in oats, and were able to determine that it causes disease susceptibility, although it looks like a resistance gene.

13. A new study published by the journal Nature Biotechnology reports the complete **genome sequence of *Bacillus amyloliquefaciens*** and discusses the genes that may contribute to its plant associated life style. *B. amyloliquefaciens* is a plant-associated bacterium that stimulates plant growth and produces secondary metabolites that suppress soil-borne plant pathogens. This bacterium, together with other rhizobacteria, is applied to a wide range of agricultural species to enhance growth, for example, by promoting seedling emergence, plant biomass and disease control. Genes responsible for signal proteins and compounds, detected by the plant during interaction, were identified. Nine gene clusters coding for numerous antibiotics and biocontrol agents have been identified. Because of the relatively small size of *B. amyloliquefaciens'* genome, as well as its similarity to the well studied *Bacillus subtilis*, researchers are now considering the application of genetic manipulation to produce secondary metabolites for developing agrobiotechnological agents such as biopesticides and plant growth stimulator. The genome sequence will also be valuable in gaining insights on the complex plant-bacteria interactions.

14. Evolution has mastered the art of turning trash to treasure - though, for scientists, witnessing the transformation can require a bit of patience. In new genetic research, scientists have traced the 170 million-year evolution of a piece of

"junk" DNA to its modern incarnation as an important regulator of energy balance in mammals. The discovery, they said, suggests that regions of the genome formerly presumed to be a genetic junkyard may actually be a hardware superstore, providing components that can be used to evolve new genes or new species. **Howard Hughes Medical Institute** international research scholar **Marcelo Rubinstein** and his colleagues reported the discovery. "The classical view has considered genomic sequences derived from mobile elements as "junk" DNA - a large accumulation of useless sequences," said Rubinstein. "However, more recent work, including the findings in this paper, is producing convincing evidence that these sequences provided raw material for the evolution of novel gene functions." Rubinstein and his colleagues had been studying one such piece of DNA, called nPE2, which enhances the activity of a gene called POMC (proopiomelanocortin). The POMC gene is expressed in cells in the brain and produces peptides that regulate a variety of behaviors, including food intake and stress-induced analgesia. Their studies showed that nPE2 is highly conserved in mammals but absent in other vertebrates, so its evolutionary origins were investigated. They found nPE2 to be highly similar to sequences present in the genomes of the marsupials opossum and wallaby.

In fact, Rubinstein and his colleagues realized that all similar sequences originated from a superfamily of mobile elements called CORE-short interspersed elements (CORE-SINES). CORE-SINES are retroposons, meaning the genetic sequence has been copied before being inserted into new sites in the genome. The researchers found a large number of other CORE-SINE superfamily members that had changed very little over evolutionary time, suggesting that nPE2's evolution from junk to regulatory DNA was not a unique event. In fact, Rubinstein suspects that thousands of currently functional elements are derived from ancient retroposon insertions - but their evolutionary history still needs to be untangled.

15. Dr. **Animesh Ray**, professor and director of Keck Graduate Institute (KGI) KGI PhD program, has shed new light on the **evolution of moveable genetic elements**, or jumping genes. The research provides evidence that movement of genes does not cause instability at the point from which the gene moves. This discovery has important implications for our understanding of molecular evolution and genetic research involving plants, including genetically modified crops. These findings can provide a precise explanation of predicting the changes a drought-resistant jumping gene from one plant put into another may cause to the DNA. Using *Arabidopsis thaliana*, Ray and his students studied the "footprint" that is left behind when a jumping gene moves to another locus.

They devised a test for examining these footprints that revealed a mechanism for the broken DNA at the launching pad region (the original location of the jumping gene) to join together to repair the vacant area. The results indicated that the DNA repaired itself in a manner that did not produce drastic abnormalities.

Ray characterized the genomic DNA as "smart" for repairing itself in a manner that doesn't produce drastic abnormalities. He also said that the process of repairing is "ancient" because the mechanism appears similar to that used by the immune system of mammals. Ancestors of plants and mammals diverged early in evolution, at least 1.5 billion years ago. The findings of Ray, his students Marybeth Langer and Lynn Sniderhan from the University of Rochester and co-author Ueli Grossniklaus, professor at the University of Zurich, were reported in the paper "Transposon Excision from an Atypical Site: A Mechanism of Evolution of Novel Transposable Elements." The work extends theories of the renowned cytogeneticist Barbara McClintock, who originally discovered moveable genetic elements. Ray's research also follows on the work of molecular geneticist Enrico Coen who has examined implications of moveable genes in plants and first proposed a similar mechanism of chromosome healing.

16. A group working at the **Swiss Institute for Experimental Cancer Research** (ISREC) in collaboration with the University of Pavia has discovered that **telomeres**, the repeated DNA-protein complexes at the end of chromosomes that progressively shorten every time a cell divides, also contain RNA. This discovery, published in Science Express, calls into question our understanding of how telomeres function, and may provide a new avenue of attack for stopping telomere renewal in cancer cells. The telomere is like a cellular clock, because every time a cell divides, the telomere shortens. After a cell has grown and divided a few dozen times, the telomeres turn on an alarm system that prevents further division. If this clock doesn't function right, cells either end up with damaged chromosomes or they become "immortal" and continue dividing endlessly – either way it's bad news and leads to cancer or disease. It was thought that telomeres were "silent" – that their DNA was not transcribed into strands of RNA. The researchers have turned this theory on its head by discovering telomeric RNA and showing that this RNA is transcribed from DNA on the telomere. A protein in the telomerase enzyme regulates the RNA in the telomere. This discovery may thus uncover key elements of telomere function.

17. Researchers at **Cold Spring Harbor Laboratory (CSHL)** have discovered three genes that interact with cancerous cells and results in 20% of lung cancers.

The three genes are located next to each other on human chromosome 14 and two are known to play key roles in fetal lung development. According to CSHL lead investigator **David Mu**, "lung cancer cells in adults can reactivate genes that are normally active in the earliest stages of lung development. We identified the mutation that triggers this abnormal re-activation of developmental genes and showed that if you turn off these genes, you stop the cancer." The CSHL research found that the three genes termed TTF1, NKX2-8, and PAX9 interact to reactivate what appears to be an early fetal gene expression pattern that results in cancer tumor growth. The collaboration of these genes and the fact that they are so close together on the chromosome may explain why this mutation is so common in lung cancer. In collaboration with Dr. William Gerald at the Memorial Sloan Kettering Cancer Center, the study finds that the mutation is more prevalent in late stage lung cancer and is possibly a risk factor for recurrence.

18. A new technique that combines **gene chip technology** with the latest generation of **gene sequencing** machines to allow fast and accurate sequencing of selected parts of the genome has been developed by researchers from the Human Genome Sequencing Center at **Baylor College of Medicine in Houston and NimbleGen Systems, Inc.**, a Wisconsin-based company recently purchased by Roche Applied Science. This new technology will replace polymerase chain reaction (PCR) for many purposes. If the aim is to sequence a whole genome for everybody, this is a huge step in that direction. The report, which appears in Nature Methods, describes the use of microarrays to enrich or increase the volume of specific genomic sequences. High throughput DNA sequencing machines made by 454 Life Sciences then determine the exact genetic code of the material. The new process is simpler, more accurate and efficient than the multiplex PCR that was previously used to sequence portions of the genome. In one experiment, more than 6,400 exons (the part of the genetic code that carries the instructions for making proteins) were analyzed. Using the old technology this would have taken at least six months.

19. **Maize Mini-chromosomes** can add stacks of functional genes to plants. A new method of constructing artificial plant chromosomes from small rings of naturally occurring plant DNA can be used to transport multiple genes at once into embryonic plants where they are expressed, duplicated as plant cells divide, and passed on to the next generation — a long-term goal for those interested in improving agricultural productivity. In the October 19, 2007, issue of PLoS-Genetics, a team of academic and commercial researchers showed that their "maize mini-chromosomes" (MMC) can introduce an entire "cassette" of novel

genes into a plant in a way that is structurally stable and functional. The MMC could be maintained indefinitely. This could be the most valuable tool that agricultural scientists, and farmers, have long dreamed of according to **Daphne Preuss**, professor of molecular genetics and cell biology at the **University of Chicago** and chief scientific officer and president of **Chromatin, Inc.**, the makers of the MMCs. The technology could be used to increase the hardiness, yield and nutritional content of crops. It could improve the production of ethanol or other biofuels. It could enable plants to make complex biochemicals, such as medicines, at very little expense. Recently, biotech giant Monsanto Company purchased non-exclusive rights to use Chromatin's mini-chromosome stacking technology in corn, cotton, soybeans, and canola. Chromatin is in discussions to license this technology to other companies, potentially capturing most of the US corn market.

PROTEOMICS

1. A team of cell biologists led by **Eugene Chin**, M.D., a research professor at The **Warren Alpert Medical School** of **Brown University** and a staff researcher at Rhode Island Hospital, found that **acetylation**, a process that modifies amino acids, plays a central role in activating interferon receptors. Interferons play a crucial role in the body's defense against infection and uncontrolled cell growth. Type 1 interferon is widely used to treat hepatitis B and C and cancers such as melanoma and leukemia. This is a major discovery in the field of signal transduction. Tyrosine phosphorylation has so far been considered the major player in signal transduction. Acetylation may be the other one. The discovery of the acetylation of cytokine receptors marks a milestone in the study of signal transduction, the process of how cells receive and respond to chemical messages.
2. The enzyme TPPII may contribute to **obesity** by stimulating the formation of fat cells, suggests a recent study in EMBO reports. The enzyme, **TPPII**, has previously been linked to making people feel hungry, but **Jonathan Graff** and colleagues now show that it may be even more deeply involved in causing obesity. The team found that TPPII actually stimulated the formation of fat cells in worms and mammalian cells and that by reducing it, fat stores decreased. Mice with lower levels of TPPII were thinner than their wild type littermates, although their food intake was comparable.
3. **Thioredoxins** (TRXs) are ubiquitous cellular workhorses that can alter the catalytic properties of specific enzymes. In these so-called redox-regulated reactions, TRXs can transfer hydrogen to convert a disulfide bridge formed

between two cysteine amino acids to their unbridged thiol forms. These TRX-responsive disulfide bridges have regulatory functions in numerous cellular processes, including regulatory cascades related to enzyme regulation, stress responses, DNA translation and transcription, protein folding and repair, cell death, and gene expression. Plants have the largest number of thioredoxins of all organisms and are found in all of the major compartments of the plant cell - cytosol, mitochondria, and chloroplasts. In chloroplasts, TRXs regulate numerous enzymes involved in the transformation of carbon into sugars. They also function to regulate the photosynthetic production of ATP, the energy currency of all living organisms and in fatty acid synthesis. Researchers have found that the **f and m type plant thioredoxins** previously thought to be localized only in chloroplasts are also found in other, **non-photosynthetic, tissues**, where they may have multiple functions. All the different types of plant TRXs are encoded in the nucleus. The f and m types were isolated from chloroplasts and function there in light-regulated reactions related to photosynthesis. h-type TRXs are localized in the cytosol and are thought to function in seeds in maturation and germination, in plant reproduction, and in other signaling pathways. Thioredoxin h is also thought to be a mobile element in plants transported through the phloem or directly from cell to cell through plasmodesmata and thus may act as a messenger molecule in signaling. Several other types of TRXs have also been found in plants.

The scientists identified regions in the control sequences of the two TRX genes, which specify whether the genes will be regulated by light, where they will be expressed, and the metabolic pathway in which they may function. Different metabolic pathways that were identified as TRX regulated include embryo and seed formation, protein storage, the circadian clock, pollen localization, and hormonal regulation.

4. The "large" G proteins function as mediators in the transfer of various signals from activated G-protein-coupled receptors (GPCR) to appropriate effectors. The first direct evidence that links the **G proteins** to salinity and heat stress tolerance has been presented by researchers at the **International Centre for Genetic Engineering and Biotechnology** in New Delhi, India. The researchers reported the isolation of the cDNAs of two isoforms of G-alpha ($G\alpha^1$ and $G\alpha^2$) and one $G\alpha$ from pea and purification of their encoded recombinant proteins. Levels of $G\alpha^1$ and $G\alpha^2$ went up following NaCl and heat treatments. Protein-protein interaction studies showed that $G\alpha$ protein interacted with the pea $G\alpha$. Consequently, transgenic tobacco plants overexpressing $G\alpha$ showed tolerance to salinity and heat, while $G\alpha$ overexpression showed only heat tolerance. This

research uncovers a new pathway to engineer to maximize crop yield in sub-optimal conditions such as high salinity and high temperature.

5. Using baker's yeast and another fungus, researchers report the first successful propagation of a prion from one organism to another. **Prions** – infectious, oddly-folded proteins that are the main suspects in fatal neurodegenerative diseases such as Cruetzfeldt-Jakob and bovine spongiform encephalopathy, or "mad cow" – remain mostly a mystery to scientists. Very few prions have been fully described. How they infect and propagate is not fully understood. New insights into prion propagation have been provided by **Susan Liebman, Professor of biological sciences at the University of Illinois at Chicago**. A key difference between the two is that the yeast prion proteins are rich in the amino acids glutamine and asparagine in the regions of the protein used to transform them into a prion. In contrast, the fungal prion lacks a rich supply of these amino acids – a characteristic it shares with the prion-forming protein in mammals, which is otherwise dissimilar.

   The researchers showed, by fusing the prion-forming domain of the fungal protein to a reporter protein, that the fungal prion could propagate in yeast. It's the first time a prion from one organism has been propagated in another organism that normally lacks that prion. It demonstrates that totally heterologous prion propagation is possible. The finding also underscores the value of the yeast model for studying factors necessary to propagate prions, now that it's been shown that propagation is not necessarily host-specific.

PLANT BIOTECHNOLOGY

1. Indian American scientists and their colleagues at Texas A and M University selectively reduced the levels of a toxic chemical in cotton, making the seeds edible and potentially transforming cotton into an important new food source. The two scientists **Keerti S. Rathore and Ganesan Sunilkumar** of Texas & their co-workers used the RNA interference technology to silence the **gossypol** gene in cottonseed only and not the rest of the plant. The expression of gossypol which is a very toxic compound is very essential for the plant because it protects it against insects and disease. Using the seed specific promoter, the cottonseed has been rendered free of gossypol and hence fit for being eaten by people. The seed typically contains about 22% protein, which is a very high quality one.
2. Government of India has launched a **National Mission on Biodiesel**. The Project involves: large scale plantation of *Jatropha curcas*; setting up of expeller units for extraction of oil from *J. curcas* seeds; setting up of biodiesel plants for making

biodiesel from Jatropha oil; and marketing such blended fuel. It is expected that this next generation technology will become the energy policy of the world by simple default. Fuel independence can be achieved by growing microalgae and other vegetable oil crops with in the greenhouse. Microalgae produce lipids (fats). Vegetable oils and fats may be converted into biodiesel through transesterification. Microalgae have the ability to produce 20 times the amount of biodiesel/hectare than soybeans.

3. Scientists led by David Mackill at the International Rice Research Institute, Philippines, have identified a gene that makes **rice flood – resistant**. Flooding causes annual losses of over $ 1 billion, with south and south-east Asia the hardest hit. The scientists have identified a gene i.e. submergence tolerant gene that enables rice to survive for up to two weeks in water. They have used genetic engineering for this task and are working on to identify other complex genes for traits such as drought- or salt – tolerance. Commercialization of the variety is expected by 2009.

4. Scientists have discovered how plant stem cells in roots detect soil structure and whether it is favourable for growth. This is a first step towards understanding how plants respond to soil compaction. Research leader Professor Liam Dolan, of the **John Innes Centre in Norwich**, UK and his team determined that the hormone ethylene regulates cell division in root stem cells. Ethylene is known to play a role in perceiving and communicating environmental cues. Stem cells in buds are activated to divide and give rise to the growth for that season. Ethylene provides signals from the environment to activate cell division when the conditions are right. The research was carried out in *Arabidopsis*, but it is thought to apply to other plant species.

5. **Salt tolerant tomato** varieties have successfully completed the first phase with experimental varieties currently undergoing field trials in target locations worldwide (Evogene Ltd., Telaviv and Zeraim Gedera Ltd., acquired by Syngenta, SA). Salt tolerant tomato collaboration advances to second phase. The use of massive irrigation practices and water sources of high salinity levels in tomato growing areas have caused a reduction in tomato yield and in severe cases the abandoning of such areas. The tomato varieties developed through this collaboration are expected to allow growers to maintain high yields under severe growing conditions displaying high salt concentrations of water and soil. Zeraim Gedera's breeding program is based on its leading tomato germplasm and utilizes advanced selection tools. These selection tools include proprietary selection assays and molecular marker for salt tolerance, developed based on computational gene discovery platform.

6. A taste modifying protein, **miraculin** has been developed by a group of **Japanese researchers**. This has been produced in transgenic tomatoes and is used as an alternative sweetener. Miraculin was first isolated from the miracle fruit, the red berries (*Richadella dulcifica*), a shrub native to West Africa. Indigenous people often use these berries to improve the palatability of their acidic maize dishes and to sweeten sour beverages. Miraculin itself is not sweet, but can turn a sour taste to a sweet taste. The sweetness induced by citric acid after exposure to miraculin has been estimated to be around 3000 times that of sucrose on a weight basis. This unique property, including the possibility of being used as an alternative low-calorie sweetener for diabetic and obese individuals, has led to increasing interests in this protein. Nine taste modifying proteins are known, but as with miraculin, the commercial feasibility of these proteins is limited, since their natural sources are tropical plants that are difficult to grow outside their normal environments.

   The miraculin gene was introduced to tomato cotyledons and recombinant miraculin accumulated at high levels in both leaves and fruits of up to 102.5 and 90.7 μg g^{-1} fresh weight, respectively. They found that the choice of plant species is an important factor in the stable production of miraculin, with tomato being a more suitable host than lettuce for miraculin production. This successful production of recombinant miraculin in transgenic tomatoes is thus a new method to make it more available for mass production of low-calorie sweetener and flavor enhancer.

7. Plant scientists have known for some time that genes from the maternal plant control **seed development**. The Oxford research, supported by the Biotechnology & Biological Sciences Research Council (BBSRC) has found an answer to the mechanism of seed development. Working in collaboration with researchers in Germany and France, Professor Hugh Dickinson's team found that only the maternal copy of a key gene responsible for delivering nutrients is active. The copy derived from the paternal plant is switched off. This gene encodes a potential signalling molecule found in the endosperm - a placenta-like layer that nourishes the developing grain, which is involved in 'calling' for nutrients from the mother plant, and so triggers an increased flow of resources. Similar mechanisms can almost certainly be expected in other cereals, and with cereal grain being a staple food across the world, the potential to harness this science to improve yields is clear. This has opened up opportunities in improving crop yield. The knowledge and molecular tools needed to harness these natural genetic processes are now available to plant breeders and could help them improve commercial varieties further. The mechanism used to switch off paternal

genes ensures supremacy of maternally derived genes. This process is known as 'imprinting' and is achieved mainly through 'methylation' - a naturally occurring chemical change in the DNA. A very similar mechanism takes place in animal embryos. However, unlike the animal imprinting systems where genes are often grouped in the chromosomal DNA, in maize imprinted genes are 'solitary' and independently regulated. This has paved way for development of better quality maize.

8. The Brazilian Enterprise of Agropecuary Research (Embrapa) developed (2007) Brazil's first **genetically modified soybean** for commercial purposes with the world's largest chemical company BASF. The transgenic soybean contains a gene of the plant Thale Cress, scientifically known as *Arabidopsis thaliana*, a member of the watercress and mustard family that is commonly grown in the lab. The gene provides the soybean with resistance to imidazolinone herbicide. Imidazolinone competes in the international herbicide market with glyphosate which is the main ingredient of herbicide Roundup developed by the U.S. Monsanto Company, Germany-based BASF's main competitor. According to Embrapa, several bio-security tests are being carried out to check the plant's impact on the environment and human feeding. The results will be sent to the National Bio-security Technical Committee in charge of authorizing the project. The new soybean is expected to take over up to 20 per cent of the Brazilian market and will enhance its competition with Monsanto. The companies expect to launch the soybean in the market by 2012.

9. The first ever 'Nina Saxena Excellence in Technology Award 2007' has been conferred on SPS Khanuja, Director, Central Institute of Medicinal and Aromatic Plants (CIMAP) and his team. The award was given for developing the plant ***Artemisia annua***, from which an **anti-malarial drug** can be extracted. The plant was produced from a variety of plants through the method of molecular breeding. The plant can now be grown in India, enhancing the rural income of the country. Earlier, *A. annua* had to be imported from China at a cost of Rs 40,000 per kg. Now through the genetic modification, the plant can be grown in the country at a cost of Rs 15,000 per kg and help in treating chloroquinone resistant cerebral malaria. At present the drug is being grown in 2,000 acres. By the next year, farming will be done over 10,000 acres.

10. The Philippines has so far approved 41 varieties of genetically-modified plants for use in the country (July 2007), with three types of corn also cleared for commercial cultivation. The corn varieties approved for commercial cultivation are ***Bt* corn**, which is resistant to corn borer, developed by U.S. agrochemical firm **Monsanto Co**. As well as corn varieties, plants that have been approved

for direct use as food, feed and for processing are soybean, canola, potato, cotton, sugarbeet and alfalfa. A genetically-modified variety of rice for food, feed and processing may also be introduced. Last year, total land area planted to *Bt* corn was estimated at around 200,000 hectares while those planted to herbicide tolerant varieties totalled around 56,000 hectares. The Philippines was the first country in Asia to approve the commercial cultivation of a genetically-modified Bt corn variety in December 2002.

11. The success of wheat as a food crop can be traced through thousands of years of genetic changes that occurred as wheat was domesticated for human use according to **University of California Davis** plant scientists Jorge Dubcovsky and Jan Dvorak. According to the authors gene mutations and the presence of multiple chromosomes – a characteristic known as "polyploidy" – enabled modern wheat to overcome several genetic bottlenecks that occurred during wheat domestication and subsequent evolution. The authors conclude that, Polyploid wheat has been able to compensate for diversity bottlenecks by capturing a relatively large proportion of the variability present in wild wheat.

    In addition, new variation is rapidly generated in the dynamic wheat genomes through gene deletions and insertions of repetitive elements into coding and regulatory gene regions. Domestication of wheat began roughly 10,000 years ago as people in western Asia began the transition from hunting and gathering to raising crops and animals. Some of the important traits that were selected for during the domestication process include increased grain size, changes in the toughness of chaff so that the wheat can be easily threshed, and retention of the grain on the plant so that it doesn't scatter in the wind before or during harvest. Globally, approximately 620 million tons of wheat are now produced each year, providing one-fifth of the calories consumed by people around the world. Ninety-five per cent of the wheat crop goes into making baked goods such as bread, cookies and pastries, while the remaining 5 per cent is durum wheat used for making pasta and related products.

12. Scientists at the **University of Missouri-Columbia** have reported molecular-level discoveries about the **mechanisms of phototropism**, the directional growth of plants toward or away from light. Phototropism is initiated when photoreceptors in a plant sense directional blue light. By understanding how phototropism works at a molecular level, work can be done toward engineering plants that produce more biomass or have increased drought tolerance.

    This information could be used to optimize plants' ability to capture light for photosynthesis, which would result in more energy capture and thus growth, or potentially agronomically useful biomass. The regulation of phototropic signaling in *Arabidopsis thaliana* has been studied. Non-phototropic hypocotyls 3 (NPH3), a protein known to be essential for phototropic responses has been studied. The scientists examined its phosphorylation, the addition or removal of a phosphate group to the protein molecule. Using a series of pharmacological treatments and immunoblot assays, the team discovered that NPH3 was a phosphorylated protein - a protein with a phosphate group attached - in seedlings grown in the darkness. When the seedlings were exposed to light, they became dephosphorylated, or lost their phosphate group. These results suggest that the absorption of light by phot1, the dominant receptor controlling phototropism, leads to NPH3's loss of a phosphate group, allowing further progression of phototropic signaling. Exposure to directional blue light stimulated NPH3's dephosphorylation. NPH3 exists as a phosphorylated protein in darkness and is rapidly dephosphorylated by a yet unidentified protein phosphatase in response to phot1 photoactivation by blue light.

13. A comprehensive review on the worldwide efforts for the **ex-situ preservation** (for example in botanical gardens and seed banks) of **vanilla genetic resources** was recently presented by researchers in France. Vanilla is a member of the orchid plant family. According to authors Severine Bory and colleagues, results of studies dealing with the taxonomy, reproductive biology and diversity of vanilla, specifically *Vanilla planifolia*, suggest the urgent need for conserving the genetic resources of the species. Commercial vanilla flavoring is extracted from the vanilla species *V. planifolia*. Various factors, which include over-collection in the wild and deforestation, led to the extinction of wild population of the species. Bory's group recommends that maintenance of germplasm in field or *in vitro* collections in genebanks is essential in perpetuating the existing genotypes to make them available for vanilla breeding and production. They added that collaborative international efforts should help protect vanilla in its area of origin, where it is highly endangered.

14. Working with Yinong Yang, a former professor of plant pathology in the Dale Bumpers College of Agricultural, Food and Life Sciences who is now at Penn State University, Tameka Bailey, Cell and molecular biology major uncovered the gene responsible for a molecular mechanism that helps regulate **rice blast resistance**. Her research focuses on certain type of proteins and the molecular mechanisms that trigger rice's response to stressful conditions, such as drought, high salinity or a biological disease called rice blast. Understanding how plants respond to these stressors will help scientists and farmers develop better ways to grow rice in less than optimal conditions. Bailey also studied proteins called mitogen-activated protein kinases. These proteins regulate the plants' response

to external stimuli, such as drought or disease. The particular type of kinase Bailey studied is the last one in a cascade of kinases that convert signals from receptors into responses from the plant. She found that these proteins regulate the plants' production of an acid called abscisic acid, which led to stress tolerance in drought and high salinity conditions, a trait that appears to be conserved in other types of plants. Bailey isolated and characterized these proteins, which are responsible for activating the plant's response to stress. To give rice plants a boost in their ability to tolerate stressful conditions genetic engineering was used to create plants that would express a great deal of the protein. To do this, she inserted extra copies of the protein kinase DNA into the DNA of a rice plant. The transgenic rice plant then expressed an abundance of that particular protein. In contrast, Bailey produced transgenic plants where the protein kinase was suppressed to see how the plants responded to stress in the absence of the protein of interest. Her studies showed that the extra boost of protein kinases led to increased drought tolerance.

15. Combining computer science, biochemistry and horticulture in the EUREKA E! 2544 E PLANTS project has resulted in a much faster and more exact model to speed the investigation and monitoring of plant behaviour and growing conditions. Farmers can make better use of resources, resulting in cheaper and better quality food for all and learn more effective crop management as they can visualize precisely and clearly what is happening in their fields. Currently, it is necessary for commercial nurseries and farmers to grow crops in real time so they can experiment and optimise plant care in terms of irrigation, spraying, temperature and nutrients. "We wanted to model how plants grow to let farmers see what is happening in a more user-friendly and advantageous way," explains Janneke Hadders of Dutch partner Dacom Plant-Service. "Our approach made it possible for the first time to integrate the behaviour and three-dimensional form on a plant component level," adds Dr Lubo Jankovic of project leader InteSys. The project started when the two UK partners – InteSys and Chembiotech Laboratories – approached Plant Research International (PRI) of Wageningen University in the Netherlands, which has extensive experience in plant modelling. PRI suggested involving Dacom as the specialist in presenting such information to the end user. Plants use simple principles of component behaviour and they interact by competing for internal and external resources. The project involved developing an analogue computer model calibrated using data from the growth of real plants. Work was carried out on chrysanthemums. This flower is an easy plant to grow in greenhouses, where farmers can control the conditions." says Hadders. Two parameters

were selected for study: temperature and radiation. They ended up with a three-dimensional model of the virtual plant growing where parameters can be entered. The resulting model allows simulation on a PC so that the farmer can observe the effect, for example of a certain temperature over a number of days.

16. **"Herculex"** is a maize variety produced by Pioneer and it is now authorised in ten countries. It is modified with a gene that makes it resistant to maize rootworm. Discussions are going on for allowing the importation of "Herculex" maize into Europe. The European Food Safety Authority (EFSA) has deemed the product to be safe.

17. Selective plant breeding and genetic engineering could be used to reduce the incidence of **iron deficiency** worldwide by improving the quality of dietary iron. Dr Michael Zimmerman, Laboratory for Human Nutrition, **Swiss Federal Institute of Technology**, Zurich, and colleagues have reviewed published literature worldwide, mainly from the last five years, to prepare a Seminar, which looks at the issue of nutritional iron deficiency in both industrialised and developing countries. The authors say: "Iron deficiency is one of the leading risk factors for disability and death worldwide, affecting an estimated 2 billion people...the high prevalence of iron deficiency in the developing world has substantial health and economic costs, including poor pregnancy outcome, impaired school performance, and decreased productivity." The World Health Organisation (WHO) estimates that 39% of children younger than five years, 48% of children between five and 14 years, 42% of all women, and 52% of pregnant women in developing countries are anaemic, with half having iron deficiency anaemia. WHO also believes that the frequency of iron deficiency in developing countries is around 2.5 times that of anaemia which is not iron deficiency related.

    The three main strategies for correcting iron deficiency are **supplementation (provision of iron without food), fortification of foods, and the relatively new approach of genetic engineering and plant breeding**. Supplementation can be targeted to high risk groups and be cost-effective; yet the logistics of distribution and absence of compliance are major limitations. Untargeted supplementation in children in tropical countries, mainly in areas of high transmission of malaria, is associated with increased infections.

    Fortification is probably the most practical, sustainable and cost-effective long-term solution to control iron deficiency at the national level. The low incidence of iron deficiency anaemia in adolescent and young women in the USA might be at least partly due to consumption of iron-fortified wheat flour. However, while fortification is common and has proven benefits, loss of iron from both wheat and rice during the milling process means that keeping the levels of iron acceptable (40 mg/kg) is difficult. This is where the authors believe genetic engineering can play a key role. The iron content in rice can be increased two- to three-fold by introduction of the ferritin gene from the soybean. Another problem – the reduction of bioavailable iron due to high phytate content – could also be solved by introducing genes which increase the activity of phytase enzymes to break down the phytate.

18. Some 10,000 years ago **white rice evolved from wild red rice** and began spreading around the globe. Researchers at **Cornell** and elsewhere have determined that 97.9 per cent of all white rice is derived from a mutation (a deletion of DNA) in a single gene originating in the *Japonica* subspecies of rice. Their report, published online in the journal PloS (Public Library of Science) Genetics, suggests that early farmers favored, bred and spread white rice around the world. The researchers report that this predominant mutation is also found in the *Indica* subspecies of white rice. They have found a second independent mutation (a single DNA substitution) in the same gene in several *Aus* varieties of rice in Bangladesh, accounting for the remaining 2.1 per cent of white rice varieties. Neither of these two mutations is found in any wild red rice species.

    Both mutations produce shortened versions of the same protein in which the missing part is responsible for activating the molecular pathway leading to grain color in rice. The researchers speculate that ancient farmers actively bred and spread white rice varieties first throughout the Himalayan region and then the rest of the world because the varieties cooked faster (requiring less fuel), their hulls were easier to remove compared with red rice, and disease and insects were easier to see amid the white grains. The farmers also may have favored one mutation over the other because it may have produced favorable grains more consistently, the researchers say. In 2006 the researchers first identified the gene that makes the rice seed's bran layer, or pericarp, white. This gave rice breeders and engineers a genetic marker to help develop new breeds. The Cornell researchers regularly introduce favorable genes from wild red rices into elite white cultivars to improve yields and provide better responses to stress, but they generally select against the gene for red pericarp because it is associated with such unfavorable "weedy" linked traits as seed dormancy and "shattering" (where seeds fall easily from the stalk).

19. The Genetic Engineering Approval Committee (**GEAC**) has approved large-scale trials of what could be India's first-ever transgenic food crop — **Bt brinjal** developed by the Maharashtra Hybrid Seed Company (Mahyco). The Expert

Committee, headed by the Vice-Chancellor of Delhi University, Dr Deepak Pental, had gone into the bio-safety data generated by Mahyco, besides suggesting detailed protocols and additional safeguards to be observed during the large scale trials. Mahyco had sought permission for large scale trials of seven genetically modified (GM) brinjal hybrids: MHB-4, 9, 10, 80, 99, 11 and 39. These hybrids contain a foreign gene cry1Ac isolated from a soil bacterium, *Bacillus thuringiensis* (Bt). The gene synthesises proteins toxic to the fruit and shoot borer (*Leucinodes orbonalis*). The cry1Ac gene's incorporation into brinjal is said to confer 'in-built' resistance to the dreaded insect pest, thereby reducing reliance on pesticides.

Brinjal is a 150-180 days crop that starts giving fruit after 60 days and yields 15 pickings over the entire cycle. Farmers often spray 2-3 times before each picking, adding up to around 50 sprays. The crop is cultivated on about 5.1-lakh hectares across the country, with annual production estimated at 80-85 lakh tonnes. The large scale trials is likely to take place over two seasons, which means consumers will not get to see the GM brinjal on their plates before 2009. So far, cotton is the only GM crop that has been approved for commercial cultivation since 2002. Mahyco had sourced the original cry1Ac gene construct from Monsanto, the US-based life sciences major that also has a 26 per cent stake in the former. The transformation process or 'event' — fitting the gene construct in the right place of the brinjal genome —was achieved at Mahyco's research centre at Jalna, Maharashtra.

20. Agricultural scientists say they have uncovered the last big secret of **vitamin C** in plants, and it will create the chance to naturally breed healthier fruits. The breakthrough in understanding just how plants manufacture vitamin C will enable state science company **Hortresearch** to identify DNA markers for individual plants naturally producing high levels of the vitamin. These plants are likely to be used in selective breeding programmes to produce fruit with more vitamin C in a form easily retained by the body, unlike large doses taken in vitamin pills.

    Existing cultivars of apples or bananas provide less than half the recommended US daily intake of 90 mg/day of vitamin C for an adult male, and even an orange doesn't quite measure up - but the new research will lead to more potent fruit. Hortresearch's science general manager, Dr Bruce Campbell said the team had isolated the last undiscovered enzyme and proved it controlled vitamin C in plants. The enzyme was the last step in a chain of research begun overseas nearly 80 years ago by scientist seeking to understand how plants produce vitamin C. NZ scientists studied kiwifruit, a plant naturally high in vitamin C, with typical green kiwifruit - bred from *Actinidia deliciosa* - containing about 100 mg in each 100 g of fruit. The scientists worked on an inedible wild kiwifruit variety called *Actinidia eriantha* - with a white, hairy skin which is easy to peel - because it contains a massive 800mg of vitamin C per 100g.

Genetically engineered forms of the small and non-vigorous vines have also been used by Hortresearch in containment to investigate how specific genes in the many different kiwifruit species affect biological processes in the plants. In these studies, transgenic plants were flowered and fruited in the greenhouse to check how inserted genetic material was expressed in the plant and inherited.

21. **Tolerance Gene** has been cloned in **sorghum**. Plants have little liking to aluminum, especially if it is aluminum in acidic soils. Because of acidity, aluminum from clay minerals dissolves in the soil, and gets converted to its toxic charged (ionic) form. Aluminum toxicity in acidic soils is a big problem in crop production in as much as half the world's arable land, mostly in developing countries in Africa, Asia and South America, are acidic. To provide plants with protection from aluminum, researchers at Cornell University have isolated a novel aluminum-tolerant gene found in some cultivars of sorghum. The gene expresses a protein in the root tip that release citric acid into the soil in response to aluminum exposure. Citric acid binds aluminum ions very effectively, preventing the toxic metal from entering the roots. The gene, they discovered, is only turned on when aluminum ions are present in the surrounding soil. The researchers have now used the sorghum gene to engineer transgenic aluminum-tolerant *Arabidopsis thaliana* and wheat plants. New genetically-engineered aluminum-tolerant sorghum lines are expected by 2008.

22. **Katholieke Universiteit Leuven**, have revealed a new mechanism demonstrating the intricate ways in which plants deal with stress. The newly discovered control system has a remarkable way of orchestrating the activity of hundreds of genes, forcing the plant into 'safety mode'; the consumption of energy is contained while the organism is stimulated to mobilize reserves. This may have a negative impact on growth, but it allows the plant to temporarily safeguard itself against pernicious stress conditions. These findings also may prove to be useful beyond the case of plants, for the results are likely to be valuable in understanding disorders such as cancer and diabetes. Plants catch sunlight and use it as an energy source to produce sugars from CO_2 and water. In doing so, they are at the very basis of the food chain. Ultimately, all life on earth depends upon this biochemical process: photosynthesis. Plants control their own energy balance. Fortunately, plants have developed different

mechanisms to detect and cope with 'stress'. Together with his American colleagues at Harvard Medical School (Boston, USA), **VIB** scientist Filip Rolland, associated with the Katholieke Universiteit Leuven, is uncovering a new system of detection and control. It is driven by KIN10 and KIN11. These 'kinases' – which are also found in human beings – react to energy shortages, when, for example, there is too little sunlight or too little sugar production. They control the activity of a broad network of genes, promoting the release of energy (catabolism) from alternative sources and blocking its assimilation (anabolism). In this way, the plant protects itself against stress conditions; like a really bad summer.

The model organism for this study was *Arabidopsis thaliana*. The scientists have tested numerous stress conditions that affect photosynthesis and energy production, such as darkness, herbicide treatment and flooding (lack of oxygen). By overexpressing the KIN10 gene, causing the plant to produce more of this protein, stress tolerance is increased and plants survive longer. By switching off these genes, their control function is eliminated. With this research, the Flemish and American scientists have succeeded for the first time in attributing KIN10 and KIN11 a key role in the control of the plant energy budget and metabolism and thus the fragile balance between growth and survival; in short, the choice between life and death.

The new insights gained by this study are not limited to the functioning of plants; they may also be important for human beings. KIN10 and KIN11, as 'fuel gauges' controlling the expression of a whole set of genes, are also found in mammals. The results with plants, therefore, may play a pioneering role in discovering new functions of these proteins in disorders such as diabetes, cancer, obesity, and aging.

23. KDML 105, a well known traditional Thai jasmine rice variety, is sweet aromatic rice. Although rice aroma combines more than 200 volatile compounds, a compound named 2-acetyl-1-pyrroline (2AP) is the main constituent. The aromatic 2AP has also been found in other cereals, pandan, bread flower, fungi and bacteria. A group of Thai scientists at the Rice Gene Discovery Rice Science Center, Kasetsart University, **Thailand has identified the DNA sequences that enhance the synthesis of 2-acetyl-1-pyrroline in plants and fungi**. A patent application has been placed at the US Patent Office.

24. **Chinese researchers** have developed a **genetically modified (GM) rice variety** that they hope can be further developed to provide an alternative oral vaccine for **preventing hepatitis B infection** in humans. The GM variety contains the gene SS1, which codes for a hepatitis B virus (HBV) surface antigen. In a feeding

experiment, the GM rice was shown to induce immunological responses in mice against the HBV surface antigen. The researchers say that the presence of the HBV antigen caused the mice to produce antibodies, which served to provide protection against the HBV virus binding to liver cells. Hepatitis B ranks among the most severe chronic diseases in the world, especially in tropical parts of Africa and in Southeast Asia. Commercial vaccines are available, but the expense of current immunization programs limits their use in developing countries. The Chinese researchers chose rice as the "carrier crop" because rice is a staple food in many developing countries.

25. **Chinese researchers** report that they have used a "**maize Ac/Ds transposon insertion mutagenesis system**" to develop a total of **20,000 mutant rice varieties**, some of which exhibit **increased yields** and **enhanced stress tolerance.** The insertional mutagenesis technology employed by the researchers involves the random insertion of specific DNA fragments throughout the rice genome. Rice varieties resulting from this process displayed a whole variety of morphological and conditional phenotypes. The researchers say this is the first time that insertional mutagenesis has been used to produce desirable traits for breeding purposes. The varieties that exhibit higher yielding potential or greater tolerance of abiotic stresses may be used in breeding programs that combine conventional breeding with marker-assisted selection. The researchers' work suggests that rice yields can be improved by the disruption of certain rice genes.
26. New research shows how the houseplant "**mother of thousands**" (*Kalanchoe diagremontiana*) makes the tiny plantlets that drop from the edges of its leaves. Having lost the ability to make viable seeds, the plant has shifted some of the processes that make seeds to the leaves, said Neelima Sinha, professor of plant biology at **UC Davis**. Many plants reproduce by throwing out long shoots or runners that can grow into new plants. But mother of thousands goes further: the plantlets are complete **miniature plants** that become disconnected from the mother plant's circulatory system and drop off, allowing them to spread rapidly and effectively. The houseplant has lost the ability to make viable seeds and only reproduces through plantlets. The researchers looked at two genes, STM and LEC, in mother of thousands and close relatives, some of which make seeds instead of plantlets. STM controls shoot growth, while LEC is involved in making seeds. Expression of STM in leaves was essential for making plantlets. In most plants LEC is expressed in seeds, but mother of thousands' version of the gene, LEC1, was expressed in leaves as well. When the researchers transferred the LEC1 variant into other plants, they were unable to make viable seeds. The findings could be useful in manipulating plant reproduction.

27. **Eucalyptus trees** genetically modified by a team of **Taiwanese and U.S. biologists** have proven capable of ingesting up to three times more carbon dioxide than normal strains, indicating a **new path to reducing greenhouse gases and global warming**, team members said yesterday. Under the auspices of National Science Council, staff members at the **Taiwan Forestry Research Institute (TFRI)** under the cabinet-level **Council of Agriculture and North Carolina State University** in the United States carried out the gene modification project that not only creates eucalyptus with a higher than normal CO_2 absorptive capacity, but also causes them to produce less lignin and more cellulose. TFRI researcher Chen Zenn-zong explained that cellulose, hemicelluloses, and lignin in trees are all created from carbon elements. However, only cellulose can be used in commercial processes of pulp manufacturing and bio-ethanol extraction. The idea behind the whole project is to increase the value of genetically-modified eucalyptus to related industries, so the ratio of cellulose and lignin was adjusted. Meanwhile, the tree's capacity in absorbing CO_2 to reduce greenhouse gases was enhanced, so that more trees planted for production, the more CO_2 are consumed. With every eucalyptus carrying 18 per cent less lignin and 4.5 per cent more cellulose, Chen estimated that a pulp factory with an annual output of 1 million tons could generate extra revenues of NT$1. 2 billion (about US$36 million).

28. **Chinese scientists** have developed a **genetically modified (GM) corn** that could help improve the nutritional value of livestock feed and reduce pollution. The research is carried out by the **Chinese Academy of Agricultural Sciences** (CAAS). The corn has now entered pre-production field trials. The GM corn produces seeds containing high levels of the **phytase enzyme**. The enzyme helps livestock to digest phosphorus which is enclosed in the indigestible form of phytate. Animals lack phytase in their system. As a result, farmers add the enzyme to animal feed to help livestock digest phosphorus. The CAAS scientists isolated the gene that produces phytase from a species of the fungus *Aspergillus*, and inserted it into corn. Preliminary test have shown that compared to regular varieties, the rate of seed germination, growth speed and yield of the GM corn were no different. The scientists said that, under current industry criteria for feed additives, adding just a few grams of the GM corn seed per kilogram of animal feed would be enough to satisfy livestock's nutritional demand for phosphorus. Chinese farmers could save up to $60 million per year in buying industrial phytase if this technology is commercialised. Phosphorus pollution caused by animal waste is a serious problem in China, resulting in widespread algal blooms in the Chinese lakes. Better phosphorous digestibility could add

to improvement of the environment. China has not yet approved any GM corn for commercial sale.

29. The first genetically modified (GM) rice is in Cebu, Philippines is gaining acceptance. Experts believe by 2011, it will become a hit to fight global malnutrition. Scientists at the **Philippine Rice Research Institute** (PhilRice) and Strive Foundation called this golden grain wonder as **"3-in-1" rice** because it is a genetically improved variety fortified with **Vitamin A, iron and Zinc**; identified by researchers as the vitamin and minerals mostly lacking in children and pregnant women around the world. Scientists, agricultural experts and researchers are excited over the new "3-in-1" GM rice because to them, this is the answer to reduce poverty and malnutrition in the country.
30. A study reveals that plants are not as passive and boring as we think, as they send warning signals to each other through their own chat network. Many herbal plants, like clover, strawberry and reed, form internal networks, by means of runners, where they can exchange information efficiently. **Josef Stuefer and his colleagues from the Radboud University in Nijmegen, the Netherlands**, were the first to demonstrate that clover plants warn each other via the network links if insects or pests are nearby. If one of the plants is attacked by caterpillars, other members of the network are warned via an internal chemical signal. As the warning is received, the intact plants strengthen their chemical and mechanical resistance so that they are less attractive for advancing caterpillars. As with computer networks, plant viruses can use the network to rapidly spread through the connected plants. The infection of one plant therefore leads to the infection of all plants within the network.
31. The **Philippine Rice Research Institute** (PhilRice) has developed a new rice variety that represents the country's first successful attempt to develop a product using marker-assisted selection. The new variety, called **Tubigan 7, can tolerate flooding** and is resistant to three of the most serious rice pests: bacterial blast, stemborer, and tungro. The Philippine National Seed Industry Council (NSIC) has officially released the rice variety. Tubigan 7 is the Philippine's second locally developed "biotech rice." The first was a tissue culture-derived variety Wagwag rice. This variety reportedly produces yields that are 15 per cent higher than the conventional harvest record in the Philippines.
32. Potato, the fourth most important food crop, is being cultivated worldwide, especially in places with relatively cool climates like the Andean tropical highlands or countries in the northern temperate zone. Sadly, potatoes are frost sensitive species incapable of cold acclimation. A brief exposure to frost can significantly reduce yield, while hard frosts can completely destroy entire

crops. Although genetic donors, particularly the cold-acclimatized wild potatoes of South America, exist, transfer of **freezing tolerance to cultivated potatoes** proved to be unsuccessful because of the complex genetics of the trait and the introduction of undesirable agronomic properties. By introducing the *AtBCF* genes for freeze tolerance from *Arabidopsis* with an activation control (promoter) solely for cold conditions, scientists successfully obtained several potato lines with increased freezing tolerance of up to -5°C. **Tony Chen and colleagues**, who authored the study, also found out that the attachment of the cold-inducible promoter minimize the expression of agronomically undesirable traits, like delayed flowering and retarded growth, previously attributed to the *AtBCF* genes.

33. In what could be an important breakthrough in agro-technology, scientists at the **Central Institute for Cotton Research (CICR)** have been able to apply **apomixis** - a technique to develop cotton hybrids that behave virtually as varieties - enabling the farmers to replicate the seeds themselves. It promises an end to the costly hybrid bargain for cotton farmers before every sowing season, and if everything goes well, farmers using other crops will also benefit. Cotton farmers across the country spend more than Rs 1,000 crore annually on seeds. If and when completed, the apomixis mission will help save the poor farmers this precious money. The Indian Council of Agriculture Research (ICAR) has already given a nod to the apomixis project submitted by the CICR under the Technology Mission for Cotton (TMC) and has even released the necessary funds. The advantages of having this variety is that it is non-contaminable and non-outcrossable (natural crossing with other varieties), thus maintaining the quality for all future generations of the variety. What is special about the apomictic hybrid is that all plants are similar to each other, with no segregation being noticed from generation to generation.

    The apomixis story actually began at the Cotton Research Station of Marathwada Agriculture University (MAU), Nanded, in the 1990s, where the then cotton specialist Anil Ansingkar crossed three cotton species - *G. arboreum* (desi) cotton known for jassids and drought resistance, *Hirsutum* (American) cotton, known for its big bolls, high yield and ginning percentage, and *Barbadense* (Egyptian) cotton known for its long staple and strength - to come out with a good variety with good economic characters. Ansingkar observed that three plants of the "abnormal trispecies crop" showed apomictic characteristics. Currently about 70 per cent of cotton is under hybrid cultivation. If an apomictic variety with the same vigour (high crop qualities) as the best hybrids is available, it will hold a great promise. Ansingkar says, "Fixing of heterosys in hybrids is great news. It will prove to be a great boon for farmers once we have commercially ready apomictic hybrids."

34. The accepted belief till date has been that starch biosynthesis is produced solely in the chloroplast. Biologist **Nora Alonso Casajús** PhD provides evidence to show that the greatest part of the precursor molecule in **starch biosynthesis** – known as ADPG (ADPglucose) – accumulates in the cytosol of the plants. This finding has meant à great advance in the race to obtain vegetables that can produce large quantities of starch, a substance the annual production of which is about a thousand million tonnes and which has become an essential raw material in multiple sectors of modern industry such as biofuels or biodegradable plastics. The thesis, entitled, Factors involved in the regulation of starch and glycogen production in plants and bacteria was recently defended at the **Public University of Navarra's Institute of Agrobiotechnology**. With the research results, it has been possible to conclude that, contrary to that proposed in the classical model, most ADPG linked to starch biosynthesis is concentrated in the cytosol. This cytosolic location of the ADPG suggests, moreover, that the enzyme responsible for the biosynthesis of the ADPG is not the plastidial AGP, but the sucrose synthase. This is why the researcher went on to produce and characterise plants that superexpressed SuSy. The research concluded, primarily, that the ADPG produced by SuSy is linked to the biosynthesis of starch; secondly, that SuSy has significant control over this biosynthesis process and, thirdly, that it is SuSy and not the AGP that catalyses the production of ADPG that accumulates in the leaves.

FOOD INDUSTRY

1. Post-Harvest technology has revolutionized the food industry and breakthroughs in large-scale storage, packaging and transportation of fruit and vegetable products using bulk aseptic food processing has been achieved. The World Food Prize 2007 has been presented to Dr. Philip E. Nelson of Purdue University. The pioneering work of Dr. Nelson has made the storage of perishable foods possible for longer periods of time without loss of nutritional quality. Food processing tanks coated with epoxy resin and sterilization of valves and filters are innovative additions to food processing industries. Hence, food products are stored and removed without reintroducing contaminants. As a result, enormous quantities of pathogen-free food could be distributed to plants around the world for final processing and packaging.

Dr. Nelson (in partnership with Scholle Corporation) developed a low-cost aseptic "bag-in-box" system for preserving and shipping foods. He (with Fran Rica Manufacturing) has also engineered a variation of the bag sealing fitment as a membrane, which ruptures during the fill and then reseals with a sterilized foil cap. This is now the standard technology used for processing and packaging of aseptically processed foods worldwide.

In the developing world, these technologies have made it affordable and convenient to transport and deliver a variety of safe food products without the need for refrigeration, averting loss due to spoilage. Citrosuco, a leading orange juice producer based in Brazil, has used the technology developed by Dr. Nelson to ship up to eight million gallons of orange juice to the United States and Europe. The technology has also been applied to bring potable water and emergency food aid to survivors of the 2004 tsunami in Southeast Asia and the victims of Hurricane Katrina in 2005, as well as to other crisis situations worldwide, and is used in school nutrition programs in developing countries.

2. The FDA and USDA approved LISTEX(TM) P100, the Natural Bacteriophage Product against *Listeria,* as GRAS (Generally Recognised as Safe), for all Food Products (July 2007). In the fight against *Listeria,* one of the most dangerous food pathogens, US food processing companies can now apply a novel yet natural tool: LISTEX(TM) bacteriophages. The FDA and USDA have approved this product from The Netherlands as GRAS, based on extensive safety and efficacy data and organoleptics tests confirming that LISTEX(TM) is safe and has no impact on taste, smell, colour, and other physical properties of treated products.

   Bacteriophages ('phage') are the most abundant micro-organisms on earth. Fresh water and seawater can contain as many as 1 billion phages per ml, while in fresh and processed meat and meat products, more than 100 million viable phages per gram are often present. Phages are harmless to humans, animals and plants, and target only bacterial cells. They are extremely specific in regard to the bacteria they recognize. The LISTEX(TM) bacteriophages target only *Listeria* bacteria (leaving desirable bacteria in place), and are easy to apply in production processes. In October 2006 the FDA had already proclaimed GRAS for LISTEX(TM) against *Listeria* in cheese. The extension to all products susceptible to *Listeria,* opens the door for the meat and fish industry to apply LISTEX(TM).

3. Application from Unilever to approve an ice structuring protein preparation as a novel food ingredient and release into European markets is under evaluation. Ice structuring proteins (ISP) are naturally occurring proteins and

peptides, which are found in a variety of living organisms (such as fish, plants, insects). ISP protect them from damage to tissues in very cold conditions by lowering the temperature at which ice crystals grow and by modifying the size and shape of ice crystals. Sourcing ISP from nature is not sustainable or economically feasible. The applicant, Unilever, therefore proposes to market a preparation of ice structuring proteins which will be obtained from the fermentation of genetically modified food grade yeast (*Saccharomyces cervisiae*) in sealed vessels. The applicant intends to use its ISP preparation in edible ices in order to influence the formation of ice structure during manufacture. The term 'edible ices' encompasses ice cream, including dairy ice cream, milk ice, water ice, fruit ice, sorbets, frozen desert and any similar products such as ice smoothies and products for which edible ice is a component. The level of ISP in edible ices will not exceed 0.01% by weight. Approval is sought under Novel Foods Regulation (EC) No. 258/97.

4. **Heinz**, one of the few food producers that breeds its own crop varieties, is stepping up efforts to develop **new tomato varieties**. It plans to sell them to its contract growers, who will be required to use the seeds to grow tomatoes for Heinz. These growers won't be allowed to sell the tomatoes to other companies. Over the past three years, Heinz increased its budget for seed research by 40%. The company says the added funding has allowed it to purchase more land adjacent to its Stockton research farm and to double the size of its seed research team to 30 people, including breeders, agronomists and a plant pathologist. Company officials predict the investment will pay off as new breeds of tomatoes help lower their ingredient costs by millions of dollars a year. The company had long focused on increasing yields, seeking varieties that grow more tomatoes per plant. But two years ago, as corn prices began rising, the company began putting an emphasis on developing sweeter tomatoes. Heinz uses traditional breeding techniques to create tomatoes with the desired traits - not genetic modification, which involves introducing DNA from another species into the seed. A parent tomato with high sugar content is selected and then breed until the right hybrid is obtained.

   In an effort to speed up the process, Heinz scientists are testing seedlings for the presence of certain DNA segments, or markers, that are associated with particular traits, such as **sweetness**. The seedlings with the right markers are then cultivated, improving the odds of producing tomatoes with the desired characteristics. The sweeter tomato varieties Heinz has already developed could be used in its ketchup in another year or two. The company is in the early stages of developing tomatoes even 5% to 10% sweeter than those.

ENVIRONMENTAL BIOTECHNOLOGY

1. A new study reports that a genetically engineered tobacco plant may be very good for the environment. It shows promise for cleaning up soil contaminated with trinitrotoluene (TNT), a widely used military explosive. TNT contamination is a major environmental problem at many World War II sites, military training areas, and explosive manufacturing sites. In addition to being explosive, TNT is toxic and a human health threat. Researchers knew that certain soil bacteria could metabolize and change TNT into non-toxic compounds. But those natural bacteria exist at levels too low to detoxify TNT. In the new study, researchers inserted a gene for a TNT-transforming bacterial enzyme into a tobacco plant. Then they tested the plant's effect on TNT-contaminated soil in comparison to regular tobacco plants grown in the same soil for several weeks. The genetically modified plants significantly reduced the toxicity of the TNT-contaminated soil. This is the first report to demonstrate that transgenic plants engineered for the phytoremediation of organic pollutants can increase the functional and genetic diversity of the bacterial community in acutely polluted soil compared to wild type plants. The findings have important implications, not only for use of genetically engineered plants for TNT remediation, but for cleaning up other sources of contamination as well.
2. Genetically modified seeds that are resistant to a low-toxicity herbicide, glyphosate, have a positive environmental impact when compared to other technologies to combat weeds, according to a recent University of Illinois study. With the emergence of **glyphosate-resistant (GR) weeds**, the environmental consequences of alternatives to the use of genetically modified seed are of increasing importance. The researchers have conducted a study that simulated the environmental effect of abandoning the glyphosate-resistant seed if weeds develop immunity to it. They utilized a well-known mammalian toxicity measure, the LD 50 dose for rats (the volume of pesticide needed to kill 50 per cent of a test population of rats), to assess one potential environmental impact. They simulated the consequences for corn, soybeans, and cotton. It was found that the use of GR seeds reduces the number of LD 50 doses applied per hectare by 17 per cent to 98 per cent, depending on the crop. If farmers switch to conventional seeds because of GR-resistant weeds but maintain the same tillage practices, this study suggests that LD 50 doses could increase by as much as 100 LD 50 doses per hectare in soybeans, and 500 LD 50 doses per hectare in cotton, or 11.4 and 19.8 per cent, respectively.
3. Scientists since the early '90s have seen the potential for cleaning up contaminated sites by growing plants able to take up nasty groundwater

pollutants through their roots. Then the plants break certain kinds of pollutants into harmless byproducts that the plants either incorporate into their roots, stems and leaves or release into the air. The problem with plants that are capable of doing this is that the process is slow and halts completely when growth stops in winter. Using plants in this way, a process called phytoremediation, often hasn't made sense given the timetables required by regulatory agencies at remediation sites. Scientists led by the **University of Washington's Sharon Doty**, genetically engineered poplar plants being grown in a laboratory and found that they were able to take as much as 91 per cent of trichloroethylene, the most common groundwater contaminant at U.S. Superfund sites, out of a liquid solution. Unaltered plants removed 3 per cent. The **poplar** plants — all cuttings just several inches tall growing in vials — also were able to break down, or metabolize, the pollutant into harmless byproducts at rates 100 times that of the control plants. The enzymes used to metabolize the contaminants are from a group called **cytochrome P450** found in both plants and animals. Poplars have a lot of P450s and Doty said scientists hope to eventually sort them to find ways to manipulate the plant's own genes to ramp up pollution degradation. In the meantime they are conducting experiments inserting a gene that produces cytochrome P450 in mammalian livers, in this case the livers of rabbits. Poplar genes producing cytochrome P450 is expressed in all their cells, but not at the rates achieved by the transgenics.

Let us now take a brief look at the Indian Scenario. Indian biotechnology has been seen as a highly promising business, considering the country's rich talent pool. **'India's Biotech Expanse'** can be summarised in the following figures:

- **1,200** in-house R&D labs in the country
- **400** non-profit research centers promoted by private players

India's rich talent pool (annual figures)

- **2.5** million graduate IT, engineering and life science professionals
- **50,000** post graduates in biosciences
- **1,500** PhDs in biosciences

(Source: Frost & Sullivan)

The industry is close to an inflection point form where much sharper growth and greater visibility will be seen. The world's biggest biotech and life science companies are attracted by India's talent and low costs and capabilities demonstrated by a host of small Indian biotech companies. Global majors like Genentech, Genzyme, Pall Life Sciences, Agilent Technologies' biotech division, Histogenetics have either just set up a base in

India, or are seen to be close to doing so. These companies will be concentrating on a range of activities, from clinical research to drug discovery.

Biogen Idec has set up an Indian subsidiary earlier this year with the objective of doing R&D and integrating India into its global clinical development programmes. Dr. Alpna Seth, who heads Biogen Idec India, says that apart from low cost of trials and R&D, the country's English-speaking population is an additional advantage. She opines, "The vast talent pool in the field of science and technology is of great significance in a highly regulated knowledge industry like ours, which involves a considerable amount of documentation and communications." The US giant, $15 billion Amgen, which is the world's largest biotech company, is planning to have its own clinical development centre in India. The company may also conduct core drug discovery related activities.

According to Alok Gupta, country head (life sciences & technology) of Yes Bank, many foreign biotech companies have shown a broad level of interest in India. This was evident at the Biotechnology Industry Organisation International Convention in Boston, US, held in May 2007. There were some 22000 participants including a strong contingent from India. He says that scientists are much cheaper in India. A scientist who in US would have to be paid between $200,000 and $300,000 a year, can be hired in India for about $60,000.

Dr. Amanda Caples, Director-biotechnology, state government of Victoria, Australia, believes that Australian drug companies will be looking to conduct clinical trials and research in India. The prime reason being low cost of trials. It is estimated that trials cost 30% less to do in India compared to Australia and 50% less compared to US. With drug development costs reaching astronomical levels — about $1 billion, of which some 70% is spent on trials and R&D, India presents a great opportunity, according to Dr. Caples.

According to Frost & Sullivan estimates the turnover of the Indian Biotechnology Industry is $2 billion in the year 2007.

REFERENCES

www.abc.net.au/ra/innovations/stories/s1729809.htm

www.academic.marist.edu/VBSC/secondfloor/biosciencesjournals.htm

www.advancedbiotech.in

www.africancrops.net/news

www.agriscape.com/journals/biotechnology

www.ausbiotech.org/.../details.asp?memberId=151624&returnUrl=%2Fabndirectory

www.bio-link.org/resFree.htm

www.biomedical-market-news.com/

www.bionews.in

www.bioone.org/perlserv/?request=get-document&doi=10.1614%2FWS-04-194

www.biosource-technologies.com

www.bspp.org.uk/ppigb/journals.htm

www.eevl.ac.uk/scitechnews/13sep01.html

www.ejcc.org.uk/.

www.epicyte.com

www.ericse.org/digests/dse96-5.html - 18k

www.eurekaalert.org

www.haccpnow.co.uk

www.intelliwareint.com/Links.html

www.internet-work.com/a2z/links/director/discrete/science/93.htm

www.jbc.org/cgi/content/full/277/25/22985

www.lib.unc.edu/prices/1999/PRIC238.HTML

www.mathtools.net/Applications_and_Industries/

www.molecularfarming.com

www.montegen.com

www.organicconsumers.org/biod/biotechind102904.cfm -

www.ourfood.com

www.pk.uni-bonn.de/ppigb/society

www.reedsmith.com/library/search_library.cfm

www.sciencedaily.com

www.slageconr.net/slsnet/slsnetnews0101.html

www.theangeljournal.com/cms/content/view/390/110/

www.theangeljournal.com/cms/index2.php

www.the-scientist.com

www.ttl.fi/Internet/English/Information/Electronic+journals/African+Newsletter

www.wifp.org/DWM/InternetPeriodicals.html

8

WHAT CONTROLS VITAMIN C LEVELS IN PLANTS?

WILLIAM LAING AND SEAN BULLEY

HortResearch scientists from New Zealand have identified the last remaining unknown enzyme in the major pathway of vitamin C (ascorbic acid) biosynthesis in plants and have shown that this enzyme controls the level of vitamin C in leaves. The results have been published as a cover story in the *Proceedings of the National Academy of Sciences (USA)* (Laing *et al.*, 2007) with an accompanying commentary from Professor Jim Giovannoni of Cornell (Giovannoni, 2007). In a near-simultaneous report, scientists from UCLA, working from a different approach based on their experience in the nematode worm *Caenorhabditis elegans*, have also reported the discovery of this enzyme (Linster *et al.*, 2007). Both approaches relied on bioinformatic analysis of the predicted protein sequence of a previously identified gene that causes low vitamin C in a mutant of the model plant *Arabidopsis thaliana*. The enzyme transfers GMP from GDP-L-galactose to an acceptor, producing L-galactose-1-phosphate (see Fig. 1). The reaction represents the first totally committed step in the pathway of ascorbic acid biosynthesis, as compounds higher in the pathway (e.g., GDP-D-mannose and GDP-L-galactose) are used in other pathways by plant cells. The acceptor is identified as phosphate (a 'phosphorylase') by one group (Linster *et al.*, 2007) and as a sugar-phosphate (a 'transferase'; although the enzyme did transfer GMP to phosphate, but at a lower rate) by the HortResearch scientists. It will be referred to as a transferase in this report. The HortResearch report also showed that transient transformation of tobacco leaves with this gene resulted in a three-fold increase in ascorbic acid levels compared to controls.

Vitamin C in Plants

Plants, especially leafy vegetables and fruit, are the major source of vitamin C for humans, as we are unable to make our own. However, many traditional fruits such as apples, bananas, and tomatoes are relatively low in this vitamin, having less than 20 mg/100 g fresh weight fruit of vitamin C. Thus, one fruit would provide less than half the recommended US daily intake of 90 mg/day for an adult male. Even an orange would only provide about this amount of vitamin C. Other non-governmental authorities propose much higher intakes of vitamin C than that recommended by the US government.

Kiwifruit are well known for their high vitamin C content, with green kiwifruit (*Actinidia deliciosa*) containing ~80 mg/100 g fruit fresh weight and yellow kiwifruit (*A. chinensis*) having ~100 mg/100 g. Thus one 90 g kiwifruit would provide close to the recommended US daily intake. However, there are other species of kiwifruit, such as the less palatable *A. eriantha* and *A. latifolia*, with over 800 mg/100 g of this vitamin, so huge genetic variation exists within this genus. Consequently, *Actinidia* is an excellent genus for studying the control of vitamin C biosynthesis. This work has been enhanced by the large germplasm collections of *Actinidia* species at HortResearch and in China, as well as HortResearch's large collection of expressed sequence tags (ESTs) and their associated physical DNA clones.

The pathway to Vitamin C

The major pathway of vitamin C biosynthesis in plants is thought to be through L-galactose, although other pathways also have been postulated, e.g., through the oxidation of the hexose sugar *myo*-inositol to glucuronate, and through galacturonate (see Fig. 1). These latter two compounds are important constituents in cell walls. The L-galactose pathway was only put on a sound footing in 1998 with the discovery of the enzyme L-galactose dehydrogenase (Wheeler, Jones, Smirnoff, 1998) and since that time, all the other enzymes in the pathway have been identified (Laing *et al.*, 2007; Conklin *et al.*, 2000).

Mutations in genes in the L-galactose pathway reduce ascorbic acid levels in whole plants (Ishikawa *et al.*, 2006). However, in the *PNAS* paper (Laing *et al.*, 2007) it was reported for the first time that overexpression of a gene for a committed enzyme in this pathway can increase ascorbic acid levels in leaves. Furthermore, fruit of *Actinidia eriantha* with high levels of ascorbic acid (800 mg/100g) also have significantly increased levels of the gene for the transferase enzyme (unpublished results).

Overseas work using genetic transformation to modify other proposed pathways of vitamin C biosynthesis has resulted in increased levels of ascorbic acid. For example, transfer of a gene for the enzyme *myo*-inositol oxygenase increased ascorbate in *Arabidopsis*, probably through increased flux through the glucuronate pathway. In strawberry, expression of the enzyme D-galacturonate reductase also increased ascorbate, presumably through the galacturonate pathway (Ishikawa *et al.*, 2006).

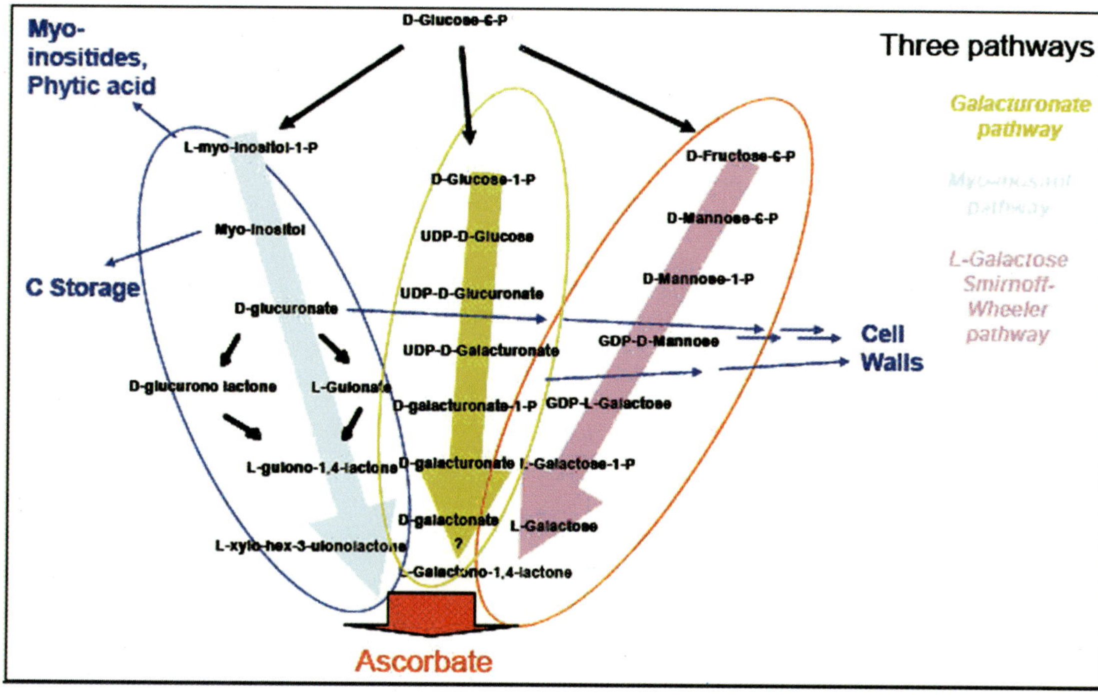

Figure 1.

Fig. 1) The three postulated pathways of ascorbic acid biosynthesis. Only the L-galactose pathway has been confirmed through mutant studies and identification of all the enzyme steps. However, some overexpression studies have provided evidence of other pathways. Blue arrows and writing represent side pathways utilizing intermediate compounds.

How to increase plant vitamin C

The worldwide aim in Vitamin C research is to increase vitamin C in fruits and leaves, and knowledge of the biochemistry of the pathways will allow us to provide more natural vitamin C for people's diets. Furthermore, high vitamin C in leaves may also make the plants more stress resistant, although this has yet to be proved. Certainly reducing vitamin C increases susceptibility to stresses, as this is how low vitamin C mutants were screened (Conklin *et al.*, 2000).

HortResearch scientists are aiming to produce fruits other than kiwifruit with much higher vitamin C content, using either traditional breeding or genetic engineering. Gene mapping of kiwifruit and apple is aimed at finding markers for high vitamin C in mapping populations, and then using this information to select for fruit with high levels of vitamin C. The transferase is expected to provide a significant lead in identifying high vitamin C markers. This map-based approach is critical in perennial crops such as tree fruit because of long generation times. In this way, it is hoped that high vitamin C apples, for example, may be selected, as most varieties of apple have significantly less than 20 mg/100 g vitamin C (http://en.wikipedia.org/wiki/Vitamin_C). Using genetic engineering to increase levels in fruit with low vitamin C is also possible (provided regulatory, IP, and consumer issues can be addressed), and it is expected that overexpression of the transferase will increase vitamin C in low vitamin C fruit, as it did in tobacco leaves.

Because of its association with scurvy and long sea voyages, vitamin C was one of the first recognized nutritional/vitamin deficiencies (http://en.wikipedia.org/wiki/Vitamin-C). However, ascorbic acid was not identified until the 20th century, and today, because of publicity from people such as Nobel Laureate Linus Pauling, vitamin C has strong public recognition. Many people know it is important to maintain a good intake of vitamin C each day, and some may supplement their intake with vitamin C from tablets. There are even people who believe that very high intakes of vitamin C (e.g., greater than 1 g per day) prevent various diseases such as cancer.

Vitamin C in such high doses has not been shown to cause significant harm. Supplemental vitamin C is likely to be rapidly excreted, because the high levels taken in one pill overwhelm the kidney. Consequently, we consider that taking the same high amount of vitamin C in the form of a fruit may slow uptake and reduce excretion. In addition, we believe this fruit form of vitamin C may also be regarded as more "natural" for those not wanting to take pills, and as pointed out by Jim Giovannoni (2007) in his commentary, "strategies to manipulate crop plants for elevated vitamin C accumulation therefore will be important in both developing and developed nations."

REFERENCES

Conklin PL Saracco SA Norris SR Last RL 2000 Identification of ascorbic acid-deficient *Arabidopsis thaliana* mutants *Genetics* **154** 847-856

Giovannoni JJ 2007 Completing a pathway to plant vitamin C synthesis *PNAS* **104**(22) 9109-9110

http://en.wikipedia.org/wiki/Vitamin_C

Ishikawa T Dowdle J Smirnoff N 2006 Progress in manipulating ascorbic acid biosynthesis and accumulation in plants *Physiologia Plantarum* **126** 343-355

Laing WA Wright MA Cooney J Bulley SM 2007 The missing step of the L-galactose pathway of ascorbate biosynthesis in plants, an L-galactose guanyltransferase, increases leaf ascorbate content *PNAS* **104** (22) 9534-95392

Linster CL *et al.* 2007 *Arabidopsis* VTC2 Encodes a GDP-L-Galactose Phosphorylase, the Last Unknown Enzyme in the Smirnoff-Wheeler Pathway to Ascorbic Acid in Plants *J Biol Chem* **282** (26) 18879-18885

Wheeler GL Jones MA Smirnoff N 1998 The biosynthetic pathway of vitamin C in higher plants *Nature* **393** 365-369

9

CATHARANTHUS ROSEUS: MOLECULAR BIOLOGY AND BIOTECHNOLOGY

CHITRA WADHWANI

Catharanthus roseus which translates as "rose-colored pure flower" (See Fig. 1) is known as the common or Madagascar Periwinkle. It is a perennial evergreen herb in the dogbane family (Apocynaceae) that was originally native to the islands of the West Indies or to Madagascar, a large island located off the southeastern coast of Africa (Stone, 1970). It has escaped cultivation and naturalized in most of the tropical world where it often becomes a rampant weed. It is established in several areas in the southern U.S. Madagascar periwinkle is grown commercially for its medicinal uses in Australia, Africa, India and Southern Europe. This fleshy perennial grows to 32 in (80 cm) in height. It has glossy, dark green, oval leaves (1-2 inches long) (See Fig.2) and flowers all summer long. The blooms of the natural wild plants are a pale pink with a purple "eye" in their centers (Heywood *et al.*, 1993) (See Fig. 3). *Catharanthus roseus* is self pollinating and carries 2n = 16 chromosomes in the somatic cells. It is self compatible with somewhat small genome size and a low chromosome number.

THE GENUS *CATHARANTHUS*

The genus *Catharanthus* belongs to the family Apocynaeceae and is closely related to the genera *Vinca* and *Amsonia*. *Catharanthus* comprises eight species; *C. coriaceus* Markgr, *C. lanceus* (Bojer ex. A.DC.) Pichon, *C. longifolius* (Pichone) Pichon, *C. trichophyllus* (Bak) Pichon are endemic in Madagaskar, *C. pusillus* (Murray) G. Don originates from India.

C. roseus, nowadays also highly appreciated as an ornamental, is a subshrub with stems of 30-100 cm long, usually upright to sometimes decumbent, containing white latex and produces a strong smell when damaged. A detailed botanical description has been published (Stearn, 1975; Pleizier, 1981; Ven Bergen, 1996).

C. roseus now has a pantropical distribution. It is naturalized in continental Africa, America, Asia, Australia and Southern Europe and on some islands in the Pacific Ocean (Pleizier, 1981; Ven Bergen, 1996). This pantropical distribution was caused by the sailors, who took the plants on board during their travels. Leaves of the plants were chewed to suppress the sensations of hunger and fatigues. Although this distribution took place many centuries ago, but it may be illustrated by the report from 1910 that *C. roseus* was used in Brazil as a remedy against haemorrhoids and scabies (Markgraf, 1976). Furthermore the earliest herbarium voucher specimens of the National Herbarium of the Netherlands (Leiden) originate from Java (1828), Brazil (1833), New Guinea (1877), Venezuela (1885), Sumatra (1899), Thailand (1900), Suriname (1901), Philippines (1904), Cuba (1905), Santo Domingo (1910), Malanaui Island (1912), Paraguay (1913), etc. (See Herbarium sheet-Fig. 4).

Linnaeus published *C. roseus,* as *Vinca rosea,* in his "Systema Naturae" in 1759. In 1828, Reichenbach used the name *Lochnera* in his "Conspectus Regni Vegetailis", but without giving a description or reference. In 1835, G. Don published *Catharanthus* in "General System of Gardening and Botany". *Lochnera* is a synonym of *Catharanthus*. Older reports on *C. roseus* go back to the botanist Flacourt, who stayed on Madagascar from 1648 to 1655 (Wijnands, 1983).

Cultivation of *C. roseus* in Europe started when seeds were sent from Madagascar to the Royal Gardens in Paris. From these plants seeds could be collected and were sent to Chelsea Physic Garden in London in 1757 by Mr. Richard, gardener of the King at Versailles and Trianon. At Chelsea, the seeds were raised by the botanist Philip Miller (1691-1771), who included a description and a plate in his "Figures of the Most Beautiful, useful and Uncommon Plants" of 1757. A herbarium specimen collected by Miller in 1758 is kept at the British Museum, and is available as species type. The plate by Miller is reproduced by Steam species type (Steam, 1975). In the meantime, David and Royen (1704-1779), director of the Hortus Botanicus in Leiden received seeds from a French diplomat. These seeds were most likely from the same French source as the seeds, which went to Chelsea. Van Royen showed Linnaeus his plant in 1758 (the start of almost 250 years of *Catharanthus* research in Leiden). In the second half of the 19th century *Catharanthus* seeds became commercially available. In the 1960's the first cultivars were offered: 'Little Blanche', 'Little Bright Eye', 'Little Delicata' and 'Little Pinkie'. In the 1980's the interest in growing *Catharanthus* cultivars as an ornamental has given rise to the breeding of plants with more colors, bigger flowers and smaller size. (See Fig. 5-8). To ensure a stable nomenclature, Division of Pharmacognosy at Leiden was appointed 'International Cultivar Registration Authority" (ICRA) by the International Society for Horticultural Science (ISHS). Accordingly, an "International Register of *Catharanthus* cultivars" was published in 1998, listing over 50 cultivars (Snoeijer, 2001). Today, over 100 cultivars are commercially available.

Habitat/ecology:

In Hawaii *Catharanthus* is commonly cultivated and frequently escaping (Wagner *et al.*, 1999). In Fiji it is often found at low elevations, escaped and naturalized in clearings coconut plantations, and waste places, and along roadsides (Smith, 1988). In Niue, it is locally very common in some plantations as well as round some houses (Sykes, 1970).

Scientific Name	:-	*Catharanthus roseus*
Common Name	:-	Vinca; Madagascar Periwinkle
Other Names	:-	Cape Periwinkle, Church Flower Red Periwinkle, Rosy Periwinkle
Native range	:-	Madagascar, grown in gardens worldwide
Habit	:-	Herb
Propagation	:-	Seed
Flowering Season	:-	Spring to fall
Light Requirements	:-	Sun to partial shade
Flower Color	:-	Pink, purple, red, salmon, white
Height	:-	6 to 18 inches
Comments	:-	Moderate to high drought tolerance

Medicinal uses:

Long before modern researchers learned of the plant's valuable and varied properties, people in faraway places were using the Madagascar periwinkle for a host of medicinal purposes.

- In India, the juice from the leaves is used to treat wasp sting.
- In Hawaii, an extract of the boiled plant is used to arrest bleeding.
- In Central America and parts of South America, the plant parts are used as a gargle to ease soar throats and chest ailments and laryngitis.
- In Cuba, Puerto Rico, Jamaica and other islands, an extract of the flower is commonly administered as eyewash for the eyes of infants.
- In Africa, leaves are used for menorrhagia and rheumatism.
- Surinamese boil ten leaves and ten flowers together for diabetes.
- Bahamians take flower decoction for asthma and flatulence, and the entire plant for tuberculosis.
- In Mauritius, the leaves infusion is given for dyspepsia and indigestion.
- In Vietnam, it is taken for diabetes and malaria.

- Curacao and Bermuda natives take the plant for high blood pressure.
- Indochinese use the stalks and leaves for dysmenorrhea. (Duke *et al.*, 1993; Dobelis, 1989)
- The leaf juice or water decoction of *Catharanthus roseus* L. (Apocyanaceae) is used as a folk medicine for the treatment of diabetes all over the world (Dobelis, 1989). Studies clearly indicated a significant antidiabetic activity with the leaf juice of *Catharanthus roseus* and support the traditional usage of the fresh leaves by Ayurvedic physicians for the control of diabetes.

NATURAL HISTORY, ORIGIN AND DISTRIBUTION

Catharanthus roseus, popularly called as Madagascar periwinkle, is a tropical, hardy and pot ornamental plant. Periwinkle is recorded as far back as 50 B.C. in folk medicine literature of Europe for its diuretic, antidysentric, hemorrhagic and wound healing properties.

Historical Uses:

- To treat diabetes, fevers, and malaria
- To regulate menstrual cycles
- To ease excessive menstrual bleeding
- As a euphoriant (feeling of well-being)

While traditional use of this plant has been ongoing for over a millennium, formal scientific research on Madagascar periwinkle began in 1952 when Dr. Clark Noble was given the leaves of the Madagascar periwinkle plant by one of his patients who had explained how popular periwinkle tea was in Jamaica for treating diabetes. Clark decided to pass the leaves onto Robert who was the Associate Director of the Collip Medical Research Laboratory at the University of Western Ontario. When Dr. Noble studied the effects of the chemicals in the leaves in rats, he had noticed that the blood sugar levels of the animals were not affected but rather their white blood cell counts had decreased. The scientists used chromatography and micro-chemical techniques to successfully separate 100 mg of vinblastine from the 400-plus alkaloids present in the Madagascar periwinkle leaves. Three years later, Dr. Robert Laing Noble and Dr. Charles T. Beer discovered vinblastine (initially named vincaleukoblastine, VLB). Clinical trials were carried out in the Princess Margaret Hospital in Toronto on patients with lymphoid cancers. When the test showed positive results, the pharmaceutical company Eli Lilly initiated production of the drug.

Origin and Geographic Distribution:

Catharanthus roseus is a perennial, evergreen herb in the dogbane family (Apocynaceae) that is native to the Indian Ocean island of Madagascar. This herb is now common in many tropical and subtropical regions worldwide and can now be found growing wild in most warm regions of the world, including the Southern U.S. *C. roseus* was introduced to many parts of the world in the 18th century. It is believed to have been introduced in India by the Portuguese missionaries in the middle of 18th century via Goa. At present, it is being cultivated in Madagascar, India, Israel and USA.

Nomenclature and Taxonomy:

Scientific name	:	*Catharanthus roseus*
Pronunciation	:	kath-uh-RANTH-us ROE-zee-us
Common name(s)	:	Periwinkle, Vinca, Madagascar Periwinkle

The botanical name for the summer flowering vinca, however, underwent a few changes from *Lochnera* to *Ammocallis* until it was finally classified as *Catharanthus*. The species name, *C. roseus*, which means rose-colored, changed only from the feminine *rosea* to the masculine *roseus* to match the gender of the Latin name of the genus. The genus name, *Catharanthus*, translates as "pure flower." The name *Catharanthus* L.G. Don is derived from the Latin words *Katharos* (pure) and *anthos* (flower), referring to the neatness and beauty of the flower; the name was assigned by George Don in 1935. Several names such as *Ammocallis rosea, Lochnera rose*a and most commonly *Vinca rosea* were used. Reichenbach, in 1935, first recognized and generically separated the existing genus *Catharanthus* from *Vinca* and designated it as *Lochnera*. However, the name did not receive nomenclature validity as he failed to provide any generic description for the genus *Lochnera*. Both *C. roseus* and *V. minor* are also known as Myrtle. Most modern cultivars are a consequence of hybrids made between *C. roseus* and other *Catharanthus* species.

Taxonomical hierarchy:

Kingdom	Plantae -Planta, Plantes, Plants, Vegetal
Subkingdom	Tracheobionta -Vascular plants
Superdivision	Spermatophyta-Seed plants
Division	Magnoliophyta -Angiospermes, Angiosperms, Flowering plants, Phanérogames, Plantes à fleurs, Plantes à fruits
Class	Magnoliopsida — Dicots, Dicotylédones, Dicotyledons
Subclass	Asteridae
Order	Gentianales
Family	Apocynaceae Adans., Nom. cons. - Apocyns, Dogbane
Sub family	Plumeroideae
Tribe	Plumerieae

Sub Tribe	Alstonrieae
Genus	*Catharanthus* G. Don – *Catharanthus*, Periwinkle
Species	*Catharanthus roseus* (L.) G. Don –

Vernacular names:

Periwinkle or Old Maid is herbaceous perennial plant once endemic to the island of Madagascar. The plant has since been widely cultivated and introduced, becoming an invasive species in some areas. During the course of its spread and naturalization over the tropics and subtropics has acquired a diversity of vernacular names as shown below in Table 1:

Table 1 : The Vernacular names of *C. roseus* being cultivated in various countries

Country		Vernacular name(s)
England	:	Cape periwinkle, Madagascar periwinkle, old-maid, rose periwinkle, rosy periwinkle.
India	:	Cape – periwinkle, Church yard blossom, Dead-man's flower, Ainskati, Rattanjor, Sadaphul, Sudukad umullai, Billagannerue (Telugu), Bara Massi and Sada Bahar (in Hindi), Gulfering (Kashmiri), Nayantara (Bengali).
Indonesia	:	Indische maagdepalm, Kembang saritijha, Kembang–dembaga.
Japan	:	Nichinchi
Phillippines	:	Chichirica
West Indies	:	Oldmaid, Ramgoat rose, Jasmine, Magdalena, Vicaria.
Spanish	:	Chatas, Chula, Pervinca de Madagascar.
French	:	Pervenche de Madagascar.

The genus *Catharanthus* is comprised of eight species of small shrubs and herbs, six of which are predominantly indigenous to Madagascar. The information about the designations, origins and distribution of the 8 species is shown in the table 2:

TABLE 2 : Origin and distribution of eight different species of *Catharanthus*

S.No.	Name of the Catharanthus species	Origin
1	*C. roseus* L.G. Don	Madagascar, now naturalized through out the tropics
2	*C. ovalis* Markgraf	Madagascar

3	*C. trichophyllous* Baker	Madagascar
4	*C. longifolius* Pichon	Madagascar
5	*C. coriaceous* Markgraf	Madagascar
6	*C. lanceous* Bojer ex ADC	Madagascar
7	*C. saitulus* pichon	Madagascar
8	*C. pusillus* Murray G. Don	India, Sri Lanka

Botanical characterization:

Catharanthus roseus is an erect, fleshy, evergreen perennial, semiwoody ever-blooming herb or sub-shrub usually grown as an annual in the flower bed. Early cultivars were low and creeping; modern cultivars may reach a height 40-80 cm high, wide at the base. In frostfree climates it develops a woody stem near the base and can get 2-3 ft (0.6-0.9 m) tall and spread out just as wide. As annuals, they are usually smaller and more prostrate. The branching starts from the base. It consists of a thick, trunk-like stem with many upright branches. The stem and leaves contain milky latex. The plant is pubescent when young and glabrous when mature. (See Fig. 9)

Root:

Root system is fibrous, spreading and shallow in nature.

Stem:

Stems are fairly rigid and woody. They are smooth with no hairs. The medium green grayish stems are upright to procumbent. The broken stem of Madagascar periwinkle exudes a milky latex sap.

Foliage:

1. *Leaf arrangement:* Opposite or alternate
2. *Stipule:* Absent
3. *Petioles:* Simple petiolate, 1.7 cm to 1.8 cm and smooth
4. *Leaf shape:* 3-8 cm long and 1.5–5 cm broad, oblong or oblong-obovate, obtuse or rounded, slightly cuspidate, finely puberulent on both surfaces, 2.5-9 cm long; midrib pale, base acute while apex is rounded to meucrunate and margin entire (Rai *et al.*, 2003).
5. *Leaf color:* The color of the upper surface of the leaves is deep green while that of lower surface is much lighter, and is deciduous. The petiole has a pinkish tinge. Leaves stay green all year in the tropics.

6. *Leaf venation:* Pinnate venation veins are prominent on the lower surface.
7. *Leaf texture:* Thick and leathery, rich glossy sheen on surface.

Inflorescence:

Flowers grow in terminal racemes, usually comprised of 4 to 5 flowers joined at one base, pedicel-3 cm, no hairs.

Flower:

1. *Floral bracts:* None
2. *Flower: Salverform,* simple, Actinomorphic violet or rose (Var. roseus), white (Var. albus) or white with red eye (Var. ocelatles), 25-37 mm in diameter, spike arrangement. Pedicels 1-2 mm long, mauve, peach, scarlet and reddish orange cultivars have also been developed. Flowering occurs from June until frost.
3. *Calyx:* Calyx lobed, linear – subulate, pubescent about 6 mm long.
4. *Corolla:* Five-petaled and pin-wheel like, each petal somewhat rhombic, with the corolla fused at the base of each of the five petals. Slender deep cylindrical corolla tube is about 2.5-3 cm long that expands to about 1.5 in (3.8 cm) across. (See Fig. 9)
5. *Stamens:* None.
6. *Ovary:* Inferior, light green, no hairs, unicarpellate, small.
7. *Style:* None.
8. Fruit: Fruits are pod like, of size 0.5 to 1 inches, green in color with dry and hard covering, ornamentally inconspicuous.
9. Seed: Many minute black cylindrical seeds with a thick seed coat. Madagascar Periwinkles are self-propagating. Seed propagation is also possible through stem cuttings. The seeds require a period of total darkness to germinate. Cuttings from mature plants also root readily.

Others:

Strong odor, grows best with a little shade and plenty of water, thrives in almost any soil, an annual in temperate areas. The plant tolerates salty soil, stormy winds and animal encroachment. It is very commonly grown in subtropical and tropical areas. Cultural usage is decoration.

Cytogenetics and genetics:

Catharanthus roseus is a diploid plant species with a karyotype comprising of 16 chromosomes, eight bivalents are visualized at meiosis. The floral morphology of *C. roseus* is conducive both for self and insect pollination. The natural populations of *C. roseus* have been observed to harbour considerable genetic variability. The natural variability and that caused by crossbreeding, induced mutagenesis and polyploidy in periwinkle has been used to develop horticulture and drug types.

Variations resulting from crossbreeding, induced mutagenesis and polyploidy have been variously employed in *C. roseus*. Initial crossing studies have shown that although the plant is self pollinated, frequent out crossing has been observed resulting into different intermediate types. Artificial tetraploids have been produced by a number of workers. Although some workers have reported that there is increase in alkaloids in tetraploids. Tetraploid plants have shorter, thicker stem, larger and thicker leaves. Tetraploids have also been reported to be less susceptible to die back disease (Rai *et al.*, 2001).

CATHARANTHUS CONTAINS ALKALOIDS AND FLAVONOIDS

The leaves have been known to contain 150 useful alkaloids among other pharmacologically active compounds. Alkaloids are part of the chemical arsenal designed to protect plants against an adverse environment. Therefore, their synthesis and accumulation are frequently induced in response to certain environmental conditions and are mediated by chemical signals, which are formed as the first responses to the external stimulus.

Indole alkaloids: Vinblastin

Vinblastine (See Fig. 10), a dimeric compound composed of catharanthine and vindoline (See Fig. 12), two monoterpene indole alkaloids, is found in *Catharanthus roseus*. The biosynthetic pathway has been elucidated. In plants, vinblastine occurs at very low concentrations (5g/2 tonnes FW). Due to its cyto-toxic effect, it may well function as phytoalexin, a chemical defense agent. Ever since 1958, when vinblastin and its analogue vincristine (See Fig. 11) were recognized as anti-cancer agents and established as important drugs in the chemotherapy of Hodgekin's disease and acute leukemia, its supply has been of great concern. Plantations of *C. roseus* could match the demand, but, for the extreme cost associated with extraction and purification. Alternative sources of higher value are the goals targeted by scientists. The development of cell cultures of *C. roseus* has been established as method in plant physiology to understand the basic processes of its biosynthesis. More recently, these efforts have been complimented with DNA technologies. Despite many years of efforts and remarkable success in increasing the production of catharanthine in cell suspensions, vinblastine has eluded all laboratories. The problem is lack of vindoline synthesis and accumulation in cell suspension cultures. Still, vindoline recurs in shoots regenerated from callus and accumulates in multiple shoot cultures. Hairy root cultures generated by employing *A. rhizogenes* have shown remarkable levels of catharanthine and traces of vinblastine. Enhanced production of catharanthine and vindoline in plants and *in vitro* is expected to be achieved by genetic manipulation of their biosynthesis or by genetically directed changes of the cytoenvironment. Cryopreservation of *C. roseus* cells has been successfully demonstrated. Finally, given sufficient yields of catharantheine from cell cultures and vindolines from shoot cultures or plants, vinblastine can be obtained through dimerization catalyzed by cell free extracts of cells cultured *in vitro*, by enzymes of such cultures, and by horseradish peroxidase with 3' 4'-anhydrovinblastine as an intermediate.

Flavonoids-anthocyanins

Anthocyanins are wide-spread metabolites among flowering plants and most conspicuous as pigments in flowers and fruits. These are potential products as marker in experimental biology and as colorant in the market. Because anthocyanins have been discovered to readily accumulate in vacuoles of cells cultured *in vitro*, it was expected that increasing its yield would only be a matter of superior cells selection and optimizing culture conditions. Culture conditions tested for improving anthocyanin synthesis and accumulation are the composition of the medium including hormones, pH and light. Remarkably, cell biomass and anthocyanin content in carrot cultures were highest on media containing a combination of galactose and glucose. An equimolar mixture of both carbohydrates, however, led to a lower performance. Stress clearly affects anthocyanin formation in plants. *In vitro* cultured material exposed to UV irradiation with supplementation of media with carbohydrate showed upward anthocyanin synthesis. With phenylalanine ammonia-lyase (PAL) and chalcone synthase (CHS) identified as key enzymes in anthocyanin biosynthesis and with genes encoding both these enzymes available as cassette, genetic manipulation of cells for improved quality and yield of production has been applied to petunia plants. PAL and CHS transgenes have not yet been expressed in cells cultured *in vitro* for production of anthocyanins. Now research is on in *Catharanthus* flavonoids.

Flavonoids in cell culture of *Catharanthus*

One of the interesting questions is whether cell cultures (callus or suspension) can synthesize the same anthocyanins as intact plants. Although the experiments could not be carried out with the sophistication as possible today, already the first early experiments showed that the formation of the backbones characteristic for *C. roseus* (petunidin, malvidin, hirsutidin) (See Fig. 13) was possible in callus cultures, and that it was induced by light (Carew and Krueger, 1976). Comparable experiments with dark-grown suspension cultures showed that the cultures turned red after dilution into a 8% sucrose solution and irradiation, and the compounds were the same as in the flowers, although the ratio was changed a bit: in petals, hirsutidin was the prominent compound, while the cultures contained mostly malvidin and petunidin (Knobloch *et al.*, 1982). It seems that there is a difference in the expression of the O-methyltransferases neccessary for the various methylation steps. Later work demonstrated that the anthocyanic profiles of flowers from both field-grown and *in vitro* regenerated plants were qualitatively similar (Piovan *et al.*, 1998).

Interestingly, a closer inspection of the induced cell cultures revealed that only about 5% of the cells turned red, indicating that the culture was very heterogeneous in its response (Knobloch *et al.*, 1982). This was confirmed in other experiments that also suggested a delay of anthocyanin formation in the lag phase of growth and the cells division phase (Hall and Yeoman, 1986b); the majority of the anthocyanins were accumulated after these stages. Additional studies confirmed that only a certain percentage of the cells (about 10%) was productive in anthocyanin formation (Hall and Yeoman, 1986a; 1987), and this was also shown in even more detailed later studies (Filippini *et al.*, 2003). The mechanism of partial differentiation with respect to capacity of flavonoid biosynthesis remains unclear.

Other experiments investigated the effects of various chemicals. One of them (Godoy-Hernández and Loyola-Vargas, 1997) found drastic increases in total anthocyanins (1476%) after application of 20 mM acetylsalicylic acid (Aspirin), but it seems a bit difficult to see a physiological relevance at such high concentration. The same might be asked for experiments with nicotinamide concentrations of 8.2 mM (Berglund *et al.*, 1993); in both cases the possible explanations sounded rather tentative. Another study showed that gibberellic acid (GA3) led to an anthocyanin increase (Ohlsson and Berglund, 2001).

What do we know about the enzymes in anthocyanin biosynthesis?

- Hotze *et al.* (1995) cloned and characterized the cytochrome P450 cinnamic acid 4-hydroxlase (C4H) and the P450 reductase (see also Meijer *et al.*, 1993) that is necessary as electron donor in the P450 reaction. The enzymes were always present in the cell cultures, including when grown in the dark. The main purpose of that work, however, was to establish a relatively simple method for functional P450 expression in *E. coli*.
- Kaltenbach *et al.* (1999) described the molecular characterization of flavonoid 3',5'-hydroxylase (F3'5'H, a cytochrome P45) and chalcone synthase, a detailed functional analysis of the F3'5'H and the tissue-specific expression in plants. The data showed that F3'5'H and flavanone hydroxylase are present in dark-grown cells; they were further induced by irradiation. The bottle-neck in flavanoid synthesis appeared to be the chalcone synthase: it was completely absent in dark-grown cultures, and had to be induced by irradiation.
- Cacace *et al.* (2003) reported the molecular and functional analysis of a new type of O-methyltransferase OMT that performs two sequential methylations at the 3'-and 5'-positions of the B-ring in myricetin (flavonol) and dihydromyricetin (dihydroflavonol). The resulting methylation pattern is

characteristic for *C. roseus* flavonol glycosides and anthocyanins. Like F3'5'G and flavanone hydroxylase, this OMT showed a quite high expression in dark-grown cell cultures.

- Schröder *et al.* (2004) described an O-methyltransferase that was not expected in *C. roseus*: The 7-OMT expected in anthocyanin biosynthesis (hirsutidin) was being looked for, but the enzyme characterized here, accepting 3'-*O*-methyl-eriodictyol (homoeriodictyol) and the corresponding flavones and flavonols as substrates, methylated them in the 4'-position of the B-ring. Flavonoids with such vicinal O-methyl groups have not yet been described from *C. roseus*, and its role *in vivo* is still an enigma.

SCIENTIFIC RESEARCH ON *CATHARANTHUS*:

Few plants have generated as much recent interest among scientists and medical communities as the Madagascar periwinkle, *Catharanthus roseus*. The interest began in the mid-1950's, when researchers, hearing of a "periwinkle tea" that was drunk in Jamaica, began to study the plant for its reported antidiabetic properties. They discovered the plant contains a motherlode of useful alkaloids. All parts of the plants are acrid and contain about 130 alkaloids, among them many show biological potency. Alkaloids belong to a class called terpenoid indole alkaloids (TIAs).

Extracts of entire dried plant contain many alkaloids of medicinal use. The principal alkaloid is vinblastine, or vincaleukoblastine (vinblastine sulfate), sold as Velban. The alkaloid has growth inhibition effects in certain human tumors. Vinblastine is used experimentally for treatment of neoplasmas, and is recommended for generalized Hodgkin's disease and resistant choricarcinoma. Another pharmacologically important alkaloid is vincristine sulfate or vincristine, sold as Oncovin. Vincristine (VCR) is used in treatment of leukemia in children. Using vinblastine and vincristine in combination chemotherapy has resulted in 80% remission in Hodgkin's disease, 99% remission in acute lymphocitic leukemia, 80% remission in Wilm's tumor, and 70% remission in gestational choricarcinoma.

There are over 70 other alkaloids that have been isolated from the plant in addition to vinblastine and vincristine. Some, such as catharanthine, leurosine sulphate, lochnerine, tetrahydroalstonine, vindoline and vindolinine lower blood sugar levels (thus easing the symptoms of diabetes). Others lower blood pressure, others act as hemostatics (arrest bleeding). Periwinkles also contain the alkaloids reserpine and serpentine, which are powerful tranquilizers (Dobelis, 1989). Ajmalicine and reserpine, present in the roots, are known to be potent antifibrillic and hypotensive agents, respectively. Because the alkaloids in this plant can have serious side effects such as nausea and hair loss, it is not recommended that people attempt to medicate themselves with periwinkles. *Catharanthus roseus* now grows wild only in Madagascar, where it is endangered by the spread of agriculture and metropolitan areas. Because it provides the anticancer alkaloids vinblastine and vincristine, *Catharanthus roseus* is cultivated on plantations in tropical countries.

Pharmacognosy of vinca alkaloids and other secondary metabolites

Alkaloids are a major class of natural products that have a physiological effect in organisms. Alkaloids are bitter-tasting plant compounds that characteristically contain carbon, hydrogen, nitrogen, and in many cases, oxygen. Because of their potent pharmacological effects, alkaloids are the basis for many drugs (Budavari & Susan, 1989). Many of them have pain - relieving or anticancer properties. More than 125 alkaloids of the indole and the dihydro-indole groups have been isolated and characterized from different plant organs.

1. SHOOT: Vindoline, Vinblastine, Vincristine, Vincaticine
2. ROOT: Ajmalicine, Serpentine

The alkaloid content of the various organs varies. Leaves, roots, stem, fruits and seeds especially their pericarp are rich in alkaloids. The flowers contain alkaloids only in small amount. Variation in the alkaloid content has been observed between the white and pink varieties.

TABLE 3 : % alkaloid content in different parts of *Catharanthus*

Plant part	% alkaloid content on dry weight basis
Root	0.12 - 9.00
Stem	0.07 - 0.46
Leaf	0.10 - 1.16
Flower	~ 0.005
Fruit	~ 0.40
Seed	~ 0.1
Pericarp	~ 0

With about 130 different compounds isolated, *C. roseus* can be regarded as a rich source of alkaloids. The alkaloids belong to the class of terpenoid indole alkaloids (TIA's). Some major products found in the plant are anhydrovinblastine, vindoline, catharanthine, ajmalicine and serpentine.

Kisakurec and Hesse in 1980, identified different alkaloid subclasses: vincosan, coryanthean, vallesiachotaman, strychnan, aspidospermatan, plumeran, ibogan, eburnan and bisindole alkaloids (about 40 compounds), most of those containing a plumeran (vindoline) and an ibogan (catharanthine) moiety.

Some of the *C. roseus* alkaloids are marketed as pharmaceuticals:

- Ajmalicine (Hydroserpan[R], Lamuran[R]) was introduced in 1957 for the treatment of hypertensia. Another important source for this alkaloid is *Rauwolfia serpentina* (L.) Bentham.
- Vinblastine (vincaleucoblastine, Velbe[R]) was introduced in 1960 and is used in the treatment of Hodgekin's disease, non-Hodgekin lymphomas, testiscarcinomas, and sometimes against breast cancer and chorio-carcinomas.
- Vincristine (leucocristine, Oncovin[R]) is an oxidized form of vinblastine and was introduced in 1963. It is used against acute leukemia, Hodgekin's disease, non-hodgekin lymphomas, rhabdomyosarcomas, Wilm's tumors in children and breast cancer.
- Anhydrovinblastine has also been claimed for use as an antineoplastic agent in the treatment of cervical and lung cancer (Schmidt *et al.*, 1998).

The antineoplastic activity of the bisindole alkaloids is attributed to their ability to disrupt microtubules, causing dissolution of mitotic spindles and metaphase arrest in dividing cells. Although the alkaloids are classified as mitotic inhibitors, their antineoplastic activity in the clinical treatment of cancer probably arises from perturbation of a variety of microtubule-dependent processes, as well as from disruption of the cell cycle and induction of programmed cell death. The alkaloids are frequently incorporated into combination chemotherapy because of their lack of cross-resistance with DNA-alkylating drugs and their different mechanisms of action.

Besides alkaloids, other secondary metabolites have been isolated from *C. roseus*, including monoterpenoid glucosides (loganin, secologanin, sweroside, deoxyloganin, dehydrologanin), steroids (catasteron, brassinolides), phenolics, flavonoids and anthocyanins.

Biochemical, cellular, pre-clinical and clinical pharmacology

The TIAs VCR and VLB act specifically on the metaphase portion of the mitotic stage of cell cycle. They produce a typical C-mitotic effect with micronuclei formation their anti-neoplastic activity is attributed to their ability to disrupt microtubules, these alkaloids bind to tubulin, the dimeric cellular proteins that play several critical roles in the cell structure and function, thus preventing the cell from making the spindles it needs to be able to move its chromosomes around as it divides. These alkaloids also seem to interfere with cells' ability to synthesize DNA and RNA (Goodman *et al.*, 1990).

VLB at 100 nM is known to completely inhibit the growth of the malarial parasite *Plasmodium falciparum* in cell cultures, presumably by affecting spindle formation, like in mammalian cells. Both VLB and VCR, at relatively low weekly doses of 1.75 mg and 14 mg, respectively, could prolong the life span of leukemic animals by 100%. They are all administered intravenously in their sulfate form once a week; these solutions are fatal if they are administered any other way, and can cause a lot of tissue irritation if they leak out of the vein.

Vinblastine sulphate is used principally in the treatment of Hodgekin's disease; it is also used against lymphosarcoma, histiocytic lymphoma, choricarcinoma, neuroblastoma, advanced testicular cancer, advanced breast cancer, and Letterer-Siwe disease. It is a component of regimen of choice for the treatment of metastatic testicular cancer and in combination with bleomycin, doxorubicin and procarbazine, in the treatment of Hodgekin's disease.

It is also in use against Kaposi's sarcoma and mycosis fungoides. In addition to the above listed diseases, vinblastine is also used to treat bladder cancer, head and neck cancer, non-small cell lung cancer, idiopathic thrombocytopenia purpura, cervical cancer, and autoimmune hemolytic anemia. Vinblastine is typically administered at a dose of 6 milligrams per square meter of body surface. It has a half-life of 24 hours in the bloodstream.

As a single agent, VCR is highly active in the treatment of childhood leukemia. It also seems to fight cancer by interfering with glutamic acid metabolism (specifically, the pathways leading from glutamic acid to the Krebs cycle and to urea formation). It is a part of famous MOPP Mechlore-thiamine, Oncovin, Procarbazine and Prednisone combination used in the treatment of acute leukemia, Hodgekin's disease, Wilm's tumor, neuroblastoma, rhabdoblastoma and reticulum cell sarcoma. Leurosine and Leurosidine act by producing ball-metaphase.

These compounds act as CNS depressants. Treatment with VLB-compounds results in inhibition of the biosynthesis of proteins, nucleic acids and lipids (Rai *et al.*, 2001). Vincristine, which is marketed as Oncovin by Eli Lilly, has a serum half-life of about 85 hours (Mutschler *et al.*, 1995). The typical dose is 1.4 milligrams per square meter of body surface once a week and neurotoxicity is the dose limiting factor (it can cause damage to the peripheral nervous system).

PRE-CLINICAL TOXICOLOGY

VLB and VCR have a highly toxic potential and show teratogenic (birth defect) effects in animals. The LD 50 for VLB and VCR in rats is 17 mg/kg and 5.2 mg/kg respectively (Rai *et al.*, 2000). The plant is reported to have caused poisoning in grazing animals in India and new South-Wales. *Cathranthus roseus* is not recommended for oral use due to possible side effects. Frequent and prolonged doses of *Catharanthus* alkaloids in cancer therapy are known to cause platelet damage. Patients usually exhibit transient side effects. Side effects can include but may not necessarily be limited to the following:

- Blood in urine, vomit, or in bowel movement (black, tarry stools)
- Dizziness, loss of balance, or trouble in hearing
- Pain, burning, redness, or swelling where the dose is given
- Pain, numbness, or tingling in the hands or feet
- Shortness of breath, trouble breathing, lung spasms
- Trouble going to the bathroom (urinating)
- Unexplained fever, chills, sore throat, or sores in mouth
- Unusual bruising or bleeding
- Yellowing of the eyes and skin
- Constipation, appetite loss
- Hair loss, headache, sensitivity to sunlight
- Muscle or jaw pain, or pain in the area where the cancer is located
- Tiredness or weakness
- Leucopenia
- Nausea, vomiting, abdominal cramps,
- Dermatitis

Because of this, people with bacterial infections should not take this drug and people with neuromuscular disorders should steer clear of this drug if possible. Likewise, people with some forms of Charcot-Marie-Tooth syndrome should avoid vincristine. Pregnant women should definitely not take it, because it causes severe birth defects in animal tests shortly after the drug is administered. People occasionally develop secondary cancers if they receive the drug along with other anticancer drugs that are known to be carcinogens.

Chemistry and uses

The plant accumulates 130 mono-terpenoid indole alkaloids (TIAs) in different organs; some of the important alkaloids are cathranthine, lochnerine, vindoline and vindolinine. Blood sugar level is lowered by these TIAs in diabetes. Vincristine and vinblastine, have anticancer properties. Periwinkle also contains the alkaloids reserpine and serpentine, which are powerful tranquilizers. Other TIAs lower blood pressure, and act as hemostatis (arrest bleeding). It also contains certain flavonoids (Brun *et al.*, 1999).

In non malignant disease

The alkaloids VCR and VLB are also useful in the treatment of a variety of thrombocytopenic disorders.

As a cardio-vascular drug

Ajmalicine (raubasine) present in root is reported to improve cerebral circulation when given intra muscularly to humans suffering from cerebral sclerosis.

Anti-diabetic activity

The 'periwinkle tea' prepared from leaf decoction of *C. roseus* is used for curing diabetes. Hypoglycemic effects of a number of other indole alkaloids such as lochnerine, tetrahydroalstonine, and Vindolidine have also been revealed.

Anti-malarial activity

Oral administration of plant extracts in doses equivalent to 4.42 g/kg have been shown to be active against malarial parasite in Leghorn chicks.

Antihelminthic activity

The plant has been reported to possess vermifungal properties. Dried leaves, incorporated into the soil produce nematicidal and ovistatic effects.

Anti bacterial and pesticidal activities

Anti bacterial action against *Vibrio cholera, Mycobacterium pyrogenes var. aureus, Salmonella typhosa and spinach mosaic virus* have been revealed by *in vitro* studies. Insect growth inhibition and insect sterilant properties have also been reported in *Heliothus armigera, Sodoptera litura, Achoea janata and Gryllodes sigillatus.*

Other uses

Folkloric uses

The plants of *C. roseus* are associated with interesting histories of use as folk–core remedies in many tropical and subtropical countries.

TABLE 4 : Cancer therapy regimes in which indole alkaloids of *Catharanthus roseus* are an essential component

Treatment Name	Combination of drugs that formed the therapy	Concerned human diseases
MOPP	Nitrogen mustard, Vincristine, Procarbazine, Prednisone	Hodgedkin's, lymphomas
CHOP	Cyclophosphoamide, Daunomycin, Vincristine Prednisone	Non-Hodgedkin's, lymphoma
ABVD	Daunomycin, Bleomycin, Vinblastine, Dacarbazine	Hodgedkin's lymphoma
PVB	Cisplatinum, Bleomycin, Vinblastine	Testicular cancer
MACOP-B	Daunomycin, Bleomycin, Vinblastine	Lymphoma

MVAC	Daunomycin, Bleomycin, Vinblastine	Bladder cancer
COP	Cyclophophoamide, Vincristine, Prednisone	Lymphoma
POC	Procarbazine, Cyclophophoamide, Vincristine	Melanoma, Small cell lung cancer
VATH	Vinblastine, Daunomycin, Thio-tepa,Halotestin	Breast cancer
VP	Vinblastine / Vindesine, Cisplatinum	Non-small cell lung cancer

Semi-Syntheic Derivative of the Bisindole Alkaloids

Despite the low abundance of the bisindole alkaloids in the plants, sufficient material has been obtained to produce several semi-synthetic derivatives, aiming at a higher specificity (lower toxicity) and a different pharmacological profile. In this respect vinblastine has become a true lead compound. With respect to biological activity, the stereochemistry of the bisindole alkaloids is of great importance. For this, seventeen VLB analogs were studied by circular dichroism (Parish *et al.*, 1998). This method allowed the assignment of the C16′ configuration (C16′-S analogs are bioactive, C16′-R analogs are inactive). Furthermore, it was found empirically that a positive Cotton effect at 305 nm was associated with biological activity.

Several semi-synthetic derivatives of the bisindole alkaloids from *C. roseus* have been studied and some are marketed {For review see (Fahy, 2000)}:

Vindesine (Systematic name: 3-(aminocarbonyl)-Î4-deacetyl-3 de(methoxycarbonyl) vinacaleukoblastine, Edisine[R]) was introduced in 1980 and is used for the treatment of leukemia's, mammacarcinomas and myelomas (Leveque, 1996).

Vinorelbine (systematic name: 3′-4′-didehydro-4′-deoxy-C′-norvincaleuko-blastine, Navelbine[R]) is used against lung and mamma-carcinomas and was introduced in 1995. In this molecule the catharanthine moiety is modified yielding an eight-membered C′-ring in place of the naturally occurring nine membered C′-ring. This structure is more easily protonated with opening of the ring and creation of an ion that will react to produce a covalent reversible bond between vinorelbine and its target protein tubulin (Potier, 1989). It is now part of several phase II clinical trials (Orecchia *et al.*, 2002).

Vinzolidine has a substituted oxazolidine-dione ring at the 4′ position of the vindoline moiety. The beta-chloroethyl side chain increased the lipophilicity of the molecule. Vinzolidine showed a greater therapeutic activity in murine tumors than either VCR or VLB, but unpredictable and severe toxicity of oral dosing schedules had led to early closure of several studies.

Vinepidine (systematic name: 4′-deoxy-22-oxo-(4′alpha)-vincaleukoblastine) was nine times less potent that vinblastine in B16 melanome cells. In L-cells, vinblastine completely inhibited growth at 40 nM, whereas vincristine and vindesine caused about

25% inhibition and vinepdine was inactive and was not further evaluated (Jordan *et al.*, 1985).

Vinflunine (systematic name: 4′deoxy-20′, 20′-difluoro-C′-norvincaleukoblastine) was obtained from vinorelbine by using superacid chemistry (Fahy *et al.*, 1997). Two fluorine atoms were selectively introduced at the 20′ position and the 3′, 4′ double bond was reduced to obtain vinflunine. It showed a superior antitumor activity *in vivo* in preclinical tumor models compared to the other *Catharanthus* alkaloids (Krucznynski *et al.*, 1998) and multidrug resistant tumor cell lines showed generally less cross resistance to vinflunine compared to VLB, VCR and vinorelbine. The inhibitory effect on the polymerization of tubuline in microtubules was determined as IC50 of 1.99 mM as compared to 1.70 mm for vinorelbine. Vinflunine acts also as an inhibitor of tubulin assembly, but exhibited quantitatively different tubulin bindings properties compared with the classic dimers. The properties of this powerful derivative have been reviewed by Kruczynski and Hill (Krucznynski and Hill, 2001). A range of other halogenated derivatives has been synthesized, e.g. 20′-chloro-4′-deoxyvinblastine (Jacquesy *et al.*, 1994; Duflos *et al.*, 1998).

Vintripol, a C23 tryptophane and a-amino-phosphonate-derivative, did not show significant antitumor activity in various clinical trials.

Vinxaltine, a C23 valine-derative, with further addition of an a-amino-phosphonate leads to a stereospecific effect with the S-epimer having significant activity and the R-epimer being relatively inactive. *In vitro*, vinxaltine is 8-370 times more active against melanoma cells compared with VCR, VLB or vindesine (Budman, 1992).

Vinglycinate (synonym: vincaleukoblastine *O* (sup4)-deacetyl-, 4-ester with *N,N*-dimethylglycine) has significant antineoplastic activity, but was never marketed (Armstrong *et al.*, 1967).

KAR-2 (synonym: KAR-2,3″-(beta-chloroethyl)-2″,4″-dioxo-3,5″-spiroxazolidino-4-deacetoxyvinblatine) showed no anticalmodulin activity in contrast to VLB. By comparative studies it was shown that the spiro-oxazolidino ring and the substitution of a formyl group to a methyl one were responsible for this effect (Orosz *et al.*, 1999).

Various other derivatives have been synthesized, which find themselves in early stages of drug development. Chemical conjugates of the bisindole alkaloids have been synthesized, which comprise oligopeptides, having amino acid sequences that are selectively proteolytically cleaved. These compounds are claimed as used in the treatment of prostate cancer and benign prostate hypertrophy (Brady *et al.*, 1999).

Biosynthetic pathway of indole alkaloids in *Catharanthus roseus*

The pathway for the synthesis of indole terpenoid alkaloids in *Catharanthus roseus* (See Fig. 14) has been worked out in considerable detail in terms of precursors, intermediates and products, localization of the enzymes involved in the process and identification and cloning of the concerned genes (See Fig. 15). A variety of biotic and abiotic stresses have been found to induce the biosynthesis and accumulation of alkaloids. Accordingly it was hypothesized that mutants of *C. roseus* tolerant / resistant to such abiotic stresses as heat and salinity may harbor constitutivity for alkaloid synthesis. Several of the salinity tolerant mutants of *C. roseus* isolated were actually found to accumulate larger than normal quantities of commercially important alkaloids.

The biosynthesis of secondary metabolites is usually discussed by starting with the description of the metabolites involved for example as in Figs. 14 and 15 where the early and late steps of alkaloid biosynthesis are presented. However these metabolites are in fact the end products of a complex process comprising the involvement of several enzymes, genes, regulatory genes and (transport through) intra-and inter-cellular compartments. The TIAs are condensation products of two biosynthetic routes, requiring coordination of the amount of intermediates supplied by both pathways. Starting from the amino acid tryptophan and the monoterpenoid geraniol, the biosynthesis of vinblastine requires the participation of at least 35 intermediates, 30 enzymes, 3 biosynthetic, 2 regulatory genes and 7 intra and intercellular compartments. Currently, the regulatory genes and (sub) cellular compartment are of particular interest, at which these process take place, which require transport of both intermediates and enzymes (targeting). The produced (secondary) metabolites are stored or either directly available for their biological function or, as in case of phytoanticipins, require a final processing for displaying their action, e.g., hydrolysis by a glucosidase.

Probably the first study on the biosynthesis of the *Catharanthus* alkaloids was performed at the end of the 1950s (Noble, 1990). Periwinkle plants were grown in an atmosphere containing $^{14}CO_2$ and after extraction of the alkaloids, column and paper chromatography, many labeled alkaloids were detected of which VLB was only a minor constituent. Since then many studies on TIA biosynthesis have been performed, allowing some detailed understanding of this, process. For practical reasons, quite a few of these studies have been performed using cell cultures. These have the advantage that they are easy to manipulate, both physiologically (rapid growth, ease of precursor feeding, etc.) and genetically. Major drawbacks of such system are that the cells normally do not accumulate vindoline (and thus no bisindole alkaloids are formed) and the limited differentiation of the cells (for example, no functional chloroplast are developed). Another difference is that while the alkaloid production in the plant seems to be constitutive, in cell cultures is inducible by various treatments. This latter property is useful in allowing a more chronological and controlled study of the biosynthesis. Here we will attempt to describe this induction process after treating cultured *C. roseus* cells with an elicitor.

BIOTECHNOLOGICAL RESEARCH ON *CATHARANTHUS ROSEUS* ALKALOIDS:

Introduction

The biotechnological part focuses on the alternative production systems for these alkaloids, for example by *in vitro* culture of *C. roseus* cells. Thus, it can be determined that to what extent the alkaloid biosynthesis pathway can be manipulated genetically ("metabolic engineering"), aiming at higher production levels of the alkaloids. Another approach is to produce the alkaloids (or their precursors) in other organisms such as yeast.

Vinblastine (VLB) and vincristine (VCR) are powerful anticancer drugs. During the last 40 years they have been used for the treatment and cure of thousands of patients, both because of their unique mode of action and their effectiveness. VLB and VCR are natural products, belonging to the group of terpenoid-indole alkaloids. They are present only as minor constituents of the complex mixture of about 130 alkaloids produced by this plant. VLB and VCR were discovered as a result of a drug screening program, performed in the late 1950's.

Due to the low levels in the plant (it takes about 500 kg of dried periwinkle leaves to isolate 1 g of VLB (Noble, 1990) and due to the large number of alkaloids present, the isolation of VLB and VCR is laborious and costly. Despite the difficulties in supply and the efficacy of the alkaloid itself, VLB has developed as a true lead compound, such that, over the years, many derivatives have been synthesized with different pharmacological profiles.

VLB and VCR are thus costly fine chemicals, which are produced in relative small volumes per year. These properties made the alkaloids excellent targets for production by biotechnological means. In the early 1980's it was shown that plant cells, as a suspension of single cells in a liquid medium, could quite readily be grown in bioreactors. Since then alkaloid producing *C. roseus* cultures have developed as a model system for the production of valuable compounds by large scale cultivation of plant cells. This made *C. roseus* one of the best studied medicinal plants.

In the late 1940's *C. roseus* was one of the species used to develop cell culture techniques using crown-gall tumors (Hildebrandt and Riker, 1949) and to unravel the processes involved in what we now know as *Agrobacterium* transformation (De Ropp, 1947). American researchers screened *C. roseus* for its antidiabetic use. Independently, *C. roseus* was also studied by a Canadian group. As a result, the anticancer alkaloids were discovered.

At present much is known about the biosynthesis of the alkaloids, despite the length and the complexity of the pathway. Quite a few enzymes have been isolated and several genes are now available for metabolic engineering, allowing the creation of transgenic cells and plants with modified alkaloid profiles.

Various aspects of *C. roseus* and its alkaloids have been reviewed before, e.g. the discovery of the alkaloids (Svoboda, 1966; Noble, 1990), the pharmacology of the alkaloids (Creasy, 1994; McCormack, 1990), their phytopharmaceutical applications (Ross, 1998), the biosynthesis of the alkaloids (Verpoorte, 1997; De Luca, 2000; De Luca, 2001) and the production of the alkaloids by biotechnological means (Verpoorte, 1997; Van er Heijden, 1989; Moreno, 1995; Verpoorte, 1991; Verpoorte, 1998; Verpoorte, 1999; Verpoorte, 2000).

Biotechnology

C. roseus and is alkaloids have become an important model system in plant biotechnological research. The aim of this research was to produce the alkaloids by large scale penicillin by large scale culture of fungi. Despite the large scientific achievements, there is no commercial biotechnical production of these alkaloids yet. Technologically it was shown that plant cells (including *C. roseus*) can be cultured on large scale, but the levels of alkaloid production are too low to allow commercial production. For increasing the production two strategies are followed:

- Manipulation of *in vitro* cultures, which comprises screening and selection for high producing cell lines and optimization of culture conditions {reviewed in (Van der Heijdin, 1989; Moreno, 1995); Verpoorte, 1991)}.
- Genetic modification or metabolic engineering, which comprises, among others, the overexpression of biosynthetic rate limiting enzyme in alkaloid biosynthesis {reviewed in (Verpoorte, 1999; Verpoorte, 2000; Verpoorte, 2002; Hughes, 2002)}.

Recent development in both strategies are discussed below:

Alkaloid production by *in vitro* culture

For the characterization of growth and alkaloid production of *C. roseus* cell cultures, detailed protocols are available (Scragg, 1999). From the early work in the late 1940's it was already shown that *C. roseus* (gall) tissue was relatively amenable to culture. Of five species tested, it was notable that only *C. roseus* was able to grow on 0.5% of methanol, ethanol, butanol, mannitol or dulcitol as the sole carbon source (Hildebrandt, 1949). Later, in the 1970's, it was shown that various medium constituents affected alkaloid production. By optimizaton, alkaloid production and induction media were obtained and some patents for alkaloid production systems were also obtained {reviewed in (Van derHeijden, 1989; Moreno, 1995).

The effects of phytohormones in *C. roseus* cells are extensively studied by Rideau and co-workers. They found that cytokinins supplied exogenously to suspsension cultured cells enhanced alkaloid accumulation. Autonomy to plant growth regulators can be either obtained via habituation or through transformation with the isopentenyltransferase (*Ipt*) gene from *Agrobacterium tumefaciens*. However, none of the above mentioned treatment lead to higher alkaloid production in suspension cultured cells. By over expression of the *Ipt* gene, transgenic cotyledons overproducing cytokinins were obtained. The non-enhanced levels of alkaloids were explained by supra-optimal concentrations of the cytokinins or by the habituated nature of the transgenic tissue (Garnier, 1996).

Plant growth regulator autonomy does not lead to total insensitivity to exogenously applied regulators. Autonomy is associated with changes in the expression of several genes (Droual, 1998). On the protein level it was shown that endogenously produced cytokinin does not mimic the effect of exogenously applied cytokinin in *C. roseus* callus. By two dimensional electrophoresis protein expression pattern of the *Ipt* transgenic line was become insensitive to exogenous cytokinin treatment (Carpin, 1997). A 28 kD polypeptide was detected, the presence of which relates to alkaloid accumulation. Both the accumulation of the peptide and ajmalicine were induced by 2,4-dichlororphenoxyacetic acid (2,4-D) depletion and NaCl treatment and both were inhibited by gibberellic acid (Carpin, 1997; Ouelhazi, 1993).

In the biosynthesis of alkaloids, but likely also in their turnover, peroxidases play an important role. It was found that addition of 2,4-D to the culture medium reduced the peroxidases activity in the medium and cellular compartments. In particular, the synthesis of a basic isoform was reduced. Cytokinins seemed to affect the acidic isoforms. No information on the effects of alkaloid accumulation was given (Limam, 1998).

Optimization of other culture conditions as light, temperature, and aeration (shaker speed, closures, of flasks) resulted in increased alkaloid levels (Verpoorte, 1991). Elicitors; both biotic and abiotic, have successfully been applied to increase alkaloid production. In this category, MeJA as a secondary messenger is widely used. Endogenous elicitors can be released by digesting the plant cell walls with fungal enzyme preparation as macerozyme (Garnier, 1996). Macerozyme was also used for the treatment of *C. roseus* hairy roots. After treatment with 1% macerozyme, TDC and phenylalanine ammonialyase were induced.

Addition of an alginate monomer to a *C. roseus* cells culture promoted the production of ajmalicine and 5'-phosphodiesterase (an antibiotic protein).

Metabolic engineering of alkaloid biosynthesis:

In the process of studying the alkaloid biosynthesis in *C. roseus* quite a few biosynthetic enzymes have been purified, characterized and their encoding genes are cloned (Verpoorte, 1997). Those genes have been expressed heterologously in order to prove the identity of the gene, to produce the encoded enzyme, to modify the secondary metabolism and to create novel metabolites (metabolic engineering, pathway transplantation). The main 'aim of homologous' over expression of these genes is to increase the production of alkaloids.

Homologous expression of Biosynthetic enzymes:

For metabolic engineering it is a prerequisite to have functional protocols for gene transfer, including efficient tissue culture methods for regeneration. For *C. roseus* the latter has always been and still is, troublesome. Regeneration of (transgenic) cells to plantlets has only been performed with low efficiency. For this reason, few transgenic *C. roseus* plants have been produced, including the generation of hairy root culture, were described by Garnier (1999). Garnier used *A. Tumefacians* strains carrying the *lpt* gene resulting in enhanced levels of cytokinins in the alkaloids.

An efficient transformation system for *C. roseus* cell cultures was developed by Hilliou (1999). Using particle bombardment, transgenic calli were recovered at a frequency, with upto 90% of transgenic calli containing both plasmids. Within three months sufficient material could be obtained for molecular and biochemical analyses. For transforming cultured *C. roseus* cell lines by particle bombardment, the activities of several promoter derivatives fused to a *Gus* (b-bluduronidase) reported gene were compared. It was shown that the cauliflower mosaic virus 35 S promoter and a deletion derivative of the Figort mosaic virus were found most useful for metabolic engineering in *C. roseus* studies (Van der Fits, 1997). Transgenic plants were obtained via particle bombardment of *C. roseus* plants through organogenesis of buds. Histological studies showed that the gene insertion method was effective with many cells and different tissues, which was judged by the presence of the reporter gene signals GUS and green fluorescent protein (Zarate, 1999).

Besides particle bombardment, suspension cultured *C. roseus* cells can be efficiently transformed by a so called ternary transformation system (Van der Fits, 2000). Co-cultivation with *A. tumefaciens* strain LBA44040 supplemented with a constitutive *vir*G mutant gene on a compatible plasmid resulted in a high frequency of transformation.

Whitmer *et al.* performed a detailed study on several aspects of alkaloid formation in transgenic *C. roseus* cells overexpressing *Tdc* and/or *Sir.* Two transformation methods have been used: Cell cultures were initiated from seedling leaf cells by *A. tumefaciens* mediated transformation and by the above mentioned co-cultivation method using a wild type *C. roseus* cell culture. The transgenic cell lines obtained from seedling leaf cells showed a wide phenotypic diversity for TIA accumulation when transferred to an alkaloid production medium (Canel, 1998). Transgenic *Str* lines showed STR activities of 10-fold higher than wild-type cells. TIA accumulation was strongly influenced by culture conditions, such as the hormonal composition of the medium and the availability of precursors. It was found that the presence of Napthyl acetic acid (NAA) during the production phase led to lower levels of TIA accumulation (Whitmer, 1998). Highly productive cells lines showed significant instability even in the first year of culture. Instability was the result of new bottlenecks occurring in the biosynthetic pathway rather than the loss of the transgenic character (Whitmer, 1999). Loganin feeding to transgenic *Str* or *Tdc* cell lines efficiently increased alkaloid accumulation, the formed strictosidine being rapidly converted to other alkaloids. Under optimized experimental conditions, including precursor feeding, a transgenic *Tdc* cell line accumulated up to 1200 mmol/L of TIA, while the wild type cell line accumulated about 580 mmol/L (Whitmer, 2002). For selection of transgenic cell lines, kanamycin was shown to be unsuitable and hygromycin was toxic for wild types cells at a low concentration (Whitmer, 1998). By co-cultivation of a non-producing *C. roseus* cells suspension culture with *A. tumefaciens*, transgenic cell lines were obtained overexpressing *Tdc, Str*, or both. The transgenic cell lines did not accumulate alkaloids. Over expression of these enzymes is thus not sufficient to open the TIA pathway (Whitmer, 1999). Addition of XAD-7 as an artificial storage site to prevent catabolism significantly increased ajmalicine accumulation in transgenic cells when they were fed with tryptamine and loganin (Whitmer, 1999).

Heterologus expression of biosynthetic enzymes

Tdc and *Str* were the first genes cloned from the TIA biosynthesis pathway in *C. roseus*. Since then these genes have been expressed in cells of various species {reviewed in (Whitmer, 1999)}. Hallard *et al.*, (1997) demonstrated the functional expression of both genes in suspension cultured tobacco cells. After feeding of secologanin, strictosidin was found in the culture medium. Strictosidin was formed from tryptamine (produced by the transgenic enzyme TDC) and subsequent condensation with secologanin catalyzed by the transgenic enzyme STR. Strictosidine was not further converted to other alkaloids, indicating that tobacco does not possess any glucosidase activity which is able to hydrolyse strictosidine. The STR from this source was also used in an asay for the determination of secologanin contents in plant tissues (Hallard, 1998). For preparative purposes *Str* has been expressed in *E.coli* (Roessner, 1992) and insect cells (Kutchan, 1994). Later on, *Sgd* was also successfully inserted and expressed in suspension cultured tobacco cells (Zarate, 2001). The SGD protein in tobacco was present in the same high molecular weight complexes as in *C. roseus*.

The efficiency of the introduction of both *Str* and *Tdc* genes in tobacco plants was studied by Leech *et al.* (1998). The genes were introduced into tobacco leaves by particle bombardment on a single transforming plasmid. In total 150 transgenic plants were analyzed by southern blotting, which indicated a 100% co-integration of *Tdc* and *Str*. However, in 41% of the plants either *Tdc* or *Str* was silenced, in another 26% both genes were silenced, with the remaining 33% of the plants expressing both genes. From seeds of self-fertilized transformants, seedlings were grown. In seven weeks old seedlings, a variation of 24-and 110-fold in levels of TDC and STR activity, respectively, was found; the maximum STR activity was 494 pkat/mg protein.

C. roseus Tdc and *Str* cDNAs were also introduced into *Cinchona officinalis* "Ledgeriana" hairy root cultures. These cultures accumulated relatively high amounts of tryptamine, strictosidine, quinine and quinidine (1.2, 2.0, 0.5 and 1 mg/g DW, respectively). Unfortunately this production proved not to be a stable trait. STR extracted from *Cinchona* did not accept 5-methoxytryptamine as a substrate indicating the enzyme's transgenic nature, since STR from *Cinchona* does accept 5-methoxytryptamine as substrate (Geerlings, 1999). *Tdc* and *Str* cDNAs were also introduced in *Weigela* 'styriaca'. The resulting hairy root and cell suspension cultures were able to produce tryptamine and some alkaloids. Unlike the wild type hairy roots, no secologanin could be detected in the transgenic hairy roots, indicating its further conversion (Hallard, 2000).

Both *Str* and *Sgd* were expressed in yeast (*Saccharomyces cerevisae*). STR activity was found inside the cells as well as in the medium, SGD was only detected inside the cells. After feeding of tryptamine ans secologanin, strictosidine (major product) and cathenamine were produced. The transgenic yeast was able to grow on a juice obtained by extraction of snowberries (*Symphoricarpus albus*), a rich source of secologanin (Geerlings, 2001). This approach is an interesting alternative source for alkaloid production (Geerlings, 1998).

From above mentioned examples it is clear that stable heterologus expression of biosynthetic genes is still a matter of chance. Various aspects of stability and instability in transgene expression were reviewed by Meyer (1998).

Further applications of *Catharanthus*-derived materials

The research on *C. roseus* was initially focused on the development of tissue culture methods and in particular on the analysis and production of its alkaloids. This work resulted in the isolation and identification of many new compounds and had a large impact on phytochemistry and organic synthesis; many new derivatives and methods have been obtained. Next to this there has been a strong interest in the plant itself; initially mainly because of the biosynthetic purposes, but later in relation to many different areas because it offered so many interesting opportunities. For example *Tdc,* which was one of the first genes cloned from *C. roseus* is used for various purposes. The *Tdc* gene is used for improving the quality of the animal feed canola by redirection of tryptophan resulting in low levels of the undesired indole glucosinolates (Chavadej, 1994). Is also has been used for selection of transgenic plants cells (Godijn, 1993) and to form the insecticide tryptamine in *Petunia,* resulting in resistant transgenic plants (Thomas, 1999). Introduction of the *Tdc* in tobacco resulted also in resistance against the sweet potato whitefly (Thomas, 1995). After expression of *Tdc* in potato it was demonstrated that the synthesis and accumulation of phenolic compounds, including lignin, was reduced due to the limited availability of phenolic monomers. This provided evidence that synthesis and accumulation of phenolic compounds in potato can be regulated by altering the substrate availability through the introduction of a single gene outside the pathway involved in substrate supply (redirection of tryptophan into tryptamine) (Yao, 1995). The *Tdc* gene from *C. roseus* was patented (Brisson, 1998).

For another application of *C. roseus* derived materials in crop protection, a *Phytophthora* resistance gene from *C. roseus* was identified and patented (Bowman, 1996). Furthermore the gene coding isochorismate synthase from *C. roseus* was recently cloned (Van Tegelen, 1999). Overexpression of this gene in *C. roseus* was used to study the pathway of 2, 3-dihydroxybenzoic acid biosynthesis (Talou, 2001). This gene may also be used for the overproduction of salicylic acid in plants to enhance pathogen resistance, as it was previously done with the *emC* gene from *E. coli.* (Linthorst, 1999; Verberne, 2000).

Other applications of *C. roseus* include the use of the cytochrome C reductase to remove metal ions from fossil fuel (Monticello, 1997) and the use of cells for various bioconversion reactions (Hamada, 1993; Hamada, 1994), e.g. of hydroquinone to arbutin (Inomata, 1991). More than 100 cultivars of *C. roseus* are commercially available at present (Snoijer, 1998).

As a result of the research performed on the alkaloid biosynthesis, *C. roseus* developed as a model system for plant biotechnology and secondary metabolism research. As such, *C. roseus* cells were intensively studied for the biosynthesis of brassinolides (Fujioka, 2000) and of 2, 3-dihydroxybenzoic acid (Andayani, 2001), a compound accumulated in the culture medium after eliciting the cells with *Pythium aphanidermatum* elicitor preparation (Moreno, 1994). The proteome of *C. roseus* is also being investigated, methods for proteomic research have been developed (Jacobs, 2001) and analysis of protein patterns in connection with alkaloid accumulation are currently being performed. Applications of proteomics on plant secondary metabolism have been reviewed by Jacobs *et al.* (Jacobs, 2000).

From the examples mentioned above it may be concluded that *C. roseus* is not only a rich source of alkaloids, but it also holds a great treasure of enzymes and genes. It may thus be expected that the number of applications will further increase over the coming years.

Isolation of indole alkaloids from *Catharanthus roseus* by centrifugal partition chromatography in the pH-zone refining mode

Centrifugal partition chromatography (CPC) in the pH-zone refining mode can be used for a preparative and efficient isolation of vindoline, vindolinine, catharanthine and vincaleukoblastine from a crude mixture of *Catharanthus roseus* alkaloids. The separation protocol was first of all tested with a synthetic mixture of vindoline, catharanthine and vincaleukoblastine. The fraction content was analyzed and the results compared with theoretical chromatograms obtained by numerical simulation. The increase in injected sample mass results in an improvement of the purity of the isolated compounds. This observation, confirmed by theory, is of prime importance for the development of preparative pH-zone refining CPC as a preparative separation method.

Catharanthus roseus G. Don (Apocynaceae) is a plant that produces indole alkaloids used in anticancer chemotherapy. The preparation and the purification of these compounds have been intensively studied (Sapi and Massiot, 1994). Aerial parts of the plant contain between 0.2 and 1% of a mixture of about 90 different alkaloids. The most abundant ones are the monomers like catharanthine and vindoline, bearing the indole and the indoline chromophore respectively (See Fig.12). The dimers resulting from the coupling of two such compounds present interesting pharmaceutical activities. Anticancer drugs like leurocristine and vincaleukoblastine (or vinblastine or VLB) belong to this category. Industrial-scale production of these molecules proceeds in two different ways. The first one follows the classical extraction and isolation scheme. This approach is hampered by the complexity of crude extracts and by the low proportion of valuable dimeric compounds. The other way resorts to the hemisynthesis or the biomimetic coupling reaction that uses monomers as starting material. Both methods require the development of an efficient, selective, and preparative chromatographic technique able to isolate dimmers and monomers from complex extraction mixtures. This goal is achieved by means of Centrifugal Partition Chromatography (CPC) when used in the pH-zone refining mode (Shinomiya *et al.*, 1995).

CPC is a liquid-liquid partition chromatographic method generally referred to as counter-current chromatography (CCC) (Foucault, 1994) even though there is, strictly speaking, no counter-current between the liquid phases. One is mobile and the other one is kept stationary inside of the column by a constant centrifugal field. The column itself is a rotor built as a disk stack in which partition cells are engraved. The CPC technique has the ability to separate fragile compounds because there is no irreversible adsorption or chemical reaction of the injected material with a solid support (Foucault and L. Chevolot, 1998). Moreover, there is generally no problem due to saturation of the solid-state phase, allowing CPC to be efficient in preparative separations (Renault *et al.*, 1997) A high selectivity is obtained by a careful choice of the biphasic solvent system, thus allowing the separation of compounds with very similar structures (Renault *et al.*, 1997). The pH-zone-refining mode in CPC was introduced as a variant of displacement chromatography (Weisz *et al.*, 1994). It is devoted to the purification of compounds whose electric charge depends on pH value. For example, a mixture of free bases is injected in the organic stationary phase along with a base stronger than all the compounds to separate. This stronger base is called the retainer base. The bases are moved along the column by pumping through an acidic aqueous solution of the displacer. Pure products are then isolated in the effluent as salts. The column is used in the descending mode, assuming that the density of the aqueous mobile phase is greater than the density of the organic stationary phase. In the ascending mode, base salts and an acidic retainer introduced in the aqueous stationary phase can be displaced by a base mixed to the mobile organic phase. Separation of organic acids or of their salts proceeds in a similar way. A characteristic of displacement chromatography is the liberation of the products from the column by contiguous blocks arranged according to their p*K*a values and also partition coefficients. In a block, the concentration of the different species (and so the pH) are constant and fixed by the acid-base and repartition equilibrium constants. The overlap between blocks is minimum and does not vary with the amount of injected sample.

The separation of alkaloids by means of pH-zone-refining CPC concerns, to date, only mixtures with a limited number of components (Ma *et al.*, 1994). The complexity of the crude alkaloid extract from *C. roseus* merited some preliminary studies. This work presents a successful application of pH-zone-refining CPC to the preparative purification of *C. roseus* alkaloids. A computer program designed for the simulation of the chromatographic process correctly interprets the experimental chromatograms. It predicts the positive effect of mass overloading on the purity of the separated products. This effect is experimentally observed, and is of prime importance for the development of pH-zone-refining CPC as a preparative purification method.

MOLECULAR BIOLOGY RESEARCH ON *CATHARANTHUS*

Work is also being carried out on a mapping population of F_2 plants of a cross made between accessions (e.g. Pink Delhi (Pd) and *gsr-8* (Id)). Differentiating polymorphic bands generated by RAPD, ISSR markers and functional genes, which are present in one or the other parent and absent in the counterpart, are examined in plants for their presence or absence. These markers are further analyzed in populations for their segregation and linkage analysis.

The overall idea of the work is to apply several complimentary approaches to reveal as yet unidentified genes, loci and sequences that determine the yield of economically useful alkaloids including serpentine, catharanthine, vindoline, vincristine and vinblastine in *C. roseus*. These are synthesized down stream of strictosidine in the TIA pathway.

The TIA pathway of *Catharanthus roseus* is highly complex. VBL is synthesized as condensation product of V and C, which are themselves synthesized via independent routes from the intermediate strictosidine, which itself is a condensation product of precursors synthesized via extensions of shikimate pathway and non-mevalonate terpenoid pathway. Three dozen intermediates, 30 enzymes, 30 biosynthetic-and 2 regulatory-genes and 7 intra-and inter-cellular compartments relating to the TIA biosynthesis have been identified. TIA production has been observed to be inducible by biotic and abiotic stresses and a couple of signal transduction pathways have been investigated. Apparently, TIA biosynthesis is far from being understood in terms of all the intermediates, enzymes, genes and regulatory mechanisms.

A detailed distributional study of alkaloids in the organs has shown that serpentine and catharanthine are accumulated in the roots, serpentine in stem, catharanthine, vindoline, vinblastine in leaves and flowers. The alkaloids accumulated in plant organs are the net stored amounts after the amount synthesized have undergone catabolism and transfer. The genes of the functions involved in these processes are yet to be marked such that these could be tracked in the breeding populations.

The contents of TIA in an organ is a quantitative character determine by many genes which can be identified in the terms of QTLs on the genetic map of the *Catharanthus roseus*. QTL analyses has been carried out for a variety of characters including metabolite yields and a number of plant systems, e.g. Capsaecine content in Capsicum species, oil content in groundnuts and sugar contents in sugar beets. Linkage maps well saturated with molecular markers have served as tools, for locating genes determining yield/ agronomic characters and for marker-assisted pyramiding of best alleles of the genes isolated from different genetic resources, in genotype improvement of a variety of crops.

The molecular markers such as restriction fragment length polymorphism (RFLP), random amplified polymorphic DNA (RAPD), simple sequence repeats (SSR) and amplified fragment length polymorphism (AFLP) have been deployed extensively to estimate recombination frequencies or linkage values between them and measurable or qualitative phenotypic traits, in the mapping populations such as F_2, double haploids and recombinant inbred lines (RILs) derived from crosses. In these studies, molecular markers effectively converted quantitative traits into a series of Medelian inherited genetic elements that could be used to select individuals with optimum combination of alleles. The bulk segregant analysis has been often used in rapid identification of molecular markers associated with major genes (genes with large effect on the expression of a measurable trait) controlling agronomic traits. The molecular maps have also provided for cloning of specific genes in diverse crop plants.

Analysis of genetic diversity and mapping of alkaloid content for marker assisted selection (MAS) in *Catharanthus roseus*:

Catharanthus roseus (Madagascar periwinkle) species, widely grown in homes and gardens, is the only source of the indispensable anticancer drugs, vincristine and vinblastine. Efforts are on to develop its pharmaceutical varieties using a number of approaches. However, not much work has been done exploiting DNA based molecular markers. Hence work is in progress for:

1. Analysis of genetic diversity in *C. roseus*
2. Construction of a genetic linkage map of *C. roseus*
3. Tagging of alkaloid content and other traits in *C. roseus*

Molecular markers are most useful for understanding the naturally occurring polymorphisms in living organisms. Markers which are most useful as indicators of diversity include those which have high polymorphic information content such as RAPD (randomly amplified polymorphic DNA) and AFLP (amplified fragment length polymorphism). Hence *C. roseus* genotypes are being collected from different geographical locations and being analyzed using RAPD, AFLP and STMS markers. The data obtained will be scored and computed to obtain dendrograms for phylogenetic analysis.

STMS markers, the most useful co-dominant markers for mapping and tagging of important traits are being developed and used. Various strategies involving the construction of genomic libraries and the use of degenerate primers are being used for cloning and characterization of microsatellite sequences from *C. roseus*. The putative microsatellite recombinants are being sequenced and the data is being used to design STMS primers for use in the analysis of genetic polymorphism in a large collection of *C. roseus* accessions. Meanwhile crosses between the contrasting *C. roseus* parents are being attempted in order to generate the segregating mapping population. The segregating population will be analysed for their alkaloid content and also with different molecular markers and the data will be used for linkage analysis and map construction.

Use of Molecular markers

Although DNA sequencing is a straight forward approach for identifying variations at a locus, it is expensive and laborious. A wide variety of techniques have, therefore, been developed in the past few years for visualizing DNA sequence polymorphism. The term DNA-fingerprinting was introduced for the first time by Alec Jeffrey in 1985 to describe bar-code-like DNA fragment patterns generated by multilocus probes after electrophoretic separation of genomic DNA fragments. The emerging patterns make up an unique feature of the analysed individual and are currently considered to be the ultimate tool for biological individualization.

The term DNA fingerprinting/profiling has been used to describe the combined use of several single locus detection systems and are being used as versatile tools for investigating various aspects of plant genomes. These include characterization of genetic variability, genome fingerprinting, genome mapping, gene localization, analysis of genome evolution, population genetics, taxonomy, plant breeding, and diagnostics.

Anonymous DNA markers are generated by a wide variety of techniques, differing greatly in their reliability (repeatability and robustness), difficulty, expense, and the nature of the polymorphism that they detect. Because of these differences, they also vary greatly in their suitability for various uses. They may be hybridization based (e.g., RFLP), or polymerase chain reaction (PCR) based (e.g., RAPD and AFLP); microsatellite or simple sequence repeat polymorphism (SSRP), cleavable amplified polymorphic sequences (CAPS); they may detect single locus or multiple locus differences, and the markers detected may be inherited in a presence/absence, dominant, or co-dominant manner.

Genetic maps have been constructed in many crop plants using these markers on a single segregating population e.g Rice, Pea, Wheat, Brassica. Each molecular marker has its advantages and drawbacks. Therefore application of molecular marker techniques to diversify questions must take into account whether or not the data derived from a technique provides the right type of information for answering the question being addressed (Karp *et al.*, 1997). The use of these molecular markers has also simplified the research on *Catharanthus roseus.*

Desirable properties of molecular markers

1. It must be polymorphic as it is the polymorphism that is measured for genetic diversity studies.
2. It should have co-dominant inheritance that is the different forms of a marker should be detectable in homo-and hetero-zygotes and these highly variable loci with codominant alleles should be scorable.
3. A marker should be evenly and frequently distributed throughout the genome.
4. It should be easy, fast and cheap to detect.
5. It should be reproducible.

It is extremely difficult to find a molecular marker which would me*et al*l the above criteria. Depending on the type of study to be undertaken, a marker system can be identified to fulfill at least a few of the above characteristics (Weising *et al.*, 1995).

Kinds of molecular markers

1) Non – PCR based detection:

Restriction fragment length polymorphism (RFLP).

2) PCR – based detection:

Such markers include random amplified polymorphic DNA [RAPD), microsatellites or simple sequence repeat polymorphism (SSRP), amplified fragment length polymorphism (AFLP), arbitrarily primed PCR (AP – PCR), etc.

3) Targeted PCR and sequencing:

This category include sequence tagged sites (STS), sequence characterized amplified region (SCARS) etc.

1. *Non PCR based marker detection*

RESTRICTION FRAGMENT LENGTH POLYMORPHISM (RFLP)

RFLPs (See Fig. 16) are simply inherited naturally occurring Mendelian characters. They have their origin in the DNA rearrangements that occur due to evolutionary processes, point mutations within the restriction enzyme recognition site sequences, insertions or deletions within the fragments, and unequal crossing over (Schlotterer *et al.*, 1992).

This hybridization-based technique requires use of a library of DNA fragments cloned into some vector. These fragments may be from the species under study or from related (even distantly related) species. The library may be based on genomic or cDNA. RFLP does not require sequencing. The DNA of the organism under study is digested with one or more restriction endonucleases, the resulting fragments separated electrophoretically according to size, and probed with DNA clones from the library (See Fig.17). Fragments matching the probe DNA are visualized by autoradiography (Southern, 1975) or the use of fluorescent labeling techniques.

The genomic libraries are easy to construct and almost all sequence types are included; however, a large number of interspersed repeats are found in inserts that detect a large number of restriction fragments forming complex patterns. In plants, this problem is overcome to some extent by using methylation-sensitive restriction enzyme *Pst*I. This helps to obtain low copy DNA sequences of small fragment sizes, which are preferred in RFLP analysis (Figdore *et al.*, 1988; Liu *et al.*, 1990; Miller *et al.*, 1990).

On the other hand cDNA libraries are difficult to construct, however, they are more popular as actual genes are analysed and they contain fewer repeat sequences. The selection of appropriate source for RFLP probe varies, with the requirement of particular application under consideration. Though genomic library probes may exhibit greater variability than gene probes from cDNA libraries, a few studies reveal the converse (Miller *et al.*, 1990). This observation may be because cDNA probes not only detect variation in coding regions of the corresponding genes but also regions flanking genes and introns of the gene.

RFLP markers were used for the first time in the construction of genetic maps by Botstein *et al.* (1980). Most RFLP markers are inherited in a co-dominant manner. RFLP markers are especially appropriate in comparative mapping studies. RFLP markers are however labor intensive and time consuming, so newer approaches, which are relatively simple, quick, and PCR based are now developed.

Advantages :-

1. It permits direct identification of genotype or cultivars, using any tissue at any development stage in environment independent manner.
2. These are co-dominant markers enabling identification of homo-and hetero-zygotes.
3. It is used in indirect selection of conventional gene using qualitative trait.
4. It is used in tagging of monogenic trait.

Limitations :-

1. Requires large amount of highly pure DNA.
2. Requires constant good supply of probes that can reliably detect variations.
3. It is laborious and expensive.
4. It is time consuming.
5. Requires expertise in autoradiography.

2. *PCR-based marker detection*

PCR-based molecular markers are well suited for questions requiring large scale surveys of plant and animal populations.

Random amplified polymorphic DNA (RAPD)

In 1991 Welsh and McClelland (Welsh *et al.*, 1991) developed a new PCR-based genetic assay namely RAPD. RAPD is an acronym for Random Amplified Polymorphic DNA, it utilises short (10 bp) primers of random sequence to amplify random sections of the genome (See Fig. 19). The oilgonucleotide serves as both forward and reverse primer and usually is able to amplify fragments from 3-10 genomic sites simultaneously. As these markers are randomly placed they can be used to make the assay useful for efficient screening of nucleotide sequence polymorphism between individuals (Tingey *et al.*, 1993). This will help to form phylogenetic relationships between species or populations using standard tree building programs.

For this is required to run RAPD-PCRs (using the same primer) for each of the samples on a gel (1.5-2.0% agarose gel is fine) and score the banding pattern, using a presence/absence score for each band position. It is noticed that over a range of species there will be some bands that are conserved and some that are unique to the species or even individual. If relationships between species have to be determined then it has to be determined which bands are conserved between the species and which are due to variation at the population level.

These polymorphisms are considered to be primarily due to variation in temperature and primer-annealing sites. Polymorphisms could be caused by differences in nucleotide sequences at the priming sites (such as point mutations), or by structural rearrangements within the amplified sequence, (e.g., insertions, deletions, inversions) (Welsh and McClelland, 1990).

RAPDs have been used for many purposes, ranging from studies at the individual level (e.g. genetic identity) to studies involving closely related species. Due to their very high genomic abundance RAPDs have also been applied in gene mapping studies.

Random amplified polymorphic DNA can provide simple and reproducible fingerprints of germplasm by employing single, arbitrary chosen primers (Welsh and McClelland, 1990). RAPD markers can detect a large number of genetic polymorphism and when linked to major genes can be potentially useful in identifying morphological traits (Williams *et al.*, 1990).

They can also be used in monitoring diversity within plant populations, for constructing linkage maps and for tracking hybrid species' origins. They are dominant markers and hence have limitations in their use as markers for mapping, which can be overcome to some extent by selecting those markers that are linked in coupling (Williams *et al.*, 1993).

Advantages:-

1. Nanogram amount of DNA is required.
2. It is based on such markers that are random distributed and abundant throughout the genome.
3. Generation of multiple bands per reaction.
4. No prior knowledge of the target sequences is required for the design of primers.
5. Simple experimental set up is required.
6. Provides quick and efficient screening for DNA sequence, and the entire genome is screened.
7. Amenable to automation.
8. Does not involve blotting or hybridization steps as results are directly read from agarose gels.
9. It is easy, less costly and quick to assay.

Applications:-

1. Construction of genetic maps.
2. Indirect selection in segregating populations during plant breeding programs.
3. Analysis of genetic structure of populations.
4. Fingerprinting of individuals.

5. Targeting markers to specific regions of genome.
6. Identification of somatic hybrids.
7. Evaluation and characterization of genetic resources.

Limitations:-

1. RAPD polymorphisms are inherited as dominant recessive characters.
2. RAPD primers are relatively short, a mismatch of even a single nucleotide can often prevent the primer from annealing; hence there is loss of band.
3. The production of non-parental bands in the offspring of known pedigrees warrants its use with caution and extreme care.
4. RAPD is sensitive to changes in PCR conditions, resulting in changes to some of the amplified fragments.

AMPLIFIED FRAGMENT LENGTH POLYMORPHISM (AFLP)

A recent approach by Zabeau *et al.* (1993) known as AFLP, is a technique based on the detection of genomic restriction fragments by PCR amplification and can be used for DNAs of any origin or complexity. The fingerprints are produced, without any prior knowledge of sequence, using a limited set of generic primers.

The number of fragments detected in a single reaction can be 'tuned' by selection of specific primer sets. AFLP technique is reliable since stringent reaction conditions are used for primer annealing. This technique thus shows an ingenious combination of RFLP and PCR techniques (Ehrlich *et al.*, 1991; Saiki *et al.*, 1988) and is extremely useful in detection of polymorphism between closely related genotypes.

Inter-simple sequence repeats (ISSRs)

A marker system called Inter-Simple Sequence Repeats (ISSRs) has only recently been developed as an anonymous, RAPDs-like approach that accesses variation in the numerous microsatellite regions dispersed throughout the various genomes (particularly the nuclear genome).

The term microsatellites was coined by Litt and Lutty (Litt and Lutty, 1989). Microsatellites are very short (usually 10-20 base-pair) stretches of DNA that are "hypervariable", expressed as different variants within populations and among different species.

These loci contain tandem repeats that vary in the number of repeat units between genotypes and are referred to as variable number of tandem repeats VNTRs i.e. a single locus that contains variable number of tandem repeats between individuals (Nakamura *et al.*, 1987) or hyper variable regions HVRs i.e. numerous loci containing tandem repeats within a genome generating high levels of polymorphism between individuals. The primers used can be either unanchored (Gupta *et al.*, 1994, Wu *et al.*, 1994) or anchored at the 3′ or 5′ end with 1 to 4 degenerate bases extended into the following sequences (Zietkiewicz *et al.*, 1994).

ISSR's are characterized by mono-, di-or trinucleotide repeats, e.g., AA..., or AG..., CAG..., that have 4-10 repeat units side-by-side. In ISSRs, we specifically target the di- and trinucleotide repeat types of microsatellite, because these are characteristic of the nuclear genome.

Microsatellites thus form an ideal marker system creating complex banding patterns by simultaneously detecting multiple DNA loci. Some of the prominent features of these markers are that they are dominant fingerprinting markers and co dominant STMS (sequence tagged micro satellites) markers. Many alleles exist in a population, the level of heterozygosity is high and they follow Mendelian inheritance.

The ISSR primers we use to generate the variation in a given DNA sample include one of these highly variable microsatellite sequences and an arbitrary pair of bases at the 3′ (rear) end. One samples for variation among DNA samples in small PCR (polymerase chain reaction) reactions using one primer at a time. Where the primer successfully locates two microsatellite regions within an amplifiable distance away on the DNA strands of the DNA sample (the "template DNA"), the PCR reaction will generate a band of a particular size (=molecular weight) for that "locus" representing the intervening stretch of DNA between the microsatellites. Usually several to many such "paired" microsatellite areas exist in a particular DNA sample, so one gets that many bands generated in the reaction, for that sample. Studies of ISSR locus heritability have demonstrated an exceedingly close approximation to classic Mendelian ratios (Tsumura *et al.* 1997).

Extremely high variability and high "mapping density" as compared with RFLP and RAPD data make these new dominant, microsatellite-based molecular markers ideal for producing genetic maps of individual species (Nagaoka and Ogihara, 1997). These features, combined with greater robustness in repeatability experiments and less prone to changing band patterns with changes in constituent or DNA template concentrations, make them superior to other readily available marker systems in investigations of genetic variation among very closely related individuals and in crop cultivar classification (Fang and Roose, 1997; Nagaoka and Ogihara, 1997). ISSR techniques are nearly identical to RAPD techniques except that ISSR primer sequences are designed from microsatellite regions and the annealing temperatures used are higher than those used for RAPD markers.

II. Targeted PCR and sequence

Sequence tagged sites (STS)

STS is a short unique sequence amplified by PCR, which identifies a known location on a chromosome. RFLP probes specifically linked to a desired trait can be converted into PCR-based STS markers based on nucleotide sequence of the probe giving polymormhic band pattern, to obtain specific amplicon. Using this technique, tedious hybridization procedures involved in RFLP analysis can be overcome.

This approach is extremely useful for studying the relationship between various species (Bustos *et al.*, 1999). When these markers are linked to some specific traits, for example in Barley the powdery mildew resistance gene (Hartl *et al.*, 1993) or stem rust resistance gene (Oh *et al.*, 1994), these can be easily integrated into plant breeding programs for marker-assisted selection of the trait of interest. Examples of STSs are:

- Sequence characterized amplified regions (SCARs)
- Cleaved amplified polymorphic sequence (CAPS)
- Expressed sequence tag markers

Sequence characterized amplified regions (SCARs)

Michelmore *et al.* (1991) and Martin *et al.* (1991) introduced this technique. A SCAR is a genomic DNA fragment at a single genetically defined locus that is defined by PCR amplification using a pair of oligonucleotide primers. Williams *et al.* (1990) converted RFLP markers into SCARS by sequencing two ends of genomic DNA clones and designing oligonucleotides primers based on the end sequences.

These primers were used directly on genomic DNA in a PCR reaction to amplify the polymorphic regions. If no amplified fragment length polymorphism is noticed then the PCR fragments can be subjected to restriction digestion to detect RFLPs with in the amplified fragments (Rafalski *et al.*, 1993). SCARs are inherited in a co dominant fashion in contrast to RAPDs, which are inherited in a dominant manner.

Cleaved amplified polymorphic sequence (CAPS)

These polymorphic patterns are generated by restriction enzyme digestion of PCR products. Such digests are compared for their differential migration during electrophoresis (Jarvis, 1994). PCR primer for this process can be synthesized based on the sequence information available in databank of genomic or cDNA sequences or cloned RAPD bands. These markers are co-dominant in nature. This is also sometimes known as PCR-RFLP.

Konieczny and Asubel (1993) first adapted the CAPS procedure for genetic mapping by developing a set of CAPS markers for use with *Arabidopsis*. They designed 18 sets of primers, which amplified products ranging in size from 0.316 to 1.728 kb from Columbia and Landsberg ecotypes These amplified products were digested with a panel of restriction endonuclease to identify restriction enzymes that generated ecotype specific patterns.

Advantages:-

1. The sizes of the cleaved and uncleaved amplification products can be adjusted arbitrarily by the appropriate placement of the PCR primers.

2. CAPS markers can be readily assayed using standard agarose gel electrophoresis.
3. CAPS markers are co-dominant
4. A small quantity of DNA is required to assay CAPS markers.
5. CAPS mapping procedure is technically simple but the results are robust because an amplification product is always obtained.
6. CAPS markers can be assayed relatively quickly.

Expressed sequence tag markers (EST)

This term was introduced by Adams *et al.* (1991). Such markers are obtained by partial sequencing of random cDNA clones. Once generated, they are useful in cloning specific genes of interest and synteny mapping of functional genes in various related organisms. ESTs are popularly used in full genome sequencing and mapping programmes underway for a number of organisms and for identifying active genes thus helping in identification of diagnostic markers. Moreover, an EST that appears to be unique helps to isolate new genes. EST markers are identified to a large extent for rice, *Arabidopsis*, etc. wherein thousands of functional cDNA clones are being converted in to EST markers (Sasaki, 1994; Cooke *et al.*, 1996).

TABLE 4 : Comparative view of properties of RAPD, RFLP, AFLP, STS

Characteristic	RAPD	RFLP	AFLP	Sequence-Tagged Micro satellites
Principle	DNA amplification	Restriction digestion	DNA amplification	DNA amplification
Detection	DNA staining	Southernblotting	DNA staining	DNA staining
DNA required	Low	High	Medium	Low quantity
DNA required	Crude	Relatively pure	Relatively pure	Crude quality
Primer requirement	Yes (random primer)	None	Yes	Yes
Probe requirement	None	Set of specific	None probe	None
Use of radioisotope	No	Yes	Yes-No	Yes-No
Part of genome surveyed	Whole genome	Generally low copy coding region	Whole	Whole
Dominant /	Dominant Co dominant	Co dominant	Dominant	Co dominant (co dominant)

Polymorphism	Medium	Medium	Medium	High
Automation	Yes		Yes	Yes
Reliability	Intermediate	High	High	High
Recurring cost	Low	High	Medium	Low

APPLICATION OF MOLECULAR BIOLOGY TOOLS IN *CATHARANTHUS*

Catharanthus roseus on account of its features is seen as a model plant for genetic investigations. In the course of this research, fingerprinting of F_2 progeny of a cross differentiated by PCR amplification using many RAPD and ISSR primers is performed. The procedure followed is presented below:-

1. Selection of 150 F_2 generation plants was done.
2. Isolation of DNA from the 150 samples and their quantification on 0.8% agarose gel was done.
3. PCR amplification of DNA preparations using various primers and their analysis on 1.2% agarose gel, 3% Metaphor gel or polyacrylamide gel was performed.
4. Scoring and analysis of data was obtained.

DNA isolation

A. Selection of plants

Leaf samples were taken from 150 F_2 generation plants for DNA isolation.

B. Isolation of DNA from leaf samples.

Principle:

Plants tissues of *Catharanthus roseus* often contain large amounts of carbohydrates and alkaloids that are not removed by phenol extraction. CTAB method is the method of choice for obtaining good quality total DNA from many plant species. CTAB (Cetyltrimethylammonium bromide) is a cationic detergent which solubilizes membranes and forms an insoluble complex with nucleic acids. When CTAB is added to a plant cell extract the nucleic acid-CTAB complex precipitates, leaving carbohydrates, proteins and other components in the supernatant. After cell disruption and incubation with hot CTAB isolation buffer, proteins are extracted by chloroform:isoamyl alcohol. The precipitate is then collected by centrifugation and resuspended in 5 M NaCI, which causes the complex to break down. The DNA pellet resulting after centrifugation is washed, dried and redissolved. The nucleic acids are now concentrated by ethanol precipitation and the RNA is removed by RNase treatment.

Polymerase chain reaction

PCR is an efficient means for *in vitro* amplification of specific DNA sequence by the simultaneous primer extension of the complementary strand of DNA. (a) PCR amplification of DNA preparation is carried out. The reaction mixture is prepared in PCR tubes kept on ice. (b) The conditions of various steps involved in PCR amplification for RAPD are:

FOR RAPDs :-

CYCLE 1(1x)..............

Step 1.............94° C for 1 min

Step 2.............36° C for 30 sec

Step 3.............72° C for 1 min

CYCLE 2(45x)..............

Step 1.............94° C for 5 sec

Step 2.............36° C for 15 sec

Step 3.............72° C for 1 min

CYCLE 3(1x)..............

............72° C for 7 min

CYCLE 4(1x)..................

...........................4° C for infinity

FOR ISSRs :-

CYCLE 1(1x)..............

Step 1.............94° C for 2 min

CYCLE 2(35x)..............

Step 1.............94° C for 20 sec

Step 2.............50° C for 50 sec

Step 3.............72° C for 1.30 min

CYCLE 3(1x)..............

.............72° C for 7 min

CYCLE 4(1x)..................

...........................4° C for infinity.

PCR reaction mixture composition varies for RAPDs and ISSRs. Certain GENE primers are also being used for mapping purpose. They follow the principle of comparative genomics as they are obtained from different plant genera like *Arabidopsis, Rice, Wheat, Pea* etc. Basically these are developed from the EST colonies. Some of them belong to the Shikimate pathway of alkaloid synthesis in *Catharanthus*. The sequences of these GENEs are first used for Primer designing and then sent for Primer synthesis.

Principle of DNA amplification:

The polymerase chain reaction is a technique for quickly "cloning" a particular piece of DNA in the test tube (rather than in living cells like *E. coli*). Thanks to this procedure, one can make virtually unlimited copies of a single DNA molecule even though it is initially present in a mixture containing many different DNA molecules (Saiki *et al.*, 1985, 1988; Mullis & Faloona, 1987). It is a method for oligo-nucleotide primer directed enzymatic amplification of a specific DNA sequence of interest. This technique is capable of amplifying a sequence 105 to 106-fold from nanogram amounts of template DNA within a large background of irrelevant sequences (e.g. from total genomic DNA). Because both strands are copied during PCR, there is an exponential increase of the number of copies of the gene. Suppose there is only one copy of the wanted gene before the cycling starts, after one cycle, there will be 2 copies, after two cycles, there will be 4 copies, three cycles will result in 8 copies and so on.

A prerequisite for amplifying a sequence using PCR is to have known, unique sequences flanking the segment of DNA to be obtained. These Primers are short oligonucleotides (containing about two dozen nucleotides) that are precisely complementary to the sequence at the 3' end of each strand of the DNA you wish to amplify. It is not necessary to know anything about the intervening sequence between the primers.

The DNA sample is heated to separate its strands and mixed with the primers. If the primers find their complementary sequences in the DNA, they bind to them. Synthesis begins (as always 5' to 3') using the original strand as the template. The reaction mixture contains all four deoxynucleotide triphosphates-dATP, dCTP, dGTP, dTTP and a heat-stable DNA polymerase from *Thermus aquaticus-Taq* DNA polymerase and using an automated thermal cycler (BIORAD icycler) to put the reaction through 30 or more cycles of denaturing, annealing of primers, and polymerization. Polymerization continues until each newly-synthesized strand has proceeded far enough to contain the site recognized by the other primer. Two DNA molecules identical to the original molecule are formed. These two molecules are heated to separate their strands, and the process is repeated. Each cycle doubles the number of DNA molecules. After amplification by PCR, the products are separated by agarose gel electrophoresis and are directly visualized after staining with ethidium bromide.

The cycling reactions:

There are three major steps in a PCR, which are repeated for 30 or 40 cycles. This is done in an automated cycler, which can heat and cool the tubes with the reaction mixture in a very short time.

1. Denaturation:

The first step in PCR amplification system is the thermal denaturation of the DNA sample by raising the temperature within a reaction tube to 90°C. During the denaturation, the double strand melts open to single stranded DNA, all enzymatic reactions stop (for example: the extension from a previous cycle). In addition to the source DNA, this reaction tube contains a vast molar excess of the two-oligonucleotide primers, a thermostable DNA polymerase (e.g., *Taq* DNA polymerase), and four deoxyribonucleotides. The temperature is maintained for about one min.

2. Annealing:

The temperature to the mixture is slowly cooled to approximately 55°C. The primers are jiggling around, caused by the Brownian motion. Ionic bonds are constantly formed and broken between the single stranded primer and the single stranded template. The more stable bonds last a little bit longer (primers that fit exactly) and on that little piece of double stranded DNA (template and primer), the polymerase can attach and starts copying the template. Once there are a few bases built in, the ionic bond is so strong between the template and the primer, that it does not break anymore.

3. Extension:

In the third step, the temperature is raised to about 75°C, which is optimum for the catalytic functioning of *Taq* DNA polymerase. The primers, where there are a few bases built in, already have a stronger ionic attraction to the template than the forces breaking these attractions. Primers that are on positions with no exact match, get loose again (because of the higher temperature) and don't give an extension of the fragment. The bases (complementary to the template) are coupled to the primer on the 3' side (the polymerase adds dNTP's from 5' to 3', reading the template from 3' to 5' side, bases are added complementary to the template). Using automated equipment, each cycle of replication can be completed in less than 5 minutes. After 30 cycles, what began as a single molecule of DNA has been amplified into more than a billion copies ($2^{30} = 1.02 \times 10^9$).

Analysis of PCR product

Agarose gel electrophoresis is an easy way to separate DNA fragments by their sizes and visualize them. In this method, DNA is forced to migrate through a highly cross-linked agarose matrix in response to an electric current. There are three factors that affect migration rate through a gel; size of the DNA, conformation of the DNA, and ionic strength of the running buffer.

Electrophoresis:

. The technique of electrophoresis is based on the fact that DNA is negatively charged at neutral pH due to its phosphate backbone. For this reason, when an electrical potential is placed on the DNA it will move toward the positive pole:

The rate at which the DNA will move toward the positive pole is slowed by making the DNA move through an agarose gel. There is a buffer solution (which maintains the proper pH and salt concentration) with 0.75% to 2.0% agarose added. Here agarose forms a porous lattice in the buffer solution and the DNA must slip through the holes in the lattice in order to move toward the positive pole. This slows the molecule down. Larger molecules will be slowed down more than smaller molecules, since the smaller molecules can fit through the holes easier. As a result, a mixture of large and small fragments of DNA that has been run through an agarose gel will be separated by size. (See Fig. 21)

Gel containing 0.8% agarose is used for separating DNA molecules in the range of 0.5 to 10 kb. 0.3% gels will separate dsDNA molecules between 5 and 60 kb size, whereas 2% gels are used for samples between 0.1 to 3 kb (Fig. 22) Since agarose gels separate DNA according to the size, the relative mobility of DNA fragments may be determined from its electrophoretic mobility by running a standard DNA marker of known relative mobility on the same gel. The DNA is visualized in the gel by addition of ethidium bromide, which fluorescence under UV light (Fig. 23).

The finished gel, after having been fully processed can be viewed for results under a UV light. Hopefully a wide range of bands or fragments of DNA can be seen at varying places down the gel. These can then be identified, distances calculated and the results used in a myriad of ways from mapping a gene to helping decide which species is the more viable to keep and try to protect.

RAPD and ISSR analysis and identification of markers:

DNA fingerprinting using randomly selected decamer oligonucleotides and ISSRs is carried out on Agarose Gel Electrophoresis to resolve polymorphic bands. A set of primers, previously selected for giving good amplification pattern, are tested in F_2 progeny of a cross. A DNA fragment was determined as a marker only when the following were satisfied:

- The fragment is found to be polymorphic between the two parents and shows segregation in F_2 population.
- It was amplified in sufficient abundance to be easily discriminated.

- It was clearly distinguishable from all other DNA fragments.
- Its amplification was repeatable in PCR reaction. Marker data was scored as 1 or A (band presence) or 0 or B (band absence) jointly with the parental origin of the marker.

Primers used in the analysis are selected according to following conditions:

1. Discernible band patterns
2. Rich polymorphism
3. Stable repeatability

These primers are used in for estimating the linkage analysis. The average rate of polymorphism is found out. Different primers produced various number of bands. Therefore, the different rates of polymorphic loci no. is seen in segregants of *Catharanthus roseus* with different primers.

INTRODUTION TO GENETIC MAPPING IN *CATHARANTHUS*

In plants three different genomes exist within most cells: nuclear, chloroplast and mitochondrial. The presence of these genomes does not interfere with the genetic analysis of the most molecular loci. In deriving a linkage map for particular species, a mapping population which will produce segregation at the loci is produced. The choice of an appropriate mapping population depends on the type of marker systems employed. Maximum genetic information is obtained from completely classified F_2 population using a codominant marker system.

The next step is to determine if the segregation at each locus is Mendelian. According to the prediction of Mendel's first law, alleles have no permanent effect on one another when present in the same plant, but segregate unchanged by passing randomly into different gametes.

In order to determine if the locus is segregating as expected, the observed segregation is compared with the expected. The statistical test use for this is the chi-square test and is calculated below.

$$X^2 = \frac{(\text{Observed} - \text{Expected})^2}{\text{Expected}}$$

Three important reservations or cautions to be noted regarding the use or calculation of x^2 values in genetics are:

1. Chi-squares can usually be applied only to numeric frequencies and not to percentages or ratios derived from frequencies.
2. Chi-squares cannot normally be applied to distributions in which the number of individuals in any class is less than 5.

3. Since in most mapping experiments the segregation ratios for many loci in the same set of individuals are calculated, multiple tests must be corrected.

After one or more markers have been checked and the genotypes determined, the next step is to determine if any of the markers are linked. Mendel's second law concerns the independent assortment of genes. When two homozygous parents are crossed, two classes of individuals will be obtained in the next generation: parental and recombinant. There are many types of genetic markers which can be used for mapping genes, such as morphological markers, isozyme markers, restriction fragment length polymorphisms (RFLPs), microsatellites or simple sequence repeats (SSRs), random amplified polymorphic DNA (RAPDs) and amplified fragment length polymorphisms (AFLPs).

Marker data are categorical data which can be classified in different categories and recorded in numerical form, such as 1 or 0 for denoting presence or absence of a particular band at a particular marker or for distinguishing the two marker genotypes (homozygote and heterozygote) for a backcross population from two in bred lines. A Genetic map is prepared on the basis of recombination data between genetic markers. Genetic map construction requires calculation of pair wise recombination frequencies, establishment of linkage groups and estimation of map distances and determination of map order. Common software package used for map construction are Linkage 1, GMendel, Mapmaker, Map manager, and Join map.

Tissue culture advances in *Catharanthus roseus*

C. roseus has a no. of features which makes it highly suitable for serving as a model plant for functional genomic analysis especially with respect to secondary metabolites. The root, shoot, leaf and anthers of *C. roseus* have been brought under tissue culture to develop calliclones to assess their alkaloid status. It is the only source of valuable anti-cancer drug, vincristine (VCR) and vinblastine (VLB).

CULTIVATION

C. roseus plant has been widely studied for its capacity to produce secondary metabolites in cell cultures. The root, shoot, leaf and anthers of *C. roseus* have been brought under tissue culture to develop calliclones to assess their alkaloid status. Presence of large amounts of auxins and gibberellins in the medium were inhibitory to alkaloid accumulation in the culture. Modifications in medium composition and addition of certain enzymes and elicitors have been shown to enhance alkaloid content of transformed root cultures. Exogenous application of growth regulators such as IAA, 2,4-D, GA show contrasting results. Application of IAA and GA3 resulted in an increased number of leave, plant height, induction of apical dominance and parthenocarpy.

Plant tissue culture for production of vinca alkaloids

The dimeric indole alkaloids, vinblastine and vincristine have become highly valued drugs in cancer chemotherapy due to their potent antitumor activity against various leukemias, Hodgkin's disease and solid tumors. They are currently produced commercially by extraction from *Catharanthus roseus* (Apocyanaceae) plants, but the process is not efficient because of very low concentrations of the alkaloids in the plant. It was reported that the concentration of both vinblastine and vincristine was only 0.0005% as a dry weight basis.

In order to produce these useful anticancer drugs much more efficiently, many scientists have tried to apply plant tissue culture technology. In fact, a large number of papers related to this approach have been presented since the first research carried out by Carew *et al.* in 1966. However, production of both alkaloids by *de novo* synthesis using the callus or the suspension cultured cells of *C. roseus* is so far not promising because the productivity of the cultured cells reported was so far very low.

Misawa and his colleagues of Allelix Inc. in Canada studied on production of vinblastine by an alternative way in collaboration with Kurz of the National Research Council of Canada and Kutney of University of British Columbia, and established an economically feasible process consisting of production of catharanthine by plant cell fermentation and a simple chemical or an enzymatic coupling.

The vinblastine molecule is derived from two monomeric alkaloids, catharanthine and vindoline as shown in (See Fig. 12).

The concentration of vindoline in the intact *C. roseus* plant is approximately 0.2% as a dry weight basis, which is much a higher level than catharanthine, and the cost of vindoline is less expensive compared to catharanthine and vinblastine. The Allelix group, therefore, investigated the production of catharanthine by a cell suspension culture process with a selected *C. roseus* cell line induced from anthers on Gamborg's B5 medium containing 2% sucrose, 1 mg/L 2,4-D and 0.1 mg/L kinetin. The cells were grown in 250 ml flasks containing 60 ml of MS liquid medium supplemented with 3% sucrose, 1 mg/L NAA and 0.1 mg/L kinetin under continuous diffuse light on a rotary shaker (250 r.p.m.) at 25° C. In experiments for optimization of catharanthine production, they transferred 7 day old cells to a test medium and subcultured for 3 passages. In the 4th passage, 60 ml cultures were harvested in triplicate after 2 or 3 weeks growth, and the cell mass and alkaloid content were determined.

The results showed that the MS medium was the most favourable for catharanthine production but the optimal levels of phytohormones for the growth and the production were varied in different cell lines. For example, one line required no phytohormones but another line required 0.1 mg/L NAA and 0.1 mg/L kinetin. Addition of various chemically defined compounds to the medium as "inducers" was found to stimulate the production efficiently. Among them effects of vanadyl sulphate, abscisic acid and NaCl on the production of catharanthine were significant. Based on the conditions optimized by using flasks, Smart *et al.*, scaled up the cultures to 10, 30 and 100 L-air lift fermentors. When abscisic acid was added to the culture as an elicitor on the 7th day of cultivation, the final titer of catharanthine was raised to 85 mg/L in a 30 L fermentor.

The second stage in this project, the Allelix's group tried to couple enzymatically or chemically catharanthine produced by the cell culture process with commercially available vindoline. As an enzyme source for the coupling, a crude preparation obtained by 70% ammonium sulphate precipitation from the cultured cells of *C. roseus* was used. The reaction mixture containing both monomeric alkaloids, Tris buffer (pH 7.0) and the enzyme preparation was incubated at 30° C and for 3 hours. It was determined that the enzyme reaction gave various dimeric alkaloids including vinamidine, 3-(R)-hydroxyvinamidine and 3'4'-anhydrovinblastine. Leurosine and catharine, oxidized derivatives of anhydrovinblastine, were also detected in the early stages of the incubation. They found that $MnCl_2$ and either FAD or FMN stimulated the coupling. Although neither vinblastine nor vincristine was detected in the mixture, it was recognized that a substantial amount of anhydrovinblastine was formed as a major coupling product when an excess amount of sodium borohydride was added to the mixture after incubation.

In order to investigate properties of the coupling enzyme(s) it was partially purified with gel filtration and isoelectric focusing and five isozymes were obtained by Endo *et al.* One of them had MW 15,000 and the other four had the same MW (37,000). All of these isozymes were shown to have peroxidase activity. Using the partially purified enzymes, anhydrovinblastine was formed with a conversion yield of about 50%.

Formation of vinblastine from vincristine as detected by Goodbody *et al.* using a crude enzyme preparation obtained from suspension cultured cells of *C. roseus*. The highest yield of conversion obtained was 13% from 0.13 mg anhydrovinblastine in 1 ml of the reaction mixture after 3 hours incubation at 30° C, pH 7.0.

During the course of these studies on coupling mechanisms, they found that ferric ion catalyzed the coupling reaction significantly in the absence of the enzyme. It is of interest that the products of the chemical coupling were not only anhydrovinblastine but also vinblastine. The yields of both alkaloids were 52.8% and 12.3%, respectively after 3 hours incubation at 30° C, pH 7.0. These products including catharanthine were analyzed by high resolution mass spectrometry as further confirmation of their identification. Circular dichroism confirmed that coupling exists between the 2 monomeric units of both vinblastine and vincristine produced either enzymatically or chemically.

This is a novel and an efficient process to produce an antitumor drug, vinblastine, and is likely to be applied commercially. The technology was transferred from the

Canadian company to a Japanese company, Mitsui Petrochemicals Industry for further development.

Hara *et al.* of Mitsui Petrochemical could increase the yield of catharanthine up to 150 mg/L in the MS medium supplemented with 1 mg/L NAA and 0.1 mg/L kinetin using the best producing cell line isolated from Allelix's cell line. The stimulating activity of NaCl and KCl on alkaloid production was also confirmed. Furthermore, the scientists of Mitsui employed high-cell density cultures and reported yields of catharanthine of 230 mg/L/week.

The yield of vinblastine by the chemical coupling reaction was also improved by the same group; addition of ferric chloride, oxalate, maleate and sodium borohydrate stimulated the yield of vinblastine from anhydrovinblastine up to 50%.

Bede *et al.*, also investigated the production of anhydrovinblastine. They employed a two-enzyme system containing horseradish peroxidase and glucose oxidase to catalyze the formation of anhydrovinblastine from catharanthine and vindoline. Although peroxidase requires hydrogen peroxide for the coupling reaction, its presence in excess in the reaction mixture may inhibit the reaction. But addition of glucose oxidase was used to allow the controlled, continuous production of hydrogen peroxide at low levels, minimizing oxidative reactions. Both enzymes were immobilized on Euperight C beads, an oxirane matrix, and the system was shown to catalyze the coupling reaction.

Transformation of *Catharanthus roseus* using a Ri plasmid vector for hairy root generation

Catharanthus roseus is a tropical/subtropical plant of family Apocynaceae. The plant is of interest due to its production of useful terpenoid indole alkaloids, including vinblastine and vincristine, which are used for blood cancer treatments (Lounasmaa and Galambos 1989). Metabolic engineering of this species by genetic transformation may lead to enhancement of the production of specific alkaloid compounds at the whole-plant level. Production of transgenic plants of *C. roseus* has been reported (Zarate *et al.* 1999). However, no evidence of genetic transformation at the molecular level is available.

Agrobacterium rhizogenes causes production of transformed hairy roots at the infection site in plants. Hairy roots are characterized by high growth rate and genetic stability. Since the early stage of hairy-root research it has been known that in vitro cultures of hairy roots are able to regenerate whole plants in tobacco (David *et al.*, 1984). This result has also been reported in other species, including sweet potato (Otani *et al.*, 1993), *Ipomoea trichocarpa* (Ohtani *et al.*, 1996), *Brassica oleracea* and *B. campestris* (Christey *et al.*, 1997), cherry (Gutirrez-Pesce *et al.*, 1998), *Crotalaria juncea* (Ohara *et al.*, 2000), *Panax ginseng* (Yang and Choi, 2000), and spinach (Ishizaki *et al.*, 2002). Plants regenerated from hairy roots are nonchimeric because they originate from single cells (Bercetche *et al.*, 1987), a condition not necessarily expected from plants transformed by infection by *A. tumefaciens*. This study describes culture conditions for plant regeneration in hairy-root cultures of *Catharanthus roseus.*

Agrobacterium rhizogenes-mediated transformation has been used to obtain transgenic plants in 89 different taxa, representing 79 species from 55 genera and 27 families. A diverse range of dicotyledonous plant families is represented, including one Gymnosperm family. In addition to the Ri plasmid, over half these plants have been transformed with foreign genes, including agronomically useful traits. Plants regenerated from hairy roots often show altered plant morphology such as dwarfing, increased rooting, altered flowering, and wrinkled leaves and/or increased branching due to *rol* gene expression. These altered phenotypic features can have potential applications for plant improvement especially in the horticultural industry where such morphological alterations may be desirable. Use of *A. rhizogenes* and *rol* gene transformation has tremendous potential for genetic manipulation of plants and has been of particular benefit for improvement of ornamental and woody plants.

Gene Transfer Methods:

The methods of gene transfer can be broadly classified into two groups:

1. Physical delivery method or DNA mediated gene transfer (DMGT) mainly involving microinjection or macroinjection, microprojectile bombardment and electroporation.

2. Vector mediated gene transfer, mainly utilizing Ti and Ri plasmid of *Agrobacterium.*

In this method, *Agrobacterium* mediated gene transfer has been successfully utilized in majority of dicotyledonous crops and microprojectile bombardment has been used in majority of monocotyledonous crops.

The cell types that have been used for the transfer of gene included:

1. Cultured cell or protoplasts

2. Meristem cell from immature embryos shoots and flower meristem.

Use of Ti and Ri plasmid for transformation:

Recently, there has been increased interest in transformation of plants by *A. rhizogenes* and the influence of Ri (root inducing) plasmid T-DNA on plant cell development. The ease with which Ri T-DNA transformed plant cells 'hairy roots' of some plant species can be cultured and regenerated to whole plants suggest that an Ri vector system might particularly Be useful for studying foreign gene expression in transgenic plants of species in which whole plants can be regenerated from transformed roots.

Agrobacterium rhizogenes incites hairy root tumors when inoculated on susceptible dicotyledonous plants. The disease is characterized by the extensive proliferation of roots from the wound site, a characteristic that distinguishes hairy root tumors from the crown gall tumor incited by strains of *A. tumefaciens*. Tumorigenicity in *Agrobacterium tumefaciens* is a trait associated with the tumor inducing (Ti) plasmid.

The molecular basis of tumorigenecity in *A. tumefaciens* has been studied extensively. In crown gall tumor, a portion of the Ti plasmid is integrated into a plant DNA. The integrated plasmid DNA or T-DNA directs the synthesis of the tumor specific compounds octapine, nopaline and most likely agropine and enables the plant cell to grow in the absence of auxin and cytokinin, phytohormones required by normal plant cell to grow.

The basis for tumorigenicity in *A. rhizogenes* is not known. Direct examination for the presence of plasmid DNA sequence in hairy root tumor DNA has not been possible since no axenic hairy root tumor cell line has been established. Therefore it remains unknown whether hairy root tumorigenesis involves the integration of specific plasmid sequence into the plant genome.

Hairy root culture

Hairy root culture is a well-established experimental system and has provided many insights into root specific metabolism and its regulation. They have been used extensively for varied purposes like root nodule research, artificial seed production, production of plant secondary metabolites, as an experimental system to study response to chemicals.

1. Hairy root disease:

Hairy root disease is caused by gram-negative soil bacterium *A. rhizogenes*. One or both of the two pieces (TL and TR) of the Ri plasmid are transferred from bacteria into nuclear genome of the host plant. The transferred T-DNA causes the plant cell to proliferate extensively and formation of adventitious roots at or near the site of infection. The transformed plant tissues are also directed by T-DNA genes to produce unusal metabolites called opines, that serve as specific nutrients for the bacteria (Chilton *et al.*, 1982). The physiological basis of the hairy root disease is not totally understood. Alteration of auxin metabolism in transformed cells has been supposed to play an important role in expression of hairy root phenotype.

TL-DNA plays the major role in hairy root induction, and the genes encoding auxin synthesis have accessory role. Physiological studies have shown that the transformed cells are more sensitive to extracellular supplied auxins than the normal roots (Shen *et al.*, 1988).

2. Characterization of *A. rhizogenes* plasmid :

A. rhizogenes Ri plasmid have been categorized into three classes; agropine, mannopine, cucumopine depending on the opine synthesized and degraded. The *A.*

rhizogenes strains contain a single copy of large Ri plasmid. In the agropine Ri plasmid, the T-DNA is composed of two parts; left T-DNA (TL-DNA) and right T-DNA (TR-DNA) physically separated on the plasmid by a non-transferred DNA sequence of 18 Kb. TR-DNA genes are homologous to Ti plasmid tumor inducing genes. Genes involved in the agropine synthesis also located in the TR-DNA region. Synthesis of auxins can be ascribed to the TR-DNA. The characteristic phenotypic changes are predominately due to TL-DNA. Four loci (rol A, B, C, D) involved in hairy root formation have been identified in the TL-DNA by insertional mutagenesis (White *et al.*, 1985). Genes of Ri TL-DNA direct the synthesis of a substance that recruits the cell to differentiate into roots under influence of endogenous auxin synthesis.

3. The genes responsible for hairy root formation:

The T-DNA of agropine type Ri plasmid, consists of two separate T-DNA regions, TL-DNA and TR-DNA. Each of the T-DNA fragments span a 15-18 Kb region, and they are separated from each other by atleast 15 Kb of non-integrated plasmid DNA. These two T-DNA fragments can be integrated independently during the infection process (Vilaine *et al.*, 1987). The TL-DNA contains 18 ORF's including the *rol* A, *rol* B, *rol* C, and *rol* D genes which correspond to ORF's 10, 11, 12 and 15 respectively. The TR-DNA contains genes (*aux 1, aux 2, rol, BTR, mas1, mas2 and ags*) for the synthesis of agropine and biosynthesis of auxin. The *rol* genes are responsible for the observed phenotype of plant regenerated after Ri mediated transformation.

The function of *rol A* protein is unknown as its sequence shares no homology with other polypeptides (Levesque *et al.*, 1998). However, it is supposed to be regulated by auxin.

rol B protein has been shown to have tyrosine phosphatase activity which suggested that kinase/ phosphatase cascade plays a major role in the signal transduction of auxin. *rol B* gene has also been shown to encode for a β-glucosidase, which hydrolyses indole-β-glucosides, suggesting a role in the release of active auxins from inactive β-glucosides (Estruch *et al.*, 1991).

rol c has a cytokinin like action, it shows a morphological change such as diminished apical dominance, increased shoot number and reduced chlorophyll are characters associated with cytokinin action. It has been shown to encode cytokinin β-glucosidas, which is capable of hydrolyzing cytokinin β-glucosides increasing the level of free cytokinins.

Ri plasmid besides having TL and TR – DNA also contains *vir* region. However the mechanism of T-DNA integration into plant genome is partially understood but it closely resembles bacterial conjugation. T-DNA transfer and processing require products of the *vir* genes (*vir* A, B, G, C, D and E).

1.

2.

3.

4.

5.

6.

7.

8.

Fig. 1. *C. roseus* translates as rose colored pure flower.
2. Glossy, dark green, oval leaves of *C. Roseus*.
3. Natural wild blooms of *C. Roseus*.
4. Herbarium sheet.
5-8. Various cultivars of *Catharanthus*.

9.

10.

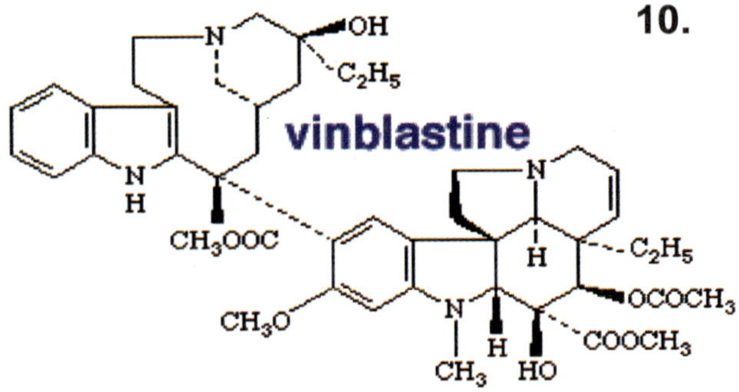

11.

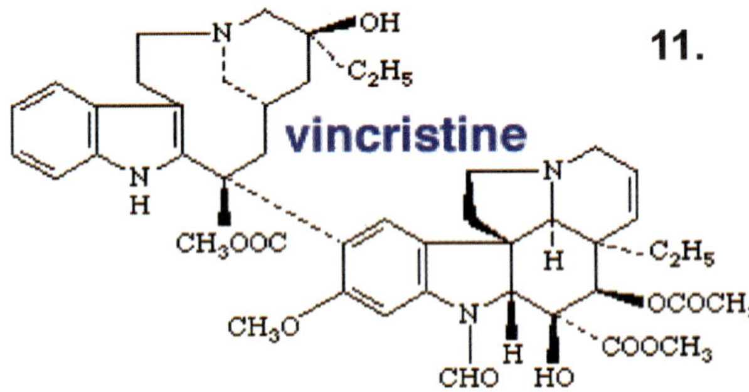

12.

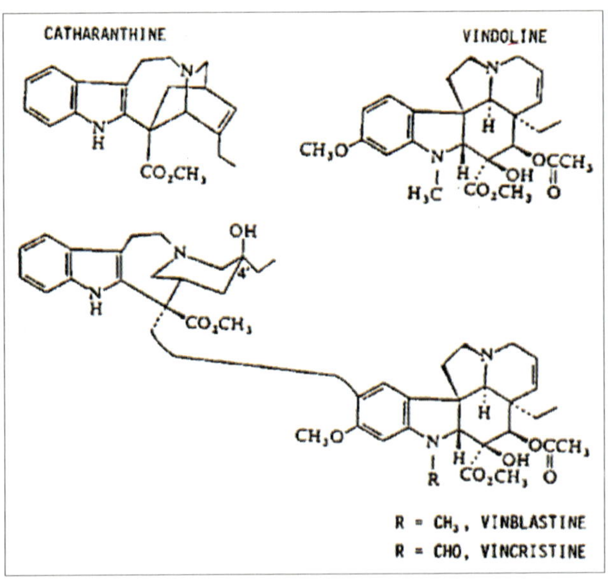

Fig. 9). Botanical characters of *C. roseus*; **10).** Vinblastine structure; **11).** Vincristine structure; **12).** Derivation of vinblastine and vincristine from catharanthine and vindoline.

13. **Anthocyanin aglyka**

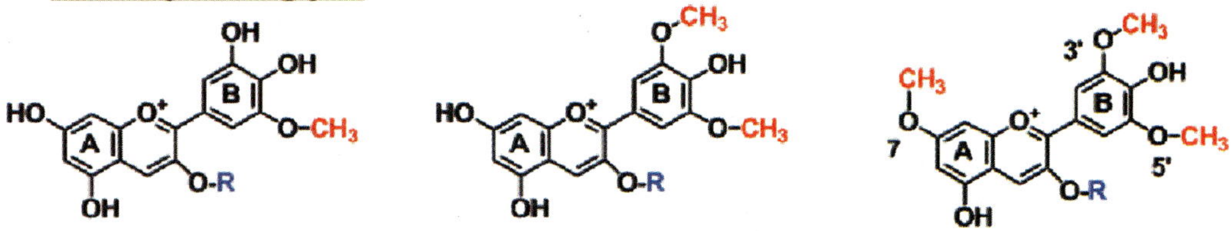

Petunidin Malvidin Hirsutidin

R = 3-O-(6-O-4-coumaroyl)glucoside (abundant)
R = 3-O-glucoside (trace to up to 10%)

Flavonol glycosides

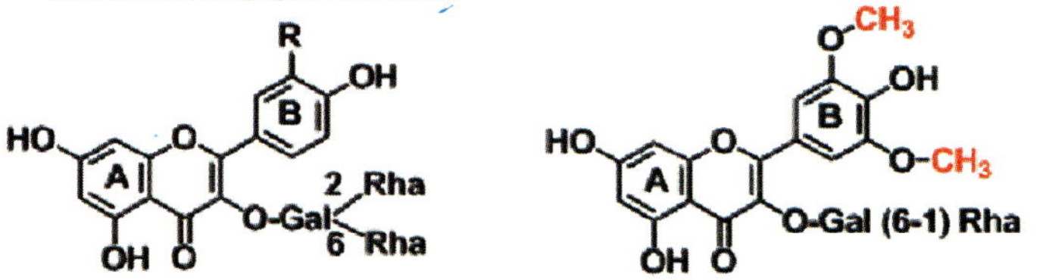

R = H backbone: Kaempferol
R = OH backbone: Quercetin

Syringetin-3-O-robinobioside

14.

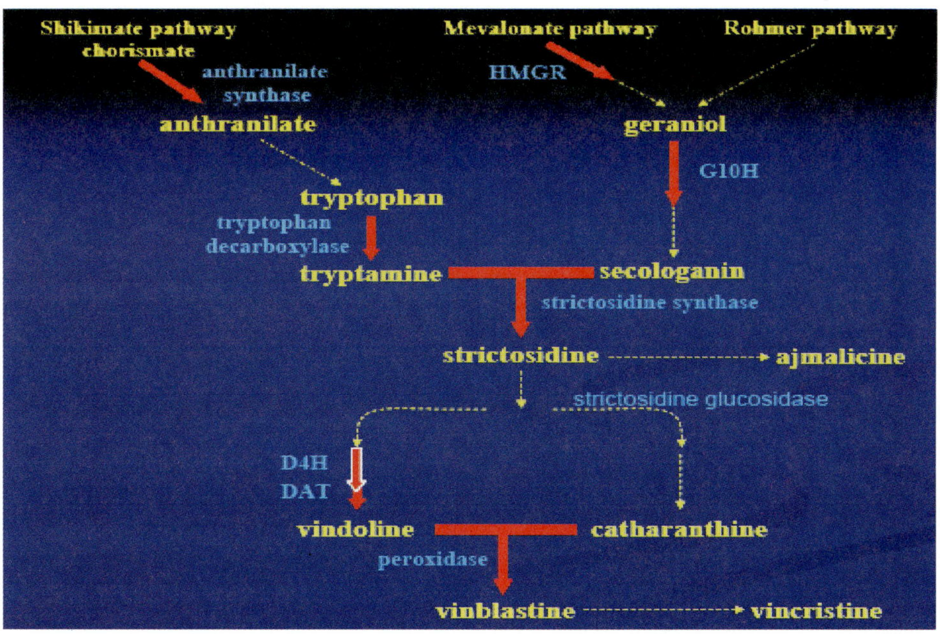

Fig. 13. Flavonoids in *Catharanthus*.

14. Pathway for the synthesis of indole terpenoid alkaloids.

15.

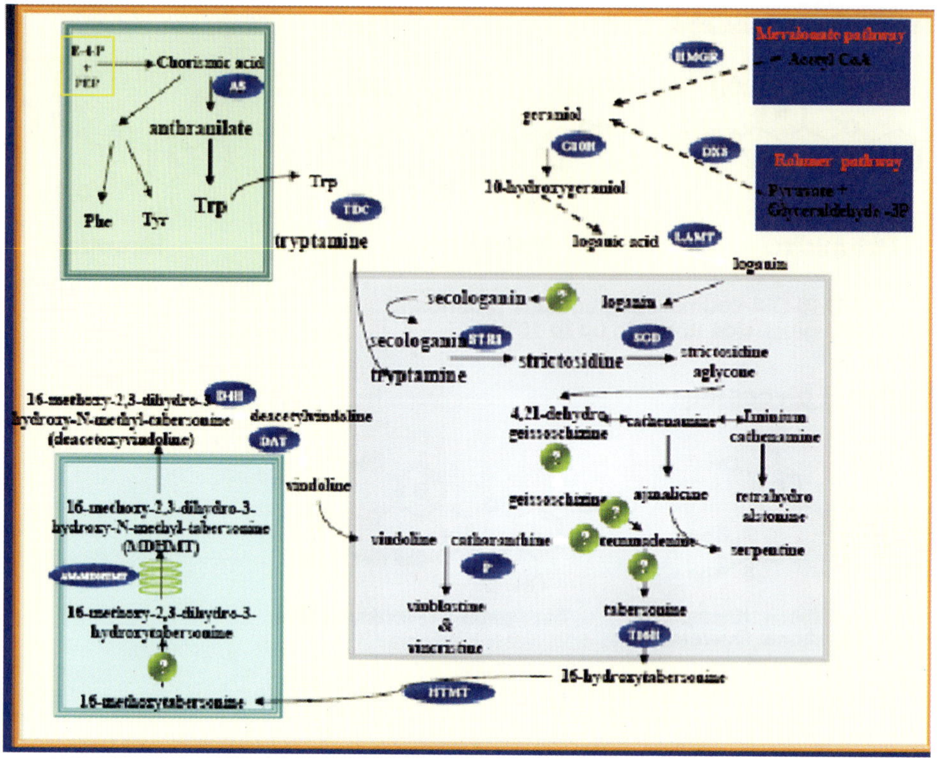

16.

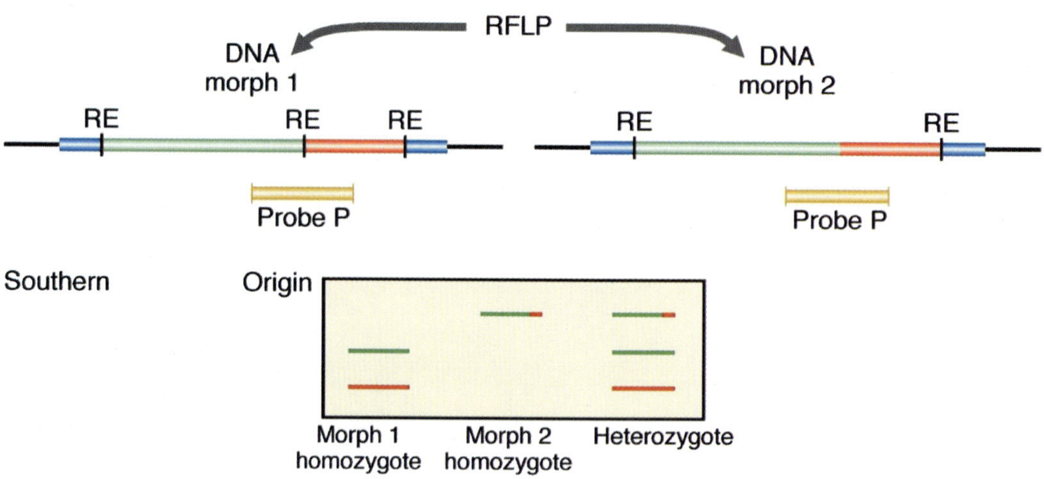

Fig. 15. Alkaloid biosynthesis in *C. Roseus*.

16. Restriction fragment length polymorphism (RFLP).

17.

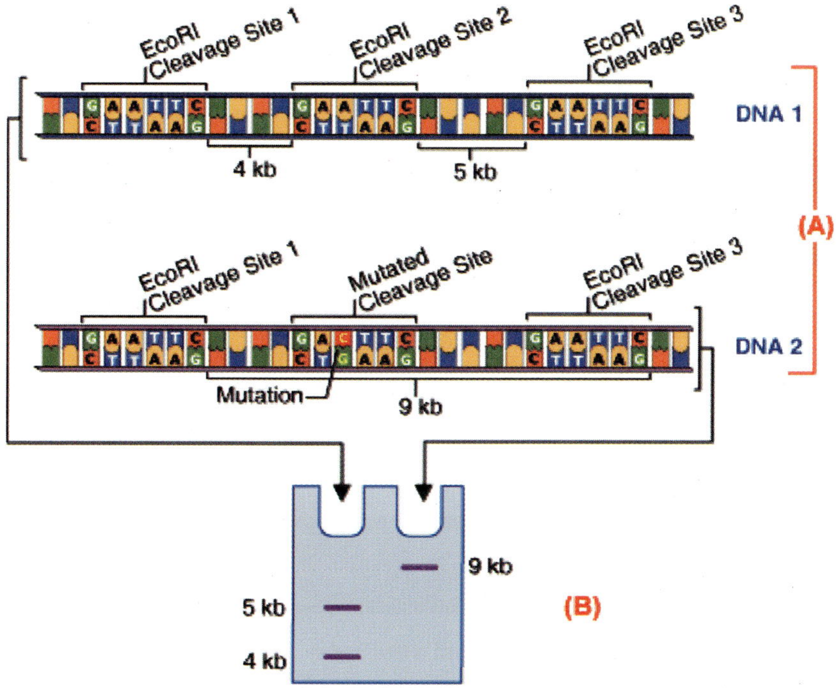

18.

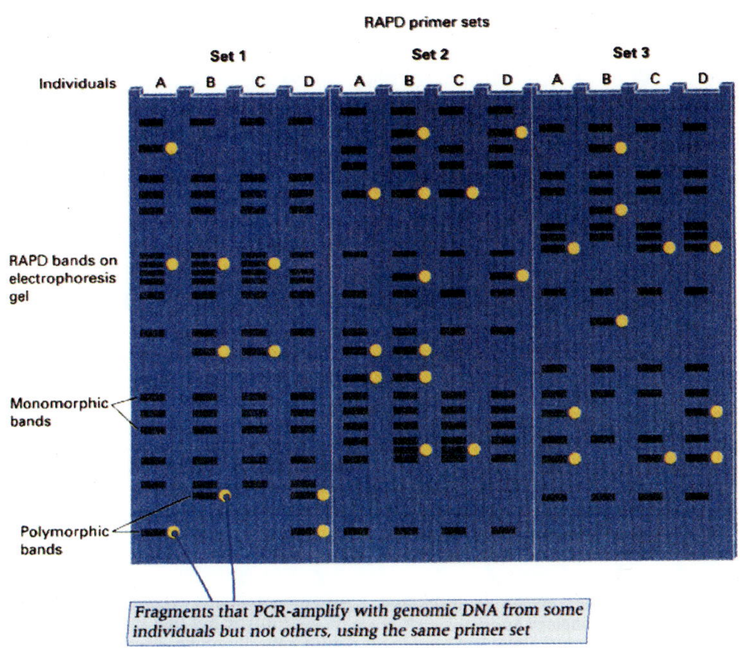

Fig. 17. Digestion of DNA with Restriction endonucleases & electrophoretic separation.
18. Random amplified polymorphic DNA (RAPD).

19.

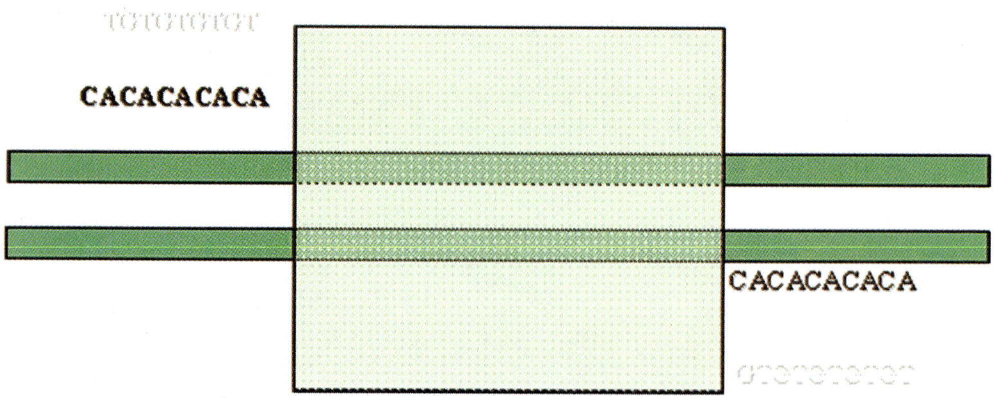

20.

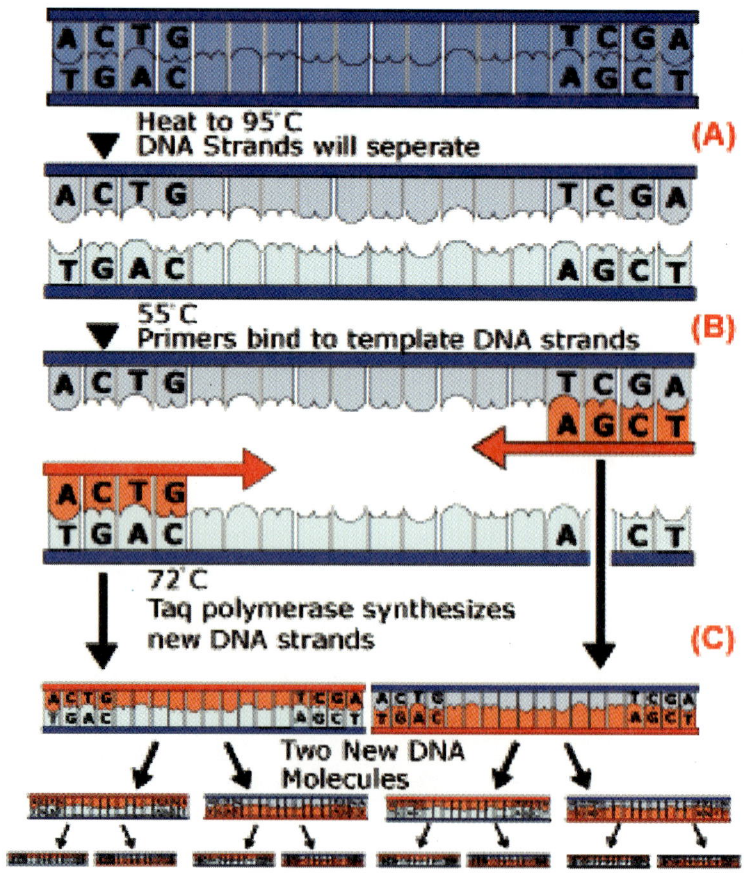

Fig. 19. Internal region of two simple sequence repeat (ISSRs).

20. Polymerase chain reaction **a)** Denaturation, **b)** Annealing & **c)** Extension

21.

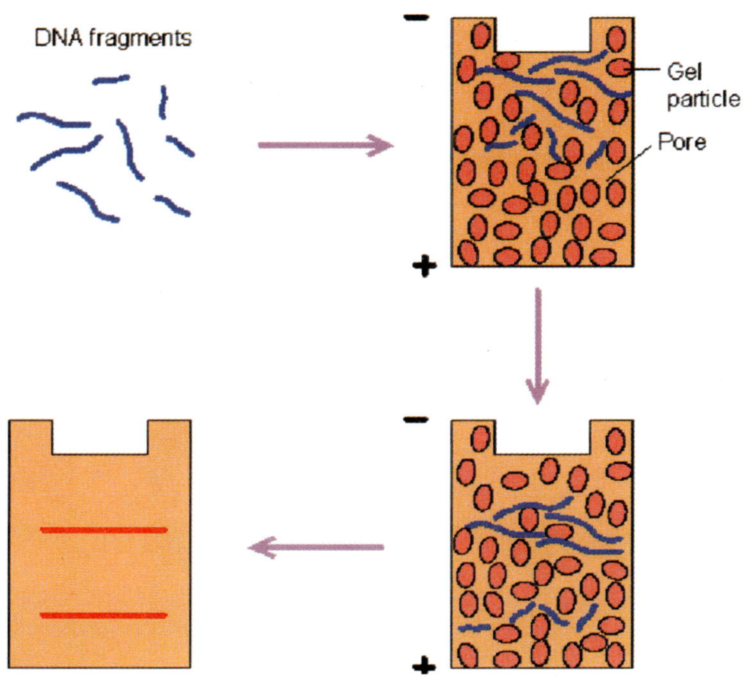

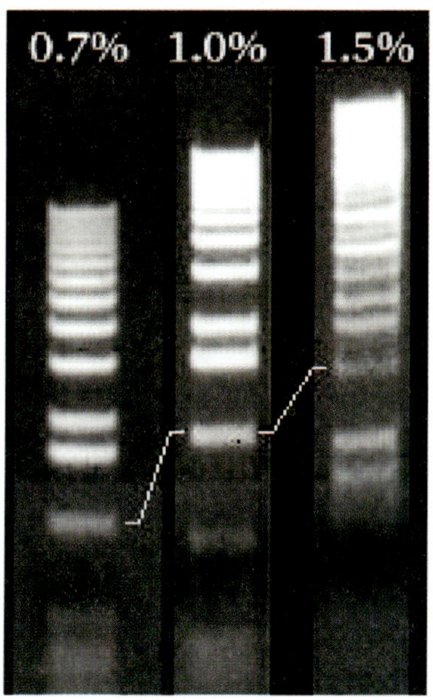

22.

23.

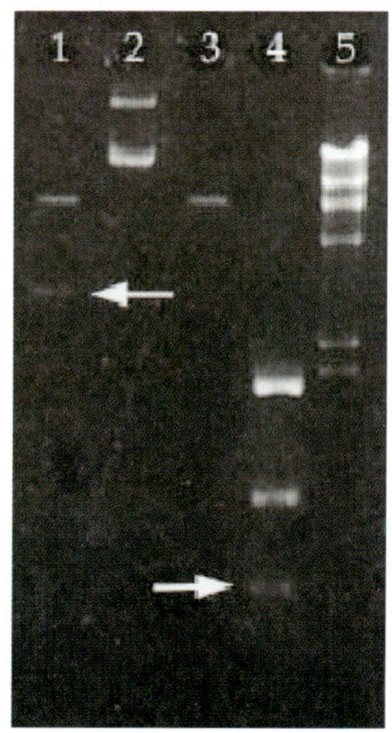

24.

Fig. 21). Principle of Electrophoresis; **22).** DNA separation depending on agarose gel concentrations; **23).** DNA visualization in the gel by addition of ethidium bromide, which fluoresce under UV light; **24).** Photograph of a gel taken by gel documentation unit.

4. Mechanism of *A. rhizogenes* - plant cell interation:

The earliest stage in the interaction is the attachment of the bacterium to the surface of the plant cell. A plant cell becomes susceptible to *Agrobacterium* when it is wounded. The wounded cells release phenolic compounds; such as acetosyringone and α-hydroxy acetosyringone that activates the vir region of the bacterial plasmid (Binns *et al.*, 1988). It has been shown that *Agrobacterium* plasmid carries three genetic components that are required by plant cell transformation (Zambryski *et al.*, 1989). The first component of Ri plasmid; T-DNA that is integrated into the plant cell, is a mobile DNA element. The second one is the virulence area *(vir)* which contains several *vir* genes. These genes do not enter the plant cell but together with the chromosomal DNA, cause the transfer of T-DNA. The third component is border sequence (25 bp) which resides in the *Agrobacterium* chromosome. The mobility of T-DNA is largely determined by these sequences, and they are the only cis elements necessary for direct T-DNA processing.

The early steps of the transfer are relatively well studied but the mechanism of interation is not completely understood. According to Zambryski it is a multistep process involving recombination, replication and repair activities, most likely mediated by host cell enzymes.

Genes encoded in T-DNA are of bacterial origin but have eukaryotic regulatory sequences enabling their expression in infected plant cells. None of the T-DNA sequences are required for transfer and integration in the host plant genome. Virulence genes that form the *vir* region of the Ri plasmid and *chv* genes found on bacterial chromosome mediate transfer of T-DNA.

Establishment of multiple shoot cultures

Multiple shoot culture of *C. roseus* produced. A node was isolated along with a piece of stem, with the purpose of forming a shoot by allowing the node to regenerate a branch. The node formed in the axil of a leaf is similar to the one in the stem tip. Shoot developed under the high cytokinin concentration was inoculated on to a fresh medium. The sole aim of experiment was to get sterile leaves in a limited time to perform *Agrobacterium* mediated transformation for the establishment of hairy root culture.

Shoot culture offer industrial alternative material for the production of useful metabolites that are not synthesized in cell culture and root cultures. However, large-scale activation shoot cultures is much more difficult than that of cell and root cultures because the shoot cultures require light irradiation and gas exchange on the leaves.

Effect of different concentrations of phytohormones as NAA and BAP on induction and growth of multiple shoot cultures was studied. 0.5 NAA / 1.0 BAP mg/lit showed the maximum growth performance with 12.2 shoots produced per node where as 0.1 NAA/0.5 BAP mg/lit showed the least response with only 4.0 shoots produced per node. Therefore, 0.5 NAA/1.0 BAP mg/lit is optimal for generation of multiple shoots in *C. roseus.*

Establishment of hairy root cultures

Roots developed on *C. roseus* leaves 14-21 days after inoculation with *Agrobacterium* strain A4 Midvein of explant showed the most vigorous response. However, when leaf explants were used it was necessary to transfer them to fresh media containing antibiotic, 3 days after inoculation, so as to prevent bacterial contamination of the supporting medium. The cultures were characterized by increased lateral branching, profusion of root hairs and absence of geotropism characteristics which are typical of transformed root cultures.

Establishment of hairy root cultures for production of valuable secondary metabolites *in vitro* has recently been reported for several plant species. Hairy root cultures formed by transformation of the plant by the soil pathogenic bacterium *A. rhizogenes* have many advantages when compared to cell suspension cultures and untransformed root cultures.

Hairy roots are potentially able to produce all root-derived products of dicotyledonous plants. Moreover, it had been possible to select highly productive hairy root clones, which grow faster than the corresponding untransformed root cultures. These cultures appeared to be genetically stable which has not been the case with high producing cell suspension cultures. Because of stable high production of secondary metabolites, hairy root cultures offer a valuable system for studying the control of gene expression and the biosynthesis of secondary metabolites.

The infection rate and the ability of transformed roots to grow further on a nutrient medium appeared to be highly dependent on the plant variety and culture condition used. On average six leaves were infected per plate by pricking method. 2-3 leaves responded per plate which was analyzed by observing characteristic callus formation at the site of infection, mostly of which were slow growing. Only 2 fast growing roots were examined which were further maintained for characterization in terms of metabolites and regeneration of transgenic plants.

The hairy roots produced represent novel material for possible production of useful alkaloids like VBL and VCR. The hairy roots produced need characterization from this point of view.

Future prospects of tissue culture in this plant

The development of *in vitro* tissue culture technique in concert with development in gene transfer technology allows the application of genetic engineering to improve the agronomic traits in crop plants. One such target is plant secondary metabolic pathway engineering. However, the yields of very few secondary metabolites have been improved in plant cell cultures. Some of the major shortcomings, which limit or prevent accumulation of secondary metabolites in cell culture, include an absence of specific enzyme activities in de-differentiated tissues and the instability of cell lines.

The hairy root cultures developed can be used further for:

- Selection of root clones.
- Quantitative and qualitative estimation of alkaloid accumulation in hairy root cultures.
- Scaling – up of alkaloid production in bioreactors for commercialization.
- Regeneration of transgenic plants.

Transgenic hairy root cultures provide an effective means to study genetic manipulations of biochemical pathways in plants. Hairy roots are also a potential production platform for a variety of secondary metabolites. As a production system they offer the potential for fast differentiated growth along with genetic and biochemical stability. For industrial production lines, a demonstration of their stability over time is needed. Although there are a limited number of reports on the long-term maintenance of hairy root cultures, there has not been any report on the long-term maintenance of transgenic hairy root cultures.

Breeding programs for *C. roseus*

Breeding programs have been set up not only for horticultural purposes, but also to modify alkaloids content. Mutation breedings were adopted to induce high alkaloid content in the dieback-resistant cultivar 'Nirmal'. Dieback disease, caused by *Pythium aphanidermatnon*, is a major limiting factor in the cultivation of *C. roseus* (Kulkarni, 1988). Two mutants, dwarf and semidwarf, with high alkaloid contents in the roots and one leaf mutant with high alkaloid content in he leaves were obtained (Kulkarni, 1998). All three mutations were found to be controlled by monogenic recessive genes.

The breeding of pure line cultivars for dual exploitation of the roots (for ajmalicine isolation) and the leaves (for isolation of bisindoles) seemed to be more appropriate than the use of hybrid cultivars (Levy, 1983). Phenolic compounds may be used as plant growth regulators: caffeic acid, gentisic acid, 3,4-dihydroxybenzaldehyde and ferulic acid enhanced growth and total alkaloid content, while *p*-coumaric acid, qurecetin and *p*-hydroxybenzoic acid inhibited the growth as well as the total alkaloid content (Sudheer, 1998). During breeding programs, evidence was obtained for phenotypic assortative mating for flower color; while self-pollination did not occur (Kulkarni, 1999). Genetic variability, heritability and genetic advance for alkaloid attributing traits in 26 genotypes of *C. roseus* were studied. In total 15 traits such as plant height, stem width and total leaf fresh weight, were studied. Both heritability and genetic advance showed that the leaf alkaloid content was under the control of additive gene action.

In various laboratories 42 *Catharanthus* specimens were analyzed for their alkaloid content. *Catharanthus trichophyllus* showed the highest accumulation of catharanthine (60 mmol/g DW) as well as two of its hybrids (crossed with *C. roseus* cultivars). The highest amount of VLB was found in dwarf cultivars (Hallard, 2000).

Economical benefits of *C. roseus*

It has a no. of features which makes it highly suitable for serving as a model plant for functional genomic analysis especially with respect to secondary metabolites. The root, shoot, leaf and anthers of *C. roseus* have been brought under tissue culture to develop calliclones to assess their alkaloid status. It is the only source of valuable anti-cancer drug, Vincristine (VCR) and Vinblastine (VLB).

PERSPECTIVES OF *CATHARANTHUS* RESEARCH

The 130 alkaloids isolated from *C. roseus* are a clear example of the unsurpassed spectrum of chemodiversity supplied by plant secondary metabolism. In general, plant secondary metabolites have found many applications, for example as drugs, flavors, fragrances and colorants. Despite the diversity, only a limited number of the 250.000 different plant species are now exploited. However, once recognized as a useful plant species, often other applications follow. Such plants become "multi-purpose" plants, e.g. the potato is an important part of the western World's diet, it supplies raw material for the industry, while it is now also studied to produce biodegradable plastics and vaccines. Other examples are Papaver and Cannabis. Also, *C. roseus* has developed into a true multi-purpose plant, it is the source for important drugs, it is a highly appreciated ornamental, its extracts are used in various dermatological preparations, and last but not least it has developed as an important model system for plant-biotechnological research.

For the realization of the initial incentive of the *C. roseus* research, i.e. the production of VLB by cell cultures, one cannot expect that this will be realized in the near future. However, analytical, enzymatic and chemical methods have been optimized allowing the production of the bisindole alkaloids in such amounts that clinical requirements can be met, and derivatives can be synthesized and tested. Although VLB has developed as a true lead compound, supply is still limited and costly. As a result of the research it became clear that besides being a source of alkaloids, *C. roseus* also contains an interesting set of enzyme and genes. Altogether the research has got a broad spin off with applications in drug development cosmetics, scientific research, and fodder quality and plant protection.

Promising perspectives of the *Catharanthus* research are found in the area of recombinatorial biochemistry, both by homologous and heterologous expression. By recombinant technology novel alkaloidal products can potentially be created. Some possibilities in this area were already mentioned by De Luca *et al*, (1998). By antisense technology one could block the vindoline biosynthesis and this may result in the accumulation of vindoline precursors, which may result in the accumulation of different bisindole alkaloids. Recombinatorial biochemistry may be of particular interest when for example basic alkaloid skeletons can be introduced into other plant species, which have their own set of reactions to decorate these skeletons, e.g. Hallard (2000) introduced *Tdc* and *Str* in weigela cells accumulating secologanin. These transgenic cells were able to produce corynanthean type of alkaloids. It is expected that in the near future more genes will become available, increasing the possibilities for metabolic engineering. The creation of new compounds in plants by introducing extra genes is no longer science fiction. If it will go as far as is seen for the polyketide biosynthetic genes in Streptomyces, where design of structures by combining the right set of genes in plants are not organized into operons. However, strictosidine can now be produced in various ways in large quantities. This opens the way to biotransformation studies, where strictosidine can be converted to more valuable compounds. It has been shown that by metabolic engineering alkaloid contents can be changed, e.g. the Cinchona cultures expression the *C. roseus Str* gene showed a different ratio of methoxylated alkaloids (Geerlings, 2000). The related exercise, producing Cinchona alkaloids by the much more easily cultivated *C. roseus* requires further understanding of the post-strictosidine steps in both species.

An important aspect for a model system is that it is characterized in great detail. In particular it is of interest to have the complete genetic information. The size of *C. roseus* genome was recently determined, 1500 Mbp, which is about 12 times the size of the Arabidopsis genome. The Arabidopsis genome is now completely sequenced (The Arabidopsis Genome Initiative, 2000). To our knowledge no *Catharanthus* genome project has been started yet. However, a start has been made with the proteome analysis of *Catharanthus* (Jacobs, 2001). Proteome analysis may lead to the identification of further biosynthetic enzymes and a better understanding of regulatory mechanisms (post translational modifications, targeting, protein-protein interactions etc.)

Almost 50 years after Garcia's publication (1994), one can summarize the perspectives of *Catharanthus* research in a single word, identical to the South-African vernacular name of periwinkle: "Kanniehood" (literally "can not die").

CONCLUSION

Living organisms are source of many valuable products of diverse chemical structures and biological activity. This is a special period of rapid progress in biotechnology. Molecular biology is finally contributing towards the development of new commercial products and new biological methodologies have spawned a new industry known as biotech industry. Research and development of biotechnologies for the production of secondary metabolites has received a major impetus by the very positive response of cells and tissues, roots and shoots, cultured *in vitro*. Synthesizing a new compound on a laboratory scale is only the first step in a long trek towards producing and marketing a new product. To realize full potential, it must be optimized and scaled up, clinical trials performed to seek approval of appropriate governmental regulatory agencies and finally the product packaged and marketed. All of these steps require several years and enormous financial costs to complete. Attention and efforts are now being shifted from enhanced production of secondary metabolites in *in vitro* systems to genetic manipulations of productions in plants like *Catharanthus roseus*, accompanied by expectations of directing secondary metabolism towards desirable compounds. Much of the support for R & D has been considered an investment for the human well being and this century will indeed be dominated by safe bio-products for the benefit of life, the environment, and the survival of human society.

REFERENCES

*Ainsworth P J Surh L C Coulter-Mackie M B 1991 Diagnostic single strand conformational polymorphism (SSCP) a simplified non-radioisotopic method as applied to a Tay-Sachs B1 variant *Nucleic Acids Research* **19** 405–406

Acton N Klayman D L 1985 vinblastine a new indole terpenoid from *Catharanthus roseus Planta Medica* **47** 442–445

Adams M D Kelley J M Gocayne J D Dubrick M Polymeropoulos M H Xiao H Merril C R Wu A Olde B Moreno R F Kerlavage A R McCombie W R Venter J C 1991 Complementary DNA sequencing expressed sequence tags and human *Genome* project *Science* **252** 1651–1656

Aerts R J Schaefer A Hesse M Baumann T W Slusarenka A 1996 *Phytochemistry* **42** 417

Andayani J M J R 2001 The isochorismate pathway as a route to 2 3 dihydrobenzoic acid in *Catharanthus roseus* cell cultures Ph D thesis Leiden University ISBN 90-73235 24-3

Armstrong J G Dyke R W Fouts P J Hawthorne J J Jansen C J Jr Peabody A M 1967 *Cancer Res* **27** 221

*Beckman J S Soller M 1990 Towards unified approach to the genetic mapping of eukaryotes based on sequence tagged microsatellite sites *Biotechnology* **8** 930–932

*Breslaur K J Frank R Blockerh Markey A 1986 Predicting DNA duplex stability from the base sequence *Proceedings of Natural Academy of Sciences of USA* **83** 3746–3750

Botstein B White R L Skolnick M Davis R W 1980 Construction of a genetic linkage map in man using restriction fragment length polymorphisms *American Journal of Human Genetics* **32** 314–331

Bowman R N 1996 *Int Appl WO* 951519698

Brady S F Feng D M Garsky V M 1999 *PCT Int Appl WO* 9928345

Brady S F Garsky V M Pawluezky J M 1999 *PCT Int Appl WO* 9902175

Brisson N De Luca V Kurz W G W 1998 *Appl US* 5814526

Brun G Dijoux M G David B Mariotte A M 1999 A new flavonol glycoside from *Catharanthus roseus* *Phytochemistry* **50** 167–169

Brun G Dijoux M G David B Mariotte A M 1999 A new flavonol glycoside from *Catharanthus roseus* *Phytochemistry* **50** 167–169

Budavari Susan ed 1989 The Merck Index An Encyclopedia of Chemicals Drugs and Biologicals Rahway NJ Merck & Co

Budman D R 1992 *Semin Oncol* **19** 639

Bustos A Soler C Jouve N 1999 Analysis by PCR based markers using designed primers to study relationships between species of Hordeum (Poaceae) *Genome* **42** 129–138

*Caetano-Anolles G Bassam B J Gresshoff P M 1991 DNA amplification fingerprinting using very short arbitrary oligonucleotide primers *Biotechnology* **9** 553–557

*Chaparoo J Werner D O'Malley D Sederoff R 1992 *Plant Genome San Diego* **21**

*Charles D J Simon J E Wood K V Heinstein P 1990 Germplasm variation in vinblastine content of *Catharanthus roseus* L using an alternative method of alkaloids analysis from crude plant extracts *Journal of Natural Products* **53** 157–160

*Charles DJ E Cebert J E Simon J 1991 Characterization of the essential oils of *Catharanthus roseus L Essential Oil Research* **3** 33–39

Canel C LopesCardoso M J Whitmer S van Der Fits L Pasquali G van der Heijden R Hoge J H C Verpoorte R 1998 *Plant Physiol* **116** 853

Carew D P Krueger R J 1976 Anthocyanidins of *Catharanthus roseus* callus cultures *Phytochemistry* **15** 442

Carlson J E Tulsieram L K Glaubitz J G Luk V W K Kauffeldt C Rutledge R 1991 Segregation of random amplified DNA markers in F1 progeny of conifers *Theoretical and applied genetics* **83** 194–200

Carpin S Garnieri F Andreu F Chenieux J-U Rideau M J 1997 *Plant Physiol Biochem* **35** 603

Carpin S Ouelhazi L Filali M Chenieux J-U Rideau M J 1997 *Plant Physiol* **150** 452

Chang C Bowman A W Lander E S Meyerowitz E W 1988 Restriction fragment length polymorphism linkage map for *Arabidopsis thaliana Proceedings of Natural Academy of Sciences of USA* **85** 6856 6860

Chavadej S Brisson N McNeil J N De Luca V 1994 *Proc Natl Acad Sci USA* **91** 2166

Cooke R *et al* 1996 Further progress towards a catalogue of all *Arabidopsis* genes analysis of a set of 5000 non-redundant ESTs *Plant Journal* **9** 101–124

Creasy W A 1994 In The monoterpenoid indole alkaloids Saxton J E Ed Wiley and sons Ltd Chichester 715–754

De Luca V St Pierre B 2000 *Trends Plant Sci* **5** 168

De Ropp R S 1947 *Am J Bot* **34** 53

Dobelis Inge N ed 1989 Magic and Medicine of Plants Pleasantville NY Reader's Digest Books

Droual M Hamdi S Creche J Keverts C Rideau M J 1998 *Plant Physiol* **153** 623

Duflos A Fahy J Thillaye D B V Barret J M Hill B 1998 *PCT Int Appl WO* 9845301

Duke S O Paul R N 1993 Development and fine structure of glandular trichomes of *Catharanthus roseus L International Journal of Plant Sciences* **154**(1) 107–118

*Epplen J T Inchrambach A Dunn N J Radola B J 1992 Advances in Electrophoresis *VCH Cambridge vol* **5** 59–112

Ehrlich H A Gelfand D H Sninsky J J 1991 Recent advances in the Polymerase Chain Reaction *Science* **252** 1643–1651

Etievant C Barret J M Kruczynski A Perrin D Hill B T 1998 *Investig New Drugs* **16** 3

*Ferreira J F S Janick J 1995 Floral morphology of *Catharanthus roseus* with special reference to trichomes *International Journal of Plant Sciences* **156** 807–815

*Ferreira J F S Janick J 1996 Roots as an enhancing factor for the production of vincristine in shoot cultures of *Catharanthus roseus Plant Cell Tissue Organ Cult* **44** 211–217

*Ferreira J F S Simon J E Janick J 1995 Developmental studies of *Catharanthus roseus* Flowering and vinblastine production under greenhouse and field conditions *Planta Medicine* **61** 167–170

*Fukuoka S Inoue T Miyao A Monna L Zhong M S Sasaki T Minobe Y 1994 Mapping of Sequence-Tagged Sites in Rice by Single Strand Conformation Polymorphism *DNA Research* **1** 271–277

Fahy J 2000 *Curr Pharm Des* **7** 1181

Fahy J Duflos A Ribet J P Jacquesy J C Berrier C Jouannetaud M P Zunino F 1997 J *Am Chem Soc* **119** 8576

Fang D Q Roose M L 1997 Identification of closely related citrus cultivars with inter-simple sequence repeat markers *Theoretical and applied genetics* **95** 408–417

Figdore S S Kennard W C Song K M Slocum M K Osborn T C 1988 Assessment of the degree of restriction fragment length polymorphism in Brassica *Theoretical and applied genetics* **75** 833–840

Filippini R Caniato R Piovan A Cappelletti E M 2003 Production of anthocyanins by *Catharanthus roseus Fitoterapia* **74** 62–67

Flavell R B 1995 *Tibtech* **13** 313–319

Fodor P Guth G Mourin E 1997 *Patent EP* 0692258

Forsyth W G C Simmonds N W 1957 Anthocyanidins of *Lochnera rosea Nature* 180–247

Foucault A P 1994 Centrifugal Partition Chromatography Chromatographic Science Series Marcel Dekker New York **68**

Foucault A P and Chevolot L 1998 *J Chromatogr A* **808** 3

Fujioka S Noguchi T Watanabe T Takatsuto S Yoshida S 2000 *Phytochemistry* **53** 549

*Gu W K Waden N F Yu J Wallace D H 1995 *Theoretical and applied genetics* **91** 465–470

Garcia F 1954 Proceedings 8th Pacific Science Congress of the National Research Council of the Phillipines IVA 182

Gardiner J M Coe E H Melia-Hancock S Hoisington D A Chao S 1993 Development of a core RFLP map using an immortalized F_2 population *Genetics* **134** 917–930

Garnier F Carpin S Label P Creche J Chenieux J-C Rideau M Hamdi S 1996 *Plant Sci* **120** 47

Garnier F Hamdi S Label P Rideau M 1999 *Biotechnol Agric For* **45** 88

Garnier F Label P Hallard J Chenieux Rideau M Hamdi S 1996 *Plant Cell Tiss Org Cult* **45** 223

Geerlings A Hallard D martinez caballero A Lopes Cardoso L van der Heijden R Verpoorte R 1999 *Plant Cell Rep* **19** 191

Geerlings A Redondo F j Comin A Memelink J van der Heijden R Verpoorte R 2001 *Appl Microbiol Biotechnol* **56** 420

Geerlings A van der Heijden R memelink J Verpoorte R 1998 *EP* 982041162

GeerlingsA Martinez-Lozano Ibanez M Memelink J van der Heijden R Verpoorte R 2000 *J Biol Chem* **275** 3051

Goddijn O J M van der Duyn Schouten P m schilperoorte R A Hoge J H C 1993 *Plant Mol Biol* **22** 907

Gupta M Chyi Y-S Romero-Severson J Owen J L 1994 Amplification of DNA markers from evolutionary diverse genomes using single primers of simple sequence repeats *Theoretical and applied genetics* **89** 998–1006

*Hearne C M Ghosh S Todd J A 1992 *Trends Genetics* **8** 288–294

*Heath D D Iwama G K Devlin R H 1993 *Nucleic Acids Research* **21** 5782–5785

*Hongtrakul V Huestis G M Knapp S J 1997 *Theoretical and applied genetics* **95** 400–407

Hallard D A C 2000 Transgenic Plant cells for production of indole alkaloids Ph D Thesis Leiden University ISBN 90-74538-29-5

Hallard D van der Heijden R Contin Jimenez E M T Snoeijer W Verpoorte R Jensen S R Cardoso M I L Pasquali G Memelink J Hoge J H C 1998 *Phytochem Anal* **9** 162

Hallard D van der Heijden R Verpoorte R Cardoso M J L Pasquali G Memelink J Hoge J H C 1997 *Plant Cell Rep* **17** 50

Hamada H Fuchikami Y Ikematso Y Hirata T Williams H J Scott A I 1994 *Phytochemistry* **37** 1037

Hamada H Fuchikami Y Jansing R L Kaminsky L S 1993 *Phytochemistry* **33** 599

Hartl L Weiss H Stephan U Zeller F J Jahoor A 1993 Use of RFLP markers for the identification of alleles of the Pm3 locus conferring Powdery mildew resistance in wheat (Triticum aestivum L) *Theoretical and applied genetics* **90** 601–606

Hayes P M 1993 Quantitative trait locus effects and environmental interaction in a sample of North American barley germplasm *Theoretical and applied genetics* **87** 392–401

Helentjaris T Slocum M Wright S Schaefer A Nienhuis J 1986 Constuction of genetic linkage maps in maize and tomato using restriction fragment length polymorphism *Theoretical and applied genetics* **72** 761–769

Heywood V H ed 1993 Flowering Plants of the World New York NY Oxford University Press

Hildebrandt A C Riker A J 1949 *Am I Bot* **36** 74

Hilliou F Christon P Leech M J 1999 *Plant Sci* **140** 179

Hughes E H Shanks J V Metab Eng 2002 **4** 41

Inomata S Yokoyama M Seto S Yanagi M 1991 *Appl Microbiol Biotechnol* **36** 315

Ito Y Shinomiya K Fales H M Weisz A and Sher A L 1995 In W D Conway and R J Petroski Editors ACS Symposium Series American Chemical Society Washington DC **593** 156–183

*Jones N H Ougham H Thomas 1997 Markers and mapping we are all geneticists now *New Phytologist* **137** 165–177

Jacobs D I van der Heijden R Verpoorte R 2000 *Phytochem Anal* **11** 277

Jacobs D I van Rijssen M S van der Heijden R Verpoorte R 2001 *Proteomics* **1** 1345

Jacquesy J C Fahy J Berrier C Bigg D Jouannetaud M P Zunino F Kruczynski A Kiss R 1994 *PCT* No *PCT* FR94/00898 1995 WO95/03312

Jarvis P Lister C Szabo V Dean C 1994 Integration of CAPS markers into the RFLP map generated using recombinant inbred lines of *Arabidopsis thaliana Plant Mol Biol* **24** 685–687

Jeffrey A J Wilson V Thein S L 1985 Individual-specific fingerprints of human DNA *Nature* **314** 67–73

Jordan M A Himes R H Wilson L 1985 *Cancer Res* **45** 2741

*Klayman D L Lin A J Acton N Scovill J P Hock J M Milhous W K Theoharides A D 1984 Isolation of vincristine from *Catharanthus roseus* growing in the United States *Journal of Natural Products* **47** 715–717

*Kolchinsky A Kolesnikova M Ananiev E 1991 *Genome* **34** 1028–1031

Karp A Kresovich S Bhat K V Ayad W G Hodgkin T 1997 Molecular tools in plant genetic resources conservation a guide to the technologies IPGRI Technical Bulletin No 2 International Plant Genetic Resources Institute Rome Italy

Kisakurek M V Hesse M 1980 In Indole and biogenetically related alkaloids Phillipson J D Zenk M H Eds Academic press London **11** 25

Knobloch K H Bast G Berlin J 1982 Medium-and light-induced formation of serpentine and anthocyanins in cell suspension cultures of *Catharanthus roseus Phytochemistry* **21** 591–594

Konieczny A Ausubel F M 1993 A procedure for mapping *Arabidopsis* mutations using recombinant ecotype-specific PCR-based markers *Plant Journal* **4** 403–410

Krucznynski A Barret J M Etievant C Colpaert F Fahy J Hill B T 1998 *Biochem Pharmacol* **55** 635

Kruczynski A Hill B T 2001 *Crit Rev Oncol Hemat* **40** 159

Kulkarni R N Baskaran K Chandrashekara R S Kumar S 1999 *Plant Breed* **118** 71

Kulkarni R N Ravindra N S 1988 *Planta Med* **54** 355

Kutchan T M Bock A Dittrich H 1994 *Phytochemistry* **35** 353

*Lagerstrom M J Parik J Malmgren H Stewart J Patterson U Landegren U 1991 Capture PCR Efficient amplification of DNA fragments adjacent to a known sequence in human and YAC DNA *PCR Methods Applications*

*Lessa E P 1992 *Molecular Biology and Evolution* **9** 323–330

Lang G Coteret J 1999 *PCT Int Appl WO* 9936042

Lang G Cotteret J 1999 *PCT Int Appl WO* 9936040

Leech M J May K Hallard D Verpoorte R De Luca V Christou P 1998 *Plant Mol Biol* **38** 765

Leveque D Wihlm J Jehl F 1996 *Bull Cancer* **83** 176

Levy A Milo J Ashri A Paleviteh D 1983 *Euphytica* **32** 557

Limam F Chahed K Ouelhaz N Ghrir R Ouelhazi L 1998 *Phytochemistry* **49** 1219

Linthorst H J M verpoorte R Verberne M C Moreno P R H van tegelen L J P Wullems G J Croes A F Stuiver M H custers J H M V Simons L H Melchers L S Bol J F 1999 *PCT Int Appl WO* 9950423

Litt M Luty J A 1989 A hypervariable microsatellite revealed by *in vitro* amplification of a dinucleotide repeat with cardiac muscle actin gene *American Journal of Human Genetics* **44** 397– 401

Liu B H Knapp S J 1990 GMENDEL a program for Mendelian segregation and linkage analysis of individual or multiple progeny populations using log-likelihood ratios *Journal of Heredity* **81** 407

*Malyshev S V Kartel N A 1997 Molecular markers in mapping of plant genomes *Molecular Biology* **31** 163–171

*Martinez B C Staba J 1988 The production of vinblastine in *Catharanthus roseus* L Tissue cultures *Adv Cell Culture* **6** 69–87

*McCormack G 2002 Cook Islands Natural Heritage Project database *Cook Islands Natural Heritage Project*

*McVaugh R 1984 Compositae Vol 12 Flora Novo-Galiciana a descriptive account of the vascular plants of Western Mexico ed Anderson W R *Pure Appl Chem* **58** 817–824

*Mizobuchi M Frohman L A 1993 Rapid amplification of genomic DNA ends *Biotechniques* **15** 214–216

*Mohan M S Nair A Bhagwat T G Krishna M Yano 1997 Genome mapping molecular markers and marker-assisted selection in crop plants *Molecular Breeding* **3** 87–103

Ma Y Ito Y Sokolovsky E and Fales H M 1994 *J Chromatogr A* **685** 259

Markgraf F Flore de Medagascar et des Comores 1976 169e Famille Aponcynaceae **139**

Martin G B Williams J G K Tanksley S D 1991 Rapid identification of markers linked to a Pseudomonas resistance gene in tomato by using random primers and near-isogenic lines *Proceedings of Natural Academy of Sciences of USA* **88** 2336–2340

McCormack J J 1990 In The Alkaloids Brossi A Suffness M Eds Academic press San Diego **37** 205–228

McCouch S R Kochert G Yu Z H Wang Z Y Khush G S Coffman W R Tanksley S D 1988 Molecular mapping of rice chromosomes *Theoretical and applied genetics* **76** 815–829

Menke F L H Parchmann S Mueller M J Kijne J W, Memelink 1999 *J Plant Physiol* **119** 1289

Meyer P 1998 In Transgenic Plant research Lindsey K Ed Harwood Academic publishers 203–275

Michelmore R W Paran I Kesseli R V 1991 Identification of marker linked to disease resistance gene by bulk segregant analysis A rapid method to detect markers in specific genomic regions using segregating populations *Proceedings of Natural Academy of Sciences of USA* **88** 9828–9832

Milbourne D Meyer R C Collins A J Ramsay L D Gebhardt C Waugh R 1998 Isolation characterization and mapping of simple sequence repeat loci in Potato *Molecular Genomics and Genetics* **259** 233–245

Miller J C Tanksley S D 1990 *Theoretical and applied genetics* **80** 437–448

Milo J Levy A Akavia N Ashri A Palevitch D 1985 Inheritance of corolla color and anthocyanin pigments in periwinkle (*Catharanthus roseus* [L] G Don) Zeitschrift für Pflanzenzüchtung **95** 352–360

Monticello D J Xu G W Mitchel K W 1997 US 57226056

Moreno P R H van der Heijden R Verpoorte R 1994 *Plant Cell Rep* **14** 188

Moreno P R H van der Heijden R Verpoorte R 1995 *Plant Cell Tiss Org Cult* **42** 1

Moreno Valenzuela O A Monforte Gonzalez M Munozsanchez J A Mendez Zeel M Loyola Vargas V M 1999 *J Plant Physiol* **155** 447

Mullis K B Faloona F A 1987 Specific synthesis of DNA *in vitro* via a polymerase-catalyzed chain reaction *Methods of Enzymology* **155** 335–350

Mustafa N R Verpoorte R 2007 Phenolic compounds in *Catharanthus roseus Phytochemistry Reviews* **6** 243–258

Mutschler Ernst Hartmut Derendorf 1995 Drug Actions Basic Principles and Therapeutic Aspects Stuttgart Germany medpharm Scientific Publishers

*Nelson D L Caskey C T 1990 PCR Technology Principles and Application for DNA Amplification (ed Erlich H A) Stockton New York 113–118

Nagaoka T Ogihara Y 1997 Applicability of inter-simple sequence repeat polymorphisms in wheat for use as DNA markers in comparison to RFLP and RAPD markers *Theoretical and applied genetics* **94** 597–602

Nakamura Y Julier C Wolff R Holm T O'Connell P Leppert M White R 1987 *Nucleic Acids Research* **15** 2537–2547

Nam H G Giraudat J Den Boer B Moonan F Loos W D B Hauge B M Goodman H M 1989 Restriction fragment length polymorphism linkage map of *Arabidopsis thaliana Plant Cell* **1** 699–705

Nishibe S Takenaka T Fujikawa T Yasukawa K Takido M Morimitsu Y Hirota A Kawamura T Noro Y 1996 Bioactive phenolic compounds from *Catharanthus roseus* and *Vinca minor Natural Medicines* (Tokyo) **50** 378–383

Noble R L 1990 *Biochem Cell Biol* **68** 1344

*Olson M Hood L Cantor C Botstein D 1989 A common language for physical mapping of human Genome *Science* **245** 1434–1435

*Orita M Suzuki Y Sekiya T Hayashi K 1989 *Genomics* **5** 874–877

Oh B J Frederiksen R A Magill C W 1994 Identification of molecular markers linked to head smut resistance gene (Shs) in Sorghum by RFLP and RAPD analysis *Phytopatho* **84** 830–833

Orecchia R Jereczek Fossa B A Catalano G Chiesa F DePas T Masci G Krengli M Vavassori A De Paoli F Robertson C Marrocco E De Braud F 2002 *Oncology* **63** 115

Orosz F Comin B Rais B Puigjaner J Kovacs J Tarkanyi G Acs T Keve T Cascante M Ovadi J 1999 *Brit J Canc* **79** 1356

Ouelhazi L Filali M Decendit A Chenieux J-C Rideau M J 1993 *Plant Physiol Biochem* **31** 42

*Paglia G P Olivieri A M Morgante M 1998 *Molecular Genomics and Genetics* **258** 466–478

*Parker J D Rabinovitch P S Burmer G C 1991 Targeted gene walking polymerase chain reaction *Nucleic Acids Research* **19** 3055–3060

*Perez T Albornoz J Dominquez A 1998 An evaluation of RAPD fragment reproducibility and nature *Molecular Ecology* **7** 1347–1357

Panaud O Chen X McCouch S R 1995 High density genetic mapping of the rice *Genome* based on sequence tagged microsatellite sites *Genome* **38** 1170–1176

Parish C A Dong J G Bommann W G Chang J Nakanishi K Berova N 1998 *Tetrahedron* **54** 15739

Piovan A Filippini R 2007 Anthocyanins in *Catharanthus roseus in vivo* and *in vitro* A review *Phytochemistry Reviews* **6** 235–242

Piovan A Filippini R Favretto D 1998 Characterization of the anthocyanins of *Catharanthus roseus*(L) G Don *in vivo* and *in vitro* by electrospray ionization ion trap mass spectrometry *Rapid Communications in Mass Spectrometry* **12** 361 367

Pleizier A C Mededelingen Landboduwbogeschool Wageningen 1981 81–91

Potier P *Semin Oncol* 1989 **16** (Suppl 4) 2

*Richardson T Cato S Ramser J Kahl G Weising K 1995 Hybridization of microsatellites to RAPD a new'source of polymorphic markers *Nucleic Acids Research* **23** 3798–3799

Rafalski J A Tingey S V 1993 Genetic diagnostics in plant breeding RAPDs microsatellites and machines *Trends Genetics* **9** 275–280

Rai S P Kumar S 2000 Induced mutation to monocotyledony in periwinkle *Catharanthus roseus* and suppression of mutant phenotype by kinetin *Journal of Genetics* **79**(3) 1–8

Rai S P Kumar S 2001 Hetrocarpous flowers resulting from a recessive mutation in periwinkle *Catharanthus roseus Current Science* (India)

Rai S P Luthra R Gupta M M Kumar S 2001 Pleiotropic morphological abiotic stress resistance of the hyper abscisic acid producing Abo-mutant in periwinkle *Catharanthus roseus Journal of Biosciences* **26** (1) 57–70

Rai S P Luthra R Kumar S 2003 Salt tolerant mutants in GLYCOPHYTIC SALINITY RESPONSE (GSR) genes in *Catharanthus roseus Theoretical and applied genetics* **106** 221–230

Renault J H Ghédira K Thépenier P Lavaud C Zèches-Hanrot M and Le Men-Olivier L 1997 *Phytochemistry* **44** 1321

Renault J H Thépenier P Zèches-Hanrot M Le Men-Olivier L Durand A Foucault A and Margraff R 1997 *J Chromatogr A* 763–345

Roessner C A Devagupta R Hasan M Williams H J Scott A I 1992 *Protein Expr Purif* **3** 295

Ross I A 1998 Medicinal plants of the World Humana Press Inc Towota N J

*Sarkar G Turner R T Bolander M E 1993 Restriction-site PCR A direct method of unknown sequence retreival adjascent to a known locus by using universal primers *PCR Methods Applications* **2** 318–322

*Schmid G W Hofheinz 1983 Total synthesis of vincristine *Journal of American Chemistry Society* **105** 624–625

*Somers D J Zhou Z Bebeli P Gustafson J P 1996 *Theoretical and applied genetics* **93** 983–989

Saiki R K Gelfand D H Stoffel S Scharf S J Higuchi R Horn G T Mullis K B Erlich H A 1988 Primer directed enzymatic amplification of DNA with a thermostable DNA polymerase *Nature* **239** 487–497

Saiki R K Scharf S Faloona F Mullis K B Horn G T Erlich H A Arnheim N 1985 Enzymatic amplification of beta-globin genomic sequences and restriction site analysis for diagnosis of sickle cell anemia *Science* **230** 1350–1354

Sambrook J Fritsch E F Maniatis T 1989 Molecular Cloning A Laboratory Manual 2nd Edition Cold Spring Harbor Laboratory Press Cold Spring Harbor New York

Sapi J and Massiot G 1994 Heterocyclic Compounds In J E Saxton Editor Monoterpenoid Indole Alkaloids Wiley Chichester **25**(4) 523–646

Sasaki T 1994 Rice Genome Newsletter for Rice Genome analysis 3/1 Tsukuba

Schlotterer C Tautz D 1992 Slippage synthesis of simple sequence DNA *Nucleic Acids Research* **20** 211–215

Schmidt B Kutney J Mayer L 1998 *PCT Int Appl WO* 9839004

Scragg AH 1999 In Methods of Molecular Biology plant cell culture protocols Hall R D Ed Humena Press Inc Totwa NJ **111** 393–402

Smith A C 1988 Flora Vitiensis nova A new flora of Fiji Lawai Kauai Hawaii National Tropical Botanical Garden **4** 85

Snoeijer W International Register of *Catharanthus* cultivars 1998 and 1st supplement 2001 Leiden/ Amsterdam Center for Drug Research Division of Pharmacognosy Leiden the Netherlands

Southern E M 1975 Detection of specific sequences among DNA fragments separated by gel electrophoresis *Journal of Molecular Biology* **98** 503 517

Stearn T 1975 In The *Catharanthus* Alkaaloides Taylor W T Farnsworth N R Eds Marcel Dekker **9** 44

Stone B C 1970 *The flora of Guam Micronesica* **6** 480 481

Sudheer B K Ram Hao S S 1998 *Indian J Plant Physiol* 302

Svoboda G H 1966 Proceedings First Symp GECA Excerpta Medial Foundation Amsterdam **9** 28

Sykes W R 1970 Contributions to the flora of Niue New Zealand Department of Scientific and Industrial Research Bulletin **200** 44-45

*Triglia T Peterson M G Kemp D J 1988 A procedure for *in vitro* amplification of DNA segments that lie outside the boundaries of known sequence *Nucleic Acids Research* **16** 81–86

Talou J R Verberne M C Muijono R A B van Tegelen L J P Bernal B G Linthorst H J M Wullems G J Bol J F Verpoorte R 2001 *Plant Physiol Biochem* **39** 595

Tanksley S D Ganal M W Martin G B Chromosome landing a paradigm for map-based gene cloning in plants with large genomes 1995 *Trends Genetics* **11** 63 68

The *Arabidopsis* Genome Initiative 2000 *Nature* **408** 796

Thomas J C Adams D G nessler C L Brown J K Bohnert H J 1995 *Plant Physiol* **109** 717

Thomas J C Akroush A M Adamus G 1999 *Plant Physiol Biochem* **37** 665

Tingey S V Rafalski J A Williams J G K 1993 Application of RAPD Technology to Plant Breeding (ed Neff M) ASHS Publishers Minnesota **3** 8

Tsumura Y Ohba K Strauss S H 1996 Diversity and inheritance of inter-simple sequence repeat polymorphisms in Douglas-fir (Pseudotsuga menziesii) and sugi (Cryptomeria japonica) *Theoretical and applied genetics* **92** 40-45

*Unkles S E Duncan J M Kinghorn J R 1992 *Current Genetics* **22** 317–318

*Vos P Hogers R Bleeker M Reijans M Van de Lee T Hornes M Frijters A Pot J Peleman T Kuiper M Zabeau M 1995 RFLP A new technique of DNA fingerprinting *Nucleic Acids Research* **23** 4407–4414

Van der Fits L Deakin E A Hoge J H C Memelink J 2000 *Plant Mol Biol* **43** 495

Van der Fits L Memelink J 1997 *Plant Mol Bio* **33** 943

Van der Heidjen R Verpoorte R Verpoorte R 1989 *Plant Cell Tiss Org Cult* **18** 231

Van Tegelen L J P Moreno P R H Croes A F Verpoorte R Wullems *J Plant Physiol* 1999 **119** 705

Ven Bergen M A 1996 Wageningen Agricult Univ Pap 1996 **96** 9

Verberne M C Verpoorte R Bol J F Mercado Blanoe J Linthorst H J M 2000 *Nature Biotechnol* **18** 779

Verpoorte R Merenlink J 2002 *Curr Opin Biotechnol* **13** 181

Verpoorte R van der Heijden R and Memelink J 2000 Transgenic Res **9** 323

Verpoorte R van der Heijden R Moreno P R H 1997 In The Alkaloides Brossi A Ed Academic Press San Deigo **49** 221 299

Verpoorte R van der Heijden R ten hoopen H J G Memelink J 1999 *Biotechnol Lett* **21** 467

Verpoorte R van der Heijden R van Gulik W J ten Hoopen H G J 1991 In The Alkaloids Brossi A Ed Academic Press San Diego **40** 187

*Weber J L May P E 1989 Abundant class of human DNA polymorphism which can be typed using the polymerase chain reaction *American Journal of Human Genetics* **44** 388–396

*Weining S Langridge P 1991 *Theoretical and applied genetics* **82** 209–216

Wagner W L Herbst D R Sohmer S H 1999 Manual of the flowering plants of Hawai'i Revised edition University of Hawai'i Press Honolulu **215** 216 1858

Weising K Nybom H Wolff K Meyer W 1995 DNA Fingerprinting in Plants and Fungi (ed Arbor A) CRC Press Boca Raton **1** 3

Weisz A Sher A L Shinomiya K Fales H M and Ito Y 1994 *J Am Chem Soc* **116** 704

Welsh J McCelland M 1990 Fingerprinting genomes using PCR with arbitrary primers *Nucleic Acids Research* **18** 7213-7218

Welsh J McClelland M 1991 *Nucleic Acids Research* **19** 6823-6831

Welsh S L 1998 Flora Societensis A summary revision of the flowering plants of the Society Islands E P S Inc Orem Utah **36**

Whitmer S 1999 Aspects of terpenoid indole alkaloid formation by transgenic cell lines of *Catharanthus roseus* overexpressing trypyophan decarboxylase and strictosidine synthase Ph D thesis Leiden University ISBN 90-74538-48-7

Whitmer S Canel C Hallard D Gonclaves C Verpoorte R 1998 *Plant Physiol* **116** 853

Whitmer S van der Heijded R Verpoorte R 2002 *Plant Cell Tiss Org Cult* **69** 85

Whitmer S van der Heijded R Verpoorte R J 2002 *Biotecthnol* 96193

Whitmer S Verpoorte R Canel C 1998 *Plant Cell Tiss Org Cult* **53** 135

Wijnands D O 1983 The botany of the Commelins A Taxonomical nomenclatural and historical account of the plants depicted in the Moninckx Atlas and in four books by Jan and Casper Commerlin on the plants in the Hortus Medicus Aristelodamensis 1682 1710 A A Balkemar Rotterdam

Williams JGK Kubelik A R Livak K J Rafalski J A Tingey S V 1990 DNA polymorphisms amplified by arbitrary primers are useful as genetic markers *Nucleic Acids Research* **18** 6531 6535

Williams M N V Pande N Nair S Mohan M Bennett J 1991 Restriction fragment length polymerase analysis of Polymerase chain reaction products amplified from mapped loci of rice (Oryza sativa L) genomic DNA *Theoretical and applied genetics* **82** 489-498

Winter and Kahl 1995 Molecular marker technologies for plant improvement World *Journal of Microbiol Biotechnology* **11** 438 448

Wu K Jones R Dannaeberger L Scolnic P A 1994 Detection of microsatellite polymorphisms without cloning *Nucleic Acids Research* **22** 3257-3258

Xue Y Q Sherman D H 2001 *Metab Eng* **3** 15

Yao K De Luca V Brisson N 1995 *Plant Cell* **7** 1787

Zabeau M Vos P 1993 Selective Restriction Fragment Amplification A general method for DNA fingerprinting European Patent Application EP 0534858

Zarate R Bonavia M Geerlings A van der Heijden R Verpoorte R 2001 *Plant Physiol Biochem* **39** 763

Zarate R Memelink J van der Heijden R Verpoorte R 1999 *Biotechnol Lett* **21** 997

Zhao X Kochert G 1992 *Molecular Genomics and Genetics* **2311** 353-359

Zietkiewicz E Rafalski A Labuda D 1994 Genome fingerprinting by simple sequence repeat (SSR) anchored polymerase chain reaction amplification *Genomics* **20** 176-183

**-Not consulted in original*

10

IDENTIFICATION OF MYCOBACTERIUM IN CLINICAL SAMPLES, PLANNING AND SURVEILLANCE

CHANDAN AGARWAL

INTRODUCTION

Trends in Tuberculosis Research

Infecting almost one-third of the human population, *Mycobacterium tuberculosis* was declared a global emergency by the World Health Organisation (WHO) in 1983. Today the bacterium is still the leading cause of death among the infectious diseases. With ninety five percent of the cases and an even higher percentage of deaths occuring in developing countries, tuberculosis (TB) is considered a disease of poverty. Although America is well above the global poverty level, TB struck New York hard in 1992. Areas around New York had incidents of TB that were twenty times the national average. It is estimated that between 2002 and 2020, 36 million people will die due to TB.

TB is not a highly contagious infectious disease, but like many of the other infectious diseases, TB spreads by droplet infection. This means that a person who has TB must sneeze or cough to get saliva containing the bacteria into the air where the healthy person can inhale the bacteria. Still the chances of inhaling enough bacteria to fall ill are low. Once inhaled, the bacteria travel to the air sacs in the lungs where they are then engulfed by special cells called macrophages. Inside the macrophages, the bacilli bacteria multiply continuously until the cell bursts open. The bacteria can either be active or inactive at this point. If the bacteria are active, the host (infected person) will get sick. If the bacteria are inactive the disease will remain hidden until a future date, the host (infected person) will then get sick.

The global impact of tuberculosis has been alarming. With an estimated annual two to three million deaths (including HIV infected individuals), over 8 million new cases and one third of the world's population infected with *Mycobacterium tuberculosis*, the WHO has declared tuberculosis an international health emergency. Hence, there has been a world wide, determined and comprehensive effort to combat tuberculosis both at the level of recruiting governmental and non governmental agencies to implement mass programmes for detection and supervised treatment, as well as encouraging and funding scientists to research and discover new improved tools for the diagnosis, treatment and prevention of human tuberculosis (World Health Organization, 2003).

There has been a qualitative change in the scenario of research in the present era of genomics, proteomics and concomitant use of bioinformatic tools to analyse, compare, predict and design appropriate targets or reagents for the detection of tuberculosis, as well as proficient vaccines and effective drugs. The availability of the mycobacterial genome, proteome and bioinformatic tools in the public domain has had a profound influence on the use of new technologies for rapid screening of a multitude of potential anti-mycobacterial compounds. Technologies with similar impact such as micro array have been used to study the complex host-pathogen interactions. However, verifying and confirming the utility and relevance of these 'critical genes' or their products requires assessment in a suitable biological system or animal model. To date, these have been the roadblocks in tuberculosis research. Further, these model systems do not match the high throughput technologies either in capacity or in determining the biological relevance of the critical gene (s) or their product (s). New technologies have enabled researchers to address age old questions such as reliable detection of tubercle bacilli or products of immune response in clinical samples in vitro or in vivo, parameters for describing protection against infection by *Mycobacterium tuberculosis*, and constructing new vaccines (Ahmed, 2004).

Several mycobacterial pathogens belonging to the tuberculosis complex have been recognized as human pathogens. They include *M. tuberculosis, M. bovis, M canneti* and *M. africanum*. These micro organisms cause identical disease in humans which can not be clinically differentiated. The commonly used first line drugs have been used for treating these clinical conditions. It has been a challenge to identify these pathogens owing to their antigenic and genetic similarities, and difficulties in identification by the classical bacteriological methods. Polymerase Chain Reaction (PCR), spoligotyping and associated technologies have been described to meet the challenge of recognizing these pathogens. Several gene targets such as hup B, dev R, rRNA, dna J and insertion sequences (1S6110, 1S990, 1S1081) have been described as reliable and relevant for differential identification of these pathogens (Prabhakar, 2004).

Besides PCR based technologies, methods are being investigated to distinguish between latent and active disease. The tuberculin test has been the mainstay in the past. However, cytokine assays that specifically detect interferon γ in whole blood and antibody assays have been shown to have the potential of becoming diagnostic tests.

Cocktails of mycobacterial antigens such as early secretory antigen 6 (ESAT-6 , Rv 3875) and culture filtrate protein 10 (CFP-10, Rv 3876) localized in the region of difference 1 (RD 1) of *Mycobacterium tuberculosis* have been shown to be suitable candidates for the detection of exposure to tubercle bacilli prior to progression towards clinical disease. Profiling antibody responses to *Mycobacterium tuberculosis* antigens at different stages of the disease (latent, non cavitary and cavitary) have shown that the gamut of antigens recognized alters with progression of the disease. Antibody based assays have an added advantage over cellular assays designed for the detection of interferon γ, in that they will not be influenced by the HIV positivity status of those being tested. In general, acceptance of a clinical diagnosis has been a contentious issue between scientists and clinicians.

Besides the previously described techniques, a new technology at the forefront for in vivo detection of active infectious foci is positron emission tomography (PET) with a glucose analogue (2-{fluorine 18} fluoro-2-deoxy-D-glucose, FDG). Granulomatous tissue has been reported to accumulate FDG. In fact, FDG uptake is more rapid by cellular infiltrates such as lymphocytes and macrophages, which are characteristically seen in early tuberculosis lesions. Activated inflammatory cells exhibit quantitatively increased glycolysis and hexose monophosphate shunt activity compared with the surrounding tissue. Taking advantage of this differential uptake of FDG by inflammatory cells, PET techniques are being applied to image tuberculomas. PET based imaging techniques along with other imaging techniques provide a hopeful means to discover cryptic and hitherto inaccessible infectious tuberculous foci (Goo, 2000).

The various means of diagnosis of tuberculosis being practised are :

DIRECT METHODS	INDIRECT METHODS
A Bacteriological	
* Smear examination for AFB	* Tuberculin Test
* Culture for mycobacteria	* CT Scan
* Detection of AFB in histopathological specimen	* Serology ELISA
* Rapid culture system :	RIA
BACTEC	PHA (Passive Haem Agglutination Test)
Others	
B Molecular biological methods	* Lymphocytic or mycobacterial products
* DNA or RNA hybridization	* ADA (Adenosine deaminase)
* PCR with DNA probe	* Tuberculostearic acid
	* HPLC (High Performance Liquid Chromatography)
C Luciferase reporter phase activity	* Detection of tubercular granuloma

The best and the most specific diagnosis is obtained by direct methods of detecting the mycobacteria or its specific genetic element from a sample.

A Comparative Status of Various Diagnostic Methods in Tuberculosis

Methods	*Sensitivity*	*Species specificity*	*Time*	*Quantity of bacilli required/ml of specimen*
AFB Smear Examination	Low	No	Hour	10^4
Conventional Culture	Moderate	Yes	Weeks	10^2
Radiometric Culture	Moderate	Yes	Days to Weeks	10^2
HPLC	Low to Moderate	Yes	Weeks	10^7-10^8
DNA Probes	Moderate to High	Yes	Weeks	10^5
RNA Probes	Moderate to high	Yes	Weeks	10^5
PCR+DNA Probe	High	Yes	Hours to days	10^1

(Glassroth, 1993)

Causative Organism

Tuberculosis is an infectious disease caused by *Mycobacterium tuberculosis* and less commonly by other organisms of the 'tuberculosis complex'. The annual incidence of new cases of all forms of TB (pulmonary and extra pulmonary) worldwide is estimated to be approximately 8 million, of which about 95% occur in developing countries. Many TB cases in developing countries remain undiscovered. Of the discovered smear positive cases, less than half complete treatment. Consequently, the estimated prevalence worldwide is 16 to 20 million, of whom about 8 to 10 million are sputum smear positive and highly infectious (Central TB Division, 2000).

The prevalence of infection signifies the number of positive reactors in respect to the total number of persons tested for tuberculin in a population. This may be present at a point or during a period of time. In India, the rate of prevalence is different at various areas and at various times. The rate was reported even upto 80% in 1949-50. However, grossly, it can be regarded as 50% as reported in 1980. This implies that 50 out of 1000 people were suffering from TB. At present about 40% of adults are infected with TB, and approximately 1.5 million cases are put on treatment every year.

An estimated 5 lakh deaths from TB occour every year (RNTCP, 2000).

Schoelein (1939) was probably the first to use the term 'tuberculosis'. He derived the simile of the small nodular lesions in the tissues of patients of tuberculosis to that of knobby protuberence (tubers) found in the roots of certain plants.

In the early sixteenth century an epidemic of tuberculosis began in Europe which spread over for the next 200 years and remained the cause of mortality of one quarter of the population. This epidemic was named as the 'white plague'. Robert Koch (1843-1910), a German physician discovered the tubercle bacilli. He presented his discovery to the Physiology Society of Berlin on the 24th of March, 1882. This is commemorated by observing the 24th of March as the **'World Tuberculosis Day'** after the centenary of Koch's discovery. The landmark developments in the history of tuberculosis are:

- Discovery of tubercle bacilli – Robert Koch in 1882
- Discovery of X-ray – Roentgen, 1895
- Discovery of streptomycin – Schatz Bugie and Walksman, 1944

Mycobacterium tuberculosis

Mycobacteria are slender rods that sometimes show branching filamentous forms resembling the fungal mycelium. In the liquid cultures they form a mold like pellicle. Hence, the name 'Mycobacteria', meaning fungus like bacteria. They do not stain readily, but once stained resist decolourisation with dilute mineral acids. They are therefore called 'acid fast bacilli' or AFB. They are aerobic, non motile, non capsulated and non sporing. The genus includes obligate parasites, opportunistic pathogens and saprophytes.

Classification of Mycobacteria

TUBERCLE BACILLI

Human	-	*M. tuberculosis*
Bovine	-	*M. bovis*
Murine	-	*M. microti*
Avian	-	*M. avium*
Cold blooded	-	*M. marinum*

LEPRA BACILLI

Human	-	*M. leprae*
Murine	-	*M. leprae murium*

A. TYPICAL MYCOBACTERIA

Photochromogens

Scotochromogens

Nonphotochromogens

Rapid growers

MYCOBACTERIA CAUSING SKIN ULCERS

M. ulcerans

M. balnei

JOHNE'S BACILLUS

M. paratuberculosis

SAPROPHYTIC MYCOBACTERIA

M. butyricum, M. phlei, M. stercoris

(Ananthanarayan, 2002)

Morphology

In the lesions it causes in the tissues, *M. tuberculosis* appears in the form of straight or curved slender rods (about 3 mm x 0. 3 mm) with parallel sides: it is often found in groups. Tubercle bacilli have been described as Gram positive, though strictly speaking, this is not correct, as after staining with basic dyes they resist decolourization by alcohol even without the mordanting effect of iodine. When stained with carbol fuchsin by the Ziehl Neelsen's method or by fluorescent dyes (auramine O, rhodamine) they resist decolourisation by 20% sulphuric acid and absolute alcohol for 10 minutes (acid and alcohol fast). Acid fastness has been ascribed variously to the presence in the bacillus of an unsaponifiable wax (mycolic acid) or to a semipermeable membrane around the cell. With fluorcochrome dye, they appear as fluorescent under blue light.

Culture Characteristics

M. tuberculosis is an aerobe, but grows more slowly than most bacteria and demands special media. Colonies appear in about two weeks and may sometimes take up to eight weeks. Optimum temperature is 37^0c and growth does not occur below 25^0c or above 40^0c. Optimum pH is 6.4-7.0.

It is usually possible to distinguish the human and bovine types by their cultural characteristics on solid egg media. The first grows luxuriantly in rough colonies and is described as 'eugonic'; the second is 'dysgonic' growing slowly in smooth colonies. These features are not invariable and may be lost on subculture. The addition of 0.5% glycerol improves the growth of *M.tuberculosis*, but has no effect or may even impair the growth of *M.bovis*. Sodium pyruvate helps the growth of both types. Human tubercle bacilli do not grow in the presence of p-nitrobenzoic acid.

Tubercle bacilli do not have exacting growth requirements but are highly susceptible even to traces of toxic substances like fatty acids in media. The toxicity is neutralized by serum albumin or charcoal.

Several media, both solid and liquid have been described for the cultivation of tubercle bacilli. The solid media are :

Lowenstein Jensen	-	Contains egg
Petragini	-	"
Dorset	-	"
Tarshis	-	Contains blood
Loeffler	-	Contains serum

The solid medium most widely employed for routine culture is Lowenstein Jensen (LJ) medium without starch, as recommended by the International Union Against Tuberculosis (IUAT). This contains coagulated hen's egg, mineral salt solution, asparagines and malachite green, the last acting as a selective agent inhibiting other bacteria.

On solid media, *M. tuberculosis* forms dry, rough, raised, irregular colonies with a wrinkled surface. They are creamy white, becoming yellowish or buff coloured on further incubation. *M. bovis* colonies, in comparison are flat, smooth, moist, white and breakup easily when touched.

Liquid media as – Dubos', Middlebrook's, Proskauer and Beck's, Sula's and Sauton's media are more common. Liquid media are not generally employed for routine cultivation, but are used for sensitivity testing, chemical analysis and preparation of antigens and vaccines.

Resistance

The tubercle bacillus is susceptible to desiccation, heat and other physical agents, but will resist some of the common antiseptics, possibly because of the hydrophobic character of bacterial cell wall. It survives exposure to 5% phenol, 15% sulphuric acid, 3% nitric acid, 5% oxalic acid and 4% sodium hydroxide. They are sensitive to formaldehyde, glutaraldehyde, tincture of iodine and ethanol.

Mycobacteria are killed at 60^0c in 15-20 minutes. Cultures may be killed by exposure to direct sunlight for two hours, but bacilli in sputum may remain alive for 20-30 hours. Bacilli in droplet nuclei may retain viability for 8-10 days under suitable conditions. Cultures remain viable at room temperature for 6-8 months and may be stored for upto two years at 20^0c (Ananthanarayan, 2002).

Antigenic Properties

The chemical composition of *M. tuberculosis* has been the subject of much inquiry. It varies qualitatively and quantitatively in different cultures depending in part on the media used. Following infection by tubercle bacilli, delayed hypersensitivity is developed to the bacillary protein (tuberculin). Human and bovine strains cannot be distinguished by skin tests.

The basic backbone of mycobacterial wall structure is formed by peptidoglycan, a polymer of N-acetyl muramic acid and N-acetyl glucosamine as in other bacteria. There are lipopolysaccharide side chains that are branched on the muramic acids. The lipopolysaccharide is made of polysaccharide arabinogalactan which is esterified at its distal end with mycolic acid that forms the major part of the outer layer of the cell wall. In addition to this, mycobacteria can have several other complex molecules associated with the outer layer of the cell wall. Apart from them, the cell wall also contains proteins and peptides which impart the quality of tuberculin like activity and often acts as receptor for iron uptake (eg. *M. smegmatis*) (William *et al.*, 2000).

Cord factor which are sulfatides or sulfated acyltrehaloses found on the outer wall of some *M.tuberculosis* may play a role in virulence because of their capacity to prevent phagosome - lysosome fusion in cultured macrophages infected with *M. tuberculosis*. Mycosides are types of glycolipids or glycosides that act as a component of the cell wall and impart phase binding properties. Further, they are serologically active and type specific. So they are used in seroagglutination technique for identification and classification of mycobacteria.

The mycolic acid in the mycobacterial wall is responsible for the acid fastness of mycobacteria. The high lipid content of the cell wall makes it hydrophobic. It imparts resistance to injury from sulfuric acid, NaOH and / or detergents but susceptibility to heat, X-ray, UV rays and alcohol.

Thus the common mycobacterial antigens are :

1. Polysaccharide Antigens :

- Arabinogalactan
- Arabino mannan
- D-glucan

They are probably identical for all mycobacterial species.

2. Protein Antigens :

- PPD (Purified Protein Derivatives) - prepared from 'old tuberculin' (OT)
- 65,000 dalton heat shock protein (65 KD hsp) - they are supposed to have an important function in maintaining the cell integrity and are produced by the organism under stress as high temperature and sub optimal culture conditions. This protein is highly antigenic and has several epitopes. Some of these epitopes appear to be species specific and some because of their similarities to epitopes

of mammalian proteins are thought to be responsible for human autoimmune conditions.

- Antigen 5: It is a species specific 38,000 dalton protein antigen. It is used extensively for serodiagnosis by ELISA. It is a potent antigen with substantial (but not absolute) degree of species specificity.
- Antigen 6 (antigen of 85 complex): This is formed by three antigens designated as 85 A, 85 B and 85 C. They appear to be encoded in separate genes and share some common epitopes. Antigen 6 (30,000 D) is a member of 85 B and is secreted by growing mycobacteria and binds to fibrinectin. It is under extensive study as a serodiagnostic antigen.

3. The 30,000 D antigen can evoke:

a. Strong skin response

b. Good IgG response

c. Elicits in vitro correlates of cell mediated hypersensitivity

Antigen p32 is another protein antigen of 85A complex.

4. Mycosides (William, 2000)

Immunity

The immune mechanisms of tuberculosis differ from those of most bacterial infections. The importance of humoral antibodies is uncertain although their presence can be detected by the usual methods.

Tubercle bacilli do not contain or secrete a toxin. The exact basis of their virulence is not understood but seems to be related to their ability to survive and multiply in macrophages. Various components of the bacillus have been shown to possess different biological activities which may influence the pathogenesis, allergy and immunity in the infection.

The various factors responsible for virulence are:

- Cord Factor
- Sulfatides
- Tubercular heat sock protein (hsp) (65KD)
- Complement activation
- LAM – This is a major hetropolysaccharide, structurally similar to the endotoxin of Gram negative bacteria. It inhibits macrophage activation by IFN γ (interferon gamma) and IL-10 (interleukin 10) that suppress mycobacteria incduced T cell proliferation. It also causes macrophages to secrete TNF responsible for fever and weight loss.

The types of immune responses found in tuberculosis are:

- Cell mediated (CD 4+) –delayed hypersensitivity
- Antibody mediated immunity and
- Autoimmunity

The various phases of type IV immune response are:

a) **Phase of Cognition:** When antigen presenting cells (macrophages and dendritic cells) present the antigenic epitope associated with class II major histocompatibility complex to the specific clone of CD4+ T lymphocytes.

b) **Phase of Activation :** After cognition the CD 4 cells are also co-stimulated by the IL-1 and IL – 6 and start secreting cytokines (IL-2) and express membrane receptors for IL-2 (IL-2R). This results in autocrine stimulation of activated CD 4 cells.

c) **Effective Phase :** As a result of autocrine and paracrine stimulation by multiple cytokines, the CD 4 cells proliferate and differentiate. They form clones of T cells that elaborate cytokines and mediate the effective phase of delayed type of hypersensitivity reactions.

This sequence completes the type IV hypersensitivity response leading to killing of mycobacteria by the activated macrophages by virtue of reactive nitrogen intermediates (NO, NO_2 and HNO_3). This is associated with the formation of epitheloid cell granuloma and clearance of the organism.

There are two different subsets of CD4+ T helper cells-called Th1 and Th2 cells as found in mice. In humans, they are called 'Th1 like' and 'Th2 like' cells. They have characteristic patterns of lymphokine secretion and they stimulate reciprocal immune functions.

Th1 cells - are responsible for delayed type hypersensitivity.

Th2 cells - provide B cell mediated reactions and augment immunoglobulin synthesis.

The B cells lie partially activated by forming combination of MHC II antigen complex on their surfaces. These partially activated B cells are stimulated by activated CD 4+ cells into fully activated B cells that further differentiate into plasma cells leading to secretion of immunoglobulins.

The important feature is the alternation in tissue reactivity which occurs as a result of infection. This was established by Koch and the phenomenon bears his name. He showed that if virulent tubercle bacilli were infected sub-cutaneously into a healthy guinea pig, the puncture wound healed in a day or two, but fourteen days later a nodule formed at the site of the injection which later ulcerated. Subsequently, the regional lymph nodes enlarged and caseated. The term 'Caseation' is used to describe a form of necrosis characteristic of tuberculosis in which the affected tissue is converted into a homogenous structureless material with the appearance and consistency of cheese. The process of autolysis is halted before the stage of liquefaction. If the injection was made into a guinea pig previously infected with tuberculosis, a nodule appears within two days at the site of injection to be rapidly followed by ulceration. No enlargement of regional lymph nodes would take place and the ulcer would usually heal rapidly.

Pathogenesis

The relationship of the infecting organism and host are more complex in tuberculosis than in any other infection, largely because the tissue reactions are modified by the development of hypersensitivity to the invading bacilli.

The changes in the tissues are of two kinds:

- Exudative and
- Proliferative

They are best followed in the lung of the experimental animal. In the earliest stages a fluid exudate is seen to surround the infecting organisms. In it are a few neutrophils some of which may contain bacilli. The exudate diminishes, most of the neutrophils perish and phagocytic monocytes appear which engulf the dead leukocytes and the bacilli. This lesion may be microscopic or may affect an entire lobe depending on the dose of bacilli and the suspectibility of the host.

If there has been previous tuberculosis infection with resulting hypersensitivity, the exudative response is great; and only moderate if the animal is not hypersensitive. There are three possible sequels:

1. First, there may be resolution and restoration of the status quo-ante: this is likely when the infecting dose is small and resistance is high.
2. Secondly, when the infecting dose is great and hypersensitivity well developed, necrosis and sloughing may result with the formation of a cavity.
3. Thirdly, a proliferative or productive lesion may develop. In this case, the monocytes are followed by epitheloid cells with pale yellow cytoplasm and vesicular nuclei which arrange themselves radially. After two or three weeks; fibroblasts, lymphocytes and plasma cells dispose themselves around the circumference of the lesions while in the centre one or more giant cells appear. These may exceed about 100 mm in diameter and they contain many dark staining nuclei arranged round their periphery. This lesion now constitutes the microscopic tubercle, the hallmark of tuberculosis infection.

Primary infection in man is commonly pulmonary and the sequence of events has been extensively studied in the lungs.

Four stages in the pathogenesis of tuberculosis have been described based on classical animal studies:

Stage 1 :	Onset	- 1st Week
Stage 2 :	Logarithmic bacillary growth	- 1st and 2nd week
Stage 3 :	Control by cell mediated immunity and DTH	- after 3 weeks
Stage 4 :	Liquifaction and lung cavity formation	

Bacilli, usually contained in droplets of infected sputum, are inhaled and carried to the periphery. A parenchymal lesion develops which may be trivial and cause no symptoms or may be of massive exudative type. Infected material usually passes by the lymphatic vessels to the regional hilar nodes where further lesions appear.

In most cases primary tuberculous infection heals rapidly, having passed unrecognized and leaves a small area of calcification at the periphery of one lung with a similar lesion in the related lymph node. The initial parechymal lesion is called a 'Ghon's focus' and it, together with the infected hilar nodes constitute the Ghon's complex.

Once the primary lesion has healed, the patient will have acquired hypersensitivity to tuberculoprotein and subsequent infection which is often termed 'post primary' will follow a course modified by this fact. Further lesions are due to reinfection from external sources or possibly by the bacilli escaping from lesions within the body.

Spread of Infection in The Body

M. tuberculosis is an intracellular organism although it is capable of extracellular multiplication in pulmonary cavities and similar situations. Nevertheless its normal habitat limits its method of spread. It can spread in the body by a number of ways :

1. The first of these is by simple extension to contiguous structures.
2. The second is by lymphatic dissemination: cells containing bacilli are carried by the lymphatic channels to regional lymph nodes which become the seat of active infection. Chains of lymph nodes may be affected in this fashion and material containing bacilli may enter the thoracic duct and thus the blood stream.
3. The third route of spread is by the blood stream itself. A caseous lymph node or other lesion ruptures into a vein discharging its content of bacilli into the circulation. This mechanism is responsible for miliary tuberculosis.
4. Finally dissemination may occour by way of the bronchi, alimentary tract or urinary passages.

a) In the first instance, the contents of the cavity may be aspirated into another part of the lung causing tuberculous bronchopneumonia.

b) In the second, infected sputum may lead to laryngeal tuberculosis or be swallowed to cause tuberculous enteritis.

c) In the third, discharges from tuberculosis of the kidney may infect the ureter and the bladder.

Epidemiology

There are two reservoirs from which tuberculous infections spread: the patient with active pulmonary tuberculosis and the tuberculous dairy cow. From the first, droplets containing mycobacteria are inhaled and gives rise to pulmonary tuberculosis; from the second infection of cervical or mesenteric nodes, perhaps with later tuberculosis of bones or joints arises. Other modes of transmission are rare. Occasionally a skin abrasion becomes infected and this was once a common occupational risk in butchers. Discharges from sinuses and contaminated excreta may generally be the vehicle of infection and oral transmission as has been recorded in infants.

Secondary factors have received much attention.The possibility of some genetic predisposition to infection exists. It has been shown, for instance that when one twin develops tuberculosis, infection of the other occurs in one of three cases when binovular and in three out of four when uniovular. The racial difference may well be due to differences in living conditions, although there is little doubt that American Negroes are more likely to develop an epidemic form of the disease. Malnutrition is commonly accompanied by overcrowding which will favour the spread of infection. It seems doubtful whether age has any great influence. It was formally believed that children and infants were more susceptible but it is clear that the age at which primary infection is acquired is steadily rising.

There are certain diseases in which an undue liability to tuberculosis has long been recognized: chief amongst these are diabetes mellitus, cirrhosis of the liver, chronic alcholism and chronic leukaemias. The incidence in these special instances has decreased with the more effective management of the disease in question.

Prolonged administration of corticosteroid drugs has been shown to reduce reistance to tuberculosis and to reactivate lesions previously considered healed. Because they often suppress the general symptoms of infection, the clinical picture may be unusual and the disease often eludes diagnosis until the process is far advanced.

Molecular Epidemiology: Achievements and Future Prospects

Our understanding of the transmission of tuberculosis (TB) has been greatly enhanced since the introduction of deoxyribonucleic acid (DNA) fingerprinting techniques for *Mycobacterium tuberculosis* in the early '90s.

In the '90s, a wide variety of genetic markers for *M. tuberculosis* were identified. However, only a minor number of these appeared to offer enough discrimination and reproducibility for wide scale implementation (Kremer, 1999; Kremer, 2005). In 1993, IS6110 Restriction Fragment Length Polymorphism (RFLP) typing was adopted as the standard method for routine typing of *M. tuberculosis* (van Embden, 1993). In this method, chromosomal DNA is digested with restriction enzyme *PvuII*. The digested DNA is separated on an agarose gel and, after Southern Blotting, hybridized with a DNA probe. This DNA probe is directed to the IS6110 insertion sequence and labelled with peroxidase, enabling enhanced chemiluminesence (ECL) detection of IS6110 containing restriction fragments. Another typing method, 'spoligotyping' has been used extensively as a secondary typing method (Bauer, 1999) and as a marker to study the phylogeny of the *M. tuberculosis* complex. Spoligotyping exploits the polymorphism in the direct repeat region of *M. tuberculosis* complex strains. This region consists of direct repeats interspersed with unique spacer sequences, and is amplified by polymerase chain reaction (PCR) with primers directed to the repeats. The PCR-product is subsequently hybridized to known spacer sequences which are immobilized on a membrane through reversed line-blotting. Because one of the primers, and hence the PCR product, is labelled with biotin, ECL detection is achieved after incubation with peroxidase labelled streptavidin. Another DNA typing method frequently used for *M. tuberculosis* is Variable Numbers of Tandem Repeats (VNTR) typing. Typing results of this method are expressed as numerical codes. Each number of the code represents the number of tandem repeats at a particular repeat locus. The number of repeats varies by strain and is determined through PCR amplification of the repeat locus with primers directed to the regions flanking that repeat locus and determination of the PCR product size. After an extended period of improvement and validation, VNTR typing is now ready to become the next gold standard for typing of *M. tuberculosis* complex isolates (Supply, 2006).

The first report on the use of IS*986* RFLP to examine transmission of TB was published in September 1990 (Hermans, 1990).

In the beginning, strain typing was mainly used to study outbreaks of TB and institutional transmission. Soon thereafter, in multiple population based studies, the rate of recent transmission and risk factors for transmission were determined. Active transmission of TB in low prevalence settings appeared to be associated for a large part to particular risk groups such as drug abusers, homeless people, and certain immigrant groups. Transmission of drug resistant bacteria could be compared to that of drug-susceptible strains (van Doorn, 2006).

Although it has become clear that the phylogeny of *M. tuberculosis* differs significantly in several geographic areas, not much is known about the dynamics of the population structure and the reasons for the genetic conservation observed among *M. tuberculosis* isolates in high prevalence areas. If particular genotypes of *M. tuberculosis* are selected, how fast does a shift towards more adapted variants occur? Are we influencing the spread of particular genotypes of *M. tuberculosis*? Best studied in this respect is the Beijing genotype family of *M. tuberculosis*. There are indications that there is indeed a dramatic and relatively fast change in the composition of the worldwide population of *M. tuberculosis*. If the current observations hold true, we may be facing a recurrent TB epidemic caused by bacteria with a higher level of evolutionary development. However, more research is needed to draw better conclusions.

Genetic Heterogeneity of M. tuberculosis and Multiple Infections

When talking about multiple *M. tuberculosis* sub populations in sputum of TB patients, two phenomenon are often confused, although they should be clearly distinguished:

- Multiple strain populations derived from a single ancestral strain displaying genetic drift and
- Multiple infections by more than one strain.

In the case of multiple (or mixed) infections, the presence of more than one *M. tuberculosis* strain is demonstrated on one occasion of culturing from clinical material. This should not be confused with re-infection, usually after curative treatment, as this refers to a new episode of the disease caused by another strain.

Numerous observations in the molecular epidemiology of TB have pointed out that bacteria are subject to evolutionary change. Sometimes minor rearrangements of IS*6110* RFLP profiles are noticed in epidemiologically related and serial patient isolates. The rate of change of IS*6110* RFLP patterns in such isolates has been studied by investigators (de Boer, 1999; Niemann, 2000). However, also within clinical *M. tuberculosis* isolates, sub populations of bacteria with minor genomic differences coexist. For example, low intensity bands in IS*6110* RFLP profiles are reliable indication of a sub population of bacteria with, for example, a one band difference in IS*6110* RFLP. Preparation of single colony cultures and subsequent IS*6110* RFLP typing of isolates with such low intensity bands showed the co-existence of separate sub populations of bacteria, either with or without a normal intensity band at the position where the low intensity band occurred in the original clinical isolate.

Though the introduction of molecular markers for *M. tuberculosis* in the early '90s has greatly facilitated our understanding of the epidemiology of TB, even the latest most optimal typing, VNTR typing, will not be completely reliable. In fact, each genetic marker only reveals a minor part of the genomic information of a bacterium. Depending on the marker, different strains will exhibit identical genotyping profiles. Furthermore, in order to be able to follow the chains of transmission in a given area and to subdivide primary, secondary, etc. sources of infection, the turnover of genotyping profiles will never be in range with the pace of transmission.

To distinguish between even genetically related strains, and to be able to follow the spread of offspring of strains in the community, more detailed multiple marker typing systems need to be developed. In fact, the most accurate typing would be whole genome sequence analysis of *M. tuberculosis* isolates. It is expected that with this information, the exact sequence in the evolutionary development of the offspring of a *M. tuberculosis* bacterium can be identified, without the interference of differences in time of incubation and confusion about the spread from primary and secondary, or even tertiary sources of infection in the same period.

Furthermore, the evolution of bacteria does not take place through whole population shifts in the genomic make up, but through mutation and multiplication of initially a single bacterium. By applying the current DNA amplification and sequence techniques, subtle genetic changes among bacterial strains are still difficult to visualize. However, creative future solutions may also deal with this phenomenon.

New Diagnostic Methods

The diagnosis of mycobacterial infections remained practically unchanged for many decades and probably would not have progressed at all without the unexpected resurgence of tuberculosis (TB) which characterized the last twenty years of the 20th century.

With microscopy lacking wide margins for improvement, the areas which most benefited from the renewed interest in TB were culture and identification, while a completely new approach emerged, aimed towards the direct detection of mycobacterial nucleic acids in clinical specimens. The following diagnostic methods can be used for the detection of the tubercle bacilli:

A) Automated Culture Methods

Although known for decades, liquid media for cultivation of mycobacteria had never attracted the attention of mycobacteriologists. In fact, the ability of a liquid medium to support a faster growth was heavily hampered by its susceptibility to contamination. The use of anti microbial combinations suitable of inhibiting the growth of the whole spectrum of potential contaminants (Gram positive and Gram negative bacteria as well as fungi) represented a turning point.

During the same period, automation was taking its first steps in microbiology, with blood cultures leading the field. The idea of exporting such technology to mycobacterial cultures evolved into selective liquid media, which were a breakthrough for diagnostic mycobacteriology.

A.1) BACTEC TB 460

The BACTEC TB 460 system (Becton Dickinson, Sparks, MD) was the first, and for many years the only, automated approach in mycobacteriology. It makes use of a radiometric instrumentation developed for blood cultures with the broth bottles replaced by vials containing a medium specific for mycobacteria.

In BACTEC TB 460 a modified Middlebrook 7H9 medium is used, in which one of the components, palmitic acid, is radiolabelled with ^{14}C. Contamination is controlled by the addition, prior to use, of a mixture of polymyxin B, amphotericin B, nalidixic acid, trimethoprim and azlocillin (PANTA) reconstituted with a poly-oxyethylene solution. The use of such a combination of antibiotics does not eliminate the decontamination step, which needs to be performed before inoculation of the samples. The vials containing the medium remain sealed through the whole culture process and the specimen is inoculated by puncturing the rubber septum with a needle.

Once the paired needles have perforated the rubber septum of the vial, the gaseous phase is aspirated and replaced with air containing 5 % CO_2. The aspirated gas is analyzed by a β-counter to quantify the eventual presence of radiolabelled CO_2.When viable mycobacteria are present in the culture vial, the radiolabelled palmitic acid is metabolized and radioactive CO_2 is liberated into the gaseous phase.

The BACTEC TB 460 was first commercialized in 1980 and soon became popular worldwide. It is not a fully automated system, as the vials, which are held in an external incubator, must be loaded into the instrument for reading. The reading is usually performed twice a week during the first 15 days of incubation, and weekly thereafter, until the 42nd day. BACTEC TB 460 is still used in many laboratories worldwide but the interest of the manufacturer to promote newly developed alternative systems is slowly taking over its still excellent performance.

From the first evaluations, BACTEC TB 460 revealed a clear superiority over solid media systems in terms of sensitivity and time of detection of positive culture (Middlebrook, 1977; Park, 1984). Many years later, despite the release of new systems, the BACTEC radiometric method is still competitive (Ganeswrie, 2004). The BACTEC TB 460 system is also suitable for mycobacterial blood culture using designated bottles.

A.2) BACTEC MGIT960

The BACTEC MGIT960 system (Becton Dickinson, Sparks MD) uses the technology of the previously developed blood culture instrument. The original system (BACTEC 9000) was first adjusted to support mycobacterial cultures but was subsequently completely redesigned to process tubes, which are much less cumbersome to handle than the original bottles.

Mycobacteria Growth Indicator Tube (MGIT) is a modified Middle Brook 7H9 medium in which a supplement is added at the moment of use. The supplement is a mixture of oleic acid, albumin, dextrose, and catalase (OADC) enrichment and the same PANTA antibiotic mixture used in the radiometric system. The presence of PANTA does not do away with the decontamination step, which needs to be done before inoculation. As the tubes containing the medium are screw capped, no needle is needed for inoculation. A silicon film embedded with a ruthenium salt is present at the bottom of the tube as a fluorescence indicator. The incubator and reader are combined in a single cabinet. The bottom of each tube, stimulated by ultraviolet light, is monitored by a fluorescence reader. Fluorescence emitting tubes are reported as positive. It is also possible to use the MGIT tubes without instrumentation, by holding the tubes in a normal incubator and observing the fluorescence under a Wood's lamp.

The oxygen normally present in the medium quenches the natural fluorescence of the ruthenium salt. If viable mycobacteria are present in the tube, oxygen is consumed due to their metabolism, the quenching effect lowers accordingly, and the bottom of the tube fluoresces when exposed to ultraviolet light. The BACTEC MGIT960 monitors the tubes at one hour intervals, alerts when they become positive and signals the end of the incubation period.

Many studies have evaluated the BACTEC MGIT960 in comparison with similar competitor systems and also with solid media. BACTEC MGIT960 turned out to be clearly faster and more sensitive than solid media, while the comparison with other automated and semi-automated systems revealed substantially overlapping performance.(Alcaide, 2000; Badak, 1996). Mycobacterial blood cultures cannot be performed with this system.

A.3) Versa TREK

The VersaTREK (previously known as the ESP system II) uses the technology of a previously developed blood culture system and is commercialized by Trek Diagnostic Systems.

It uses a modified Middlebrook 7H9 medium to which the OADC enrichment must be added. Two different antimicrobial mixtures are available. The first one, also known as AS, includes polymyxin B, azlocillin, fosfomycin, nalidixic acid, and amphotericin B. The second contains polymyxin B, vancomycin, nalidixic acid, and amphotericin B (PVNA). Usually, AS is used, while PVNA is used for heavily contaminated samples. The presence of such antimicrobial mixtures for contamination control does not eliminate the decontamination step, which needs to be performed before inoculating the sample. The bottles of medium hold a cellulose sponge whose large surface area allegedly improves growth. Bottles are inoculated through a rubber septum by means of a syringe.

Incubator and reader are combined in a single cabinet, which also shakes the bottles during the incubation. The pressure within each bottle is monitored by a manometer through a proper connector. Cultures presenting a decreased headspace pressure are reported as positive. If viable mycobacteria are present in the bottle, the oxygen consumption due to their metabolism reduces the internal pressure.

Several studies have evaluated VersaTREK in comparison with solid media and similar automated and semi-automated competitor systems. VersaTREK clearly performs better than solid media but shows no substantial advantage over other systems (Tortoli, 1998; Williams Bouyer *et al.*, 2000).

Mycobacterial blood cultures can also be performed with the VersaTREK system. However, whole blood cannot be used and a previous treatment is required to obtain sediment for inoculation. Either a buffy coat or sediment obtained with the lysis centrifugation method is suitable to inoculate the bottles. The lysis centrifugation method consists of saponin containing tubes to lyse blood cells, a proper centrifugation procedure, and special pipettes for elimination of supernatant and collection of the sediment.

A.4) BacT/Alert 3D

BacT/Alert 3D (previously known as MB/BacT) is commercialized by bioMérieux and uses the technology of a previously developed blood culture system.

A modified Middlebrook 7H9 medium is used in which a supplement, a mixture of OADC enrichment and polymyxin B, amphotericin B, nalidixic acid, trimethoprim, vancomycin and azlocillin, is added at the moment of use. The presence of such contamination controlling antibiotics does not eliminate the decontamination step needed before inoculation. The bottles of medium have a CO_2 sensor at the bottom and are inoculated through a rubber septum by means of a syringe. The incubator and reader are combined in a single machine which does not shake the bottles during incubation.

The CO_2 sensor is impacted by a light whose reflected ray is monitored by a photodiode. Bottles producing specific changes in the intensity of the reflected light are reported as positive. If viable mycobacteria are present in the bottle, the CO_2 produced by their metabolism causes a change in the color of the sensor from green to yellow, which alters the intensity of the reflected light ray.

BacT/ALERT 3D monitors the bottles at 10 min intervals, alerts when they become positive, and signals the end of the incubation period. Many studies have evaluated the BacT/ALERT 3D. The system turned out to be clearly faster and more sensitive than conventional media, while the comparison with other automated and semi-automated systems did not reveal significant differences (Alcaide, 2000; Piersimoni, 2001). The system is also suitable for mycobacterial blood cultures, provided proper bottles are used; no previous treatment of the blood is required.

B) Nucleic Acid Amplification Methods

When the polymerase chain reaction (PCR) methodology took its first steps into diagnostic microbiology, a restricted number of microorganisms appeared to have the potential to benefit from the novel technique. *M. tuberculosis* was among them, and the dream of the rapid diagnosis of TB appeared to be about to come true.

One of the first findings on the way to developing a PCR technique aimed at *M. tuberculosis* detection was that, although different targets were investigated, none of them were suitable for differentiating *M. tuberculosis* from other species belonging to the *M. tuberculosis* complex. Such limitation, due to the extremely high genome similarity (close to 100 %) among the members of the *M.tuberculosis* complex, did not, however, cool the enthusiasm. In fact, the differentiation of such species is of very limited relevance from the clinical and therapeutic point of view.

Among the first proposed genomic targets for diagnostic PCR was the newly detected insertion element IS*6110* which being present in multiple copies (from 4 to 20 in more than 95 % of *M. tuberculosis* strains), appeared to have the potential for an enhanced sensitivity. Other successfully used deoxyribonucleic acid (DNA) regions include the 65 kiloDalton (kDa) heat shock protein gene, the gene encoding the 126 kDa fusion protein, and the gene encoding the β-subunit of ribonucleic acid (RNA) polymerase; all of them are present in single copies in *M. tuberculosis* complex genomes.

In the 90's, many laboratories developed various *in house* PCR protocols and soon the aspiration of an increased sensitivity led to the adoption of nested PCR. In this procedure, the amplification of a large region of DNA is followed by a second amplification targeted to a shorter internal stretch. Most of such *in house* PCR procedures achieved a sensitivity never matched by commercial systems but were often burdened by the high incidence of false positive results due to amplicon cross contamination of specimens.

In the last few years, several amplification methods have been commercialized.

B.1) Amplified MTD

Amplified *Mycobacterium tuberculosis* Direct Test (AMTD) was developed by Gen-Probe (San Diego, CA, USA). A *M. tuberculosis* complex specific region of the 16S ribosomal RNA gene produces double stranded ribosomal DNA, due to the combined action of reverse transcriptase and ribonuclease. In turn, RNA polymerase catalyzes the synthesis of multiple stretches of ribosomal RNA from the ribosomal DNA synthesized before. A new cycle starts when the newly produced ribosomal RNA undergoes further transcription by reverse transcriptase.

The sensitivity of the method is increased by the presence, in each bacterium, of a high number of 16S ribosomal RNA target molecules (about 2,000) compared to only one copy of 16S ribosomal DNA. Another advantage of the amplification from RNA relies on the low stability of such a molecule; this minimizes both the risk of contamination and the incidence of false positive results due to the persistency of stable nucleic acids (DNA) in the host organism, even after the complete eradication of the infection.

The detection of amplification products relies on hybridization with a specific, single-strand DNA probe labelled with a chemiluminescent molecule. The whole process is performed manually, starting with the extraction by means of sonication, continuing with the addition of different reagents until the final reading with the luminometer. Thermal cyclers are not needed and the whole amplification step is carried out on a heating block at 42°C. The turnaround time is 2.5 hours. No internal control is provided in the kit to monitor the presence of inhibitors. The method is approved by the Food and Drug Administration of the United States of America for testing smear positive and smear negative respiratory samples.

The sensitivity of this method had been reported to be in the range 91.7 % to 100 % in smear positive samples and from 65.5 % to 92.9 % in acid fast bacilli (AFB) smear negative samples by (Alcala, 2001; Piersimoni, 2003).

B.2) Amplicor MTB Test

The Amplicor MTB Test (Roche Molecular Systems, Basel, Switzerland) relies on standard PCR. A 584 bp fragment of the 16S ribosomal RNA gene, comprising a species-specific region flanked by genus specific sequences, is amplified using biotinylated primers. In the master mix, an unusual combination of nucleotides is present-as an adjunct to adenine, guanine and cytosine, uracil is used in place of thymine. As a consequence, the amplification product differs from the target DNA in that it contains uracil instead of thymine. This device is part of a contamination control system based on the use of uracil-N-glycosylase, an enzyme that fragments DNA wherever uracil is present. The enzyme, added to the samples before amplification, destroys any amplicon resulting from previous amplifications without damaging the uracil free target DNA. Because of the genus specific nature of the annealing regions, 16S ribosomal DNA belonging to any mycobacterial species is amplified by this PCR. The use, in the revealing phase, of magnetic beads coated with *M. tuberculosis* complex specific probes allows the removal, by washing, of any other DNA. The detection of the specific amplification product is performed by adding an avidin-enzyme conjugate and a chromogenic substrate.

The specificity of this method is close to 100 % while sensitivity ranges from 90 % to 100 % in smear positive samples and from 50 % to 95.9 % in smear negative ones (Bogard, 2001; Piersimoni, 2003)

B.3) BD ProbeTec ET

The BD ProbeTec ET (Becton Dickinson, Sparks, MD) uses DNA polymerase and isothermal strand displacement amplification to produce multiple copies of IS*6110*, an insertion element unique to *M. tuberculosis* complex. The rationale of strand displacement amplification is extremely complex.

Some manipulation is required before introduction of the sample into the automatic instrument; each sample is first inactivated at 105°C, and then sonicated to extract the DNA, transferred into a priming well at 72.5°C, and subsequently into an amplification well at 54°C. In the BD ProbeTec ET instrument, the microplate containing the samples and the amplification reagents is incubated at 52.5°C and the fluorescence emitted is continuously monitored. A thermal cycler is not required. The turnaround time is 3.5 to 4 hours.

This method has a sensitivity ranging from 98.5 % to 100 % for smear positive samples and is very variable (0.33 %-100 %) for smear negative ones (Barrett, 2002).

Although direct amplification methods are used worldwide, they are far from having revolutionized clinical mycobacteriology. Culture, supported by microscopy, still remains the gold standard, and molecular methods only represent a useful support in some cases, to speed up the diagnosis of TB. The unsatisfactory sensitivity is the major limitation of amplification based methods. It is now evident that paucibacillary specimens have little chance of being detected by molecular amplification. Factors that contribute to the reduction of the sensitivity are an uneven distribution of bacilli in the sample, the suboptimal extraction of nucleic acids and sometimes the presence of inhibitors. The phenol chloroform extraction unquestionably provides the best yield but being cumbersome and time consuming, also raises the risk of contamination. To minimize this risk and to make the technique user friendly, the commercial systems have probably oversimplified this step by reducing it to sonication or heat treatment only. The sediment of a number of samples contains substances inhibiting the amplification process. The reason for their presence is unknown and at present there is no known method for neutralizing them. The use of an internal control represents a major feature to be taken into account at the moment of choosing an amplification method.

Although the specificity of amplification methods is substantially good, the possibility of false positive results should be borne in mind by both microbiologists and clinicians. A therapy like the one against TB, which is long and not exempt from side effects, should not be undertaken only on the basis of a positive result of a single amplification test. The major reason for false positive nucleic acid amplification results is the contamination of samples, possibly in the pre-analytic, but mostly in the analytic phase. The application of dedicated procedures such as the one employing uracil N glycosylase or the adoption of sealed amplification chambers is useful. More important still are general precautions such as the frequent decontamination of the work environment with 10 % bleach and the exposure of pipettes, tips and bench surfaces to ultraviolet light when not in use.

Finally, a major role is played by the training and the expertise of the operator. A particular category of false positive results is that concerning samples obtained from patients under treatment. In these patients, the detectability of mycobacterial DNA over a long period of time, despite the effectiveness of the treatment, is well known and clearly makes DNA amplification useless for treatment monitoring. The above limitations and recommendations have also been stressed by the US Centers for Disease Control and Prevention.

C) Genetic Identification Methods

Following the extraordinary development of molecular methods, the identification of mycobacteria, previously based on phenotypic investigations, suddenly started to rely on genotypic methods. Different genetic approaches developed in research laboratories became rapidly popular in diagnostic laboratories and some of them were transformed into commercial diagnostic kits.

C.1) PCR Restriction Enzyme Analysis

The PCR restriction enzyme analysis (PRA) method is based on the amplification of a 441-bp fragment of the *hsp65* gene by PCR, followed by the digestion of the amplified product with two restriction enzymes *Bst*EII and *Hae*III according to the procedure first described by Telenti, 1993. The products of the digestion reaction are then separated and visualized by agarose gel electrophoresis. The restriction pattern thus obtained is compared to an algorithm present in the PRASITE. This database comprises 74 PRA patterns corresponding to 38 defined species of mycobacteria. In recent years, several novel PRA patterns from newly characterized species have been described. The PRA method has also been used with other genomic regions for the identification of mycobacteria such as the *rpoB* and *gyrB* genes with good results (Lee, 2000; Goh, 2006).

There are many studies on the application of the PRA-*hsp65* method for the rapid identification of mycobacteria. In general, the method has proved to be practical, cost-effective, and highly accurate for mycobacterial identification. The accuracy of the identification can be further improved when combined with minimal microbiological characteristics such as growth rate and pigmentation. Nevertheless, attention should be paid to a few technical details such as gel preparation and running, and some training is needed in the interpretation of patterns.

C.2) DNA Probes

The DNA probe technology for identification of fastidious organisms is still one of the most successful molecular diagnostic procedures worldwide. In this sense, the role played by commercial DNA probes in the quality improvement of mycobacterial identification in clinical laboratories cannot be disregarded.

C.2 (a) AccuProbe

The precursor system, AccuProbe (Gen-Probe, San Diego, CA), was developed almost 20 years ago. Different AccuProbe kits are available for identifying mycobacteria belonging to the *M. tuberculosis* complex, the *M. avium* complex and the species *M. kansasii, M. bacterium gordonae, M. avium* and *M. intracellulare*. They can be used on cultures grown either on solid or in liquid medium. The only equipment required is a sonicator for cell lysis and a luminometer for the final reading.

The sensitivity and specificity of AccuProbe are widely acknowledged in the literature relating to TB (Bull, 1992; Tortoli, 1996). Only a few unspecific reactions have been reported, mainly for the *M. avium* complex probe, which also hybridizes with the recently described species *M. palustre* (Torkko, 2002), *M. parascrofulaceum* and *M. saskatchewanense*.

C.2 (b) Line Probe Assay

The line probe assay uses the reverse hybridization technology with differently specific DNA probes immobilized in parallel lines on a paper strip. The target DNA, previously extracted by boiling, is PCR amplified using biotinylated primers and finally incubated with the strip. Once the hybridization has been carried out under highly stringent conditions and the unbound amplicons have been washed out, the hybridized probe is revealed as a colored band, developed following the addition of a streptavidin-labelled enzyme and a chromogenic substrate. The specificity of the hybridized line probe is inferred by the position of the colored band on the strip. Three commercial methods are available, INNO-LiPA MYCOBACTERIA (Innogenetics, Ghent, Belgium), GenoType Mycobacterium (Hain, Germany), and GenoType MTBC (Hain, Germany).

C.3) Genetic Sequencing

Every genetic region which is highly conserved and at the same time includes moderately variable sequences is a potential target for identification. A number of such regions are known in the genome of living organisms; among them the best known include genes for 16S ribosomal DNA, 23S ribosomal DNA, *hsp65* and the non encoding stretch ITS.

The 16S ribosomal DNA, which is about 1,500 bp long, is by far the most popular target for sequencing and the one for which the largest database is available. In this gene, universal sequences shared by practically every living organism co-exist with genus-specific sequences common to the organisms belonging to the same genus (e.g. the genus *Mycobacterium*) and with species specific sequences that differentiate between species. With regard to the mycobacteria, almost all the sequences characterized by species specific variability are concentrated in the first third of the gene. A practical consequence is that the determination of the nucleotide sequence of the first 500 bp of the gene allows the differentiation of almost all mycobacterial species known at present. Genetic sequencing is nowadays performed using automatic sequencers that render it highly reproducible. Once a sequence has been determined, its comparison with known sequences is required.

D) Non Conventional Phenotypic Diagnostic Methods

In addition to the so called conventional methods for TB diagnosis and besides the automated and molecular diagnostic methods described above, some new technologies have been proposed, such as phage based assays and rapid detection of growth by microscopic observation of microcolonies in solid or liquid media and analysis of cell wall mycolic acids etc.

D.1) Thin Layer Chromatography (TLC)

TLC uses silica plates (stationary phase) on the surface of which the mycolic acids extracted from the mycobacterial strain are separated as a result of their different affinity for a solvent (mobile phase) advancing by capillarity. Once the plate has been stained, each species displays a particular dot pattern according to its mycolic acid content that can be identified by comparison with patterns of reference strains with known mycolic acid composition run in parallel (Minnikin, 1975)

The high number of mycobacterial species, already over 130, has substantially scaled down the relevance of TLC for identification at the species level. With only seven types of mycolic acids and with most mycobacteria including no more than two or three of them, the number of TLC patterns shared by more than one, and often by many species is high.

D.2) Gas Liquid Chromatography (GLC)

In GLC, a gas (mobile phase) is used to carry the sample through a liquid (stationary phase) contained in a column. Once the lipids extracted from a mycobacterial strain have been injected, the high operating temperature of the column (about 300°C) produces the cleavage of the mycolic acids in saturated methyl esters 22, 24 and 26-carbon atoms long. Along with such mycolic acid fragments, saturated and unsaturated fatty acids (including tuberculostearic acid) and alcohols are eluted.

The recognition of different elution products is usually obtained by mass spectrometry. Cleavage products which are unvaried within single mycolic acid types, produce along with fatty acids and alcohols, patterns consistent to single species and suitable for their differentiation (Guerrant; 1981; Lambert, 1986).

The major problems concerning the GLC are the limited inter laboratory reproducibility and the insufficient discriminative power related to the high number of *Mycobacterium* species.

D.3) High Performance Liquid Chromatography (HPLC)

The HPLC uses high pressure to carry a liquid (mobile phase) containing the extracted sample through the particulate (stationary phase) present in the column. The various types of mycolic acids, previously saponified, extracted and derivatized to bromophenacyl esters are separated in the column and eluted at different times. On the basis of individual ultraviolet absorbance, the detector plots single fractions as peaks arranged in a profile. The profile of each species is sufficiently different from those of other species to provide identification when visually compared with profiles of known mycobacteria (Butler, 1991). A fluorescence based detection system may also be used that is more sensitive than the ultraviolet based system.

For many years, HPLC has been considered the only phenotypic method suitable for differentiating almost all mycobacterial species. The number of species hardly differentiable or not distinguishable at all, which was insignificant until a few years ago, has recently increased due to the continuous description of new species, in particular, of rapidly growing mycobacteria.

E) Smear Examination For AFB

This is of use in distinguishing a few bacterial species as tubercle bacilli from all other. The acid fast bacteria are relatively impermeable and resistant to simple stain but when stained with a strong reagent (basic fuchsin 5% phenol with heat) can resist decolorization by strong acid. The decolorized non acid fast organisms are counter stained with a contrasting colour eg. malachite green or methylene blue. These bacteria are exceptionally rich in lipid fatty acids in their cell wall and because of this, the bacteria become acid fast. When lipids are removed by treatment with solvents the cells are no longer acid fast (*M. leprae*).

Some other acid fast organisms are :

- *Corynebacterium spp.*
- *Nocardia spp.*
- *Rhodococcus equi.*
- *Isospora spp.*
- *Cryptosporidium spp.*
- *Legionella micdadci.*

The acid fast smear remains of enormous value in the primary care setting because in the hands of a trained microbiologist, it is easy to perform and reliably interpreted. A flow chart for the preparation of Kinyoun (cold carbol fuchsin), Ziehl Neelsen and flurochrome stained smears is as follows:

Steps	Kinyoun	Ziehl-Neelsen	Fluorochrome
Apply specimen	Apply specimen to central area of slide and allow to air dry	Apply specimen to central area of slide and allow to air dry	Apply specimen to central area of slide and allow to air dry
For all 3 methods	Fix by immersing in absolute methanol for 1 min.	Fix by immersing in absolute methanol for 1 min.	Fix by immersing in absolute methanol for 1 min.
Primary Stain	Kinyoun carbolfuchsin, 2 min	Carbolfuchsin, cover with filter paper, heat to steaming, let it stand for 5 min, remove paper	Auramine-O or auramine-rhodamine, 15 min.
Rinse	Rinse gently with water and drain excess from side	Rinse gently with water and drain excess from side	Rinse gently with water and drain excess from side
Decolorize	3% acid alcohol until color no longer runs from slide	3% acid alcohol until color no longer runs from slide	0.5% acid-alcohol, 2 min.
Rinse	Rinse gently with water and drain excess from side	Rinse gently with water and drain excess from side	Rinse gently with water and drain excess from side
Counterstain	Methylene blue or brilliant green, 30 sec	Methylene blue or brilliant green, 30 sec	Acridine orange or potassium permanganate, 2 min.
Rinse	Rinse gently with water and drain excess from side	Rinse gently with water and drain excess from side	Rinse gently with water and drain excess from side
Dry	Allow slide to air dry without blotting	Allow slide to air dry without blotting	Allow slide to air dry without blotting
Examine thoroughly	1,000X	1,000X	Screen 200-400 X, Confirm 600-1,000X

Every patient who has cough for three weeks or more, with or without other symptoms, should have three sputum samples examined for AFB (RNTCP, 2000). When pulmonary TB is suspected, at least three sputum specimens (SPOT-EARLY MORNING-SPOT) should be collected and examined by microscopy. Ideally all the three specimens should be collected within two days.

If the patient is attending a Peripheral Health Institute (PHI) which is also a designated microscopy centre, sputum will be examined at the same facility. If the patient is attending a PHI which is not a designated microscopy centre, there are two options:

1) either the sputum container can be transported to the microscopy centre, or
2) the patient himself can be referred to the microscopy centre.

What ever arrangement is most convenient to the patient and ensures prompt diagnosis should be used.

Guidelines For Collecting Sputum

The following guidelines (RNTCP, 2000) should be followed while collecting the sputum samples:

1. A specimen is collected on the spot when a patient is suspected of having TB. The patient is given the sputum container with his laboratory serial number written on it. The person collecting the sputum demonstrates how to open and close the container , takes the patient to an open space far away from other people and demonstrates with actual actions how to bring out sputum. The patient is instructed to inhale deeply 2-3 times with his mouth open, cough out deeply from the chest, open the container and spit out the sputum into it, and close the container. This is called a **spot specimen.**
2. The patient is then given a sputum container with his laboratory serial number written on it (not on the lid) to collect an early morning specimen before his second meeting with the laboratory technician. This is called an **early morning specimen.**
3. When the patient returns with the early morning specimen, a second **spot specimen** is collected in another container under the supervision of a staff member (Central TB Division, 2000).

A good sputum sample is

- thick (semi-solid), coughed out deeply from the lungs
- purulent (yellowish mucus)
- sufficient in amount (at least 2 ml)

A poor quality sputum sample

- contains only saliva or nasal mucus
- is small in quantity (less than 2 ml)

A good sputum container is

- Disposable

- Clean
- Wide mouthed
- Unbreakable
- Easily burnt
- Made of clear thin plastic
- Leak proof
- Provided with tight fitting lid (RNTCP, 2000)

The smear is examined by oil emersion method and at least 100 fields should be examined systematically.

If the examination is negative after looking at 100 fields, then a more thorough search is recommended for another new 100 fields. A smear size of 10 x 20 mm consists of 10,000 oil emersion fields.

If the slide has	**Results**	**Positive Grading**	**No. of fields to be examined**
More than 10 AFB per oil immersion field	Positive	3+	20
1-10 AFB per oil emersion field	Positive	2+	50
10-99 AFB per 100 oil immersion fields	Positive	1+	100
1-9 AFB per 100 oil immersion fields	Scanty	Record exact figure	200
No AFB in 100 oil immersion fields	Negative	—	100

The various causes of false positive sputum smear are :

- Food particles
- Precipitated stain
- Saprophytic acid fast bacilli
- *Mycobacterium kansasii* or *Nocardia*
- Spores of *Bacillus subtilis,* fibres and pollens
- Contamination of bacilli from one smear to another
- Scratches on the slide (RNTCP, 2000).

The causes of false negative sputum smear are :

- Inadequate sputum collection
- Improper storage

- Failure to select suitable portion of sputum for examination
- Inadequate preparation of smear or staining of the slide
- Inadequate examination of the smear
- Reading error – observer related
- False identification, labelling or recording (RNTCP, 2000)

Drugs and Drug Interactions

Drug treatment is fundamental for controlling TB, promoting the cure of the patients and breaking the chain of transmission when the antituberculosis drug regimen is completely and correctly followed.

Antituberculosis drug treatment started in 1944, when streptomycin (SM) and para-aminosalicylic acid (PAS) were discovered. In 1950, the first trial was performed comparing the efficacy of SM and PAS both as monotherapy or combined. The study demonstrated that combined therapy was more effective and resulted in the first multidrug antituberculosis treatment that consisted of a long course of both drugs. In 1952, a third drug, isoniazid (INH) was added to the previous combination greatly improving the efficacy of treatment. In 1960, ethambutol (EMB) substituted PAS and the treatment course was reduced to 18 months. In the '70s, with the introduction of rifampicin (RIF) into the combination, treatment was shortened to just nine months. Finally, in 1980, pyrazinamide (PZA) was introduced into the antituberculosis treatment, which could be reduced further to only six months.

Two biological features explain why combined drug therapy is more effective at curing TB than monotherapy. One is that treatment of active TB with a single drug results in the selection of drug resistant bacilli and failure to eliminate the disease. The other is that different populations of tubercle bacilli-each of them showing a distinct pattern of susceptibility for antituberculosis drugs may co-exist in a TB patient (Shamputa, 2006).

Since 1970, no new drug has been discovered for antituberculosis treatment, which today seems insufficient to confront the disease. Fortunately, research efforts have been successful in providing a wide range of new molecules with promising antituberculosis activity.

Antituberculosis treatment has two main objectives (Onyebujoh, 2005). First, there is a need to rapidly kill those bacilli living extracellularly in lung cavities, which are metabolically active and are dividing continuously; this is required in order to attain the negativization of sputum and therefore to prevent further transmission of the disease. Second, it is necessary to achieve complete sterilization and elimination of those bacilli replicating less actively in acidic and anoxic closed lesions, and to kill semi dormant bacilli living intracellularly in other host tissues, otherwise these bacilli may persist and will be responsible for subsequent TB relapses. INH is the drug with the highest activity against rapidly dividing bacilli, whereas RIF and PZA have the greatest sterilizing activity against bacteria that are not dividing. These reasons, along with the prevention of drug resistance, support the use of a combination therapy for the treatment of TB.

Drugs for treating TB are usually classified as first and second line drugs. Traditionally, there are five first line drugs: INH, RIF, PZA, EMB, and SM. Second line drugs include the aminoglycosides kanamycin and amikacin, the polypeptide capreomycin, PAS, cycloserine, the thioamides ethionamide and prothionamide and several fluoroquinolones such as moxifloxacin, levofloxacin and gatifloxacin. Some reports, however, include SM among the second line drugs, since its use has declined in recent years due to the high rates of resistance and also because other more effective drugs have been incorporated into the anti tuberculosis treatment. Similarly, new drugs such as the rifamycin derivatives rifapentine and rifabutin can be considered among the first line drugs, and in the near future, it is quite likely that some fluoroquinolones could be incorporated into the standard antituberculosis treatment, thus being considered as first line drugs. The current short course treatment for the complete elimination of active and dormant bacilli involves two phases:

- **initial phase**: three or more drugs (usually isoniazid, rifampicin, pyrazinamide and ethambutol or streptomycin) are used for two months and allow a rapid killing of actively dividing bacteria, resulting in the negativization of sputum
- **continuation phase**: fewer drugs (usually isoniazid and rifampicin) are used for 4 to 7 months, aimed at killing any remaining or dormant bacilli and preventing recurrence.

When resistance to any of the first line drugs is found or highly suspected, or when adverse effects to first line drugs develop during therapy, the treatment should include other drugs known as second line drugs. The World Health Organization (WHO) has established four TB diagnostic categories, assuming from a public health perspective that the highest priority of national TB programs is to identify and cure those patients with sputum smear positive pulmonary TB, i.e. infectious TB patients (World Health Organization, 2003). Category I comprises those patients with a high priority for treatment who are new smear positive patients, new smear negative pulmonary TB patients with extensive parenchymal involvement, patients with concomitant HIV/acquired immunodeficiency syndrome (AIDS) disease or severe forms of extrapulmonary TB. Patients with a lower priority for treatment are classified as follows: Category II (relapse, treatment failure or default), Category III (new smear negative pulmonary TB other than in Category I and less severe forms of extrapulmonary TB) and Category IV (chronic sputum positive TB after re-treatment and proven or suspected MDR-TB).

In addition to these guidelines for TB treatment, there are other alternatives. For example, the Center for Disease Control and Prevention of the United States also suggests continuation phases consisting of INH and rifapentine once per week for four months for patients in Category I (Centers for Disease Control and Prevention, 2003). This treatment can be used when sputum is negative for acidfast bacilli (AFB) after the first two months of treatment but should be extended to nine months if the result of the culture at that time point is still positive. These guidelines apply only to HIV negative patients as regimens containing rifapentine should not be administered to HIV/AIDS patients.

In general, the duration of the continuation phase must be estimated once the first two months of treatment (initial phase) have been completed. If the patient had cavitations on initial chest radiography and cultures are still positive after two months of treatment, the continuation phase should be extended to 31 weeks (seven months). When drug resistance develops, patients should be treated with a new combination containing at least three drugs that they had never received before (or that do not show cross resistance with those to which resistance is suspected). In these conditions the treatment is longer, more toxic, more expensive and less effective than regimens containing first line drugs and should be directly observed. In children, drug regimens similar to those for adults can be given, although EMB is not recommended because of its ocular toxicity. Rifapentine has not been approved for pediatric use.

In case of pregnancy, similar drug regimens can be prescribed, although SM and other second line aminoglycosides must not be given because they are ototoxic for the fetus. Also, there has been concern about the use of PZA. Then, a drug regimen of nine months of INH and RIF supplemented with EMB during the first month has been proposed.

The WHO recommends not using SM or thiacetazone in HIV positive patients in order to prevent the adverse effects of these drugs, often enhanced by antiretroviral drugs; EMB can be used instead.

Chemoprophylaxis of TB is indicated for asymptomatic patients having a positive tuberculin skin test (TST) but not showing active disease (latent TB infection), especially when they are at risk of developing the disease : for example, HIV positive patients (Balcells, 2006). This is aimed at preventing the occurrence of TB. Prophylaxis is most frequently achieved by the administration of INH only at doses of 300 mg daily for six to nine months (although there is a risk of developing INH resistance). When resistance to INH is suspected, other regimens including RIF, PZA or EMB, can be administered, although there is a greater chance of having adverse effects. In TB prophylaxis, RIF can be given concurrently with INH, reducing the prophylaxis treatment to three months.

It is of prime importance to ensure the patient's adherence to the antituberculosis treatment in order to achieve complete elimination of the bacilli (and hence avoid disease relapse) and also to prevent the emergence of drug resistance. For this reason, the antituberculosis treatment has to be supervised. Both the patient's adherence and supervision are often difficult to manage when the antituberculosis treatment has to be administered on a daily basis. Alternative treatments based on an intermittent administration of drugs (three times, twice and even once per week) facilitate the patient's adherence and the supervision of treatment. Intermittent treatment is possible because antituberculosis drugs have a marked post antibiotic effect. After the tuberculous bacillus has been exposed to drugs, there is a lag period (up to several days) during which its growth is interrupted even after the drug concentration has fallen to sub-inhibitory levels. Thus, there is no need to maintain a continuous inhibitory drug concentration to kill the bacilli or prevent growth.

Drug Resistance Mechanisms

The natural drug resistance of *M. tuberculosis* is an important obstacle for the treatment and control of TB. This resistance has traditionally been attributed to the unusual multi layer cell envelope and active multidrug efflux pumps (De Rossi, 2006; Jarlier, 1994). Recent insights into mechanisms that neutralize the toxicity of antibiotics in the cytoplasm have revealed other systems that function in synergy with the permeability barrier and efflux systems to provide natural resistance. Drugs inhibiting these intrinsic systems would enable many antibiotics which are already available but have not been used for TB to gain a new potential use against *M. tuberculosis* (Lomovskaya, 2006; Nguyen, 2006).

Acquired Drug Resistance

Knowledge of the molecular basis of drug resistance in *M. tuberculosis* increased with the sequencing of the genome and the development of molecular tools. In other bacterial species, acquired drug resistance is mediated by plasmids or transposons, but in contrast, *M. tuberculosis* acquired drug resistance is caused by mutations in chromosomal genes (Heym 1994). So far, no single pleiotropic mutation has been found in *M. tuberculosis* to cause a MDR phenotype. The MDR phenotype is caused by sequential accumulation of mutations in different genes involved in resistance to individual drugs, due to inappropriate treatment or poor adherence to treatment (Zhang, 2000). However, it is important to observe that some resistant strains do not present these classic mutations, suggesting the possibility of the existence of other mechanisms such as efflux pumps and alterations in the permeability of the cell wall.

New Developments and Perspectives

"The history of tuberculosis (TB) has been one of scientific, medical and political failure." According to the Global TB Control Report released by the World Health Organization (WHO), the good news is that *"the worldwide TB epidemic has leveled off for the first time since the disease was declared a public health emergency in 1993."* The bad news is that *"at the current rate of progress, the 1990 prevalence and mortality rates will not be halved worldwide by 2015."* The Global Plan to Stop TB needs to triple investment in order to achieve such a goal (World Health Organization, 2007).

TB is the only disease ever declared a global emergency by the WHO. Paradoxically, although we count on effective and proven cost effective interventions for its control, TB continues to cause great mortality and suffering, especially in poor and less developed countries. In association with the HIV/AIDS pandemic it forms a lethal combination. In addition, multidrug resistant (MDR) TB and the recently described extensively drug resistant (XDR) TB with further resistance to key second line drugs and which are virtually incurable severely complicate the management and control of the disease worldwide (Dorman, 2007; Shah, 2007). Almost 8.8 million new cases of TB were reported in 2005 and 1.6 million deaths were attributed to the disease. Asia and Sub-Saharan Africa accounted for 7.4 million new cases of TB worldwide (World Health Organization, 2007). Substantial scientific advances were made in knowledge about the agent and the disease in the '90s. First, the complete genome sequence of the tubercle bacillus was deciphered. Second, molecular epidemiology came into being, shedding a new light on mechanisms of TB transmission. Third, cellular mechanisms involved in *M. tuberculosis* resistance to several drugs were discovered. Fourth, new tools for speeding TB diagnosis and assessing drug susceptibility were sought while old methods were re-discovered and/or re-formulated. Lastly, research on drug and vaccine development exploded.

As for more recent advances, an analysis published in *Nature Medicine* highlights the most interesting scientific findings made in TB over the last three years, as identified by TB researchers. The three articles that were considered to have the highest impact in the field were:

- the description for the first time in 40 years of a new drug, a diarylquinoline that acts on an entirely new mycobacterial target, the proton pump of adenosine triphosphate (ATP) synthase (Andries, 2005)
- the report from a rural South African area of a deadly outbreak of what thereafter became known as XDR-TB (Gandhi, 2006) and
- the description of a novel recombinant Bacille Calmette-Guérin (BCG) vaccine candidate that works by over expressing not an antigen, but a membrane perforating enzyme (Grode, 2005).

The main topics of these three papers underline the most pressing gaps in the translation of TB research.

Bacillus and The Disease Under The Light of Molecular Epidemiology

The accelerated expansion that TB research underwent during the '90s is reflected by the way in which molecular epidemiology is regarded nowadays. In fact, molecular strain typing of *M. tuberculosis* is no longer a novelty.

In less than two decades, it has changed our view of TB transmission dynamics, challenging traditional dogmas and answering unsolved epidemiological questions (Mathema, 2006). Nowadays, molecular epidemiology is embedded within almost every aspect of TB research, laboratory diagnosis, clinical management and control interventions.Soon after standardization was agreed on DNA restriction fragment length polymorphism (RFLP) with IS6110, its use was generalized worldwide (van Embden, 1993). Other DNA strain typing methods were developed, IS6110 RFLP became the gold standard of *M. tuberculosis* strain genotyping and a new discipline-molecular epidemiology came into being. The considerable amount of information gathered in national and international *M. tuberculosis* genotype databases throughout the world enabled the analysis of global TB dissemination and promoted phylogeographical studies (Brudey, 2006) that in turn, incorporated more sophisticated and accurate markers of *M. tuberculosis* evolution (Arnold, 2007).

A common conviction of previous times was that the genome of the tubercle bacillus was extremely stable and homogeneous. Phenotypical discrimination between strains and even between species within the *M. tuberculosis* complex was limited to phage typing and comparison of drug resistance patterns, which were the only tools available for differentiation. Molecular epidemiology refuted such belief by revealing the existence of wide *M. tuberculosis* polymorphisms in population based studies (van Soolingen, 2001).

Molecular epidemiology tools also enabled the identification, description and differentiation of rare species within the *M. tuberculosis* complex such as *M. canettii, M. microti, M. caprae,* and *M. pinnipedii*. These species had previously been overlooked, mainly because they were difficult to distinguish by conventional biochemical tests. The differentiation between these *M. tuberculosis* complex taxons contributed to the triggering of evolutionary studies (Gutierrez, 2005). In turn, differentiation to the species level by spoligotyping (Kamerbeek, 1997) a user friendly genotyping tool applied worldwide turned out to have practical implications on medical management and epidemiology.

Genotyping tools have gained a well deserved place in national TB control programs. Indeed, molecular epidemiology studies have a direct impact on the surveillance of TB transmission and serve to adjust control strategies, not only in industrialized countries (van Doorn 2006) but also in medium and poor resource countries (Godreuil, 2007; Prodinger, 2007).

Some challenging issues raised by molecular epidemiology studies are related to reinfection (Shen, 2006) and multiple infection (Shamputa, 2006; van Rie, 2005), loss of strain fitness associated with drug resistance (Gagneux, 2006; Toungoussova, 2004), differential virulence and immunopathogenesis (Dormans, 2004; Manca, 2005), tissue or organ affinity (Caws, 2006), vaccine development and protection (Abebe, 2006; Grode, 2005). Even the results of genomics rely upon the lineage of the few strains than have so far been sequenced.

A reformulated version of the Variable Number of Tandem Repeats (VNTR) typing based on 15 Mycobacterial Interspersed Repetitive Units (MIRU) loci is emerging as the tool of choice and probably the next gold standard for *M. tuberculosis* genotyping at least for the near future (Supply, 2006). Unfortunately, this stimulating prospect poses a practical problem for laboratories in medium and low resource countries that managed to perform strain typing during the '90s. The manual VNTR-MIRUs procedure is still too labor demanding, while high throughput techniques entirely or partially automated are not affordable for most laboratories in such countries, at least in the short or medium term. In addition, these laboratories have little chance of translating the data gathered in existing databases on RFLP and spoligotype information into the new MIRUs language, at a reasonable pace. Thus, as long as science continues to advance, the scientific gap between industrialized and developing countries will widen.

In spite of the impressive advances made in the field with the existing tools, the ideal method for strain typing has not yet been achieved. Microarray techniques are envisaged as the true future tools for DNA strain typing. Hopefully, VNTR-MIRUs and/or microarray genotyping methods will become standardized and their use generalized so that the increased demand for these techniques will contribute towards lowering their cost to become affordable worldwide.

New Perspectives in Diagnosis

It is now 125 years since the tubercle bacillus was described by Robert Koch. Disappointingly, the diagnosis of the disease still relies on the same microscopy technique based on the specific Ziehl Neelsen staining of the bacillus, which was already available soon after that fundamental discovery. Much more progress needs to be made in obtaining better and faster diagnostic methods. Indeed, in high burden resource poor countries, where TB is a major public health problem, the diagnosis of active disease is mainly performed by direct microscopic examination of sputum smear samples. This technique, although simple and inexpensive, lacks sensitivity in comparison to *M. tuberculosis* culture. Several modifications mainly based on concentration and centrifugation techniques have been proposed to improve the sensitivity of sputum smear microscopy, with varying results. A recent systematic review and meta-analysis has shown that specificity does not vary substantially between different methods, but sensitivity can be improved. By comparison with direct smears, centrifugation and overnight sedimentation are more sensitive when preceded by any of several chemical methods (Steingart, 2006a). No other major improvement has been obtained in the classical staining method based on the Ziehl Neelsen technique developed many years ago.

Fluorescent microscopy proved to be faster and more sensitive than conventional microscopy based on Ziehl Neelsen staining and is the standard diagnostic method in high income countries (Steingart, 2006b). It has the additional advantage of demanding less effort from the laboratorist, thus reducing fatigue and human error. As for low income countries, the eventual introduction of fluorescent microscopy should be evaluated carefully because it requires a more expensive microscope and a more complex technique. It should be noted, however, that the main burden of fluorescent microscopy lies in the maintenance of the mercury lamp, rather than in the initial cost of the equipment. Lately, an inexpensive device has been released into the market that can be adapted to any fluorescence microscope. It has a long life, low power consumption 'Royal Blue' Luxeon light emitting diode that is used in place of the high cost, short lived, and environment unfriendly mercury vapor lamp (Anthony, 2006). In fact, this lamp lasts 200 times as long and costs 10 times less than the vapor mercury lamp. This form of illumination is suitable for the detection of auramine-O stained bacilli and may become an affordable alternative for improving diagnostic microscopy in laboratories serving poor-income settings with a high load of smear examinations (Van Hung, 2007).

Fluorescein diacetate staining was recently evaluated for assessing bacilli viability in sputum smears. It has been proposed for the early and accurate detection of TB treatment failure in poor income settings with a high TB burden (Hamid Salim, 2006).

Cultivation of *M. tuberculosis* is the gold standard for the diagnosis of active TB in the laboratory. This has been traditionally performed in egg or agar based solid media. Although slow and time consuming, it is relatively simple to perform and rather inexpensive in most settings. Newer alternative methods based on liquid culture media and giving faster results such as the BACTEC radiometric method, BACTEC MGIT960 and BactT/Alert have proven useful especially in medium and high income countries with the necessary resources to use them routinely (Scarparo, 2002). It is now a standard recommendation that the combination of a solid and a liquid culture medium gives the best sensitivity in recovering mycobacteria in primary culture.

The development of many new nucleic acid amplification techniques, such as the polymerase chain reaction (PCR) opened new possibilities for the rapid diagnosis of many infectious diseases. However, when applied to TB diagnosis, only two methods, the Amplicor *Mycobacterium Tuberculosis* Test (Amplicor) (Roche Diagnostic Systems,

Inc., New Jersey) and the Amplified *Mycobacterium Tuberculosis* Direct Test (MTD) (Gen-Probe, California), both commercially available, have received approval by the US Food and Drug Administration (FDA) for direct application in clinical samples (Centers for Disease Control, 2000). Compared to culture and clinical status, nucleic acid amplification tests have high sensitivity and specificity in smear positive samples. The current recommendation is that molecular tests should always be interpreted in conjunction with the patient's clinical data (Pfyffer, 2003). More recent approaches combining PCR amplification and fluorescent probe detection in the same tube, such as real time PCR technology, are promising to give improved sensitivities and specificities. More evaluations in target populations are needed to assess the real impact on the diagnosis of the disease (Espy, 2006).

Until recently, the tuberculin skin test was the only available test for the diagnosis of latent TB infection. Based on the detection of delayed type hypersensitivity to purified protein derivative (PPD) obtained from *M. tuberculosis*, it measures the size of the induration produced after sub-cutaneous inoculation of a standardized dose. Since PPD is actually a raw mixture of several antigens shared by *M. tuberculosis*, *M. bovis* BCG and several non tuberculous mycobacteria (NTM), a positive tuberculin skin test result could be due to latent TB infection, previous BCG vaccination or previous exposure to NTM. Additionally, it has various disadvantages, such as variability in the interpretation by different readers, the need of some experience to correctly interpret the result and the requirement for the patient to return after 48-72 hours for test reading.

In order to overcome these disadvantages, immune based blood tests have been recently introduced to detect latent TB infection. Interferon γ (IFNγ) assays measure the amount of IFN γ produced or the actual number of IFNγ producing T lymphocytes in response to specific antigens. Both approaches are based on the fact that T cells sensitized with tuberculous antigens will produce IFN γ when they are re-exposed ex vivo to mycobacterial antigens; a high amount of IFN γ production is then presumed to correlate with TB infection (Pai, 2004). The current IFN γ assays use more specific antigens to *M. tuberculosis* than PPD: the early secretory antigen target 6 (ESAT-6), and the culture filtrate protein 10 (CFP-10). There are many studies evaluating IFN γ assays in different populations. Based on published evidence, the T-SPOT TB assay seems to be more sensitive than the QuantiFERON-TB Gold test in detecting latent as well as active TB infection. However, the absence of a real gold standard for the diagnosis of latent TB infection prevents a more definitive conclusion. Further studies comparing these two assays are needed, especially in immunosuppressed patients (Richeldi, 2006).

It should be remarked that TB diagnosis in endemic countries has so far gained little benefit from scientific progress, namely biotechnological developments (Perkins, 2006a;

Perkins, 2006b). There is an urgent demand for a field friendly test, a point of care which enables to diagnose the disease on the spot in order to avoid delays in diagnosis, thus, preventing further transmission and reducing complications. This type of test is particularly useful when patients do not return for care and would greatly benefit people in settings such as prisons, homeless shelters and clinics for migrant workers who have no ready access to, or do not seek, public health service assistance. The ideal diagnostics for TB should be available in one hour and should require no electricity, refrigeration, fresh water supply or highly trained personnel (Keeler, 2006).

Serological tests aimed at the detection of either antigens specific to, or antibodies directed against *M. tuberculosis* would in principle provide a choice platform for a point-of-care diagnostic tool. In the development of tests for antibody detection, careful attention should be paid to the selection of the target group and the control population groups for performance evaluation. Inclusion and exclusion criteria should be quite stringent regarding age range, geographical location, previous exposure to *M. tuberculosis* and environmental mycobacteria, tuberculin skin test status, previous BCG vaccination and unrelated pathologies for differential diagnosis. Also, the sensitivity of the test should be evaluated in the actual target population, namely patients with (pulmonary and/or extrapulmonary) paucibacillary TB rather than in acid fast bacilli smear positive pulmonary cases. Another issue to consider is the genetic diversity, not only between hosts but also between *M. tuberculosis* lineages, which might render a promising test only suitable for a restricted geographical region (Lopez, 2003).

A quite different approach that utterly fulfills the requirements of the point-of-care diagnosis is based on the electronic nose technology, which is able to detect and identify tiny amounts of virtually every substance in a few minutes. The device can be assembled as the sensory part of a portable artificial intelligence system, able to detect several microbes simultaneously through their specific "odors". Such a system could be used to investigate the agent either directly in the patient's breath or in a swab of a specimen obtained from any bodily site. This highly suited technology has already been reported for other bacterial human pathogens but is still awaiting development for TB diagnosis (Dutta, 2006).

Lastly, diagnostic methods currently under development and expected to be available in two to four years include a dipstick PCR, the detection of mycobacterial proteins in urine and a blood antibody test, among others (Marris, 2007).

The Problem of Drug Resistance Detection

Traditionally, drug resistance in TB was assessed by culturing *M. tuberculosis* on solid media in the presence of antibiotics and measuring growth by enumeration of colonies. This methodology although simple to perform and rather inexpensive, is quite slow and laborious, requiring several weeks to give the final results (Heifets, 1999).

Several molecular tools have also been developed and proposed as rapid methods to detect drug resistance. They search for genetic determinants of resistance rather than for the resistance phenotype and involve molecular nucleic acid amplification by PCR and detection of amplified products for specific mutations correlating with drug resistance. Molecular methods have several advantages over culture based techniques: shorter turnaround time, no need for growth of the organism, the possibility for direct application in clinical samples, less biohazard risks and feasibility for automation. However, not all molecular mechanisms of drug resistance are known. In most cases, molecular methods have been directed towards detecting resistance to rifampicin for two major reasons. First, rifampicin resistance is a good surrogate marker for treatment failure and in settings with a high prevalence of drug resistance, for multidrug resistance. Second, the associated mutations are well defined, restricted to a short chromosomal segment and their prevalence is sufficiently known worldwide. Current developments aim at the simultaneous identification of *M. tuberculosis* and detection of resistance to two or more key drugs. The aim would be to achieve identification and multiple drug resistance detection directly on clinical specimens, thus avoiding the delay implied in culturing the bacilli (Cavusoglu, 2006; Sekiguchi, 2007).

Recent Advances in Drug Development

Associated with the problem of drug resistance is the search for new anti tuberculosis drugs. Almost no new anti tuberculosis drug classes have been developed over the last 40 years. This lack of investment became evident when the HIV/AIDS pandemic emerged, soon followed by MDR-TB and the unavoidable interactions between anti tuberculosis and anti retroviral drugs. Many candidate compounds have been considered in the last decade, but very few of them have entered into further evaluations.

The Global Alliance for TB Drug Development was established in 2000 to promote the development of new anti tuberculosis compounds. Its goal is to bring a new anti tuberculosis drug into the market by 2010. The most advanced program on new drugs is examining moxifloxacin, and is about to enter phase III clinical trials in multiple centers. The program aims at using this fluoroquinolone instead of ethambutol or isoniazid in the first line drug scheme of anti tuberculosis treatment, in order to shorten the current 6-month duration of the treatment (Burman, 2006). A similar program is being carried out in Africa, where gatifloxacin, another fluoroquinolone, is also substituted for ethambutol. This combination treatment is currently in Phase III clinical trials aimed at shortening the standard regimen to four months. PA-824 is another pro-drug of the nitroimidazole class that has already passed animal model testing in combined therapy, and is currently undergoing phase I clinical trials (Nuermberger, 2006). Interestingly, a protein was described in *M. tuberculosis* that is involved in both intracellular drug activation and resistance to this drug candidate. Other promising drugs are in the preclinical phase of evaluation. OPC-67683 is a nitro dihydro-imidazooxazole derivative that inhibits mycolic acid biosynthesis, is free of mutagenicity and is highly active against TB in vitro and in mice. The diarylquinoline known as R207910, TMC207 or compound J is a new compound exhibiting a completely novel mode of action : inhibition of ATP synthase and very high activity against *M. tuberculosis, M. leprae* and *M. ulcerans.* This promising compound is being tested in phase IIa trials on TB treatment (Lounis, 2006).

The availability of such a spectrum of new drug candidates offers great promise but also entails a great challenge. The role played by each drug must be explored within the frame of a multidrug treatment regimen. In the immediate future, a complete series of clinical trials will be needed to find the optimal treatment scheme of ultra short duration, i.e. 2 months or even shorter. Innovative strategies must be designed in order to meet this challenge.

Research and development is also needed on innovative drug formulations and drug delivery systems aimed at increasing compliance and achieving a high local drug concentration while minimizing systemic side effects. In this respect, the growing field of inhalation therapy offers a very promising new prospect (Chow, 2007). A technology based on porous particles for pulmonary drug delivery is already in use for insulin. This technology presented as a simple, low cost, disposable, dry powder inhaler can be applied to the delivery of anti tuberculosis drugs.

Recent Advances in Vaccine Development

Vaccine development is a problematic issue for many reasons. It is not appealing for the industry because it demands a huge investment, takes a long time and the risk of failure is high.

Major obstacles also lie in the initial vaccine design itself, such as the difficulty in inducing a potent and long lasting cellular immune response in humans, due to our poor understanding of host parasite interactions. Vaccines that promise to be more potent than BCG were designed following a rational approach: they are recombinant BCGs over expressing major *M. tuberculosis* antigens. Grode et al. used an entirely novel approach : Instead of using an antigen, these authors selected the membrane perforating listeriolysin to construct a recombinant BCG. This hemolytic enzyme, produced by *Listeria monocytogenes,* allows the agent to escape from the phagosomes of infected host cells. This improves the access of mycobacterial antigens to the major histocompatibility complex Type I pathway, thus resulting in better CD8+ T cell stimulation. At least in the mouse model, recombinant BCG secreting listeriolysin elicited an enhanced protection against a strain of the *M. tuberculosis* Beijing/W genotype family, while parental BCG failed to do so consistently (Grode, 2005). Future vaccines must be able to protect against the most prevalent, transmissible and/or virulent lineages worldwide and not merely against laboratory domesticated strains. A provocative finding was reported by researchers at the Institut Pasteur, where the BCG vaccine was first developed in 1924. They gathered a large body of evidence on the evolutionary sequence of events leading to modern vaccine variants. By comparing various genetic markers, the authors were able to construct an evolutionary tree, showing that distinct variants of the original BCG strain underwent different deletions during in vitro passages, mainly in the process of adaptation to the broth medium used for vaccine production. These mutations might have been responsible for a gradual loss in immunogenicity and protection ability. Oddly enough, their findings suggest that the early BCG Japan strain may even be superior to the later ones that are currently much more widely used, such as the Danish BCG strain. It is speculated that, in general, vaccine manufacturers tended to prefer BCG strains that cause less serious local reactions over those provoking more intense inflammation at the site of injection. As a result, less protective strains for vaccine production might have unintentionally been selected through time (Brosch, 2007).

The tubercle bacillus is both an amazing creature and a formidable enemy that has proven hard to conquer. It is pushing science to its limits. Medical research itself has developed into a complex and engaging living creature whose evolution is driven by selective pressure. Synergic efforts, interdisciplinary approaches and translational research are expressions of its adaptive response. The greater the challenge the more eager the endeavor must be. We just can not afford to lose this battle.

REFERENCES

Abebe F Bjune G 2006 The emergence of Beijing family genotypes of *Mycobacterium tuberculosis* and low level protection by bacille Calmette Guerin (BCG) vaccines: is there a link? *Clin Exp Immunol* **145**: 389-97

Ahmed N Hasnain S E 2004 Genomics of *Mycobacterium tuberculosis*: old threats and new trends *Indian J Med Res* **120**: 207-12

Alcaide F Benítez MA Martín R Escribà JM 2000 Evaluation of the BACTEC MGIT 960 and MB BAC/T system for routine detection of *Mycobacterium tuberculosis J Clin Microbiol* **38**: 3131-2

Alcalà L Ruiz-Serrano MJ Hernangomez S 2001 Evaluation of the upgraded amplified *Mycobacterium tuberculosis* direct test (Gen-Probe) for direct detection of *Mycobacterium tuberculosis* in respiratory and non respiratory specimens *Diagn Microbiol Infect Dis* **41**: 51-6

Ananthanarayan R Panikar CKJ 2002 Textbook of Microbiology Sixth edition

Andries K Verhasselt P Guillemont J 2005 A diarylquinoline drug active on the ATP synthase of *Mycobacterium tuberculosis Science* **307**: 223-7

Anthony RM Kolk AH Kuijper S Klatser PR 2006 Light emitting diodes for auramine-O fluorescence microscopic screening of *Mycobacterium tuberculosis Int J Tuberc Lung Dis* **10**: 1060-2

Arnold C 2007 Molecular evolution of *Mycobacterium tuberculosis Clin Microbiol Infect* **13**: 120-8

Badak FZ Kiska DL Setterquist S Hartley C O'Connell MA Hopfer RL 1996 Comparison of Mycobacteria Growth Indicator Tube with BACTEC 460 for detection and recovery of mycobacteria from clinical specimens *J Clin Microbiol* **34**: 2236-9

Balcells ME Thomas SL Godfrey-Faussett P Grant AD 2006 Isoniazid preventive therapy and risk for resistant tuberculosis *Emerg Infect Dis* **12**: 744-51

Barrett A Magee JG Freeman R 2002 An evaluation of the BD ProbeTec ET system for the direct detection of *Mycobacterium tuberculosis* in respiratory samples *J Med Microbiol* **51**: 895-8

Bauer J Andersen AB Kremer K Miorner H 1999 Usefulness of spoligotyping to discriminate IS*6110* low copy number *Mycobacterium tuberculosis* complex strains cultured in Denmark *J Clin Microbiol* **37**: 2602-6

Bogard M Vincelette J Antinozzi R 2001 Multicentre study of a commercial automated polymerase chain reaction system for the rapid detection of *Mycobacterium tuberculosis* in respiratory specimens in routine clinical practice *Eur J Clin Microbiol Infect Dis* **20**: 724-31

Brosch R Gordon SV Garnier T 2007 Genome plasticity of BCG and impact on vaccine efficacy *Proc Natl Acad Sci U S A* 2007; 104: 5596-601

Brudey K Driscoll JR Rigouts L 2006 *Mycobacterium tuberculosis* complex genetic diversity: mining the fourth international spoligotyping database (SpolDB4) for classification population genetics and epidemiology *BMC Microbiol* **6**: 23

Bull TJ Shanson DC 1992 Evaluation of a commercial chemiluminescent Gen Probe system "AccuProbe" for the rapid identification of mycobacteria including "MAIC X" isolated from blood and other sites from patients with AIDS *J Hosp Infect* **21**: 143-9

Burman WJ Goldberg S Johnson JL 2006 Moxifloxacin versus ethambutol in the first 2 months of treatment for pulmonary tuberculosis *Am J Respir Crit Care Med* **174**: 331-8

Butler WR Jost KC Kilburn JO 1991 Identification of mycobacteria by high performance liquid chromatography *J Clin Microbiol* **29**: 2468-72

Cavusoglu C Turhan A Akinci P Soyler I 2006 Evaluation of the Genotype MTBDR assay for rapid detection of rifampin and isoniazid resistance in *Mycobacterium tuberculosis* isolates *J Clin Microbiol* **44**: 2338-42

Caws M Thwaites G Stepniewska K 2006 Beijing genotype of *Mycobacterium tuberculosis* is significantly associated with human immunodeficiency virus infection and multidrug resistance in cases of tuberculous meningitis *J Clin Microbiol* **44**: 3934-9

Centers for Disease Control and Prevention (CDC) 2003 Treatment of tuberculosis American Thoracic Society CDC and Infectious Diseases Society of America Morbidity and Mortality Weekly Report 52 (No RR-11)

Chow AH Tong HH Chattopadhyay P Shekunov BY 2007 Particle Engineering for Pulmonary Drug Delivery *Pharm Res* Jan 24 de Boer AS Borgdorff MW de Haas PE Nagelkerke NJ van Embden JD van Soolingen D 1999 Analysis of rate of change of IS*6110* RFLP patterns of *Mycobacterium tuberculosis* based on serial patient isolates *J Infect Dis* **180**: 1238-44

Dormans J Burger M Aguilar D 2004 Correlation of virulence lung pathology bacterial load and delayed type hypersensitivity responses after infection with different *Mycobacterium tuberculosis* genotypes in a BALB/c mouse model *Clin Exp Immunol* **137**: 460-8

Dorman SE Chaisson RE 2007 From magic bullets back to the Magic Mountain: the rise of extensively drug-resistant tuberculosis *Nat Med* **13**: 295-8

Dutta R Dutta R 2006 Intelligent Bayes Classifier (IBC) for ENT infection classification in hospital environment *Biomed Eng Online* **5**: 65

De Rossi E Ainsa JA Riccardi G 2006 Role of mycobacterial efflux transporters in drug resistance: an unresolved question *FEMS Microbiol Rev* **30**: 36-52

Espy MJ Uhl JR Sloan LM 2006 Real time PCR in clinical microbiology: applications for routine laboratory testing *Clin Microbiol Rev* **19**: 165-256

Gagneux S Burgos MV DeRiemer K 2006 Impact of bacterial genetics on the transmission of isoniazid resistant *Mycobacterium tuberculosis Pathog* **2**: e61

Gandhi NR Moll A Sturm AW 2006 Extensively drug resistant tuberculosis as a cause of death in patients co-infected with tuberculosis and HIV in a rural area of South Africa *Lancet* **368**: 1575-80

Ganeswrie R Chui CS Balan S Puthucheary SD 2004 Comparison of BACTEC MGIT 960 system and BACTEC 460 TB system for growth and detection of Mycobacteria from clinical specimens *Malays J Pathol* **26**: 99-103

Glassroth J Richrman C B and Mersfeild ES 1993 Tuberculosis a comprehensive international approach marcel dekker New York p 159

Godreuil S Torrea G Terru D 2007 First molecular epidemiology study of *Mycobacterium tuberculosis* in Burkina Faso *J Clin Microbiol* **45**: 921-7

Goh KS Fabre M Huard RC Schmid S Sola C Rastogi N 2006 Study of the gyrB gene polymorphism as a tool to differentiate among *Mycobacterium tuberculosis* complex subspecies further underlines the older evolutionary age of 'Mycobacterium canettii' *Mol Cell Probes* **20**: 182-90

Goo JM Yeo JS Kim Hy 2000 Pulmonary tuberculoma evaluated by means of FDG-PET *Radiology* **216**: 117-21

Grode L Seiler P Baumann S 2005 Increased vaccine efficacy against tuberculosis of recombinant *Mycobacterium bovis* bacille Calmette-Guerin mutants that secrete listeriolysin *J Clin Invest* **115**: 2472-9

Guerrant GO Lambert MA Moss CW 1981 Gas chromatographic analysis of mycolic acid cleavage products in mycobacteria *J Clin Microbiol* **13**: 899-907

Gutierrez MC Brisse S Brosch R 2005 Ancient origin and gene mosaicism of the progenitor of Mycobacterium tuberculosis *PLoS Pathog* **1**: e5

Hamid Salim A Aung KJ Hossain MA Van Deun A 2006 Early and rapid microscopy based diagnosis of true treatment failure and MDR-TB *Int J Tuberc Lung Dis* **10**: 1248-54

Heifets LB Cangelosi GA 1999 Drug susceptibility testing of *Mycobacterium tuberculosis*: a neglected problem at the turn of the century *Int J Tuberc Lung Dis* **3**: 564-81

Hermans PW van Soolingen D Dale JW 1990 Insertion element IS*986* from *Mycobacterium tuberculosis*: a useful tool for diagnosis and epidemiology of tuberculosis *J Clin Microbiol* **28**: 2051-8

Heym B Honore N Truffot-Pernot C 1994 Implications of multidrug resistance for the future of short course chemotherapy of tuberculosis: a molecular study *Lancet* 344: 293-8

Jarlier V Nikaido H 1994 Mycobacterial cell wall: structure and role in natural resistance to antibiotics *FEMS Microbiol Lett* **123**: 11-8

Kamerbeek J Schouls L Kolk A 1997 Simultaneous detection and strain differentiation of *Mycobacterium tuberculosis* for diagnosis and epidemiology *J Clin Microbiol* **35**: 907-14

Keeler E Perkins MD Small P 2006 Reducing the global burden of tuberculosis: the contribution of improved diagnostics *Nature* 444 Suppl **1**: 49-57

Kremer K Arnold C Cataldi A 2005 Discriminatory power and reproducibility of novel DNA typing methods for *Mycobacterium tuberculosis* complex strains *J Clin Microbiol* **43**: 5628-38

Kremer K van Soolingen D Frothingham R 1999 Comparison of methods based on different molecular epidemiological markers for typing of *Mycobacterium tuberculosis* complex strains: interlaboratory study of discriminatory power and reproducibility *J Clin Microbiol* **37**: 2607-18

Lambert MA Moss CW Silcox VA Good RC 1986 Analysis of mycolic acid cleavage products and cellular fatty acids of *Mycobacterium* species by capillary gas chromatography *J Clin Microbiol* **23**: 731-6

Lee H Park HJ Cho SN Bai GH Kim SJ 2000 Species identification of mycobacteria by PCR-restriction fragment length polymorphism of the *rpo*B gene *J Clin Microbiol* **38**: 2966-71

Lomovskaya O Bostian KA 2006 Practical applications and feasibility of efflux pump inhibitors in the clinic—a vision for applied use *Biochem Pharmacol* **71**: 910-8

Lopez B Aguilar D Orozco H 2003 A marked difference in pathogenesis and immune response induced by different *Mycobacterium tuberculosis* genotypes *Clin Exp Immunol* **133**: 30-7

Lounis N Veziris N Chauffour A Truffot Pernot C Andries K Jarlier V 2006 Combinations of R207910 with drugs used to treat multidrug resistant tuberculosis have the potential to shorten treatment duration *Antimicrob Agents Chemother* **50**: 3543-7

McShane H Pathan AA Sander CR 2004 Recombinant modified vaccinia virus Ankara expressing antigen 85A boosts BCG primed and naturally acquired antimycobacterial immunity in humans *Nat Med* **10**: 1240-4

Manca C Reed MB Freeman S 2004 Differential monocyte activation underlies strain specific *Mycobacterium tuberculosis* pathogenesis *Infect Immun* **72**: 5511-4

Marris E 2007 From TB tests just a 'yes or no' answer please *Nature Medicine* **13**: 267 doi:10 1038/ nm0307-267

Mathema B Kurepina NE Bifani PJ Kreiswirth BN 2006 Molecular epidemiology of tuberculosis: current insights. *Clin Microbiol Rev* 2006; **19**: 658-85

Middlebrook G Reggiardo Z Tigert WD 1977 Automatable radiometric detection of growth of *Mycobacterium tuberculosis* in selective media *Am Rev Respir Dis*; **115**: 1066-9

Minnikin DE Al Shamaony L Goodfellow M 1975 Differentiation of *Mycobacterium Nocardia* and related taxa by thin layer chromatographic analysis of whole organism methanolysates *J Gen Microbiol* **88**: 200-4

Nguyen L Thompson CJ 2006 Foundations of antibiotic resistance in bacterial physiology: the mycobacterial paradigm *Trends Microbiol* **14**: 304-12

Niemann S Rusch Gerdes S Richter E Thielen H Heykes Uden H Diel R 2000 Stability of IS6110 restriction fragment length polymorphism patterns of *Mycobacterium tuberculosis* strains in actual chains of transmission *J Clin Microbiol* **38**:2563-7

Nuermberger E Rosenthal I Tyagi S 2006 Combination chemotherapy with the nitroimidazopyran PA-824 and first line drugs in a murine model of tuberculosis *Antimicrob Agents Chemother* **50**: 2621-5

Onyebujoh P Zumla A Ribeiro I 2005 Treatment of tuberculosis: present status and future prospects *Bull World Health Organ* **83**: 857-65

Pai M Riley LW Colford JM Jr 2004 Interferon gamma assays in the immunodiagnosis of tuberculosis: a systematic review *Lancet Infect Dis* **4**: 761-76

Park CH Hixon DL Ferguson CB Hall SL Risheim CC Cook CB (1984) Rapid recovery of mycobacteria from clinical specimens using automated radiometric technique *Am J Clin Pathol* **81**: 341-5

Perkins MD Roscigno G Zumla A 2006a Progress towards improved tuberculosis diagnostics for developing countries *Lancet* **367**: 942-3

Perkins MD Small PM 2006b Admitting defeat *Int J Tuberc Lung Dis* **10**: 1

Pfyffer GE Kaufmann SHE Hahn H 2003 Mycobacteria and TB Basel Karger Switzerland pp 67-83

Piersimoni C Scarparo C Callegaro A 2001 Comparison of MB/BacT ALERT 3D System with radiometric BACTEC system and Lowenstein Jensen medium for recovery and Identification of mycobacteria from clinical specimens: a multicentre study *J Clin Microbiol* **39**: 651-7

Piersimoni C Scarparo C 2003 Relevance of commercial amplification methods for direct detection of *Mycobacterium tuberculosis* complex in clinical samples *J Clin Microbiol* **41**: 5355-65

Prabhakar S Mishra A Singhal A Tyagi J S 2004 Use of hup B gene encoding a histone like protein of *Mycobacterium tuberculosis* and *M bovis J Clin Microbiol* **42**: 2724-32

Prodinger WM 2007 Molecular epidemiology of tuberculosis: toy or tool? A review of the literature and examples from Central Europe *Wien Klin Wochenschr* **119**: 80-9

Price's Text book of the Practice of Medicine 1973 Oxford University Press Eleventh Edition Sir Ronal Bodley Scott

Revised National Tuberculosis Control Programme 2000 Ministry of Health and Family Welfare-New Delhi

Richeldi L 2006 An update on the diagnosis of tuberculosis infection *Am J Respir Crit Care Med* **174**: 736-42

Scarparo C Piccoli P Rigon A Ruggiero G Ricordi P Piersimoni C 2002 Evaluation of the BACTEC MGIT 960 in comparison with BACTEC 460 TB for detection and recovery of mycobacteria from clinical specimens *Diagn Microbiol Infect Dis* **44**: 157-61

Sekiguchi J Miyoshi Akiyama T Augustynowicz Kopec E 2007 Detection of multidrug resistance in *Mycobacterium tuberculosis J Clin Microbiol* **45**: 179-92

Shah NS Wright A Bai G H 2007 Worldwide emergence of extensively drug resistant tuberculosis *Emerg Infect Dis* **13**: 380-7

Shamputa IC Jugheli L Sadradze N 2006 Mixed infection and clonal representativeness of a single sputum sample in tuberculosis patients from a penitentiary hospital in Georgia *Respir Res* **7**: 99

Shen G Xue Z Shen X 2006 The study recurrent tuberculosis and exogenous reinfection Shanghai China *Emerg Infect Dis* **12**: 1776-8

Supply P Allix C Lesjean S 2006 Proposal for standardization of optimized mycobacterial interspersed repetitive unit variable number tandem repeat typing of *Mycobacterium tuberculosis J Clin Microbiol* **44**: 4498-510

Steingart KR Henry M Ng V 2006a Fluorescence versus conventional sputum smear microscopy for tuberculosis: a systematic review *Lancet Infect Dis* **6**: 570-81

Steingart KR Ng V Henry M 2006b Sputum processing methods to improve the sensitivity of smear microscopy for tuberculosis: a systematic review *Lancet Infect Dis* **6**: 664-74

Telenti A Marchesi F Balz M Bally F Böttger EC Bodmer T 1993 Rapid identification of mycobacteria to the species level by polymerase chain reaction and restriction enzyme analysis *J Clin Microbiol* **31**: 175-8

Torkko P Suomalainen S Iiavanainen E 2002 *Mycobacterium palustre* sp nov a potentially pathogenic slow-growing mycobacterium isolated from veterinary and clinical specimens and Finnish stream water *Int J Syst Evol Microbiol* **52**: 1519-25

Tortoli E Simonetti MT Lavinia F 1996 Evaluation of reformulated chemiluminescent DNA probe (AccuProbe) for culture identification of *Mycobacterium kansasii J Clin Microbiol* **34**: 2838-40

Tortoli E Cichero P Chirillo MG 1998 Multicentre comparison of ESP Culture System II with BACTEC 460TB and with Lowenstein Jensen medium for recovery of mycobacteria from different clinical specimens including blood *J Clin Microbiol* **36**: 1378-81

Toungoussova OS Caugant DA Sandven P Mariandyshev AO Bjune G 2004 Impact of drug resistance on fitness of *Mycobacterium tuberculosis* strains of the W-Beijing genotype *FEMS Immunol Med Microbiol* **42**: 281-90

Tuberculosis : A text book by William N Rom Stuart M Garaus 2000 : Oxford University Press p : 167-169

van Doorn HR de Haas PE Kremer K Vandenbroucke Grauls CM Borgdorff MW van Soolingen D 2006 Public health impact of isoniazid resistant *Mycobacterium tuberculosis* strains with a mutation at amino acid position 315 of katG: a decade of experience in The Netherlands *Clin Microbiol Infect* **12**: 769-75

van Embden JD Cave MD Crawford JT 1993 Strain identification of *Mycobacterium tuberculosis* by DNA fingerprinting: recommendations for a standardized methodology *J Clin Microbiol* **31**: 406-9

Van Hung N Sy DN Anthony RM Cobelens FG van Soolingen D 2007 Fluorescence microscopy for tuberculosis diagnosis *Lancet Infect Dis* **7**: 238-9

van Rie A Victor TC Richardson M 2005 Reinfection and mixed infection cause changing *Mycobacterium tuberculosis* drug resistance patterns *Am J Respir Crit Care Med* **172**: 636-42

van Soolingen D 2001 Molecular epidemiology of tuberculosis and other mycobacterial infections: main methodologies and achievements *J Intern Med* **249**: 1-26

Williams Bouyer N Yorke R Lee HI Woods GL 2000 Comparison of the BACTEC MGIT 960 and ESP Culture System II for growth and detection of mycobacteria *J Clin Microbiol* **38**: 4167-70

World Health Organization (WHO) 2003 Treatment of tuberculosis: guideline for national programmers 3rd edition Geneva

World Health Organization 2007 Global tuberculosis Control – surveillance Planning Financing WHO report

Zhang Y Telenti A 2000 Genetics of drug resistance in *Mycobacterium tuberculosis* in : Molecular genetics of Mycobacteria

INDEX

A

B

C

D

E

F

G

H

I

J

K

L

M

N

O

P

R

S

T

U

V

W

X